The plants of Prince Edward Island

Reprinted
D.S. Erskine
formerly of Plant Research Institute
Ottawa, Ontario
Publication 1088 1960

with new records, nomenclatural changes, and corrections and deletions

P.M. Catling
Biosystematics Research Institute
Ottawa, Ontario

D.S. Erskine
71 Green Bush Road
Willowdale, Ontario

R.B. MacLaren
5 Rosemount Drive
Charlottetown, Prince Edward Island

Research Branch
Agriculture Canada
1985

Publication 1798

Available in Canada through

Authorized Bookstore Agents
and other bookstores

or by mail from

Canadian Government Publishing Centre
Supply and Services Canada
Ottawa, Canada K1A 0S9

Catalogue No. A53-1088/1985E — Canada: $16.50
ISBN 0-660-11794-0 — Other Countries $19.80

Price subject to change without notice

Canadian Cataloguing in Publication Data

Erskine, D. S.

The plants of Prince Edward Island

(Publication ; 1798)

Originally issued as: Publication ; 1088.
Includes bibliographical references.

1. Botany — Prince Edward Island. I. Catling, Paul M. II. MacLaren, R. B. III. Canada. Agriculture Canada. Research Branch. IV. Title. V. Series: Publication (Canada. Agriculture Canada). English ; 1798.

QK203.P7E67 1985 581.9717 C85-097204-3

CONTENTS

Cover: Stemless lady's-slipper, *Cypripedium acaule* Ait., the floral emblem of Prince Edward Island. Drawing by Michael Sydor. See the correction to page 118 on page xxi.

PREFACE

The Plants of Prince Edward Island, originally published in 1960, is a very popular publication. Since it has been out of print we have received many requests for a reprinting. After some deliberation it was decided that the most expedient method of making the information available again was to reprint the 1960 publication and to add a new introduction to the reprint to bring it up to date as much as possible and to improve its utility. This 1985 introductory update reports 60 taxa that are additions to the island's flora, new localities for 11 rare taxa, name changes for 116 taxa, 16 corrections to previous listings, and 33 miscellaneous corrections as well as other information.

P.M. Catling
1985

INTRODUCTION TO THE PLANTS OF PRINCE EDWARD ISLAND, SECOND EDITION, WITH NEW RECORDS, NOMENCLATURAL CHANGES, AND CORRECTIONS AND DELETIONS

Since 1960 a lot of research has been conducted that is relevant to the flora of Prince Edward Island. Information on vegetational history and geology has recently become available. Vascular plant species that were not known there before have been discovered on the island, and additional localities have been discovered for some rare species. The status of some plants has changed. Taxonomic research has resulted in changes in a number of scientific names and authorities. Reexamination of some of the justifying specimens, along with more recent information, has disclosed identification errors that can now be corrected. Other errors and misprints have also come to our notice. To make this reprint as useful as possible, we have dealt with this new material under the appropriate headings that follow.

HISTORY AND GEOLOGY

In 1973, the surficial deposits of the island were mapped and descriptions of the distribution and nature of glacial and postglacial deposits and events, were published (Prest 1973).

It was previously noted that "the vegetational history of the island is at least hypothetical, until analyses may be made of peat and lake-sediment samples for fossil pollen." A recent study (Anderson 1980) has outlined the vegetational history. Following the melting of the Wisconsin glacier 10 000 years ago, Prince Edward Island had a short-lived tundra vegetation followed by a combination of spruce forest-tundra from 10 000 to 8 000 before present (BP). Pine forests developed after this period. Between 6 500 and 4 500 BP hemlock and pine were prevalent. Beech appeared about 3 400 BP and beech, hemlock and birch forests were dominant until modern times. The fossil pollen record suggests that the climate became warmer 8 000 BP and may have been much warmer than present 3 000–1 500 BP, but became cooler after this period. European settlement 200–250 BP had a marked effect on the vegetation. Substantial reduction of forests occurred along with the introduction of Eurasian weedy species and an increase in grass cover (Anderson 1980).

NEW RECORDS

Taxa new to the flora

New records for the flora of Prince Edward Island are a result of both new discoveries and correction of names on material that was previously misidentified. In the latter case the new record, i.e., the revised identification, is listed below and the incorrect identification to be deleted is listed under "Corrections and deletions" with the appropriate cross-reference. Name (nomenclatural) corrections also result in new names (list follows), but these are not included as "new records" unless the application of the name has been altered somewhat. The new records listed here include both records published after 1960 and various unpublished records. Square brackets indicate reports for which we have been unable to locate justifying material.

Herbarium acronyms, used under this heading and elsewhere to indicate the location of voucher material, follow: ACAD, Herbarium of Acadia University, Wolfville, Nova Scotia; DAO, Herbarium of the Biosystematics Research Institute, Agriculture Canada, Ottawa, Ontario; GH, Gray Herbarium of Harvard University, Cambridge, Massachusetts; NSPM, Herbarium of the Nova Scotia Provincial Museum, Halifax, Nova Scotia; NSAC, Herbarium of the Nova Scotia Agricultural College, Truro, Nova Scotia; UNB, Connell Memorial Herbarium, University of New Brunswick, Fredericton, New Brunswick; UPEI, Herbarium of the University of Prince Edward Island, Charlottetown, Prince Edward Island.

ISOETACEAE

Isoetes macrospora Dur. The previous mapping of *I. riparia* from Lake Verde and Glenfinnan Lake (Erskine 1960, see reprint following this introduction) is based on specimens of *I. macrospora* at ACAD, DAO and NSPM. It has been shown in this area of Queens Co. by Roland and Smith (1969).

OPHIOGLOSSACEAE

[***Botrychium dissectum*** Spreng.]. Shown by Roland and Smith (1969) in the region of West Point, Prince Co.

Botrychium simplex E. Hitchc. Moist turfy ground about 300 m south of the lighthouse and in heath on tops of the east cliffs, North Point, Prince Co. (28 July 1984, *P.M. Catling, V.R. Brownell, S. Varga* and *D.A. Metsger*, DAO, UNB). This species was previously rejected because of misidentification.

SPARGANIACEAE

[*Sparganium minimum* (Hartm.) Fries]. Shown from the Brackley Point Rd. area and North Point by Roland and Smith (1969).

JUNCAGINACEAE

Triglochin gaspensis Rieth & D. Love. Reported in 1960 as the small form of *T. elata* and because it was not recognized with formal taxonomic rank, it was not mapped or discussed. The only place where this taxon has been collected in Prince Edward Island is Souris, Kings Co. (4 Aug. 1952, *D.S. Erskine*, DAO).

POACEAE

Oryzopsis asperifolia Michx. Greenwich, Kings Co. (22 July 1971, *D. Griffin*, UPEI).

Dichanthelium ovale (Ell.) Gould & Clark var. *addissonii* (Nash) Gould & Clark. Collections previously referred to *P. subvillosum* Ashe and *P. lanuginosum* Ell. var. *implicatum* (Scribn.) Fern. belong here: Bothwell, Kings Co. (10 Aug. 1945, *A.E. Roland, W.G. Dore 45-2160A*, DAO); Alberton, Prince Co. (27 June 1952, *D. Erskine 1010, W.G. Dore*, DAO); Hunter River, Queens Co. (7 July 1952, *D. Erskine 1228*, DAO); Wood Islands, Queens Co. (21 July 1952, *D. Erskine 1333*, DAO); Wellington, Prince Co. (30 July 1952, *D. Erskine 1430*, DAO); Bothwell, Kings Co. (16 July 1953, *D. Erskine 2009, A.J. Smith*, DAO); North Point, Prince Co. (28 July 1984, *P.M. Catling, V.R. Brownell, S. Varga* and *D.A. Metsger*, DAO). See also under "Corrections and deletions."

CYPERACEAE

Eleocharis nitida Fern. Muddy depression above the shore, Cavendish Narrows, Lot 11, Prince Co. (15 Aug. 1970, *D. Erskine*, DAO). A poorly understood taxon lumped with *E. tenuis* by some recent authors.

[*Carex lupulina* Muhl.]. Shown from the area west of Charlottetown in Queens Co. by Roland and Smith (1969), but there are no justifying specimens at DAO, ACAD, NSPM or NSAC, and it was not mapped in Prince Edward Island by Reznicek and Ball *in* Can. J. Bot. 52: 2387-2399. 1974.

[*Carex lurida* Wahl.]. Shown from the Rollo Bay area in Kings Co. by Roland and Smith (1969), but there are no justifying specimens at ACAD, DAO, NSPM or NSAC.

Carex rugosperma Mackenzie. Brackley Beach and Brackley Point in Queens Co. See under "Corrections and deletions" *Carex umbellata*.

Carex wiegandii Mackenzie. 1.6 km (1 mile) south of Murray River, Kings County (22 July 1953, *D. Erskine 2133* and *A.J. Smith*, DAO). Previously determined and mapped as *Carex atlantica* (see under "Corrections and deletions").

JUNCACEAE

Juncus militaris Bigel. Keefe's Lake along route 5 east of Avondale, Queens Co. (29 July 1956, *D. Erskine*, DAO). Mapped by Roland and Smith (1969) and shown in Keefe's Lake (Erskine 1960, p. 22) but not included in the annotated list.

ORCHIDACEAE

Cypripedium calceolus L. var. ***pubescens*** (Willd.) Corell. Open swampy tamarack woods, ca. 4 km (2.5 miles) WNW of Miscouche, Prince Co. (29 July 1984, *P.M. Catling, V.R. Brownell, S. Varga* and *D.A. Metsger*, DAO, UNB).

Spiranthes cernua (L.) L.M.C. Rich. The specimen reported as *S. lucida* from Heatherdale in Kings Co. (6 Sept. 1957, *D. Erskine*, DAO) is actually this species. Also it has been found 1.6 km (1 mile) north of Caledonia in Kings Co. (25 Sept. 1954, *D. Erskine 2507*, DAO).

Spiranthes ochroleuca (Rydb. *ex* Britton) Rydb. East Point, Kings Co. (15 Sept. 1980, *D. Erskine*, TRT).

JUGLANDACEAE

[***Juglans cinerea*** L.]. Not mapped in Prince Edward Island by Little (*in* USDA Misc. Publ. 1146: map 133-E. 1971) and probably not native but now spreading from cultivation (D. McAskill, pers. comm.). In the mid-1970's this species was propagated and distributed on the island by the previously named Department of Agriculture and Forestry.

BETULACEAE

Betula cordifolia Regel. At least several localities on the island (Roland and Smith 1969; Brittain and Grant *in* Can. Field-Nat. 79: 253-257. 1965).

Betula pumila L. var. ***pumila***. Previously listed incorrectly as var. *renifolia* Fern.

[***Ostrya virginiana*** (Mill.) K. Koch.]. One location in Prince Co. and two locations in Queens Co. (D. McAskill, pers. comm.). Listed earlier in J.W. MacLeod's *The forests of Prince Edward Island* (Forestry Chron. 23: 190-193. 1947).

URTICACEAE

[***Laportea canadensis*** (L.) Gaudich.]. Dunk River, Lower Newton, Prince Co. (July–Aug. 1967, *D.S. Erskine*). Specimen apparently lost.

POLYGONACEAE

Polygonum arenastrum Boreau. See McNeill *in* Can. J. Bot. 59(12): 2744-2751. 1981.

CHENOPODIACEAE

Atriplex franktonii Taschereau. Launching Place, near Cardigan, Kings Co. (6 Sept. 1957, *D. Erskine*, DAO). Victoria, Queens Co. (22 Sept. 1968, *I. J. Bassett* and *K. W. Spicer, 5154, 5156*, DAO).

Atriplex subspicata (Nutt.) Rydb. Charlottetown, Queens Co. (21 Sept. 1968, *I.J. Bassett* and *K.W. Spicer 5146*, DAO). Wood's Island, Kings Co. (21 Sept. 1968, *I.J. Bassett* and *K.W. Spicer 5137*, DAO). East Beach, Wood Island Ferry Terminal, Kings Co. (22 Sept. 1979, *D.F. Brunton* and *H.L. Dickson*, DAO).

Chenopodium strictum Roth. var. ***glaucophyllum*** (Aellen) Wahl. See Bassett and Crompton *in* Can. J. Bot. 60: 586-610. 1982.

FUMARIACEAE

Dicentra cucullaria (L.) Bernh. Dunk River below Stanchel (5 June 1975, *R.B. MacLaren*, DAO). This species was listed in 1960 but in square brackets because there were no substantiating specimens.

BRASSICACEAE

Dentaria diphylla Michx. In woods in boggy soil 3–4 m from the edge of the Morell River, about 200 m below Indian Bridge, Kings Co. (24 May 1983, *R.B. MacLaren s.n.*, DAO).

Conringia orientalis (L.) Dumort. Murray River, Kings Co. (9 July 1974, *R.B. MacLaren*, DAO, P.E.I. Dept. Agric. Herb.). Reported in 1960 but without a voucher (Cody and MacLaren 1976).

Nasturtium microphyllum (Boenn.) Reichenb. Still water of large stream at western outskirts of Kensington, Prince Co. (10 Sept. 1954, *D. Erskine 2487*, DAO). In marshy clay at roadside, Mount Herbert, 6.4 km (4 miles) E of Charlottetown, Queens Co. (11 July 1956, *D. Erskine s.n.*, DAO). Both of these collections were initially determined as *N. officinale* and are the only justifying specimens for the occurrence of the latter taxon in Prince Edward Island.

HAMAMELIDACEAE

Hamamelis virginiana L. Murray River, shrubs in pine wood near Millpond (Aug. 1982, *D.S. Erskine*, DAO). Previously reported from the site by Stewart in 1971 in an International Biological Program report.

ROSACEAE

Crataegus holmesiana Ashe var. ***villipes*** Ashe. Specimens collected in 1952 and 1953 and identified later. Hunter River, route 13, Queens Co. (21 June 1953, *A.J. Smith 129*, DAO). Brackley Beach, National Park, Queens Co. (5 Sept. 1952, *D. Erskine 1680*, DAO). Brackley Point, Queens Co., 25 June 1953, *D. Erskine* 1777, DAO).

Geum canadense Jacq. Edge of wood, east side, Savage Harbour (16 Aug. 1967, *D.S. Erskine 005.67*, ACAD). Mapped for this area by Roland and Smith (1969, p. 452).

FABACEAE

Lotus corniculatus L. Wilmot Valley, Prince Co. (Aug. 1982, *D.S. Erskine*, DAO). Now widely used as a fodder crop and naturalized along roads and in pastures.

[*Coronilla varia* L.] Planted along roadsides to control erosion. Persisting but no evidence of it spreading.

Medicago sativa L. ssp. *falcata* (L.) Arcangeli. St. Felix (8 July 1975, *R.B. MacLaren*, DAO). Recent authors recognize two subspecies, this one not reported previously as growing without cultivation.

OXALIDACEAE

Oxalis corniculata L. A specimen from Charlottetown (Fernald and St. John, 14 Aug. 1914) was previously reported as *O. dillenii*.

GERANIACEAE

Erodium cicutarium (L.) L'Hér. Machon Point, Kings Co. (6 Aug. 1974, *R.B. MacLaren* (Cody and MacLaren 1976)).

EUPHORBIACEAE

Euphorbia glyptosperma Engelm. Several plants scattered along railroad at Montague, Kings Co. (7 Sept. 1976, *R.B. MacLaren*, DAO).

EMPETRACEAE

Empetrum rubrum Vahl ssp. *eamesii* (Fern. & Wieg.) R. Good. Roadside bank, Stanhope, Queens Co. (27 June 1945, *W.G. Dore* and *E. Gorham 45.322*, ACAD). Mapped by Roland and Smith (1969).

ACERACEAE

[*Acer platanoides* L.]. Sporadically escaped from cultivation (D. McAskill, pers. comm.).

MALVACEAE

[*Hibiscus trionum* L.]. This was overlooked by Erskine (1960) in his compilation, despite its listing by Campbell (1952).

CISTACEAE

Lechea maritima Leggett var. *subcylindrica* Hodgdon. Open dry sand of old and stable dune crests with *Hudsonia*, east end of Hog Island, Prince Co. (29 July 1984, *P.M. Catling, V.R. Brownell, S. Varga* and *D.A. Metsger*, DAO, UNB). This material has the leaves on the basal shoots (of the previous year) thick and hairy between the midrib and margins. Also, the seeds are smooth and lacking the reticulate membranaceous coat that is characteristic of the seeds of *L. intermedia* Leggett. The var. *subcylindrica* is restricted to the north coast of New Brunswick and this location on Prince Edward Island (Hodgdon *in* Rhodora 40: 29-69, 87-131. 1938).

VIOLACEAE

Viola conspersa Reichenb. Moist spruce forest south of the lighthouse on the west side, about 1 km south of the lighthouse, North Point, Prince Co. (28 July 1984, *P.M. Catling, V.R. Brownell, S. Varga* and *D.A. Metsger*, DAO, UNB).

Viola lanceolata L. Moist open ground with *Carex cannescens*, at the edge of the swamp south of the lighthouse, North Point, Prince Co. (28 July 1984, *P.M. Catling, V.R. Brownell, S. Varga* and *D.A. Metsger*, DAO, UNB).

ONAGRACEAE

Circaea lutetiana L. ssp. *canadensis* (L.) Asch. & Mag. under woods, east side, Savage Harbour (16 Aug. 1967, *J.S. Erskine 001.67*, ACAD). Mapped by Roland and Smith (1969), Cavendish (Aug. 1982, *D.S. Erskine*, DAO). See Boufford *in* Ann. Mo. Bot. Gard. 69: 985-986. 1982.

Circaea × *intermedia* Ehrh. (*C. alpina* L. × *C. lutetiana* L.). Lower Newton, Prince Co. (1967, *D.S. Erskine*, DAO). Scales Pond (1975, *D.S. Erskine*, DAO). See Boufford *in* Ann. Mo. Bot. Gard. 69: 985-986. 1982.

APIACEAE

Osmorhiza longistylis (Torr.) DC. Under woods, east side, Savage Harbour, (16 Aug. 1967, *J.S. Erskine 002.67*, ACAD). This specimen was collected after Erskine's flora was published but was mapped for the island by Roland and Smith (1969, p. 545).

BORAGINACEAE

Borago officinalis L. Three plants found at the end of an abandoned garden plot at Charlottetown Research Station (27 Sept. 1983, *R.B. MacLaren s.n.*, DAO).

SCROPHULARIACEAE

Verbascum phlomoides L. Dry field, Surrey, Queens Co. (20 July 1914, *M.L. Fernald* and *H. St. John 1174*, GH).

Euphrasia borealis (Towns.) Wettst. (*E. brevipila* in part, *E. stricta* in part, × *E. aequalis*). By a lobster cannery in turf on top of the sea cliff, 1.6 km (1 mile) SW of North Point, Prince Co. (28 July 1953, *D. Erskine 2166*, DAO).

Euphrasia stricta Wolf *ex* J.F. Lehmann. This taxon was dubiously reported in 1960 under *E. rigidula*. Listed from Prince Edward Island in a recent monograph by Sell and Yeo (*in* Bot. J. Linn. Soc. 63: 189-234. 1970) based on a sheet from Bideford (8 Aug. 1948, *A.R.A. Taylor*, DAO).

Scrophularia nodosa L. Vernon River, Queens Co. (3 July 1974, *R.B. MacLaren*, see Cody and MacLaren 1976). Also reported from Pownal (R. Curley, P.E.I. Dep. Agric., pers. comm.).

RUBIACEAE

Galium verum L. Port Hill, Prince Co. (5 Aug. 1975, *R.B. MacLaren*, DAO), and seen by MacLaren at Cove Head, Queens Co.

VALERIANACEAE

Valeriana officinalis L. Abundant and spreading at edge of churchyard, Rollo Bay, Kings Co. (5 July 1952, *D.S. Erskine 1191* and *J.E. Campbell*, DAO), seen also at Murray River and Charlottetown. This species was reported from Rollo Bay and Murray River in the 1960 edition (p. 43) but was left out of the annotated list.

LAMIACEAE

Teucrium canadense L. In eelgrass wrack on upper salt marsh, Cavendish Narrows in the vicinity of Milligan's Wharf southeast of Freeland (Aug. 1982, *D.S. Erskine*, DAO).

ASTERACEAE

Aster* × *blakei (Porter) House. Open boggy swampland above the spruce slope, 8 km (5 miles) west (by south) of Tignish, Prince Co. (29 July 1953, *D. Erskine* and *A.J. Smith, 2222 2223*, DAO).

Antennaria neodioica Greene ssp. ***canadensis*** (Greene) Bayer & Stebbins (*A. canadensis, A. neglecta* Greene var. *randii* (Fern.) Cronq.). Along fence, Heatherdale, Kings Co. (16 June 1948, *J.M. Bruce*, ACAD). Most other *Antennaria* collections from Prince Edward Island appear to be referrable to *A. neodioica* Greene ssp. *neodioica* (Bayer and Stebbins *in* Syst. Bot. 7(3): 300-313. 1982).

Crepis tectorum L. Previously in square brackets because in 1961 it was doubtful that it was naturalized. Since then it has been found in several places and is apparently well established. The most recent collections are from Rustico Island (Aug. 1982, *D.S. Erskine*, DAO).

Bidens connata Muhl. Village of O'Leary (*Fernald et al.*, GH). Included in the 1961 edition, but without a specimen. Also seen near Cherry Valley by D.S. Erskine.

[***Hieracium acuminatum*** Jordan]. Shown (as *Hieracium lachenalii* C.C. Gmel.) from the area of Dunk River in Prince Co. (Roland and Smith 1969).

New localities for certain rare species

TYPHACEAE

Typha angustifolia L. Ross Corner, Prince Co. (1960, *G.C. Cunningham*).

POACEAE

Phragmites australis (Cav.) Trin. *ex* Steudel. Previously known only from the Dunk River estuary, this species has recently been found at Cavendish Narrows and Lennox Island, both in Prince Co. (Aug. 1982, *D.S. Erskine*, DAO).

Phalaris canariensis L. Cavendish, Prince Edward Island National Park (5 Sept. 1982, *W.J. Cody 32189*, DAO).

FAGACEAE

[***Quercus robur*** L.] Spreading from cultivation at De Sable, Dunstaffnage, Orwell and Brudenell (D. McAskill, pers. comm.).

CHENOPODIACEAE

Atriplex laciniata L. Miminegash (Aug. 1982, *D.S. Erskine*, DAO).

CARYOPHYLLACEAE

Dianthus armeria L. Brackley Beach (Aug. 1982, *D.S. Erskine*, DAO).

MALVACEAE

Malva verticillata L. Cornwall (7 Oct. 1981, *R.B. MacLaren*, DAO). At edge of grainfield, Alexandra, Queens Co. (1976, *F. Houston*, P.E.I. Dep. Agric. Herb.).

APIACEAE

Angelica atropurpurea L. Murray River, Kings Co. (13 Aug. 1974, *R.B. MacLaren*, P.E.I. Dep. Agric. Herb., photo DAO, see Cody and MacLaren 1976).

SOLANACEAE

Hyoscyamus niger L. Miminegash (1919, *Perry*, ACAD). The listing of this species had been questioned and material evidence was lacking until the specimen at ACAD was recently located and confirmed.

SCROPHULARIACEAE

Veronica persica Poiret. Malpeque, Prince Co. (22 July 1904, *J. Fowler*, DAO) and Chepstow, Kings Co. (15 Sept. 1980, *D.S. Erskine*, DAO).

ASTERACEAE

Tussilago farfara L. Indian Bridge, Bangor, Kings Co. (9 May 1974, *R.B. MacLaren*, DAO, P.E.I. Dep. Agric. Herb., Cody and MacLaren 1976) and abundantly naturalized on Lennox Island, Prince Co. (Aug. 1983, *D.S. Erskine*, DAO).

Changes in status

In addition to new records for the province and new records for certain rare species, the status of other species has changed appreciably since 1960. For example, *Polygonum sachalinense* (p. 140) is only an occasional escape and has not become established since the extensive colonies at Souris were destroyed. In contrast, *Bromus inermis* (p. 66) and *Daucus carota* (p. 208) have become much more common and are now common throughout the province.

With regard to cultivated vegetation (described previously on p. 31), the practice of making silage has become more common; the cultivation of corn and tobacco in the southeast (and locally elsewhere) has become common. Barley and mixed grain are now more widely grown.

NOMENCLATURAL CHANGES

There were two new combinations published in the first printing, namely *Toxicodendron radicans* (L.) Ktze. var. *rydbergii* (Small) Erskine and *Hudsonia* × *intermedia* (Peck) Erskine. The effective publication date for these is 7 Feb. 1961. The combination *Plantago juncoides* Lam. var. *decipiens* (Barnéoud) was incorrectly indicated as new (on p. 236).

It has been recommended that certain family names be changed and the recommendation has been accepted by some recent authors. Included here are Poaceae to replace Gramineae, Brassicaceae to replace Cruciferae, Fabaceae to replace Leguminosae, Apiaceae to replace Umbelliferae and Asteraceae to replace Compositae.

Since 1960, many systematists or a specific author (as indicated) have recommended various changes in scientific names and authorities. These recommended name changes are not all readily available and do not appear in many more recent publications. They are listed below alphabetically with the old name and authority first, followed by the proposed name and authority. References are included where they are particularly useful or explanatory. Although no distinction is made here between scientific name changes, which represent corrections (involving the *International Code of Botanical Nomenclature*), and changes reflecting a general change in taxonomic judgment, it is possible to follow up many of these changes using the references provided.

Alisma triviale Pursh - *Alisma plantago-aquatica* L. var. *americana* Schultes & Schultes, but *A. triviale* has been maintained by Pogan *in* Can. J. Bot. 41: 1011-1013. 1963 and by Bjorquist *in* Opera Bot. 19: 104-105. 1968.

Alnus crispa (Ait.) Pursh var. *mollis* Fern. - *Alnus viridis* (Villars) Lamarck & DeCandolle ssp. *crispa* (Aiton) Turrill. See Furlow *in* Rhodora 81: 209. 1979.

Alnus rugosa (Du Roi) Spreng. var. *americana* (Regel) Fern. - *Alnus incana* (Linnaeus) Moench ssp. *rugosa* (Du Roi) Clausen. See Furlow *in* Rhodora 81: 209. 1979.

Ambrosia psilostachya DC. var. *coronopifolia* (T. & G.) Farwell - *Ambrosia psilostachya* DC. See Bassett and Crompton *in* Can. J. Plant Sci. 55: 463-476. 1975.

Antennaria petaloidea Fern. var. *subcorymbosa* Fern. - *Antennaria neodioica* Greene ssp. *petaloidea* (Fern.) Bayer & Stebbins. See Bayer and Stebbins *in* Syst. Bot. 7(3): 300-313. 1982.

Antennaria neodioica Greene var. *attenuata* Fern. - *Antennaria neodioica* Greene ssp. *neodioica*. See Bayer and Stebbins *in* Syst. Bot. 7(3): 300-313. 1982.

Antennaria neodioica Greene var. *chlorophylla* Fern. - *Antennaria neodioica* Greene ssp. *neodioica*. See Bayer and Stebbins *in* Syst. Bot. 7(3): 300-313. 1982.

Antennaria neodioica Greene var. *neodioica* - *Antennaria neodioica* Greene ssp. *neodioica*. See Bayer and Stebbins *in* Syst. Bot. 7(3): 300-313. 1982.

Agropyron repens (L.) Beauv. - Some recent authors use *Elymus repens* (L.) Gould or *Elytrigia repens* (L.) Nevski.

Agropyron trachycaulum (Link) Malte. - Some recent authors use *Elymus trachycaulis* (Link) Gould *ex* Shinners. The P.E.I. taxa are then subsp. *trachycaulis*, subsp. *major* (Vasey) Tsvelev (= *A. trachycaulum* var. *majus*) and subsp. *novae-angliae* (Scribner) Tsvelev (= *A. trachycaulum* var. *novae-angliae*).

Agrostis palustris Huds. - *Agrostis stolonifera* L.

Agrostis tenuis Sibth. - *Agrostis capillaris* L.

Arenaria lateriflora L. - *Moehringia lateriflora* (L.) Fenzl.

Arenaria groenlandica (Retz.) Spreng. - *Minuartia groenlandica* (Retz.) Ostenf.

Arenaria peploides L. var. *robusta* Fern. - *Honckenya peploides* (L.) Ehrh. subsp. *robusta* (Fern.) Hultén. See McNeill *in* Rhodora 82: 495. 1980.

Arenaria stricta Michx. - *Minuartia michauxii* (Retz.) Ostenf.

Arctium minus (Hill) Bernh. - *Arctium minus* Berhn.

Aster junciformis Rydb.- Some recent authors use *Aster borealis* (T. & G.) Provancher.

Athyrium thelypterioides (Michx.) Desv. Some recent authors use *Diplazium acrostichoides* (Sw.) Butters.

Atriplex patula L. var. *hastata* (L.) Gray - *A. prostrata* Buch. See Bassett, Crompton, McNeill and Taschereau *in* Agric. Can. Monogr. 31: 72 pp. 1983.

Atriplex patula L. var. *littoralis* (L.) Gray - *Atriplex littoralis* L. See Bassett, Crompton, McNeill and Taschereau *in* Agric. Can. Monogr. 31: 72 pp. 1983.

Atriplex sabulosa Rouy - *Atriplex laciniata* L. See Bassett, Crompton, McNeill and Taschereau *in* Agric. Can. Monogr. 31: 72 pp. 1983.

Azalea canadensis Ktze. - *Rhododendron canadense* (L.) BSP.

Betula lutea Michx. f. - *Betula alleghenienss* Britton.

Brassica hirta Moench - *Sinapis alba* L.

Brassica kaber (DC.) Wheeler - *Sinapis arvensis* L.

Calamagrostis neglecta (Ehrh.) P. Gaertner, B. Meyer, & Scherb. - *Calamagrostis stricta* (Timm) Koeler.

Calopogon pulchellus (Salisb.) R. Br. - *Calopogon tuberosus* (L.) BSP. See Voss *in* Rhodora 68: 460. 1966.

Carex abdita Bickn. - *C. umbellata* Willd.

Carex angustior Mackenzie - *Carex echinata* Murray. See Reznicek and Ball *in* Univ. Mich. Herb. Contrib. 14: 153-203. 1980.

Carex bebbii Olney - *Carex bebbii* (Bailey) Fern. See Voss *in* Rhodora 68: 451. 1966.

Carex canescens L. var. *disjuncta* Fern. - *Carex canescens* L. ssp. *disjuncta* (Fern.) Toivonen. See Toivonen *in* Ann. Bot. Fenn. 18: 94. 1981.

Carex cephalantha (Bailey) Bick. - *Carex echinata* Murray ssp. *echinata*. See Reznicek *in* Univ. Mich. Herb. Contrib. 14: 153-203. 1980.

Carex hystricina Muhl. - *Carex hystericina* Willd. See Voss *in* Rhodora 68: 452. 1966.

Carex tonsa (Fern.) Bickn. - *Carex rugosperma* Mack. var. *tonsa* (Fernald) E.G. Voss. See Voss *in* Rhodora 68: 452. 1966.

Chrysanthemum leucanthemum L. - *Leucanthemum vulgare* Lam.

Chrysanthemum parthenium (L.) Bernh. - *Tanacetum parthenium* (L.) Schultz.

Cinna latifolia (Trev.) Griseb. - *Cinna latifolia* (Goepp.) Griseb.

Comandra richardsiana Fern. - sometimes included with *C. umbellata* (L.) Nutt. or *C. umbellata* (L.) Nutt. var. *decumbens* E.J. Hill.

Comptonia peregrina (L.) Coult. - *Myrica asplenifolia* L.

Convolvulus sepium L. - *Calystegia sepium* (L.) R. Br. See Brumitt and Groves *in* Kew Bull. 36(2): 422. 1981.

Cornus stolonifera Michx. - *Cornus sericea* L., possibly the correct name.

Crataegus succulenta Schrad. - *Crataegus succulenta* Link.

Digitaria ischaemum (Schreb.) Muhl. - *Digitaria ischaemum* (Schreb.) Schreb. *ex* Muhl.

Dryopteris spinulosa (O.F. Muell.) Watt. var. *americana* (Fisch.) Fern. - *Dryopteris campyloptera* (Kze.) Clarkson.

Dryopteris spinulosa (O.F. Muell.) Watt. var. *spinulosa* - *Dryopteris carthusiana* (Vill.) H.P. Fuchs.

Dryopteris spinulosa (O.F. Muell.) Watt. var. *fructuosa* (Gilbert) Trudell. - *Dryopteris* × *triploidea* Wherry.

Dryopteris spinulosa (O.F. Muell.) Watt. var. *intermedia* (Muhl.) Underw. - *Dryopteris intermedia* (Muhl.) A. Gray.

Empetrum atropurpureum Fern. & Wieg. - *Empetrum rubrum* Vahl. var. *atropurpureum* (Fern. & Wieg.) R. Good.

Eragrostis poaeoides Beauv. - *Eragrostis minor* Host. See McNeill and Dore *in* Nat. Can. (Que.) 103: 533-567. 1976.

Erigeron canadensis L. - *Conyza canadensis* (L.) Cronq.

Euphorbia supina Raf. - *Euphorbia maculata* L. See Mulligan and Lindsay *in* Nat. Can. (Que.) 105: 37-40. 1978.

× *Euphrasia aequalis* Collen - *Euphrasia borealis* (Towns.) Wettst.

Euphrasia americana Wettst. - *Euphrasia nemorosa* (Pers.) Wallr.

Euphrasia randii Rob. var. *reeksii* Fern. - *Euphrasia randii* Rob. var. *randii*.

Euphrasia rigidula Jordan - *Euphrasia stricta* Wolf *ex* J.F. Lehmann.

Festuca capillata Lam. - *Festuca tenuiflora* Sibth.

Galinsoga ciliata (Raf.) Blake - *G. quadriradiata* Ruiz et Pavon. See Canne *in* Rhodora 79: 319-389. 1977.

Gnaphalium macounii Greene - *Gnaphalium viscosum* HBK.

Habenaria blephariglottis (Willd.) Torr. - *Platanthera blephariglottis* (Willd.) Lindley.

Habenaria clavellata (Michx.) Spreng. - *Platanthera clavellata* (Michx.) Luer.

Habenaria dilatata (Pursh) Gray - *Platanthera dilatata* (Pursh) Lindley.

Habenaria fimbriata (Ait.) R. Br. (not yet confirmed in P.E.I.) - *Platanthera grandiflora* (Bigelow) Lindley.

Habenaria hookeri Torrey - *Platanthera hookeri* (Torrey) Lindley.

Habenaria hyperborea (L.) R. Br. - *Platanthera hyperborea* (L.) Lindley.

Habenaria lacera (Michx.) Lodd. - *Platanthera lacera* (Michx.) G. Don.

Habenaria obtusata (Pursh) Richards - *Platanthera obtusata* (Banks *ex.* Pursh) Lindley.

Habenaria orbiculata (Pursh) Torrey - *Platanthera orbiculata* (Pursh) Lindley.
Habenaria psycodes (L.) Spreng. - *Platanthera psycodes* (L.) Lindley.
Heracleum lanatum Michx. - *Heracleum sphondylium* L. ssp. *montanum* (Schleicher *ex* Gaudin) Briquet *in* Shiny & R. Keller. See Brummit *in* Rhodora 73: 578-584. 1971.
Hieracium pratense Tausch - *Hieracium caespitosum* Dumort.
Hypericum boreale (Britton) Bickn. - *Hypericum mutilum* L. ssp. *boreale* (Britton) J.M. Gillett. See Gillett *in* Natl. Mus. Can. Publ. Bot. No. 11. 1981.
Hypericum virginicum L. var. *fraseri* (Spach) Fern. - *Triadenum virginicum* (L.) Raf. ssp. *fraseri* (Spach) J.M. Gillett. See Gillett *in* Natl. Mus. Can. Publ. Bot. No. 11. 1981.
Juncus alpinus Vill. - *Juncus alpinoarticulatus* Chaix *in* Villars. See Hamet-Ahti *in* Ann. Bot. Fenn. 17: 341-342. 1980.
Juncus effusus L. var. *effusus* - *Juncus effusus* L. ssp. *effusus*. See Hamet-Ahti *in* Ann. Bot. Fenn. 17: 183-191. 1980.
Juncus effusus L. var. *solutus* Fern. & Wieg. - *Juncus effusus* L. ssp. *solutus* (Fern. & Wieg.) Hamet-Ahti. See Hamet-Ahti *in* Ann. Bot. Fenn. 17: 183-191. 1980.
Juncus effusus L. var. *costulatus* Fern. - *Juncus pylaei* Leharpe. See Hamet-Ahti *in* Ann. Bot. Fenn. 17: 183-191. 1980.
Juncus effusus var. *compactus* Lej. & Court. - *Juncus effusus* L. ssp. *effusus*. See Hamet-Ahti *in* Ann. Bot. Fenn. 17: 183-191. 1980.
Juncus effusus L. var. *pylaie* (Leharpe) Fern. & Wieg. - *Juncus pylaie* Leharpe. See Hamet-Ahti *in* Ann. Bot. Fenn. 17: 183-191. 1980.
Lactuca scariola L. - *Lactuca serriola* L.
Lappula myosotis Moench - *Lappula squarrosa* (Retz.) Dumort.
Lychnis alba Mill. - *Silene pratensis* (Raf.) Godron & Gren. See McNeill and Prentice *in* Taxon 30(1): 27-32. 1981.
Lycopodium complanatum L. var. *flabelliforme* Fern. - *Lycopodium digitatum* A. Braun.
Lycopodium complanatum L. - *Diphasium complanatum* (L.) Rothm.
Lycopodium inundatum L. - *Lepidotis inundata* (L.) C. Borner.
Lycopodium tristachyum Pursh - *Diphasium tristachyum* (Pursh) Rothm.
Lycopsis arvensis L. - *Anchusa arvensis* (L.) Bieb.
Matricaria matricarioides (Less.) Porter - *Chamomilla suaveolens* (Pursh) Rydb.
Microstylis unifolia (Michx.) BSP. - *Malaxis unifolia* Michx.
Nicandra physalodes (L.) Pers. - *Nicandra physalodes* (L.) Gaertn.
Nuphar variegata Engelm. - *N. variegata* Durand *in* Clinton. See Voss *in* Taxon 14(5): 154-160. 1965.
Oxalis dillenii Jacq. ssp. *dillenii sensu* Eiten (*in* Am. Midl. Nat. 69(2): 257-309. 1963) - *Oxalis stricta* L. *sensu* Young (*in* Flora Europaea 2: 192. 1968) and Lourtieg (*in* Phytologia 42(2): 57-198. 1979).
Oxalis stricta L. *sensu* Eiten (*in* Am. Midl. Nat. 69(2): 257-309. 1963) - *Oxalis europaea* Jordan *in* F.W. Schultz *sensu* Young *in* Flora Europaea 2: 192. 1968. *Oxalis fontana* Bunge var. *fontana sensu* Lourtieg *in* Phytologia 42(2): 57-198. 1979.
Panicum boreale Nash - *Dichanthelium boreale* (Nash) Freckmann (Phytologia 39: 269. 1978).

Panicum depauperatum Muhl. var. *psilophyllum* Fern. - *Panicum depauperatum* Muhl. var. *involutum* (Torr.) Wood. See Voss *in* Rhodora 68: 443. 1966. *Dichanthelium depauperatum* (Muhl.) See Gould *in* Brittonia 26: 59. 1975.

Panicum lanuginosum Ell. var. *implicatum* (Scribn.) Fern. - *Dichanthelium acuminatum* (Swartz) Gould & Clark var. *implicatum* (Scribn.) Gould & Clark. See Gould and Clark *in* Ann. Mo. Bot. Gard. 65: 1088-1132. 1978.

Panicum subvillosum Ashe - *Dichanthelium acuminatum* (Swartz) Gould & Clark var. *acuminatum*. See Gould and Clark *in* Ann. Mo. Bot. Gard. 65: 1088-1132. 1978.

Phragmites communis Trin. var. *berlandieri* (Fourn.) Fern. - *Phragmites australis* (Cav.) Trin. *ex* Steudel. See Voss *in* Mich. Bot. 11: 31. 1972.

Plantago juncoides Lam. var. *decipiens* (Lam.) Bernéoud - *Plantago maritima* L. See Bassett *in* Agric. Can. Monogr. 7: 25. 1973.

Plantago oliganthos Roem. & Schult. - *Plantago maritima* L. See Bassett *in* Agric. Can. Monogr. 7: 25. 1973.

Polygonum allocarpum Blake is included with *Polygonum fowleri* Robins. by some recent authors.

Polygonum aviculare L. Some authors refer to the eastern Canadian plants as *P. monspeliense*. See McNeill *in* Can. J. Bot. 59(12): 2744-2751. 1981.

Rhinanthus crista-galli L. - *Rhinanthus minor* L.

Rosa cinnamomea L. - *Rosa majales* J. Herrmann

Salix gracilis Anders. var. *textoris* Fern. - *Salix petiolaris* Smith. See Ball *in* Bull. Torrey Bot. Club 75: 178-187. 1948.

Salix rigida Muhl. - *Salix eriocephala* Michx. See Argus *in* Brittonia 32(2): 170-177. 1980.

Satureja acinos (L.) Scheele - *Acinos arvensis* (Lam.) Dandy.

Scirpus americanus Pers. - *Scirpus pungens* Vahl. See Schuyler *in* Rhodora 76: 51-52. 1974.

Silene cucubalus Wibel - *Silene vulgaris* (Moench) Garcke.

Sonchus arvensis L. var. *glabrescens* Gunth., Grab. & Wimm. - *Sonchus uliginosus* Bieb.

Stellaria media (L.) Cyrillo. - *Stellaria media* (L.) Vill. See Turkington *in* Can. J. Plant Sci. 60: 981. 1980.

Suaeda americana (Pers.) Fern. - *Suaeda calceoliformis* (Hook.) Moq. - See McNeill et. al. *in* Rhodora 79: 133-138. 1977.

Thelypteris dryopteris (L.) Slosson. - *Gymnocarpium dryopteris* (L.) Newm. ssp. *dryopteris*

Thelypteris phegopteris (L.) Slosson. - *Phegopteris connectilis* (Michx.) Watt.

Torreyochloa fernaldii (Hitchc.) Church. - Some recent authors include this taxon in *Puccinellia*, the name then being *Puccinellia fernaldii* (Hitchc.) E.G. Voss. See Voss *in* Rhodora 68: 445. 1966.

Toxicodendron radicans (L.) Ktze. - *Rhus radicans* L. See Mulligan and Junkins *in* Nat. Can. (Que.) 105: 291-293. 1978 and McNeill *in* Nat. Can. (Que.) 108: 237-244. 1981.

Trifolium agrarium L. - *Trifolium aureum* Pollich.

Trifolium procumbens L. - *Trifolium campestre* Schreber.

Urtica gracilis Ait. - *Urtica dioica* L. ssp. *gracilis* (Ait.) Selander. See Bassett, Crompton and Woodland *in* Can. J. Bot. 52: 503-516. 1974.

Urtica procera Muhl. - *Urtica dioica* L. spp. *gracilis* (Ait.) Selander. See Bassett, Crompton and Woodland *in* Can. J. Bot. 52: 503-516. 1974.

Zizania aquatica L. var. *interior* Fassett - *Zizania palustris* L. var. *interior* (Fassett) Dore.

CORRECTIONS AND DELETIONS

ISOETACEAE

Isoetes riparia Engelm. The justifying specimens from Lake Verde and Glenfinnan Lake have been revised to *I. macrospora* Dur.

ASPIDIACEAE

Thelypteris simulata (Davenp.) Nieuwl. Included in the 1960 edition on the basis of a report by Roland (1947) but not included in the more up-to-date Roland and Smith (1969). No justifying specimen could be found.

POTAMOGETONACEAE

Potamogeton gramineus L. Several attempts to relocate the marsh at Cherry Valley, where this species occurred, have failed. Because there are no voucher specimens it should be treated as unconfirmed and placed in square brackets.

POACEAE

Panicum subvillosum Ashe.

Panicum lanuginosum Ell. var. ***implicatum*** (Scribn.) Fern. The justifying specimens have been revised to *Dichanthelium ovale* (Ell.) Gould & Clark var. *addissonii* (Nash) Gould & Clark. See Gould and Clark *in* Ann. Mo. Bot. Gard. 65: 1088-1132. 1978. See also under "New records."

CYPERACEAE

Carex atlantica Bailey (presumably ssp. *atlantica*). The justifying specimen from a swampy roadside ditch 1.6 km (1 mile) south of Murray River, Kings Co. (22 July 1953, *D. Erskine* and *A.J. Smith 2133*, DAO) has been revised to *Carex wiegandii* Mackenzie. See Reznicek and Ball *in* Univ. Mich. Herb. Contrib. 14: 153-203. 1980.

Carex atlantica (Bailey) ssp. ***capillacea*** (Bailey) Reznicek. (= *C. howei* Mack). The mapping of *C. howei* in Prince Co. by Roland and Smith (1969, p. 158) is apparently based on a sheet at NSPM, collected at McNeill's Mills on Western Road (13 Aug. 1953, *D. Erskine 2381* and *A.J. Smith*). The plants on this sheet, however, are clearly referrable to *Carex interior* Bailey.

Carex umbellata Schkuhr. The only justifying specimens at DAO are from Brackley Beach and Brackley Point in Queens Co. In conformity with the trend toward restricting the name *umbellata* to the taxon with the relatively short-beaked, smaller perigynia (Voss *in* Mich. Flora, p. 287. 1972), these specimens are best referred to *Carex rugosperma* Mackenzie.

JUNCACEAE

Juncus compressus Jacq. This report is based on specimens of *Juncus gerardii* Loisel. See Stuckey *in* Can. Field-Nat. 95(2): 169. 1981.

ORCHIDACEAE

Platanthera grandiflora (Bigel.) Lindl. (= *Habenaria fimbriata* (Ait.) R. Br., *H. grandiflora* (Bigel.) Torr.). Erskine (1960) rejected previous reports of this species in P.E.I., explaining that they were based on specimens of *P. psycodes*. The mapping for the Bay Fortune area by Roland and Smith (1969, p. 224) apparently rests on a specimen at ACAD collected in mixed woods at Bay Fortune, Kings Co. (29 July 1954, *H.E. Aitken 56*). The relatively small sepals, deeply lobed and narrow lobes of the lip and oblong (instead of ovate) petals suggest that this plant is better referred to *Platanthera* × *andrewsii* (Niles) Luer (*P. psycodes* × *P. lacera*).

Spiranthes lucida (H.H. Eaton) Ames. This report is based on specimens of *S. cernua* (DAO).

CHENOPODIACEAE

Chenopodium rubrum L. This report is based on specimens of *C. glaucum* (DAO).

NYMPHAEACEAE

Nymphaea alba Ait. The collection from Charlottetown, Queens Co. (5 July 1952, *D.S. Erskine*, DAO) is lacking rhizomes. In the European plant, the leaves are reported to be crowded on the rhizome, whereas the North American plant has the leaves scattered. Other differences are less convincing. The reference to "leaves crowded" on the label refers to the surface of the water (D.S. Erskine, pers. comm.) rather than the rhizome. Until the Prince Edward Island material is studied in much more detail, we prefer to treat the plants previously referred to *N. alba* in Prince Edward Island as a larger cultivated form of *N. odorata*.

BRASSICACEAE

Nasturtium officinale R. Br. There are only two justifying specimens at DAO, both of which have been revised to *N. microphyllum*. See under "New records."

HALORAGIDACEAE

Myriophyllum tenellum Bigel. Although not included in the annotated list, this species was listed in descriptions of freshwater communities on page 23. It should be deleted due to the absence of material evidence.

GENTIANACEAE

Nymphoides cordata (Ell.) Fern. (= *Limnanthemun cordatum* (Ell.)). The mapping of this species in Queens Co. by Roland and Smith (1969, p. 582) is based on a sheet at ACAD (30 Aug. 1953, *D. Erskine 2477*), which has been revised to *Nymphaea odorata* Ait. based on the thin roundish and purple-tinged basal leaves and lack of glandular dots on the undersurface of the floating leaves.

SCROPHULARIACEAE

Euphrasia tetraquetra (Brébisson) Arrondeau (*E. canadensis* Towns.) Not listed for Prince Edward Island in recent revision by Sell and Yeo *in* Bot. J. Linn. Soc. 63: 189-234. 1970. Only one specimen so labeled was found at DAO *(Erskine 2491)*. This has been revised to *E. nemorosa*.

OTHER CORRECTIONS

Title page: the actual date of publication was 7 February 1961.

page 7, lines 32 and 33: delete "Watson died about 1908" and replace with "Watson's botanical work ceased about 1908."

page 12, line 40: The New Brunswick clayey till is not the parent material for the O'Leary clay loam; the latter is based on till from deeper shalier beds of the local bedrock exposed in the shallower western part of the sedimentary basin.

page 19, line 9: The percentages were derived from figures for potential agricultural land (farmland) and are incorrect because this was not equivalent to total land area, even in the Central Upland.

page 29, line 11: delete "*nashu*"; insert "*nashii.*"

page 33, line 13 and following: The fact of a probable land bridge connecting Prince Edward Island with the mainland (Kranck 1970) represents another means of post-Wisconsin migration.

page 35, line 38: The statement relating to *Pilea pumila* must be dropped because it has now been found in northern Nova Scotia.

page 36, line 10: The sentence beginning "Thus ..." is not entirely reasonable because it does not rule out the possibility of relicts after the destruction of presettlement forest habitat.

page 37, line 2: insert "do not" after "which."

page 37, line 2: delete "35"; insert "135."

page 37, line 5: delete "*Pilea pumila.*"

page 43, line 33: delete "*Teucrium canadense*" because it has recently been found. See under "New records."

page 50, line 19: delete "*marginale*"; insert "*marginalis.*"
page 54, between lines 22 and 23: insert "PINACEAE."
page 56, line 6: delete "and"; insert "an."
page 56, line 33: delete "Scoria"; insert "Scotia"; delete second "a" of "Hackamatack."
page 74, line 5: delete "FUBETTE"; insert "FINETTE."
page 118, line 32: delete "provincial emblem" under "*Cypripedium reginae* Walt." With the floral emblem act in 1947 the provincial legislative assembly of Prince Edward Island established the lady's-slipper orchid as the province's floral emblem. Of the three species of lady's-slipper found in the province, the floral emblem became the showy lady's-slipper, *Cypripedium reginae*. This plant was shown on the cover of the 1960 edition of the *Plants of Prince Edward Island*. A few years later not enough of this species could be found in the province to make a display for the Fathers of Confederation Building in Charlottetown. The showy lady's-slipper had always been scarce and largely restricted to the western part of Prince County. Therefore it became apparent that the showy lady's-slipper was not the best choice for the provincial floral emblem. In 1965 a more precise and appropriate botanical name was included in an amendment to the floral emblem act: the stemless lady's-slipper, also called the moccasin flower, *Cypripedium acaule* Ait., shown on the front cover of this edition.
page 128, line 31: "Nuts...;" replace with "Nuts formerly gathered and eaten."
page 142, line 15: delete "Launching."
page 152, line 11: delete "*variegatum*"; insert "*variegata.*"
page 154, line 23: transfer Map 408 to line 26.
page 176, line 18: delete "Bailey)"; insert "Blanchard; *R. canadensis* var. *pergratus* (Blanchard) Bailey, Roland)."
page 176, lines 19 and 20: delete "*R. canadensis*...Roland)."
page 185, map 517: delete "var. *linearis*."
page 192, line 14: delete square brackets.
page 194, line 5 from bottom: delete square brackets.
page 210, line 28: delete comma from "(DC.), Domin."
page 220, line 22: delete square brackets and apply to line 27.
page 234, line 11: delete "southwestern"; insert "southeastern."
page 236, line 13: delete "n. comb., based"; insert "Fern..."
page 254, line 23: remove square brackets; delete "if present"; insert "West Prince."
page 262, line 12: delete second "a" of "Tragopogan" and replace with "o."

ACKNOWLEDGMENTS

We appreciate correspondence from D. McAskill and I. MacQuarrie. A.A. Reznicek and B. Crins provided many helpful comments. V.R. Brownell, S. Varga and D.A. Metsger assisted in field studies.

REFERENCES

Anderson, T.W. 1980. Holocene vegetation and climatic history of Prince Edward Island, Canada. Can. J. Earth Sci. 17: 1152-1165.

Cody, W.J. and R.B. MacLaren. 1976. Additions and rediscoveries of five plant species in Prince Edward Island. Can. Field-Nat. 90: 53-54.

Kranck, K. 1972. Geomorphological development and post-Pleistocene sea level changes, Northumberland Strait, Maritime Provinces. Can. J. Earth Sci. 9: 835-844.

Prest, V.K. 1973. Surficial deposits of Prince Edward Island. Geol. Surv. Can. Map 1366A (with descriptive notes).

Roland, A.E. and E.C. Smith. 1969. The flora of Nova Scotia. Nova Scotia Museum. 746 pp.

P.M. Catling

D.S. Erskine

R.B. MacLaren

PUBLICATION 1088 DECEMBER 1960

THE PLANTS

of

PRINCE EDWARD ISLAND

by

David S. Erskine

Plant Research Institute

Research Branch

CANADA DEPARTMENT OF AGRICULTURE

CONTENTS

THE PLANTS OF PRINCE EDWARD ISLAND

INTRODUCTION

An annotated list of the plants of Prince Edward Island, besides fulfilling a long-felt want, may serve many interests. Sources of information on the flora are at present limited to very incomplete, privately published lists without indication of the distribution of the species. "The Island," as inhabitants call the province, is a predominantly agricultural area; 87% of its land surface is farmland. Thus, the survey may serve the needs of those who want to know what wild plants are present that may serve as hosts to fungus or insect diseases of crop plants, or what noxious weeds are present that should be stamped out. At the same time, a survey of the plants serves the broader interests of plant geographers who wish to know what species occur and how far agriculture has modified the natural flora.

As only a dozen of the species present are not included in Roland's *Flora of Nova Scotia* (1945), and the *Grasses of Nova Scotia* by Dore and Roland (1942), keys for identification and descriptive notes are superfluous. Victorin's *Flore Laurentienne* (1935) is also applicable to the area, but, however, it includes a smaller proportion of the Island's species. Comprehensive for the entire maritime area are the eighth edition of *Gray's Manual of Botany* (1950), and Gleason's edition of *Britton and Brown's Illustrated Flora* (1952), either of which must be resorted to for new additions to the flora, but which seldom give any indication of the presence or absence of a species from the Island itself.

Procedure of Field Survey

The survey was projected and initiated by the Botany Division, of the Canada Department of Agriculture, Ottawa, and carried out in the summers of 1952 and 1953 by the author under the guidance of W.G. Dore, and in 1953 with the field assistance of A.J. Smith of Macdonald College. A base of operations was set up at the Science Service Laboratory at Charlottetown, but limited transportation and the expanse of the Island made the establishment of field bases essential. These were set up for periods of two to seven days at towns and villages in the country: at Alberton, Hunter River, Wood Islands, Wellington, Souris (twice), Dalvay, O'Leary and Brackley Beach in the first year; at Tignish (twice), O'Leary (twice), Bideford, Wellington, DeSable, Sea View, Hunter River, Cavendish, Brackley Beach, Souris (twice) and Murray River, the second. It was planned as far as possible to cover each apparently significant region twice at different parts of the season. Although walking is the only means of progression permitting collection of plants, bus, train and car were used to expand the radius of exploration at each center; all-day trips from Charlottetown through the country or to a given locality often proved very profitable. On car or train, a list of sight records was kept in the manner described by Groh (1927) for his weed survey. At first, collecting of specimens took precedence; our objective was to obtain a complete set of the Island species in triplicate. As this phase approached completion, sight records and novelties became more important, the former filling in the natural distribution pattern of common species in the Island, the latter adding rare species.

Collected specimens were made to conform to the size of the standard herbarium sheet, 17 by 11 inches. Plants of a herbaceous nature were excavated to show underground structures. The specimens were dried in newsprint folders between blotters in a plant press aerated by corrugates and set over infra-red bulbs on a collapsible aluminum support. This device did a thorough job of drying in 24 hours in all but the most succulent plants. Records as to locality, habitat, distinctive features and date of collection of the specimens were kept on numbered field labels. The sight records were kept by locality and transferred to taxonomic order by using initials of locality against a checklist. On the distribution maps, collections were indicated by solid discs, sight records by open circles. The prepared specimens resulting from the work of the survey, some 1800 collections in all, were distributed in sets as follows: first or "master" set, retained for preservation in the herbarium of the Botany Division, Ottawa; second set, to Acadia University, Wolfville, Nova Scotia.; third, fourth and fifth sets, less complete, to the University of Montreal, the British Museum of Natural History and the New York Botanic Garden. Small maps similar to those in this publication were reproduced photographically in five copies and affixed to the first five sets of specimens. A representative set (one of each species) was placed in the Science Service Laboratory, Charlottetown. Excess replicates have been distributed variously.

Previously, knowledge of the flora of Prince Edward Island was summarized from three sources, herbarium specimens, published lists and conversations with the local amateurs of botany. All herbaria of significance were consulted. The information, kept in loose-leaf files, recorded all data on locality, habitat and collection. These loose-leaf files are being preserved at Ottawa. The list at the end of the historical section presents the distribution of Prince Edward Island specimens in herbaria.

History of Botanical Investigations

The development of botanical knowledge of the Island has taken two separate courses: that acquired by botanists of universities and museums outside the province and that by local amateurs. It began late and developed haltingly. The first native worker in botany was the general naturalist Francis Bain of York Point near Charlottetown, better known as a geologist and author of ***The Natural History of Prince Edward Island*** (1890), a minute school book which treats the plants in 44 pages, family by family, with occasional details of occurrence and habitat for conspicuous species and with entries such as "five asters" for less easily distinguished species. To discover which "five asters" were intended, one has to turn to the privately published list of the Island flora by John MacSwain and Francis Bain (1891). This list included 430 vascular plants. A supplement was printed in 1892 and another in 1894, the year of Bain's death. In 1888, John Macoun, Botanist to the Geological Survey, made the first major plant collection from the Island while based at Brackley Beach with his honeymooning son. He collected at Tignish, Charlottetown, Mount Stewart and Tracadie, East Point and Lake Verde, concentrating however at Brackley Point. The specimens are now at the National Museum, Ottawa. The published records of this collection are included in Macoun's ***Catalogue of Canadian Plants***, Vol. 5, 1890, with the pteridophytes treated by Burgess who credited Bain with the Island records. The son, James M. Macoun, published additional notes on some of these specimens in the Canadian Journal of Science (1895). A few notable collections were made for the National Museum by the geologist W.J. Wilson, as assistant to Robert Chalmers on the first Pleistocene geology survey of the Island in 1892-3.

The next chapter of exploration is delineated by the Island botanist Lawrence W. Watson, of Charlottetown. The interest in violets arising from the species-splitting of the American botanist E.L. Greene spread to Ottawa, to the younger Macoun and to James Fletcher of the Department of Agriculture, and so to Watson on whom they called for Maritime material. Live material of ***Viola adunca*** supplied to the experimental student of violets, Ezra Brainerd in Vermont, by Watson became the type of ***V. adunca*** var. ***glabra;*** an odd white form of a blue violet was named ***V. Watsoni*** by Greene, and the ***V. incognita*** of the Island was named ***V. nesiotica*** by Greene, all names since discarded. In 1901, the visit of the amateur J.R. Churchill of Massachusetts to Summerside and Tracadie included a collecting trip with Watson and added several species to the herbarium of the New England Botanical Club. In 1904 Fletcher stopped briefly in Charlottetown to collect a few violets. For two summers (1902 and 1903) Watson was employed by the National Museum as a collector of fossils and plants in the Island, but his projected flora of the Island was never prepared and his collections sent to Macoun for identification were discarded. Letters containing some of Macoun's replies are in the possession of H.A. Messervy of Charlottetown. Watson died about 1908. About this time, the vogue of Spotton's High School Botany in Toronto led to its adaptation for Island use by MacSwain (1907). Descriptions were taken over, plants not known from the Island were dropped from the MacSwain and Bain list and the few added by Watson included, to make a net total of 430 vascular plants. Copies of MacSwain's list were kept and annotated by the amateurs who later were to revise it.

Meanwhile, the rising tide of amateur botany in New England produced its synthesist in M.L. Fernald of Harvard University. The theory of persistence of plants through the later Ice Ages in the Gulf of St. Lawrence area on unglaciated "nunataks" or refugia was already forming in his mind when Robert Chalmers' conclusion that at least central Prince Edward Island had escaped glaciation drew his attention to this province. In 1912 and 1914 with Harold St. John as assistant and, for a brief space in 1912 with Long and Bartram of Philadelphia as companions, he made the most extensive collection for the Island, 713 duplicate specimens of which were deposited at the National Museum. It seems likely that the Island rather failed to fulfill Fernald's hopes of finding relict preglacial elements, for apart from incidental notes on certain coastal or marsh-dwelling species, no account of the trip was published. Thus, until the eighth edition of Gray's Manual appeared (1950), his finds were unknown to Island botanists.

Blythe Hurst, the sage of Smelt Creek near Brackley Point, was the next naturalist of note. To him as the columnist Agricola of the Charlottetown *Guardian* came the notes of the younger collectors: first, H.A. Messervy of Charlottetown, whose travels covered most of the Island, and R.R. Hurst, plant pathologist in charge of the Science Service Laboratory, recording many new introductions, then others. The MacSwain list, shorn of all descriptions and records of occurrence, was revised by Blythe Hurst to include these additions and "corrected" according to the nomenclature of Gray's Manual, seventh edition of 1907, as "A New Flora of Prince Edward Island", published in the *Transactions of the Royal Canadian Institute* (1933). Of particular significance to the Island botanists was the weed survey of the Island made by Herbert Groh of the federal Department of Agriculture in 1926. His survey, published in *Scientific Agriculture* for 1927, listed from sight records the occurrence of introduced, weedy or poisonous plants, by the number out of 55 Lots visited in which he had seen them. Some 200 specimens were added to the herbarium of the Department of Agriculture, Ottawa. At the same time, a visit from M.O. Malte, the botanist of the National Museum, added only specimens. Malte, like Fernald, projected a flora of the Gulf of St. Lawrence – Maritime region, but both abandoned them. John Adams of the Division of Botany, Ottawa, published in 1937 an article with additions to Hurst's 1933 list, increased by the sedges (records mostly taken from Mackenzie's monograph) and a few grasses. These groups remained the weakest links in Hurst's privately printed 1940 revision of his own list. He supplemented this last "New Flora" in 1941, and listed other new discoveries in his newspaper column, "Newsy Notes by Agricola," down to 1946. Blythe Hurst died in 1951. Specimens collected by Blythe Hurst, R.R. Hurst and Messervy which had not made their way to the Department of Agriculture herbarium at Ottawa were destroyed by the fire that razed the Science Service Laboratory at Charlottetown in that year. A revision, including as new these records and the results of a week's weed survey by J. Bassett in 1950, and with nomenclature corrected by the eighth edition of Gray's Manual (1950) was mimeographed by the Science Service Laboratory in 1952, and is referred to as "Campbell 1952" in the synonymy.

In 1945, first Dore and Gorham in June collected 122 specimens in the Island during the course of a botanical investigation of pastures through the Maritimes, then in August Dore and Roland spent three intense days gathering 338 unicates in the eastern part of the province. Only six of these latter, kept separate, escaped the fire which destroyed Roland's herbarium at Truro soon afterwards. Perhaps the impression gained of the inadequacy of Hurst's list coupled with this stroke of fate prompted the conception of the present project. Then Acadia University at Wolfville received Warren's early herbarium (1902-6) from North River built up with encouragement of Watson. Under E.C. Smith's curatorship, Acadia continued to receive specimens: of McGowan from Kilmuir, of Bruce from Heatherdale, of Aitken from Bay Fortune, etc. Some data on these and other preserved collections and their disposition in herbaria may be tabulated as follows:

Adams, John (1936), Summerside and Brackley Beach; basis for additions to Hurst's 1933 list (DAO) 34

Aitken, H.E. (1950), Bay Fortune, Kings County, a student collection (AU) 33

Anderson, E.G. (1951), Charlottetown and Summerside; weed collection (DAO) 12

Bassett, J. (1950), a week's wide-ranging ragweed survey and general collection (DAO) 186

Blanchard, W.H. (1909), materials for a study of the Blackberries (Can) 7

Bruce, J.M. (1948-9), a general collection from southern Kings County (AU) 129

Campbell, J.E. (1952-3), additions to the Floristic Survey (DAO) 28

Campbell, Sterling (1938), wild plants (private collection) 18

Churchill, J.R. (1901), Summerside and Tracadie, general, mainly at Harvard University (DAO, Can) 7

Dore, W.G. (1945), two general collections, one of 122 specimens with E. Gorham, one of 338 with A.E. Roland, all but 6 of the latter lost by fire (DAO, Dal, AU, etc.) 128

Eastham, J.W. (1912), near Orwell, Queens County (DAO) 3

Erskine, D.S. (1951 to 1954, 1956), two large general collections, some of the first with W.G. Dore, most of the latter with A.J. Smith, the basis of the present report (DAO, AU, Mtl. NY, BM, PEI, NSMS, etc.) 1847

Erskine, J.S. (1955), two-day general collection (NSMS) 36

Fernald, M.L. (1912, 1914), two large general collections made with H. St. John and for part of 1912 with B. Long, the major part at Harvard University, many cited specimens (Can) 713

Fletcher, James (1904), casual collection made at Charlottetown, Brackley Beach and Summerside on official Trip (DAO) 13

Fyles, Faith (1915), small set of weedy or poisonous plants, mostly from Summerside as part of Government project (DAO) 24

Groh, Herbert (1926), 56 weed-survey collections and (1929, 1930, 1932, 1934, 1937, 1940) 24 small casual additions (DAO) 80

Haviland, Margaret (née Grubbe) (1853), the first collection, casual (Kew) 30

Hurst, Blythe (1926-44), occasional specimens sent for determination (DAO) 21

Hurst, R.R. (1928-45), occasional specimens sent for determination (DAO) 9

Kerr, Robert (1949), student collection at Ellerslie (AU) 21

Leard, G.A. (1951), specimens from around Souris sent for identification (DAO) 7

Macoun, John (1888), first major general collection (Can) 395

Malte, M.O. (1926), general collection made in preparation for flora of the Maritimes (Can) 99

McGowan, L.J. (1950), student collection from Kilmuir area (AU) 20

Messervy, H.A. (1938), orchids sent to Ottawa (DAO) 6

Perry, H.G. (1919), incidental to Fisheries work at large (AU) 6

Power, E.E. (1950), York and Bideford, incidental to nematode work (DAO) 26

Scott, R.V. (1948), student collection from Ellerslie, incidental to fisheries work (AU) 19

Smith, M.W. (1940, 1953), mainly aquatics, in preparation for fisheries ecology (DAO) 33

Taylor, A.R.A. (1948), small general teaching collection, mainly from near Ellerslie, incidental to fisheries work (UNB) 55

Warren, A. Emerson (1911), New Glasgow area (DAO) 7

Warren, G.C. (1902-06), North River, Queens County general herbarium (AU) 136

Watson, L.W. (1901-04), Charlottetown area and violets (Can) 9

Wilson, W.J. (1893), small set incidental to geological work (Can) 5

Miscellaneous: collected by visiting firemen, Science Service employees, or sent in for advice by farmers (DAO) 39

Total specimens available in Canadian herbaria 4241

Acknowledgments

The list of acknowledgments can never be complete, but for the kindness of all those who helped in the project the author extends many thanks. To the systematic staff of the Botany Division at Ottawa thanks for the assistance with taxonomic questions, and to Doctors H.A. Senn and W.G. Dore for the continuance of the project when it seemed most unlikely to reach fruition (1954); to the staff of the National Museum for allowing me to catalogue and examine their set of Fernald's and Macoun's Prince Edward Island specimens at weekends and Christmas in 1952; to Dr. Smith at Acadia and Dr. Taylor at the University of New Brunswick for access to their herbaria; and to Doctors Sampson and Wood for looking up records at Harvard for me. To the staff of the Science Service Laboratory of Charlottetown, most of all, thanks for their interest in the work, such that there was no one who did not at one time or another assist in transporting us to the field, and for the use of their facilities, in particular to Mr. Hurst (whose photographs illustrate the report) and Mr. Campbell on whom the previous burden fell heaviest. To Mr. H.A. Messervy of Charlottetown, who gave me access to all his manuscript notes and to Watson's letters from Macoun, and drove us on several of our collecting trips; to Blythe Hurst, Jr., for giving me access to his father's manuscripts; the late Dr. G.C. Warren of Sydney, Nova Scotia, for the use of his MacSwain-Spotton with the MacSwain & Bain list, and for helpful correspondence; thanks. To Mr. G.A. Leard of Souris, Mr. & Mrs. Alton Webb of O'Leary, Mr. Sterling Campbell of Cavendish who drove and guided us to good localities and rare plants; to Mr. R. Found of the Fisheries Station at Ellerslie, Mr. V. Henderson of the Entomology Division, Mr. W. Profitt, the Assistant Provincial Forester, for assistance with transportation and exploration, many thanks. For the general guidance of the project, the preparation of the plot maps, selection of photographs and reading of the manuscript, Dr. Dore is deserving of much credit.

GENERAL FEATURES OF THE ISLAND AND ITS VEGETATION

Physical Features

Prince Edward Island, a crescent-shaped island lying to the south of the Gulf of St. Lawrence, has a length of about 145 miles and an area of about 2100 square miles, and lies about 9 miles from New Brunswick and 14 from Nova Scotia at the nearest points. Its situation in latitudes from 45° 57'N to 47° gives a cool temperate climate with freezing winters and moderately warm summers, means ranging from 17°F for February to 67°F for July at the Experimental Farm, Charlottetown. Proximity to the sea modifies temperatures from the extremes of New Brunswick, but less so than if Nova Scotia were not interposed between it and open ocean. Drift ice accumulating in the Gulf circulates through the Northumberland Strait and, more especially, around the northern shores of the Island before finding its exit at Cabot Strait. This cooling delays the spring; the average date of the last frost at Charlottetown is May 13 but there is variation of as much as a fortnight either way. However, the sea also delays autumn frost till mid-October (average), giving the Island the longest frost-free season (180 days) of the Maritimes except the southwestern parts of Nova Scotia. Locally, climate varies from being more maritime in the east to more continental in the west. Summer heat is higher in the west.

Rainfall (including snowfall reduced to equivalent rainfall) is fairly even throughout the year, least in spring and heaviest in fall (from 2.8 inches in April to 4.6 inches in December). The average precipitation of 43 inches is sufficient to class the climate as humid, but above-average summer temperatures make for some summer drought. Snow falls throughout the frost season, lying from the beginning of December to late March, but more or less complete thaws in January are of almost annual occurrence.

Geologically, the Island consists of a more or less uniform Permo-Carboniferous bedrock, slightly folded, eroded and overlain by deposits of Pleistocene (Ice) age which form the basis of the present soils. The bedrock is predominantly sandstone, but with interbedded shales and a very few small pockets of limestone (Park Corner, Crown Point, Miminegash). The shales are most conspicuous where the sandstone of a fold-produced dome has been eroded away to form Hillsborough Bay, so that a sandstone escarpment from Nine Mile Creek to Point Prim surrounds the shales of greater age. Many of the springs in the uplands are located at the contact of sandstone with impervious clay. A small Triassic igneous rock area occurs at Hog Island off Malpeque.

Pleistocene ice, though it hardly shaped the form of Prince Edward Island, brought foreign clay and stones, distributed local material into new forms and disrupted drainage. As the historical geology of this period in the Island is now being studied intensively by the Canadian Geological Survey, only tentative hypotheses accounting for the changes wrought by ice may be made. Ice from New Brunswick brought some calcareous and clay till from far beyond the Northumberland Strait; ice possibly from Cape Breton and of late Wisconsin age may be responsible for foreign material in the till of the eastern coast. The major ice movement must have been from the west onto the Island; following the long axis of the Island, it would rapidly have become heavy with the soft easily eroded sandstone and shale of the Island, so that most till east of the Northumberland Strait slope of western Prince County is sandy and of local origin. Western Prince was part of an area depressed by the weight of ice which became submerged by the sea to heights of 75 feet above present sea level between the time melting ice left it bare and the time the earth's crust resumed its normal level. The present uplands there are thus flanked by terraces, often of beach gravels, while former barrier-beach ponds have become present day bogs (e.g. the large East Bideford Bog and the "Black Marsh" at North Point).

Submerged peat of postglacial age is found along the southeastern coast at localities from Charlottetown round to Launching, and with the submerged forest stumps of the Queens Clay coast, indicates that for a time after the retreat of the ice the Strait was at least narrower and shallower by some 20 feet than at present. Assuming that the depression of western Prince by ice was accompanied by a corresponding rise in the eastern end of the island, there may have been a land bridge to Nova Scotia. In the section on phytogeography, we conclude that it perhaps had very slight significance in the formation of the present flora.

The soils of the Island have developed since the Ice Age on these diverse geological types. On the Central Uplands and their outliers, deeper, more finely grained soils have been formed, the Alberry fine sandy loam and the Charlottetown fine sandy loam. The New Brunswick clayey till produced the O'Leary clay loam; the exposed clays of the Hillsborough Bay area, the Queens clay. On the kames and eskers, dry sandy soils developed (Dunstaffnage sandy loam), and on the postglacial beaches of western Prince, the Kildare sandy loam. Where the Cape Breton lobe scraped off the old weathered layer, the dry thin Culloden sandy loam developed. The poorly drained phases occupy the only other large areas, the Egmont clay loam corresponding to the O'Leary, the Armedale to the Charlottetown and the sand loams.

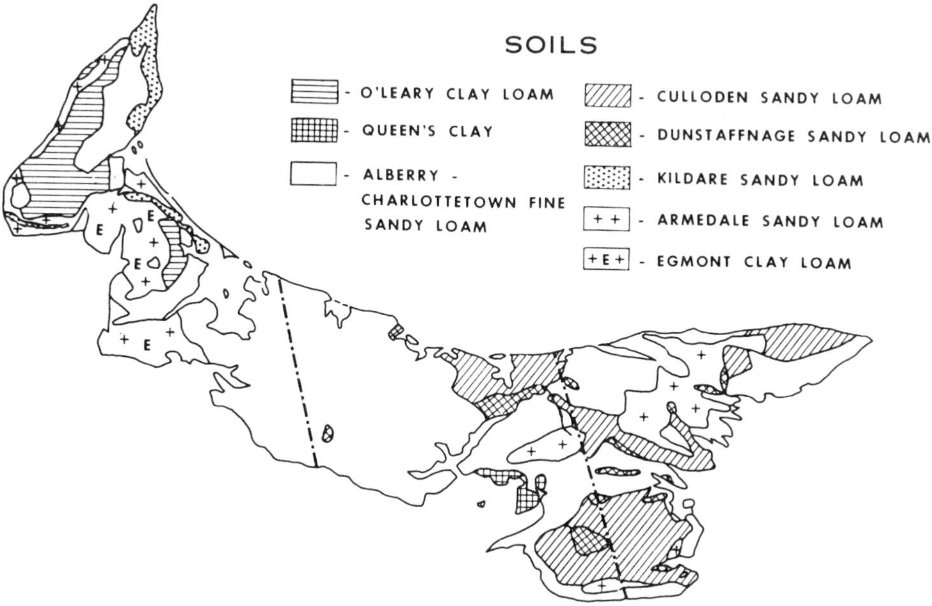

Generalized soil map of Prince Edward Island (Based on Whiteside, 1950).

Drainage, controlling as it does both the type of soil and the types of plant cover developed at each site, is of first importance as an ecological factor. The Island has low summits of under 500 feet above sea level, but short slopes to the sea or rivers give most areas adequate drainage. Thus we may with reason speak of a Central Upland although its summits south of Fredericton lie at 480 feet at most. From this center the highest divides lie in a line almost parallel to the Prince-Queens county line from Park Corner to Bonshaw. These are the well-drained hardwood hills. As the long slopes, consequently, lie to the east and west, it is on the lower broader divides running in these directions, that the nearest small swamps or bogs occur. East of the Covehead Road the upland shades off altogether at heights of 100 to 125 feet.

South of Montague River and Orwell Bay, hilly country rises to an upland with summits at 400 feet near Iona and 425 at Caledonia; there are short slopes to the south and west, longer ones toward the north and east. A fringe of lowland surrounds the hills; extensive swamps parallel Murray River, both to north and south.

East of the Hillsborough River the low sandy lands are cut off from the sea by the Pownal escarpment which, with heights of 150 to 300 feet, diverts drainage to the long slopes toward the river. The lakes of this area are scarcely drained at all, their waters filtering through swamps toward the headwaters of creeks tributary to the Hillsborough. From Fort Augustus and Pisquid eastward, the hills form a belt subdivided by the valleys of northward-draining rivers, while from Souris to Boughton River or even Montague River, the only well-drained land is on the seaward ends of the southern peninsulas separated by southeastward-draining valleys. Thus, within these fringes of hills, there are extensive swamps and large poorly-drained areas. The way in which the southeastern-directed valleys are the site of the Dunstaffnage kames and eskers suggests that drainage toward this side may have been impeded by glacial deposits. The large northeastern peninsula resembles the smaller southeastward ones in having better drained areas nearer the tip and large swamps on the gradual slopes of its broader base.

In western Prince (i.e., west of Summerside and the Miscouche isthmus) two NE-SW trending divides are the main feature; one, the backbone of West Prince (i.e., west of Portage), lies near and parallel to the northwest coast and reaches heights of 175 feet; the other, reaching only 125 feet, lies transversely across Central Prince from Abrams Village to Bideford. Central Prince north of this ridge includes the largest area of poorly drained Egmont soils, the second large area of the Egmont lying south of it toward the Miscouche isthmus. In western Prince the lower summits make for poor drainage except where the seaward ends of valleys have cut ravines. Lesser streams have often cut down fairly easily into the postglacial clays found below 75 feet, becoming wide, slow, and bordered by an alluvial flat suitable for alder thicket. Eastward, stonier alluvium is found along the larger north-flowing streams (Hunter River, Winter River, Indian River).

Coastlines in the Island appear widely different, the North Shore with its wide crescentic sweep and great bays all barred by sand islands built by the Gulf currents, the South Shore with its three great southwest-facing bays of muddy shores. The east coast with its many estuaries tending to be barred by sand spits is an open Gulf coast like the North, though more sheltered. Any appreciable depth of sand along the shore tends to become windblown into dunes, usually by onshore winds and so parallel to the coast, but also by longshore winds that cross-pile them.

Vegetational Features

The vegetational history of the Island is, at best, hypothetical until analyses may be made of peat and lake-sediment samples for fossil pollen. However, the changing distribution of land and sea areas and the fluctuations of the climate of eastern North America have been the most significant factors, and the waning of the fourth (Wisconsin) ice sheet the point of departure. In Gaspé and western Newfoundland, mountains <u>may</u> have stood above the ice at its maximum extent; but on the Island, with its greatest relief only 500 feet, nunataks cannot have existed. The possibility of plants having survived the glaciation on the Island depends on whether the Central Upland really was unglaciated. On the Island occur none of those species whose characteristic distribution suggests localized survival. Secondly, Hultén propounded a hypothesis that the continental shelf, bare during glaciation, has served as a refugium during the Ice Age. If so, the Laurentian (Gulf) endemics of coastal habitat may have reached the Island thence as the ice melted away and the sea advanced.

As the ice accumulated, it depressed the continent beneath it. As it shrank through melting, the continent returned to its normal level strip by strip, each released along what is called a 'hinge-line.' One hinge lay across Prince Edward Island: consequently as the ice melted back toward New Brunswick, the sea temporarily inundated the depressed portions of Prince County to as much as 75 feet above present sea level. At the same time, by reason of a contemporaneous hinge, the sea covered southern New Brunswick and Maine. But with so much of the world's water in the form of ice, the continental shelf was extensive above sea level, and a migration route along the Atlantic Coastal Plain brought southern plants to Nova Scotia. Here too, the effect on the flora of the Island has been very slight, but a few of the bog plants must have used this route. The relatively fresh waters of this postglacial sea gave its coasts as a migration route to estuarine plants: only one (***Samolus***) occurs of those discussed in this connection by Fassett. However, a few of the coastal plants which entered the Great Lakes along postglacial shores must have arrived at the Island about this time. Interestingly enough, two of the salt marsh plants of the Bay of Fundy reach their northern limits in Bedeque and Hillsborough Bays, suggesting migration by such a continuous inner coastal route.

The forests of Prince Edward Island, part of the 'Acadian Forest' of the present, seem to have arrived in two waves of invasion. The Boreal element is general and well represented: predominantly of conifers, white and black spruce, jack pine, larch, and balsam fir, with white birch and mountain ash, it is the only forest of the North Shore and the bogs. It may have had 7000 years to cross the waters of the postglacial sea and its shrunken remnants, or have entered by land from Nova Scotia. At least its representation in the Island is rather complete. The northern deciduous forest element, represented by the "northern hardwoods" (beech, sugar maple, yellow birch) and by white pine, hemlock and red spruce, which forms the climax forest of the uplands, shows a full quota of tree species but very little of the herbaceous "spring flora" so characteristic of the mainland hardwoods. This reached its greatest extent during the 'thermal maximum' about 3000 years ago, almost certainly after the Island had become isolated by Northumberland Strait.

VEGETATIONAL ASSOCIATIONS

Previous Knowledge of Vegetation

Our source of information regarding the nature of the vegetation before 1800 is confined mainly to the kind of notes made by travelers. Since 1820, visitors to Prince Edward Island have seldom felt that their observations could add anything to our knowledge of its plant cover.

The early French records are presented by Harvey (1926). Jacques Cartier, the discoverer, in 1534 found the "treeless lands" of the North Shore covered by "pease. . . and wild oats like rye, one would say sown and tilled," the *Lathyrus japonicus* and *Elymus mollis* of sandy beaches, by "white [?] and red gooseberry bushes, strawberries, raspberries," *Ribes hirtellum, Fragaria virginiana* and *Rubus idaeus* ssp. *strigosus* indicating an early stage in succession on cleared or burnt land. The forest was composed of "cedars, yew trees [balsam fir?], pines, white elms, ash trees, willows, others unknown to us [Europeans]." This association of *Thuja occidentalis, Abies balsamea* (?), *Pinus strobus, Ulmus americana, Fraxinus americana, Salix* spp., marks his landing as in western Prince. Nicolas Denys in the middle seventeenth century regarded "firs" as dominant, as conifers are on most of the lowlands, and remarked on the presence of beech and birch. These latter, *Fagus grandifolia* and probably its associate *Betula lutea*, are part of the characteristic upland vegetation. Gotteville, founder in 1720 of Fort LaJoie near Rocky Point, noted the local association of oak," cherry, beech and pines of mast size" (*Quercus rubra* var. *borealis, Prunus* sp., *Fagus, Pinus strobus*), which may be found even now in the woods on the grounds at Riverside near Charlottetown. *Quercus rubra* was known from the North Shore as large groves near Tracadie and mixed with *Pinus strobus* at Malpeque.

Gaudet (1956) summarizes some later observations beginning with that of Rochon (1759). Rochon noted a large grove of cedar between Malpeque and Cascumpeque, made up of "two kinds, red and white" – a valid distinction among spruce and pine, common though invalid in hemlock, but locally unheard of in cedar. Probably he referred to the spruce, since he maintained that the "incense" (gum) of the red was chewed among the Acadians. Lord Selkirk in 1792 distinguished the hardwood and the pine communities, too: "The most common species of timber are beech and maple, . . . frequently intermixed [with] birch of different kinds, spruce, firs and other species of the pine tribe," but "in some places, the pines entirely predominate, this . . . indicating a soil of an inferior quality." Johnstone (1820) remarked that Prince Edward Island was "one entire forest of wood," listed the prevalent hardwoods (with elm included, oddly enough) and softwoods, and noted that "promiscuous" mixtures of hard and softwood were to be found in places while in other parts clumps of a particular kind were found by themselves. Major Pollard (1898) spoke of virgin forest of evergreen "firs" mixed with oak, birch, maple, ash and poplar, and of majestic pine and hemlock growing amid an undergrowth of hazel, alder, aspen, juniper (i.e. larch), cedar, and tangled brambles.

Stewart (1806), classifying land by its vegetation type as to degree of suitability for settlement, generalized the ecology of the central upland and its eastern borders:

"*on the best land*, maple [*Acer saccharum. A. rubrum*], beech [*Fagus*], black and yellow birch one species, [*Betula lutea*], mixed with firs [*Picea rubens, Abies balsamea*] and pine [*Pinus strobus*], with an undergrowth of yew [*Taxus canadensis*]".
"*on the next best land*, no evergreens mixed with hardwoods, and more yew".
"*on the third grade*, poor land, with a thin upper stratum and the sub-soil cold and hard ... the hardwoods are not mixed with evergreens and there is no yew".
"*the worst land* is covered with spruce [*Picea glauca, P. mariana*], small white birch [*Betula populifolia*] and scrubby pines".
"*swamps* are covered with black spruce [*Picea mariana*]".

From relics of the original vegetation, it would appear that the best land lay on the gentler slopes of the central upland, the next best on the steeper slopes, north slopes and ravines, and the third undoubtedly on the dry hilltops. The poor land comprised the sandy parts of the valleys and lowlands near the shore, the morainic lands of Tracadie, and perhaps much of the east-central lowland (e.g., the pine groves of Mermaid).

Associations

From the vestiges of forest in the woodlots, swamps and ravines, the virgin forest sketchily described in the records gathered by Harvey and Gaudet and in Stewart's land classification may be reconstructed. (A more uniform picture is thus arrived at than today's patchwork remnants permit.) Except for dunes, salt marshes and bogs, the land was entirely forested.

1. Forest. On the uplands the dominant forest, usually of mixed forest aspect, was of the "northern hardwoods" (beech, sugar maple and yellow birch) and their coniferous associates (Braun, 1950), white pine and hemlock, their proportions varying with the nature of the site. Red spruce, considered by Halliday (1937) the unique species of the Acadian Forest Region which covered the Maritimes, made a poor third to the other upland conifers. Halliday's generalization (after Macoun) by the words "in spite of the flat topography and low elevation, maple occurs generally throughout, although in other parts of the [Central Acadian] section [including southern New Brunswick and the Gulf and Fundy slopes of Nova Scotia] it is confined to higher and better-drained positions" fails to appreciate Island topography.

The differentiation of the upland forest by site, already indicated roughly by Stewart's first three land types, corresponds closely to that found by Long (1952) in the western portion of the Central Acadian Section, near Fredericton, N.B. Pure stands of beech (Stewart's third type) dominated the dry hilltops of the central and southeastern uplands; sugar maple with some yellow birch predominated on the flatter summit areas of the north-eastern peninsula and the low uplands of western Prince, occurring also in the central uplands on the more gradual slopes. On gentle north-facing slopes or in shady ravines, red maple and hemlock predominated, particularly in the central and southeastern uplands; red maple alone or with yellow birch grew on the almost imperceptible O'Leary clay slopes. On the Island red spruce is less characteristic of low sites, probably because it here approaches its northern summer-temperature limit.

Typically, the ground cover of such sites is sparse and a shrub layer almost lacking. The striped maple *(Acer pensylvanicum)* and mountain maple *(A. spicatum)*, small and straggly but essentially trees, form a subsidiary element in the stands dominated by sugar maple. The herbs are mostly spring-flora perennials: *Smilacina racemosa, Ranunculus abortivus, Aralia nudicaulis, Trillium undulatum, T. cernuum, Medeola virginiana, Trientalis borealis,* and such localized species as *Claytonia caroliniana, Viola pensylvanica, Osmorhiza claytonii, Panax trifolius,* semi-saprophytes like *Listera convallarioides, Pyrola elliptica,* saprophytes like *Monotropa uniflora* and *Corallorhiza maculata* and parasites like *Epifagus virginiana.* Where hemlock was abundant a few saprophytes alone grew; under red spruce woods, the mossy floor would harbor *Oxalis montana, Viola incognita,* and various Pyroleae *(Chimaphila, Orthilia).*

On the lowlands a wider range of sites made for a more heterogeneous forest. Deviating least from the hardwood forest is that of the stream-valley clays in western Prince. Dominated by red maple, now mainly in alders and always with the occasional elm, and interstitial species, elder *(Sambucus canadensis)* and less invariably, *Clematis virginiana* and *Viburnum trilobum,* with *Salix lucida* around open wet spaces, it corresponds closely to the Acereto-ulmetum (maple-elm community) of the upper clay plain of postglacial terraces in Kamouraska County, Que., described by Hamel (1955). Along streams cedar may form pure stands, usually with a limited undergrowth of herbs: *Viola* spp. most typical. White and black ash are also scattered along streams here, where the tall herbs *Eupatorium maculatum, Aster puniceus* and *A. umbellatus, Urtica* spp. and *Geum laciniatum* grow. Without ash, the same association is found along the larger rivers of the north slope, far to the east of Prince. The herbaceous plants of the red maple forest vary with the wetness of the site: with poor drainage, the ferns *Onoclea sensibilis* and *Osmunda cinnamomea* with scattered *Rubus pubescens* are most common; with better but still damp sites, *Aralia nudicaulis, Thelypteris noveboracensis, Dryopteris spinulosa, Aster acuminatus* and *Carex debilis* var. *rudgei* are more typical. Red maple forest without elm is characteristic of the depressions of the sandy east-central lowland.

The abundance of glacial and marine sands in the lowlands makes for a contrasting forest, probably then dominated by white pine or white pine and oak, associated with red pine and wire birch in the most gravelly sites. Where these sands are little above the water table, the most extensive cedar woods of west Prince occur. The undergrowth of the pine woods consists of widely scattered Pyroleae, but under wire birch *(Betula populifolia),* the shrub *Comptonia peregrina,* the goldenrods *Solidago puberula* and *S. nemoralis,* and the shrubby Ericaceae are admitted. Black spruce *(Picea mariana)*, an occasional element in these low dry sites, dominated the true swamps of the lowlands almost exclusively, save for larches *(Larix laricina)* where they bordered on bogs. Along the North Shore, exposed sites and older fixed dunes allowed white spruce *(P. glauca)* forest to hold its own.

Forest changes. The major change in the forest since settlement began has been the wholesale clearing of land for crops and pasture. As well-drained fine sandy loam soils were most favorable to agriculture, the uplands were cleared more completely than the lowlands; some 72% of the central upland is cleared at present. In the west, too, the most extensive clearing has been in these areas, though the large areas of poorly drained soil reduce the average cleared area to 50%. A somewhat similar situation exists in the southeast, where the fringe of lowlands is less completely cleared but the upland soil is less favorable. Here again some 50% is cleared and in use. In the northeast, because of the large swamps the highest proportion of woodland remains; some 40% is cleared. By the combined average, 60% of the Island is devoted to agriculture, an exceptionally high proportion for Canada east of Quebec City. The remaining 40% is by no means all in woods, for some 8% of the total is in unimproved waste land, marsh or barren. The result has been the reduction of hardwood-pine-hemlock forest to a greater extent than black spruce and red maple, which form the most extensive forest areas.

Economic demand has made for selective cutting of much of the woodland that remains. In the 18th century all visitors noted that contemporary "defence resource", the mast timber of white pines. This was cut and exported to England; later white pine and oak were used in shipbuilding, and the former in woodworking. White pine stumps rotted from the back pastures within the last 70 years in many areas. The great size of the hemlock made it useful for boards, often only for barn siding because of its relative weakness, and its bark was sought for tanning. Sugar maple, used till fifty years ago as a source of sugar, was often maintained in a grove near the farmhouse, especially in the west; most of the birch from these groves was undoubtedly removed for fuel, though the paper birch has often replaced it. During the past twenty years, the demand for pit props and pulpwood has led to the cutting of much second-growth white spruce in Kings County.

One of the earliest forces released by the settler was fire. About 1738, when the North Shore was only a base for drying fish, fire burnt from Tracadie eastward to East Point, devastating the entire north-eastern peninsula. It must have spread easily in the dry spruce and pine forest. One result of this burning was the encouragement of the heath shrubs; the resulting blueberry barrens of Tracadie have been perpetuated by accidental and deliberate use of fire ever since. The fire of 1840 in central Prince produced a great blueberry barren at Conway which persisted for over thirty years. (This the biggest fire of Prince County history burned from Campbellton to Bideford.)

The opening up of the hardwood stands by selective cutting has brought disease to several forest trees. Yellow birch and white birch (as a member of the forest) have been devastated by dieback. The beech of hilltops has often suffered from the recently (1934) introduced aphid-borne canker *(Nectria coccinea)* so badly that only low thickets of stump-sprout trees persist; however, there are many more undamaged stands than in Nova Scotia. Very little white pine remained to suffer from or spread the blister rust when it struck early in this century. The larch of bogs, a species apt to grow in extensive pure stands, was also reduced by an introduced insect, the sawfly, in the 1890's. However, the role of native diseases in the case of successional species such as the larch, with their large pure stands, is probably less permanently significant; destruction and recovery may take place cyclically, as it seems to with the marine eelgrass *Zostera* attacked by wasting disease.

Forest succession. New plant communities (modifications of old ones) have been brought about by human intervention in the forest. Fire was not an unknown thing in 1534; even the discoverer Cartier found the community of raspberries and gooseberries in "treeless lands", suggesting fires set by lightning or Indians. And there were plants adapted to survival of fire: as in Nova Scotia (Martin, 1955) the perennials with deep rhizomes, the bracken and sheep laurel, sprout following a fire in the same summer. Other heaths such as the blueberries are close behind; the fireweed *(Epilobium angustifolium)* with its cork-protected rhizome flourishes in this community. The seeds which drift into the bare spaces are usually grasses such as *Agrostis scabra* and *Danthonia spicata,* or fleshy fruits carried by birds: raspberries *(Rubus idaeus)* and gooseberries or currants (*Ribes* spp.). The jack pine *(Pinus banksiana)* whose slowly opening cones are burst open by the heat of fire, formerly a rare element of bogs and Kildare beaches, has become abundant near Tignish and East Bideford, following fire. The hardwood forest opened up by cutting and disease takes over a century to regenerate. Consequently, much of the upland forest at present is in various stages of succession. Openings often permit a dense growth of fern, *Dennstaedtia punctilobula* with clumps of *Dryopteris spinulosa,* to spring up; such partial openings are colonized by plants with berry fruits: *Actaea rubra* and the red elder *Sambucus pubens* are characteristic. Red spruce and red maple are apt to follow, or paper birch and fir. More complete cutting brings forth an earlier stage characterized by sedges *(Carex emmonsii, C. deflexa)* and red maple stump sprouts, sometimes with aspens and wire birch. However, the maples become the shade trees soon, and the species of partial openings appear, with the new sedges *Carex deweyana, C. arctata* and *C. communis.* With deeper shade, the herbs *Clintonia borealis, Maianthemum canadense, Smilacina racemosa* and *Cornus* appear.

On the poorly drained soils, cutting has been less extensive. However, fire has not been less frequent. In either case the barrens that result persist longer than in other sites. The typical heaths are there, but rhodora *(Azalea canadensis),* wintergreen *(Gaultheria procumbens)* and leatherleaf *(Chamaedaphne calyculata)* are the most important, with sheep laurel next to them, and *Spiraea tomentosa* accompanying them. Such sites are apt to become peaty, with the tall shrubs *Nemopanthus mucronata* or *Vibrunum cassinoides* in clumps and the lichen-coated larch and black spruce growing up very slowly. The orchids *Spiranthes lacera, S. cernua* and the club moss *Lycopodium tristachyum*, all typical of exposed sites, find a place on the hummocks in such barrens. On the heavier lowland soils, partial red maple cutting has given room to aspens and fir with some spruce. *Osmunda cinnamomea* replaces bracken as the common fern. As a mossy ground cover is established in increasing shade, the sedges *Carex brunnescens, C. debilis* var. *rudgei* in damp, *C. leptalea* and *C. disperma* in wetter spots become typical; *Cornus canadensis* and *Coptis groenlandica* are very characteristic herbs. Along the wooded margins of swales or rivers these sedges may be accompanied by *Athyrium filix-femina, Streptopus amplexifolius* and *Ribes lacustre.* The cutting of red or white spruce woods on damp soils usually produces vegetation first in the damp pockets with peat and bog plants such as asters and *Epilobium* spp. and sedges; then fireweed, raspberries, elder, among the slash; then, red maple with balsam fir and paper birch following closely. Cutting in the pine woods produces a heath barren much like the habitat some ten years after a fire: same goldenrods, bracken, blueberries, sheep laurel and *Pyrola rotundifolia*, under wire birch and aspen with scant regeneration of pine – except in central Prince, where good examples may be seen near Freeland and Perceval River.

Reversion from pasture follows a different course, comparable to the succession on dunes. The grass is colonized directly by sun-tolerant young white spruce, with some alder and bayberry rapidly (15 years) shaded out. Until natural

White spruce quickly invades neglected pastures. Springton.

thinning occurs, only saprophytes (fungi) appear beneath; then moss (*Pleurozium*) and *Maianthemum* with *Cornus canadensis*. On the steeper slopes, *Corylus cornuta* and *Diervilla lonicera* appear at the sunlit edges of the spruce woods. In the southeastern uplands, the pastures may be colonized by wire birch and red maple rather than white spruce if these are close at hand. On damp or heavy soil the regeneration of forest from pasture is conducted by marsh fern (*Thelypteris palustris*) and larch, rather than white spruce; in flat sandy loam soil, the pasture succession parallels that of the dunes most closely, goldenrods (*Solidago canadensis, S. graminifolia*) giving way to bayberry, and that in turn to white spruce. Thus, forest succession on the lowland soils has previewed the later stages of succession in the habitats adverse to forest: the bogs, dunes and salt marshes, the lakes and ponds.

2. Fresh-water communities. Spring and rain-fed lakes form a very different plant community from that of stream-fed lakes. They are low in mineral nutrients and at first sandy-bottomed for lack of sediments. Thus, with the growth of plants they have a tendency to become acid from the accumulation of only partially decayed plant debris; consequently, they have a tendency to become bog lakes. The larger, Keefe's Lake, Glenfinnan Lake and Lake Verde, though partly

Keefe's Lake with zone of *Juncus militaris* in the two to three-foot depths.

bordered by peat, show little tendency this way; the smaller Mermaid Lake more; the ponds at Village Green and Murray River and the Lot 10 Lake at Portage are definitely bog-lakes.

The larger lakes have now acquired in their middle depths one or two feet of muck, in which the water lilies are rooted. *Nuphar variegatum* is found in all the lakes, boggy ones as well, in this medium; the rare *Nymphaea odorata* only in Lake Verde and one of the Murray River ponds. The scarcely encumbered sands of the shallower water are the medium of *Eriocaulon septangulare* and *Lobelia dortmanna* with emergent flowers, and of the completely submerged *Isoëtes riparia*, *Myriophyllum tenellum*, *Eleocharis acicularis* and *Elatine minima*. The smaller Mermaid and Verde Lakes have a border of floating-leaved *Sparganium multipedunculatum* and *S. angustifolium* (species which reappear in the sandy barrier-beach ponds of the North Shore with larger Myriophylla and *Potamogeton natans*). The three larger lakes have an inshore border of various rushes, all with *Eleocharis palustris*, in Glenfinnan and Keefe's Lakes overshadowed by tall *Juncus militaris*. Mermaid Lake has *Scirpus americanus* and *S. subterminalis* in its sandy shallows. The peaty bank bordering open water at Mermaid and Glenfinnan Lakes is dominated by *S. cyperinus;* sandy borders are usually

Shallows of the sandier shore of peat-bordered Mermaid Lake, showing *Nuphar variegatum* lily pads among the *Scirpus subterminalis* rushes.

dominated by *Myrica gale* and clumps of *Ilex verticillata,* species which likewise persist to the small pond or even swamp stage. With the transition to bog-pond, a floating mat of peat and heath roots encroaches on the water in an encircling ring. Its outer edge is the characteristic site of *Woodwardia virginica,* the distinctive characteristic heaths include *Gaylussacia dumosa, Kalmia polifolia* and even *G. baccata* (also found on barrens;) the wet peat is the site of *Habenaria blephariglottis, Sarracenia purpurea, Drosera rotundifolia, Eriophorum virginicum, E. tenellum* and the other more tolerant cotton grasses. To the rear, the heaths of peaty barrens become more common, and the succession follows the same course.

Most characteristic of the northern and eastern coasts is the formation of barrier-beach ponds. These show a succession on the whole similar to that of stream-fed lakes, which indeed they are. The artificial millponds, mostly a century old at least, likewise belong to this type. The distinctive appearance of these lakes is given by the far greater abundance of 'bulrushes,' *Typha latifolia,* which extend well out into the much greater (more quickly accumulated) organic debris of the shallows. One upland lake, Pisquid Pond, also stream-fed, answers equally to this description. The cattail (*Typha*) border of all these ponds is filled in by *Alisma triviale, Polygonum amphibium, Lysimachia thyrsiflora, Potentilla palustris,* fringed by a separate border of *Sparganium eurycarpum* with scattered *Bidens cernua, Sagittaria latifolia, Polygonum hydropiper* and *Bidens frondosa.* The influx of mineral nutrient (often fertilizer from the hill farms) stimulates the formation of a waterbloom of *Lemna minor* and no doubt of the other Lemnaceae. Pisquid Pond and Fernald's "pond at Grand Tracadie" are characterized by the dense belt of *Ceratophyllum demersum* and floating *Spirodela polyrhiza* outside the *Typha* zone. These ponds are far richer in pondweeds than the lakes: *Potamogeton foliosus* and *P. friesii* are the most characteristic. Where manure reaches the ponds, *Acorus calamus* typically occurs. These pond shores show some diversity: M.W. Smith (1946) notes that on firm clay margins *Scirpus validus* is more abundant than *Typha,* and that *Hippuris vulgaris* occurs only around the influx of cold streams, However, the *S. validus* persists into the foul-smelling quaking-marsh stage that follows the filling of such ponds by vegetation. This stage often shows large quantities of *Equisetum fluviatile,* a minor zone-former on open pond shores, and much *Glyceria grandis.* The first shrub to follow is hoary alder (*Alnus rugosa*) with *Juncus effusus* and *J. balticus* filling in. As the alder thickets become more dense, *Chrysosplenium americanum* and *Hydrocotyle americana* occupy the mucky pools; asters, *Calamagrostis canadensis, Rosa virginiana,* with *Habenaria psycodes* and *Polygonum sagittatum* cover the ground. Black spruce would commonly follow. Such an association (without the pools) is also found on damp headlands in western Prince.

Differences between upland and lowland streams are apparent in their vegetation, perhaps largely due to the difference in oxygen content between swift-running streams and slow. The sandy-bottomed upland streams and rivers usually have *Ranunculus trichophyllus* as the only submerged species, joined by *Myriophyllum exalbescens* in the southeastern upland. In slower water long-trailing beds of *Potamogeton praelongus* become common, and nearer the lowlands this species is replaced by *P. perfoliatus* var. *bupleuroides*, or more rarely *P. alpinus*. In the Dunk, *Zannichellia palustris* (often halophytic) penetrates for two miles above salt water.

3. Bog communities. Although peat (usually of *Sphagnum subsecundum*) is found in pockets in the lowland woods of red maple, and bog sphagna may take over a poorly drained fire barren, the bog community is determined by peat and

A boggy depression with sedges and cotton grass in the center, heath plants (here mainly *Azalea canadensis*) surrounding, and black spruce in the background. Near Mermaid Pond.

is essentially a later stage in lake succession. The vegetation of floating mats described previously is that of bogs, with the exception of species of the water's edge like *Woodwardia virginica*. All the other members of that community occur in the great bog of Black Banks, oddly enough the only one of the great bogs whose lake-basin origin might be questioned. Like many smaller pond-basin bogs, the Black Marsh at North Point is a sea of leatherleaf (*Chamaedaphne calyculata*) and cotton grass (*Eriophorum angustifolium*) interspersed with islands of stunted larch and black spruce, each with its "beach" of *Scirpus cespitosus*

The heath of *Chamaedaphne calyculata* at Black Marsh with islands of black spruce.

var. *callosus.* These clumps include *Nemopanthus mucronata, Aronia prunifolia,* and *Amelanchier ? fernaldii.* In the hummocks of peat, *Rubus chamaemorus* and *Vaccinium oxycoccos* are characteristic. Pools in the peat are surrounded by *Sarracenia, Drosera rotundifolia* and occasionally *Utricularia cornuta.* Damp depressions harbor the magenta-flowered orchids *Calopogon pulchellus, Pogonia ophioglossoides* and rarely *Arethusa bulbosa.* The great bogs of Portage and Miscouche are largely wooded, as the red maple and wire birch and black spruce have encroached on the open peat. Smaller bogs are essentially similar but lack the full variety of species.

4. Maritime associations. Various communities, fresh as well as salt, range along the coast usually in series of narrow bands in the treeless zone occasioned by the immediate proximity of the sea. Eroding cliffs, windswept sand of beaches and dunes, waterlogged hollows, estuarine flats, brackish lagoons, salty tidal marshes, and exposed headlands present a great diversity of habitats with corresponding diversity of floras that can be sketched only briefly.

The yielding shale and slumping sandstone of the Island's capes erode too rapidly to support a permanent or even a consistent community of plants. Weeds are the typical opportunists: *Senecio sylvaticus, Sonchus asper,* and those halophytes of weedy genera, *Atriplex patula* and *Plantago juncoides.* The wetter and more permanent grassy clay cliffs may support plants such as *Triglochin palustris.* (Inland cliffs are equally limited: the only ones worth the name are those of the Dunk which support the Island's only *Polypodium virginianum* and *Sphenopholis intermedia.*)

The sandy beaches and wet stretches of sandy soil favor *Elymus mollis* rather than the marram grass *Ammophila breviligulata,* but its associates are the beach pea (*Lathyrus japonicus*), *Cakile edentula* and the silverweed *Potentilla anserina,* often with *Arenaria peploides* and *Glaux maritima,* or, on the south shore, *Xanthium echinatum.* Deeper sand is found windblown into ridges of dunes, usually parallel to the coast and thus sheltering alternating hollows, the dune slacks. The dunes are stabilized by the rhizomatous *Ammophila,* the dune grass par excellence. Spaces among the grass are filled by *Lathyrus japonicus*

Sand dunes tend to become stabilized by dune grass (*Ammophila breviligulata*) and later colonized by white spruce (*Picea glauca*).

(itself rhizomatous), by *Carex silicea, Iris hookeri* or *Artemisia stelleriana* or by small annuals such as *Trifolium procumbens* or *Euphorbia polygonifolia.* On the lower dunes *Sonchus arvensis* is usual. At Bothwell two species of *Hudsonia* cover the dunes, but even there as on all North Shore dunes, the prevalent species is *H. tomentosa.* Small dicranoid mosses, filling space, become the site of *Sagina nodosa* colonies. Blowouts in the dunes are first colonized by the *Cakile* of sandy beaches. The first woody species to arrive is the bay-

A lingering patch of marram grass in the white spruce forest on the inner dunes. Brackley Beach.

berry among which *Anaphalis margaritacea* and *Solidago graminifolia* are the perennial weeds; the second is the white spruce which, except for rare replacement by cedar in west Prince, becomes the forest. However, while the spruce is scattered and the sand loose, the shrubs *Empetrum atropurpureum, Juniperus horizontalis,* or *Toxicodendron radicans* (poison ivy) may appear, and the perennial herb *Smilacina stellata.*

The slacks may be wet and carpeted with the moss *Aulacomnium palustre,* in which case the characteristic vascular plants are *Epilobium palustre, Cicuta bulbifera, Juncus pelocarpus* and *Vaccinium macrocarpon. Solidago sempervirens* never extends to slacks within the first ridge of dunes, *S. graminifolia* and *Aster novi-belgii* replacing it. Drier slacks are colonized by *Juncus balticus* and *J. alpinus,* often with the orchid *Liparis loeselii* filling gaps. On dry or wet slacks, the asters, goldenrods and everlasting only slightly precede the arrival of bayberry. Sometimes the sand may blow inland for a short space, favoring the growth of the warmth-loving grass *Poa compressa,* or into barrier-beach ponds creating sand flats colonized by *Limosella subulata, Rumex maritimus, Scirpus americanus, S. maritimus* and *S. paludosus.* The very local species *Aster laurentianus* belongs to this habitat. Salt spray blown over the cliffs on to sandy ground may create the association of a sandy beach at a higher level; this is particularly true on the Kildare sands of western Prince.

Muddy shores, in this predominantly sandstone land, are mainly located around estuaries, the largest salt marshes being those of the Dunk and Hillsborough Rivers. So few species colonize these saline habitats that their characteristic coloration marks the zones, bright-green *Spartina alterniflora* on mud submerged at every tide, reddish-green *S. patens* above it and blue-green *Puccinellia pumila* replacing the latter where hay-cutting or tramping have reduced it. Zones from lowest to uppermost are dotted with *Plantago juncoides* and bushy *Limonium nashu*, and occasional clumps of *Triglochin elata*. Bare patches of mud on the marsh are colonized by *Salicornia europaea* and *Suaeda maritima*, or in upper zones by *Spergularia marina* or *S. canadensis*. A weedy belt of *Atriplex patula* follows the high-water mark with the biggest roll of eelgrass wrack. The upper limit of salt-water influence is a weedy tension-zone of which *Solidago sempervirens* and *Spartina pectinata* are the most constant members, and *Ligusticum scothicum* one of the less. Pools on the salt marsh are often occupied by submerged *Ruppia maritima*. The muddy banks of estuarine creeks are the characteristic habitat of *Carex paleacea* and *C. salina*, while a carpet of *Eleocharis parvula* may here grow even below the *Spartina alterniflora* belt. The sea water affords a home to only one vascular plant, the rhizomatous eelgrass *Zostera marina*, which forms great beds off sandy shores in sheltered bays or estuaries.

Besides the dune succession which leads to white spruce forest, the latter may appear on exposed land above the beaches of headlands. These headlands may even be treeless as in the case of extreme North Point, covered by *Festuca rubra*, *Plantago juncoides* var. *laurentiana* and *Trifolium arvense*, or East Point covered by hummocky *Empetrum nigrum* with sandy patches of *Plantago juncoides* var. *decipiens*. At Stanhope, for instance, where the dune sand lends force to the effects of the west wind, there are large areas of *Juniperus communis*, *J. horizontalis* and *Empetrum nigrum* with scattered *Lechea intermedia*, *Vaccinium vitis-idaea* and introduced *Thymus serpyllum*. The borders of taller spruce woods in such sites are characterized by *Deschampsia flexuosa*, while pasture edges grow abundant *Euphrasia americana*.

5. *Cultivated vegetation.* The types of use made of the farmland (87% of the Island) is shown by this separation (1951):

Cultivated hay	203,783 acres
Improved pasture	197,937 acres
Grains	177,636 acres
Potatoes & turnips	37,156 acres
Farm wood lots	346,191 acres
Unimproved	103,318 acres
Miscellaneous crops	39,283 acres
Total farmland	1,095,304 acres

North Point, an eroding cape exposed to the full force of the Gulf storms, is essentially barren of coastal vegetation.

Unimproved land may include scrub and heath used as pasture together with swamp or bogland. All this land, in hay, oats, other grains, potatoes and turnips, is of interest botanically as a source of annual weeds, which grow in the disturbed ground as space fillers among the crop plants. Here one finds in particular the mustards and stitchworts, knotgrasses and pigweeds. But cultivation varies with the crop and has different effects. Potatoes, grown in the usual rotation, are relatively weed-free except for a little lamb's quarters (*Chenopodium album*) or knotgrass (*Polygonum aviculare*). The ridging of rows in the cultivation of root crops helps 'clean' the soil, and oats, the usual grain, grows too closely for cultivation and supports an abundant crop of devil's paintbrush (*Hieracium floribundum, H. pratense* and *H. aurantiacum*) and mustards, especially wild radish (*Raphanus raphanistrum*). The practice of chemical weed-killing will alter the proportions greatly, while nct extinguishing any species. Just so in the 1900's, the clean grain seed enforced by inspection eliminated the corn cockle (*Agrostemma githago*) but has reduced the abundance of only a few of the other species affected. Mustard seed lives so long in the soil that, once established, even crop rotation proved no way to counteract it. The planted hay crop, following in the common rotation, is a timothy-clover mixture, either red clover or alsike being usual. The practice of making silage has not yet become widespread but, especially in the central upland, one sees sweet clover (*Melilotus officinalis*) or alfalfa (*Medicago sativa*) abundant in the grass planted for this purpose and as pasturage. In the second year of the hay crop the clovers have largely died out, but have served their purpose in nitrifying the soil; this year it is a timothy hay crop and the spaces for weeds are still small. The fifth year, it is usual to pasture the field. The rotation may vary from this (the usual for mixed farming) according to the specialty of the farm. Farms with more land in potatoes (*Solanum tuberosum*) rotate fields every three years through potatoes, grain and clover hay; cattle farms, to assure more land in fodder crops, use a grain, clover hay, timothy, and pasture rotation. Because of the rotation the establishment of perennial weeds is unusual, but a biennial such as the daisy (*Chrysanthemum leucanthemum*), establishing itself in the oatfields as the

Perhaps nowhere else in the world do the daisies (*Chrysanthemum leucauthemum*) grow in as great profusion to brighten the landscape in early July; hay fields are solid white with them. Hunter River.

rosette, survives the first mowing and flowers next June so abundantly as to whiten the hayfields of the Central Uplands.

While rotation alters the tillage and the vegetation of cropland every year, 'permanent pasture' supports a different vegetation. Although the original pasture grasses must have been introduced from Europe, those which now dominate the pastures are essentially naturalized and seldom planted. Among these the bent, called browntop or P.E.I. bent (*Agrostis tenuis*) is most abundant in the Island as in other comparably cool humid temperate regions. It is replaced behind the dunes or on moraines by the Canada bluegrass (*Poa compressa*), another European, favored by warm well-aerated soils here where, generally, the climate tends to be too cool for it; in damp or clayey pastures by the creeping bent (*Agrostis palustris*), or where these are too much trodden for perennials to survive, by the annual bluegrass (*Poa annua*). The weeds of pastures are, like the grasses they must compete with, perennials: black-eyed Susans (*Rudbeckia serotina*) and native goldenrods (*Solidago canadensis, S. graminifolia*) where the grass is heavy; poisonous ragwort (*Senecio jacobaea*) and prickly thistles (*Cirsium arvense*) avoided by the stock even where the grass is overgrazed. Where pasture is neglected, forest succession is initiated. On the sandy Culloden soils, as on the fixed dunes, goldenrod and hawkweed (*Hieracium pilosella*) are replaced by a dense shrubbery of bayberry (*Myrica pensylvanica*) which is soon overtopped by white spruce or wire birch and white birch. In wet places, it is the soft rush (*Juncus effusus*) and marsh fern (*Thelypteris palustris*) which thrive on overgrazed lands, and the larch which replaces the white spruce as the sun-tolerant weed tree. On the upland, fir and red maple succeed the white spruce in whose shade they germinate, and thus regenerate the climax forest.

Areas in "nonagricultural human use" have, besides various weeds not found on agricultural land, also the deliberately planted shade trees and garden shrubs and herbs. Residential Charlottetown streets are lined with European lindens (*Tilia europaea*) and Scotch elms (*Ulmus glabra*) to the exclusion of other species. The former are planted, usually for avenues, at many of the older estates in the country; the latter are not. The English oak (*Quercus robur*), never seen in towns, is grown at a few old farms near navigable estuaries. In the western villages where shade trees are few, balm of Gilead (*Populus gileadensis*) or the native balsam poplar (*P. balsamifera*) or the "English" (Lombardy) poplar (*P. nigra* var. *italica*) are most often planted; in Wellington and Alberton box elder (*Acer negundo*) gives the streets their bright dusty green. Lilac (*Syringa vulgaris*) is the favorite ornamental shrub, followed perhaps by old double roses (*Rosa cinnamomea*). The use of dwarf elm (*Ulmus pumila*) or Siberian pea (*Caragana arborescens*) for hedges is a sure sign of recent architecture and of Experimental Farm leadership.

Along the railroads wild barley (*Hordeum jubatum*) and barnyard daisy (*Chrysanthemum leucanthemum*) and various annuals such as small toadflax (*Chaenorrhinum*), small yellow clovers (*Trifolium procumbens*), crab grasses (*Digitaria ischaemum*) and foxtail grasses (*Setaria glauca, S. viridis*), and, more recently, prostrate pigweeds (*Amaranthus albus*) and spurges (*Euphorbia supina*), used to be common. But by the use of weed killer these are being greatly reduced and in some cases may be prevented from establishing themselves along rights-of-way. Used along roadsides to save the effort of brush-cutting, weed killer spray has killed a band of alder, elder, wire birch and escaped box elder, as well as the weeds, leaving only grass and dead sticks. The use of western grain as feed for stock or poultry has led to the arrival of western weeds at sidings and farmyards: pineapple weed (*Matricaria matricarioides*, already naturalized), occasional wild oats (*Avena fatua*), prickly lettuce (*Lactuca *scariola*), great ragweed (*Ambrosia trifida*, whose establishment could ruin much of the tourist industry), cockleburs (*Xanthium italicum*), sunflowers (*Helianthus* spp.) and flax (*Linum usitatissimum*) signify such locations. The waste land around houses has long supported its vile crop of couch grass (*Agropyron repens*) and burdocks (*Arctium minus*), or catnip (*Nepeta cataria*) and motherwort (*Leonurus cardiaca*) where hens have scratched in the shade of hedges.

PHYTCGEOGRAPHY

Absence and Presence

In a discussion of species absent from Prince Edward Island, it is necessary to take into account present and past factors in plant distribution. The simplest hypothesis to explain the presence of the present native flora is by natural diffusion from eastern New Brunswick and northern Nova Scotia across the waters of Northumberland Strait, varying in bredth from 9 to 40 miles.

To test this hypothesis, a list was drawn up of the indigenous flora of the adjacent mainland within the drainage area of Northumberland Strait. Comparison of this Island list with Roland's Flora of Nova Scotia shows that 98% of the Island flora is found in Nova Scotia, in some part or other; only one or two species in the Island are not represented in either Nova Scotia or New Brunswick (*Rumex persicarioides*, *Bidens heterodoxa*). Of the flora of northern Nova Scotia, some 650 indigenous species, a full 80%, cross to the Island. The absences lie mainly in one of three ecological communities: (1) the aquatics, probably because the lakes of the Island are so few and small, (2) the intervale plants, for lack of river-alluvium areas in the Island, (3) the geophytes of the spring flora under hardwood forest. In the face of abundant hardwood forest in the Island, historical factors may serve to explain this last absence.

That diffusion across the waters of the Strait has taken place, there need be no doubt. Prince County, it seems, has never been linked by land to New Brunswick since the retreat of the ice, yet there are over 30 species there which are absent from the rest of Prince Edward Island and present in New Brunswick. Secondly, there are 15 species in the Island that do not occur in Nova Scotia at all, but which do occur in New Brunswick. And, if one postulates that it was the saline waters of the Strait which barred most of the "spring flora geophytes" from the Island, how can one account for the presence of a few of them (*Viola pensylvanica*, *Claytonia caroliniana*, *Panax trifolius*) in localized areas without assuming that each colony represents diffusion across the Straits, either from the west or south? It may be argued that ecological necessities have confined some of these local species to western Prince, but their migration via a Nova Scotian land bridge (see Geology Chapter 2) is out of the question.

Of the species that may have migrated by a land bridge formerly connecting the Island to Nova Scotia, the greater part can give no evidence. The intervale and hardwood flora of Nova Scotia by its distribution pattern in Nova Scotia (from the New Brunswick border to "the hub", thence southwest to Kings County and northeast to Cape Breton), let alone by its fossil history in the United States, shows that it has entered Nova Scotia via the Chignecto Isthmus and therefore since the recession of the postglacial sea that submerged the the Isthmus and the presumably earlier submergence of any land bridge to Prince Edward Island. The species that have entered Nova Scotia from the southwest rather than from New Brunswick are referred to as the "Coastal Plain Element" (not "Coastal Plain flora," for even in Nova Scotia they are off the Atlantic Coastal Plain). However, none of these reach the Island. But, other species of southwestern affinities, occurring both in Nova Scotia and in southern andeastern New Brunswick, appear in Prince Edward Island. These are either plants of sandy lands, such as the *Panica, Hudsonia ericoides* and *Viola fimbriatula,* or of pond and bog, such as *Woodwardia virginica, Habenaria blephariglottis* and *Gaylussacia dumosa.* Possibly these species colonized sandy deposits, bog and swampy pond hollows in the new land left by the receding postglacial sea, arriving from nearby Nova Scotia rather than from New England to the southwest. Several such species stop at Northumberland Strait, but of the dozen or so which Nova Scotia and the Island have in common, many (e.g., *Panicum depauperatum, Lactuca hirsuta, Viola fimbriatula*) seem absent from northern Nova Scotia, their ranges broken between the Annapolis Valley and the Island. Thus, there is little or no positive evidence of the significance of the land bridge in the migration of the present flora to the Island. One can only hope that pollen analysis will resolve some of the past history now obscured by wider diffusion.

Another group whose distribution may show evidence of the past continuity of Prince Edward Island with the mainland are those which follow the North Shore. The simply psammophilous or halophilous species follow round from the Gulf coasts of the Island to the Northumberland Strait coast. But other species confined to the Gulf shore show ranges extending no farther south than the Gulf shore of New Brunswick (e.g., *Juncus alpinus,* the Gulf endemic *Aster laurentianus)* or that coast and northern Cape Breton (*Parnassia parviflora, Empetrum atropurpureum, Eleocharis pauciflora*) or also with outposts on the cold coasts of the Bay of Fundy (*Juniperus horizontalis* and *Iris hookeri*) or the Bay and the outer coasts of Nova Scotia (*Euphrasia randii*). While ecological requirements (cooler summers, exposed sites) may have kept them from the Strait shores, their disrupted ranges at these their southern limits on the coast suggest a continuity dating back to a colder postglacial stage.

An interesting range pattern is shown by some eight species which occur in New Brunswick, Prince Edward Island and Cape Breton but not on the Nova Scotian mainland. Most of them (viz., *Aster junciformis, Betula pumila, Caltha palustris, Galium labradoricum, Vaccinium uliginosum*), are near their southern limit but others such as *Comandra richardsiana, Anemone canadensis* and *Hypericum majus,* while surely reaching their farthest east in the Gulf area, are not boreal types at all. If these migrated overland by way of northern Nova Scotia, those near their southern limit may have lost mainland stations through warmer summers; if so, the three latter must have migrated via Prince Edward Island and probably after the Gulf assumed its present form, or else they would have mainland stations. And the Prince Edward Island summer except along the Gulf shores is as warm as that of the mainland.

Local Distribution.

Within the Island the flora tends to be fairly uniform. Some 250 of the 625 native species, at least, may be seen from the distribution maps to be wide-ranging in the province. Some of these seem to be rare in Kings County. However, as that area was the least completely investigated during the survey, many apparent gaps there may eventually be filled. Others show the distribution: western Prince and east-central lowlands, their absence from the central uplands and the southeast being due to the absence of suitable habitats. The same might, of course, be said of some 40 common halophytes-psammophytes restricted to the coastline. Another 30-odd species (in *Carex*, section *Stellulatae*, in *Amelanchier*, *Crataegus* and *Rubus* particularly) were too imperfectly understood for their ranges as mapped to be reliably complete. However, in spite of the exclusion of at least 320 of the native flora from consideration, 23% showing very restricted ranges in the Island may be of phytogeographic interest.

Some 16 show restriction to the eastern part of the Island, some 44 to Prince west of Summerside, another 23 so widely scattered and scarce east of that isthmus as to suggest a late spread from it, some 14 a restriction to the central uplands, some 25 to the North Shore, 22 to the east-central lowlands, altogether 144 of narrow range. Of these groups none can be quite regarded as homogeneous. Some of the North Shore species are of relict status (e.g., *Aster laurentianus*), possibly related to the unstable shore habitat. Many of these are species (e.g., *Iris hookeri*, *Euphrasia randii*) restricted, possibly by the cooler summers, to the Gulf area and the coasts of the adjacent Maritime Provinces and New England. One is an Atlantic element, of the Coastal Plain type (*Corema conradii*). The group restricted to western Prince includes many species abundant in that area (*Carex gracillima*, *Salix rigida*, *Populus balsamifera*, *Thuja occidentalis*, *Pinus banksiana*, *Geum aleppicum*, *Fraxinus americana*, *Veronica scutellata*, *Eupatorium perfoliatum*). The equal failure of *Thuja* to pass the Chignecto Isthmus into Nova Scotia, almost due south of the Miscouche Isthmus, suggests some historic barrier. Among the seven or more native species of the Island that occur also in New Brunswick but not in Nova Scotia most are of this coastal endemic relict status. *Pilea pumila*, however, stands out as one that reaches as far east as the upper Hillsborough River but does not enter Nova Scotia though its range presents a hardwood forest pattern. The fact that some 23 (e.g., *Viburnum trilobum*, *Juncus dudleyi*) which are more abundant in western Prince have stations east of the central uplands suggests that species of more efficient dispersal may cross the barrier while others must very gradually make their way round it by way of the coastal streams. Furthermore, many of the western Prince group are confined west of the isthmus at Portage (e.g., *Angelica atropurpurea, Solidago gigantea*) or to the Tignish area (e.g., *Ranunculus pensylvanicus, Senecio aureus*), which reduces the significance of the Miscouche isthmus to that of one of several bottlenecks to the slower migrants.

But the most striking thing about the species of restricted range, on the whole, is how few their stations are. This is particularly true of these species in the central upland, for which I have advanced the argument of chance arrival after the formation of Northumberland Strait accompanied by ineffective dispersal. Thus, we may look upon all species local in Prince Edward Island, except for the few relict survivors of the North Shore, as late arrivals that have been slow to fill their scattered ecological niches.

The introduced flora (316 species) includes many diverse elements, which may be separated historically. The most important is the European element. Of 226 introduced species mapped in this list, 200 are of European origin, only 26 of American origin. The reasons for this are twofold: the source of the cultivated plants of Prince Edward Island (including the South American potato) was western Europe, hence the weeds brought in with them are mostly European; secondly, most of the American species which would be considered as weeds (e.g., the goldenrods) were native to the area and are classified as indigenous. Of the European species, at least 28 were introduced as crop plants (the degree of intent is difficult to assess with pasture grasses), another 30 as ornamentals and shade trees, while 142 would be classed, at least in the present age, as weeds. (Many of the mustards, for instance, have lost their status as cultivated plants). Historically, the introduction of European plants belongs to the period of settlement from the British Isles, with attendant clearing and cultivation. After Confederation, the building of railways, trade with mainland Canada, and the establishment of a train-ferry link with New Brunswick all helped to introduce weeds from North American sources. The railways have become the route of *Chaenorrhinum minus, Tragopogon pratensis, Lactuca scariola, Euphorbia supina, Collomia linearis*, and many other weeds. Though the source of the first three is originally European, all have entered the Island from American sources. Since the period 1890 to 1910, this has been true, for the modern centralized overseas trade does not come to Island ports. Apart from Blythe Hurst's sporadic and continual introductions, the only recent European weed to arrive by sea was *Arnoseris minima*. Weed seeds have been brought in with feed grain and these have given rise, particularly at railway sidings, to an element from western Canada, often species originally Russian.

One question that has often concerned phytogeographers is the degree to which plants find themselves at the limit of their range within an area. The insular position of Prince Edward Island makes such decisions rather arbitrary, since the limit imposed by the Gulf is much farther south on its North Shore than in Gaspé or Cape Breton, which have many more plants near their southern limits. Tentatively, then, one may divide the native species of the flora into two groups: those which cross the Gulf – about 35 species (or some 22 percent), and those that are found on either side beyond it – some 490. Of the former which come in from the southwest, practically all are at the northern margin of their range, only a few (e.g., ***Thuja occidentalis, Pilea pumila,*** discussed above) are at the eastern margin and do not spread on into Nova Scotia.

Those that are at a southern limit, some 30 species, mostly of the North Shore only, make up a mere 5%; those at their western limit another 5%. The latter are the 'Atlantic' species which reach eastern New Brunswick and also Newfoundland, but seem restricted to a relatively coastal migration pathway. Half of the Atlantic group of the Maritimes do not extend as far inland as Prince Edward Island.

Thus, the great majority of those 68% of the Island's species that also range across the Gulf are species which in the Maritimes stop at nothing but the Atlantic, the ubiquitous species of this area. That many of these are far from common in the Island shows the significance of the other factors (e.g. cooler summer temperature, late time of arrival, slow rate of dispersal, the few lakes, the wide separation of marshes) in limiting the distribution of species there.

Summarizing Statistics

Although no final summary of the flora can be expected, as additional introductions and adventives find their way into the flora and occasional native species continue to be discovered, a tabulation of the species now listed will give some idea of the composition of the flora.

Taxonomic rank	Native	Introd.	Total
Species	624	316	940
Varieties	69	13	82
Hybrids	12	2	14

When compared with that of Nova Scotia it is a small flora: Nova Scotia (Roland, 1947, modified by later discoveries) has 1012 native species and hybrids and 243 varieties, but the province has an area ten times as great as Prince Edward Island and has a greater variety of habitats. The significance of agriculture and settlement in the flora of Prince Edward Island may be shown by comparison of the introduced flora with that of Nova Scotia; the larger province has 392 introduced species and hybrids, and 30 introduced varieties, only a little more than the Island. For the Nova Scotian flora, Roland has not distinguished the adventive (recently or occasionally arriving species which as yet do not perpetuate themselves in the flora) from the fully naturalized species, but at least 101 "introduced species" of his list are not naturalized in Nova Scotia. Prince Edward Island has a naturalized flora of 199 to Nova Scotia's less than 291 species, for 117 Island species must be regarded as merely adventive.

A system often used in comparing floras as to the structure of the vegetation they make up is the "Raunkiaer spectrum."

Area	F	Ch	H	Cr	T
Connecticut	15.0%	1.9%	49.4%	21.7%	11.7%
Prince Edward Island	15.2%	4.0%	47.1%	24.4%	9.0%
Gaspé	12.1%	9.0%	49.0%	22.1%	7.5%

These figures for the native flora show a decided similarity between all these temperate forested regions of the northeast. However, the trees and shrubs (F - phanerophytes), form a smaller proportion of the flora in Gaspé where there are alpine areas. For the chamaephytes (Ch) or generally woody plants with low horizontal or decumbent stems, the proportion rises from Connecticut north to Gaspé, which with its alpine areas has many more in the flora. For the hemicryptophytes (H), herbaceous plants with stolons or rosettes or tillers near ground level, the proportion shows no significant trend; likewise for the cryptophytes (Cr), rhizomatous and/or aquatic plants (no bulb plants in P.E.I.). The proportion of annuals (T - therophytes) falls off to the northward, showing the Island flora once again in the logical intermediate position. Addition of the naturalized and adventive plants reduces the proportion of all classes relative to the therophytes, which rise to 13.5% in Connecticut, 14.1% in Gaspé, and overwhelmingly in the Island to 18%, probably because the native flora is small relative to Connecticut and the cultivated area large relative to Gaspé.

The significance of the recent survey can be seen from the fact that, although over 80 species have been excluded from the list of Hurst and the one of MacSwain and Bain, the 716-species list of Hurst in 1940 has been expanded to a 939-species list (or, excluding adventives often listed by Hurst an 822-species list).

REFERENCES

This list includes both major reference works and articles dealing specifically with Prince Edward Island or its plants. For species mentioned only incidentally the appropriate reference is included in the body of the list. The important records in contributions of a serial nature, such as J.M. Macoun's "Contributions to Canadian Botany" and Groh's "Canadian Weed Survey," have been treated in this way to facilitate reference.

Adams, John. 1937. Some additions to the flora of Prince Edward Island. Can. Field Nat. 51: 105-107.

Auer, Vaino. 1930. Peat bogs in southeastern Canada. Canada Dept. of Mines, Mem. 162.

Bain, Francis. 1890. The Natural History of Prince Edward Island. Charlottetown, Haszard.

Bain, Francis. 1892. Additions to "Plants of Prince Edward Island". 3 p. Charlottetown, Prince Edward Island Nat. Hist. Soc.

Braun, E. Lucy. 1950. Deciduous forests of eastern North America. Philadelphia, Blakiston.

Britton, N.L. and Addison Brown. 1913. Illustrated Flora of the Northern United States and Canada. 2nd ed. New York, Scribner, 3 vols.

Burgess, W.H.L. 1890. Pteridophyta. In Macoun, John. Catalogue of Canadian plants. Montreal.

Campbell, J. Ewen. 1952. A New Flora of Prince Edward Island. rev. ed. Charlottetown, Laboratory of Plant Pathology, mimeographed.

Chalmers, Robert. 1895. Report on the surface geology of eastern New Brunswick, north-western Nova Scotia and a portion of Prince Edward Island. Geol. Surv. Can. Ann. Rep. 7(M): 149 p.

Churchill, J.R. 1902. Some plants from Prince Edward Island. Rhodora 4: 31-36.

Clarke, J.A. 1906. Weeds and their eradication. In Prince Edward Island Dept. of Agriculture, Annual rept. ...1906. Charlottetown, p. XXVIII.

Copeland, E.B. 1947. Genera Filicum. Waltham, Chronica Botanica.

Dore, W.G. Phytogeography of the Maritime provinces (unpublished notes).

Dore, W.G. and A.E. Roland. 1942. The grasses of Nova Scotia. Proc. N.S. Inst. Sci. 20: 177-288.

Ennis, B. 1928. The life forms of Connecticut plants and their significance in relation to climate. Conn. Geol. Nat. Hist. Bul. 43: 1-100.

Fassett, N.C. 1928. The vegetation of the estuaries of northeastern North America. Proc. Boston Nat. Hist. Soc. 39: 73-130.

Fernald, M.L. 1925. Persistence of plants in unglaciated areas of boreal America. Am. Acad. Arts Sci. Mem. 15: 239-342

Fernald, M.L. 1950. Gray's manual of botany. 8th ed. New York, American Book Co.

Gaudet, J.F. 1956. Forestry past and present on Prince Edward Island. Maritime A.I.C. Conference, Charlottetown, 3 p. (mimeographed).

Gleason, H.A. 1952. The new Britton and Brown illustrated flora of the northeastern United States and adjacent Canada. New York, New York Botanical Garden. 3 vol.

Gorham, R.P. 1944. The known distribution of the buckthorns in the Maritime provinces. Acadian Nat. 1: 118-124.

Groh, H. 1927. A weed survey of Prince Edward Island. Sci. Agr. 7: 388-395.

Halliday, W.E.D. 1937. A forest classification for Canada. For. Serv. Bul. 83.

Hamel, Aubert. 1955. Esquisse écologique des comtés de l'Islet et de Kamouraska. Can. J. Bot. 33: 223-250.

Harvey, D.C. 1926. The French régime in Prince Edward Island. New Haven, Yale University Press.

Hultén, Erik. 1937. Outline of the history of arctic and boreal biota during the Quaternary period. Stockholm.

Hurst, Blythe (Sr.). 1933. Flowering plants and ferns of Prince Edward Island. Trans. Roy. Can. Inst. 19: 251-273.

Hurst, Blythe (Sr.). 1940. A new flora of Prince Edward Island. Charlottetown, Guardian Printing Office.

Hurst, Blythe (Sr.). 1941. Addenda to 1941. In later reprinting of Hurst (1940).

Lochhead, William. 1902. The weeds of Prince Edward Island. In Prince Edward Dept. of Agriculture, Annual rept. ...1906. Charlottetown, p. I - XXVII.

Long, H.D. 1952. Forest types and sites of the Acadia Forest Experimental Station. Fredericton, Acadia For. Exp. Sta. (M 226), mimeographed.

MacLeod, J.W. 1947. The forests of Prince Edward Island. Forestry Chron. 23: 190-193.

Macoun, John. 1883-1903. Catalogue of Canadian plants. Montreal and Ottawa, 7 parts. (Except for two specimens cited in earlier volumes, all Island vascular plant records are published in Part 5 (1890), the last volume on vascular plants.)

MacSwain, John. 1907. The flora of Prince Edward Island. In: Spotton, H.B. The Elements of Structural Botany; P.E.I. ed. Toronto, Gage, p. 1-104.

MacSwain, J.F., and F. Bain. 1891. List of Prince Edward Island plants. Charlottetown, P.E.I. Nat. Hist. Soc., 8 pp.

Marie-Victorin, Frère. 1935. Flore laurientienne. Montreal, Imprimerie de la Salle.

Martin, Lynton J. 1955. Observations on the origin and early development ot a plant community following a forest fire. Forestry Chron. 31: 154-161.

Owen, E.B. 1949. Pleistocene deposits of O'Leary map area, Prince County, Prince Edward Island. Geol. Surv. Can., Paper 49-6. map.

Robinson, B.L. and M.L. Fernald. 1907. Gray's new manual of botany. 7th rev. ed. New York, American Book Co.

Roland, A.E. 1947. The flora of Nova Scotia. Truro, Truro Printing & Publishing Co. Reprinted from Proc. N.S. Inst. Sci. 21: 95-642, 1944-1946.

Scoggan, H.J. 1950. The flora of Bic and the Gaspe Peninsula, Quebec. Nat. Mus. Can. Bul. 115.

Smith, M.W. 1946. A biological reconnaissance of ponds in the Prince Edward Island National Park. Acadian Nat. 2(6): 81-101.

Stewart, John. 1806. An account of Prince Edward Island. London, Winchester. p. 29-58.

Watson, L.W. 1899. Among our orchids. P.E.I. Mag. 1: 220-224.

Watson, L.W. 1900. Wolves in sheep's clothing. P.E.I. Mag. 2: 275-278.

Watson, L.W. 1907. Violaceae. In MacSwain (1907).

Whiteside, G.B. 1950. Soil survey of Prince Edward Island. Charlottetown, Experimental Farms Service. 3 maps.

Whiting, M.M. 1949. Early collection of plants in Prince Edward Island. Kew Bull. 1948: 236.

The Annotated List of Species with Maps

ANNOTATED LIST OF SPECIES

A completely definitive enumeration of the flora of an area is seldom possible: new introductions appear, stations of old ones are eradicated or die out, a few rare native species remain to be discovered, and revisions of taxonomic concepts place former records in doubt. Nevertheless, all the original plants and those thought now to be permanent members of the flora of Prince Edward Island have been listed; they are set in **boldface** type. These include native species if any specimen exists, and introduced species if their persistence as perennials or by repeated advents seems assured. [Brackets] have been used to set off species not regarded as members of the permanent flora: species now extinct, very casually adventive, very infrequent as garden escapes, or to be rejected from record.

Under each species, its distribution and habitat is set forth, in fairly general terms for the sake of brevity, the maps providing more specific details of distribution. For weedy species an effort has been made to note first recorded appearances. Groh (1927) in his report of a weed survey gives only the number of the Lots (the local units of land survey), out of the 55 he visited, in which he detected introduced, poisonous or weedy species. His figures indicate frequency at the time, though not distribution.

The taxonomy is essentially that of Gray's Manual, eighth edition (Fernald, 1950). Deviations from this are found in the true ferns (Filices) in which it follows Copeland's ***Genera Filicum*** (1947), with some reservations, and in the Ericaceae where the taxonomic concepts of H.F. Copeland* have been followed. Where some generic segregations have appeared attractive (e.g., in ***Habenaria, Potentilla, Polygonum***), the synonymy includes the alternative generic name with author of the appropriate specific combination, in the fashion of Gray's Manual (e.g., under ***Habenaria clavellata*** (Michx.) Spreng. will be found the name ***Gymnadeniopsis*** Rydb., meaning that the alternative name ***Gymnadeniopsis clavellata*** (Michx.) Rydb. is available to those who find the concept ***Habenaria*** too inclusive.)

The earlier references to each species in literature on the Island flora are appended as briefly as possible. Several examples will show the method:-

"***Polygonum sagittatum*** L. ... MacSwain and Bain, MacSwain, Hurst," means that the name ***P. sagittatum*** appears correctly in the lists of 1891, 1907, 1933, 1940 and 1952 (understood to agree with the 1940 list unless otherwise noticed). "***Cypripedium reginae*** Walt. ... (***C. spectabile*** Salisb., MacSwain and Bain, MacSwain; ***C. hirsutum***, Hurst), Campbell," means that the apparently correct

* **Copeland, Madrona 6: 113, 1941, on Monotropoideae, and in Am. Midl. Nat. 30: 533, 1943, on Rhododendroideae.**

name *C. reginae* was not used until 1952, but that earlier records of the species appeared under the names of *C. spectabile* Salisb., the latter name correctly applied to this species, as is shown here by the use of its author's name, not preceded but only followed by a comma – in 1891 and 1907, and in 1933 and 1940 under the incorrectly applied name of *C. hirsutum*, its authority omitted). Such changes and misapplications have resulted from changing rules of nomenclature in the period from 1867 to 1935, developments in Europe and the United States for which the Island botanists cannot be held responsible.

"*Stellaria graminea* L. ... (*S. longifolia*, MacSwain and Bain, MacSwain, Hurst; *S. longipes*, MacSwain and Bain, MacSwain, Hurst); MacSwain, Hurst," shows that the same species was carried from 1891 under two names and since 1907 under three. *S. longifolia* and *S. longipes* have never been found in the Island. If they were ever questioned, it must have been assumed by each that his predecessor had found it. The semicolons separate synonymy deriving from different nomenclatural sources.

"*Potentilla rolandii* Boivin, Can. Field-Nat. 65:21, 1951. ... (*P. pacifica*, Roland, not Howell)". As it is assumed that Roland's Flora will serve for identification of species, it is necessary to point out that the name *P. pacifica* used there is taxonomically incorrect, Roland's species not being the original *P. pacifica* of Howell. The source of the new name is given.

Asterisks (*) indicate species not previously reported from the Island or known from the large herbaria. The sign X precedes the name of a hybrid, thus: "X *Solidago asperula* Desf.", the name given to hybrids of *S. rugosa* and *S. sempervirens*.

The following records from the 1891 list of MacSwain and Bain must be excluded in addition to those specifically rejected in the annotated list:

Panicum agrostoides
Carex sychnocephala Carey - mistake for *C. crawfordii* or other young plants of *Carex* sect. *Ovales*.
Polygonella articulata (L.) Meisn. (*Polygonum* L.)
Ribes trifidum Michx. - (*R. glandulosum*).
Aesculus hippocastanum L. - introduced only.
Teucrium canadense L.
Monarda angustifolia - no such name.
Mentha borealis Michx. - (*M. arvensis* var. *canadensis*).
Scrophularia marilandica L.
Galium boreale L.
G. latifolium Torr.

Overlooked in Campbell's list is a record of *Hibiscus trionum* L., evidently an adventive, not established. Omitted among common garden escapes is *Valeriana officinalis* L., CHERRY PIE, seen at Rollo Bay, Murray River and elsewhere.

EQUISETACEAE

Equisetum arvense L. Map 1. Common throughout. Damp meadows, in damp wood-edges has elongated branches, railway embankments, damp sand behind dunes as a depressed sterile state, f. ***decumbens*** (Mey.) Klinge. Weed and poisonous plant. 28 Lots (Groh). MacSwain and Bain, Hurst.

Equisetum sylvaticum L., var. ***pauciramosum*** Milde Map 2. Throughout in low, poorly drained sites, more commonly in partial shade than *E. arvense.* 18 Lots (Groh). MacSwain and Bain, Hurst.

[*Equisetum palustre* L. Not seen, and absent from Nova Scotia. One station on the Tignish railroad (Groh, in error for *E. fluviatile* ?), Hurst.]

Equisetum fluviatile L. Map 3. Frequent in western Prince County, eastward scattered, most common along North Shore and in east-central lowland. Shallow standing water of streams, ponds and marshes. (*E. limosum* L., Macoun), Hurst.

[*Equisetum hyemale* L. Not seen. Brackley Point (Macoun), no specimen at National Herbarium. Wrongly ascribed to Groh by Hurst.]

Equisetum scirpoides Michx. Map 4. Mount Stewart, bogs and boggy spots (*Fernald et al.* in 1912). Brackley Point (Macoun), without specimen. Hurst.

LYCOPODIACEAE

Lycopodium lucidulum Michx. Map 5. Scattered throughout, in leaf mould of mixed woods. Macoun, MacSwain and Bain, Hurst.

Lycopodium inundatum L. Map 6. Scattered. Peat bogs, especially in mossy pockets or on bare damp sand or mud in full light. Macoun, Hurst.

Lycopodium annotinum L. Map 7. Scattered throughout. Dry mixed woods. Macoun, MacSwain and Bain, Hurst.

Var. ***acrifolium*** Fern. occurs in similar habitats at least at Rocky Point (cited in original description Rhodora 17: 124, 1915), and at Bear River. Intergradations occur as in Macoun's material from Cantire.

Lycopodium clavatum L. Map 8. Scattered throughout, on light soil, in dry open woods. MacSwain and Bain, Hurst.

In old fields on sunny banks the prevalent form is var. *megastachyon* Fern. & Bissell, possibly the growth-form developed in response to those conditions, as single and paired spikes may be borne on alternate peduncles of one rhizome. Colville Station, Royalty Junction, Peakes Station. (Var. *monostachyon* Macoun, Hurst, Campbell). Used as a Christmas decoration (Bain).

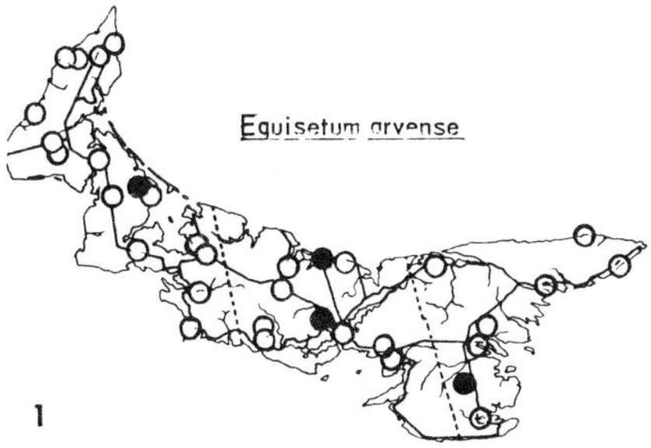
Equisetum arvense

1

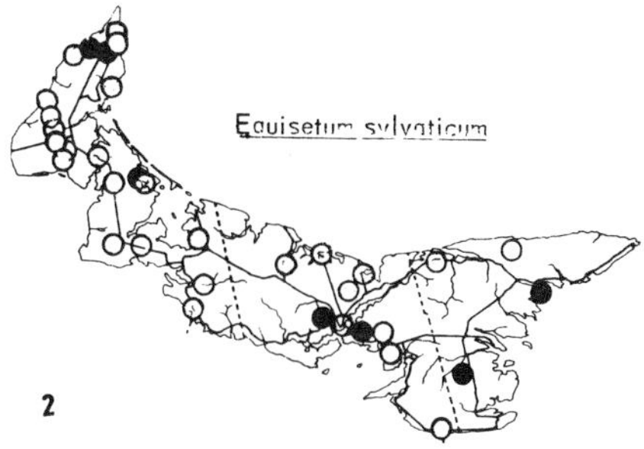
Equisetum sylvaticum

2

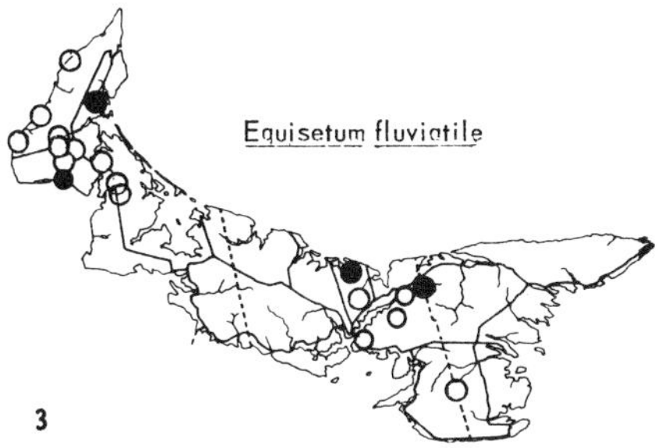
Equisetum fluviatile

3

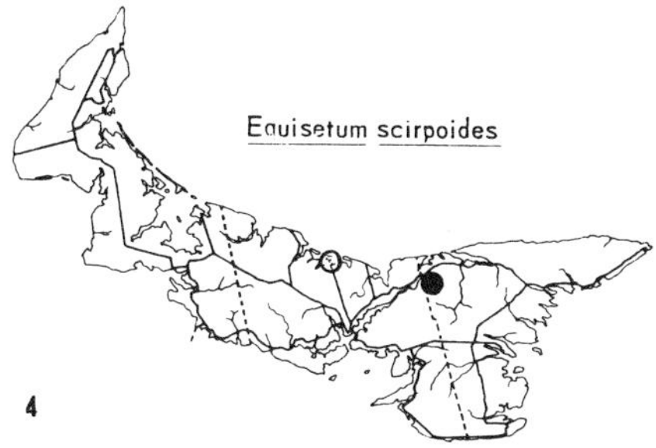
Equisetum scirpoides

4

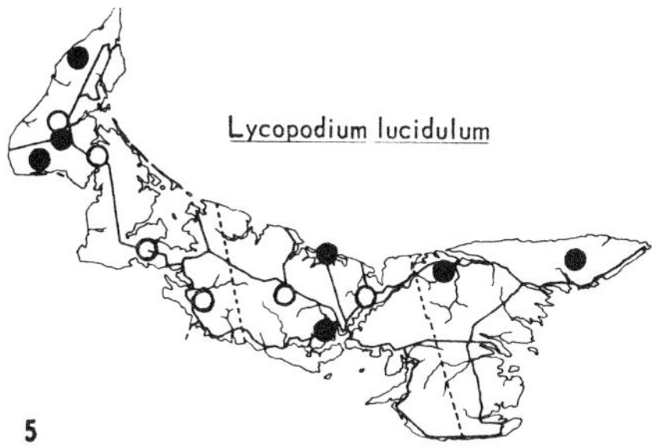
Lycopodium lucidulum

5

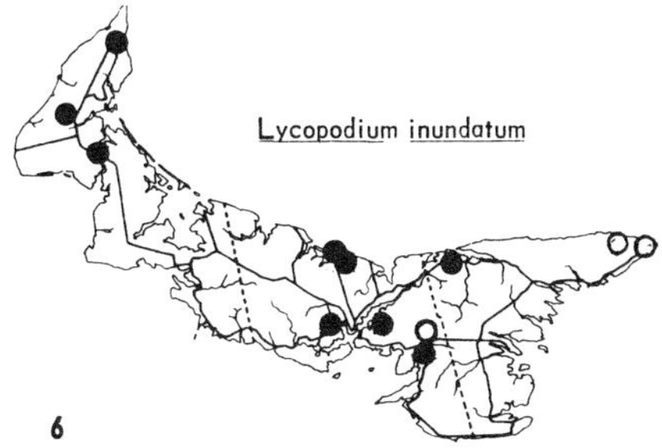
Lycopodium inundatum

6

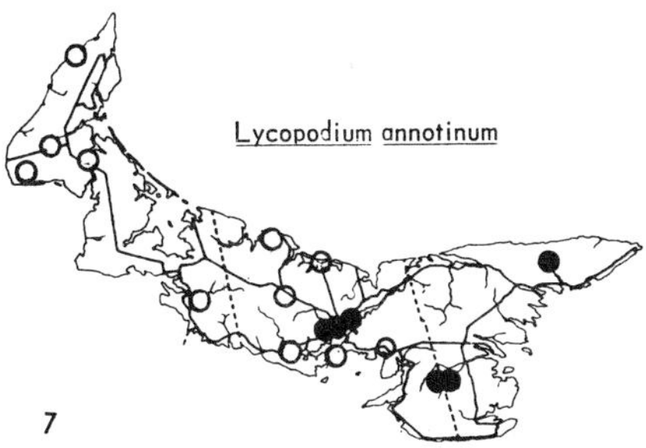
Lycopodium annotinum

7

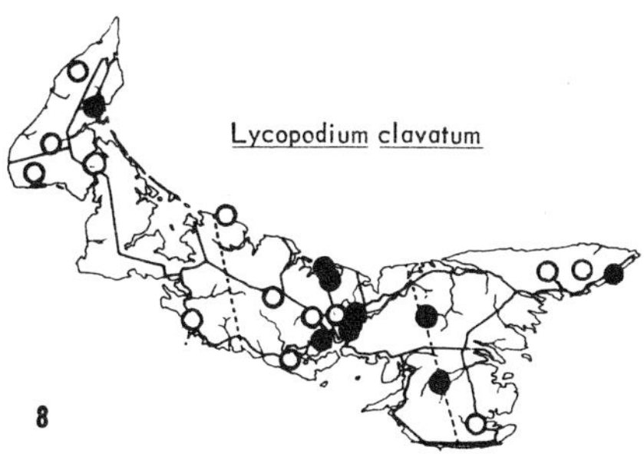
Lycopodium clavatum

8

Lycopodium obscurum L. Map 9. "GROUND SPRUCE" (Watson) Scattered throughout, in dry woods on hills and bluffs. Sun-forms have relatively erect branches with appressed leaves, a development comparable to the var. *megastachyon* of *L. clavatum*. Macoun, Hurst. (*L. dendroideum* Michx., MacSwain & Bain).

Lycopodium sabinaefolium Willd. Map 10. Old fields and land cleared from spruce woods. Alberton, Stanhope, Souris, Murray Harbour North. At Souris, the var. *sabinaefolium* and var. *sitchense* (Rupr.) Fern. grow side by side. Macoun, Hurst.

Lycopodium complanatum L. Map 11. Western Prince County and around Charlottetown, to be expected elsewhere. Dry woods on sandy or light soil. Macoun, and MacSwain and Bain referred to the following as *L. complanatum.*

Var. ***flabelliforme*** Fern. Map 12. Scattered throughout. Light soils, in open woods and on cleared banks. This may well be recognized, following Marie-Victorin, as a species, *L. flabelliforme* (Fern.) Blanchard. Hurst.

Lycopodium tristachyum Pursh Map 13. Scattered in Western Prince County; Murray River. Dry heathy clearings in open sites on light acid soil. Excluded by Hurst.

ISOËTACEAE

****Isoëtes riparia*** Engelm. Map 14. Lake Verde, Glenfinnan Lake. Shallow water on a bottom of mud-covered sand. Characteristic of acid waters low in nutriment. Determined by Rao.

OPHIOGLOSSACEAE

[***Botrychium simplex*** E. Hitchc. Reported by Burgess in Macoun (1890), on the authority of Bain. In a letter to L.W. Watson, Macoun identified Bain's specimen as a "small *B. virginianum*". MacSwain and Bain, Hurst.]

Botrychium matricariaefolium A. Br. Map 15. Hamilton near Malpeque, Hunter River, Black River, Valleyfield East near Montague. In hardwood leaf mold, and in cleared land reverting to white spruce. (*B. ramosum*, Hurst). Burgess, in Macoun.

****Botrychium lanceolatum*** (Gmel.) Ångström var. ***angustisegmentum*** Pease & Moore Map 16. Hamilton near Malpeque, Hunter River. Always found in association with *B. matricariaefolium.*

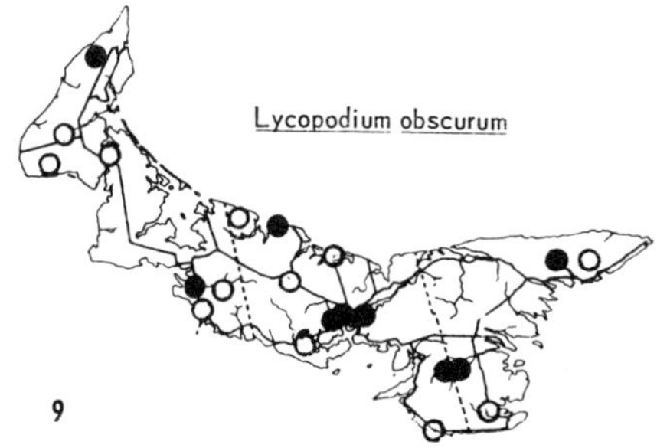

9

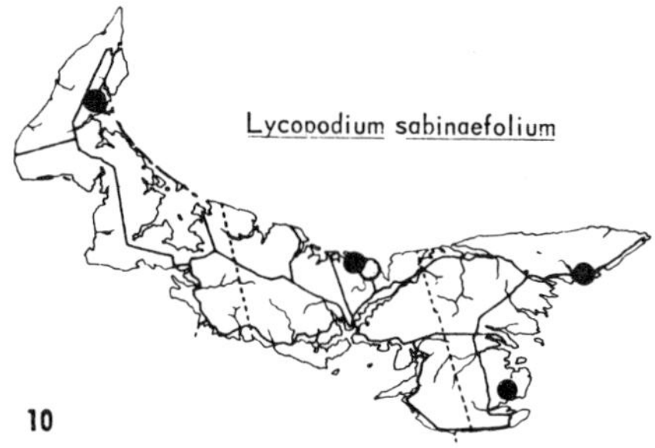

10

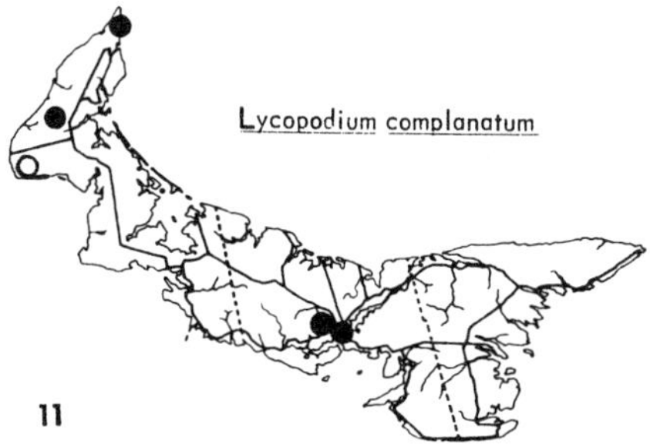

11

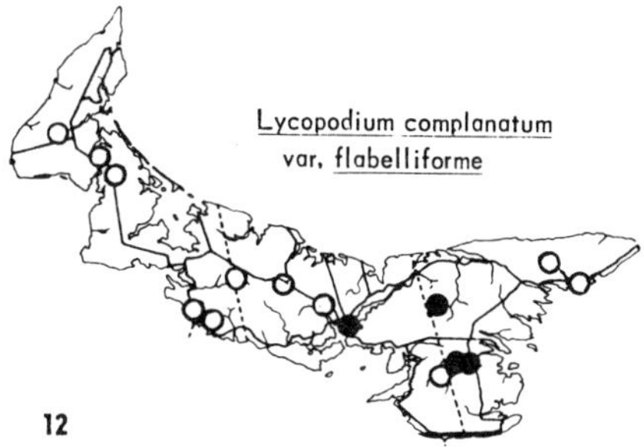

12

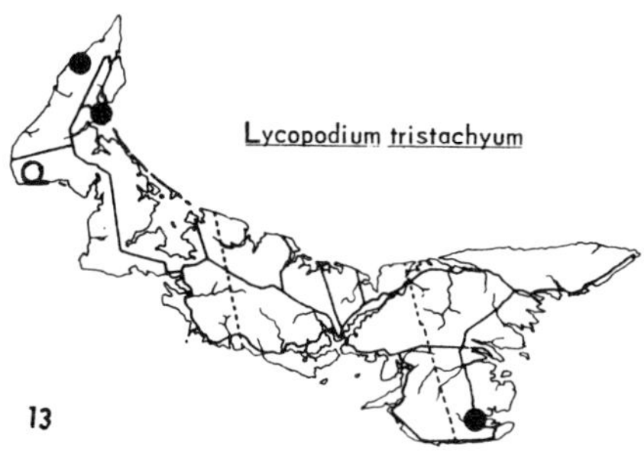

13

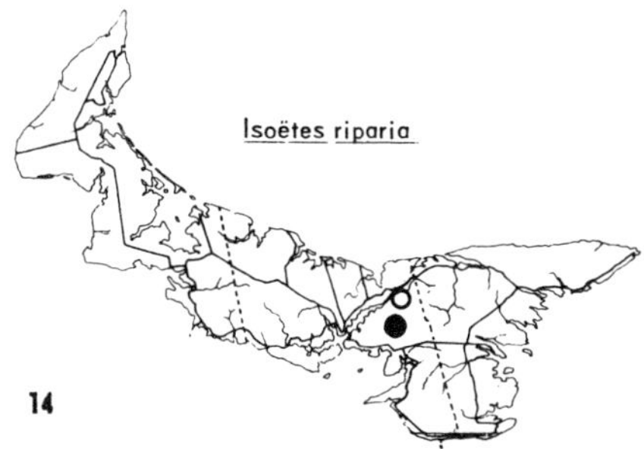

14

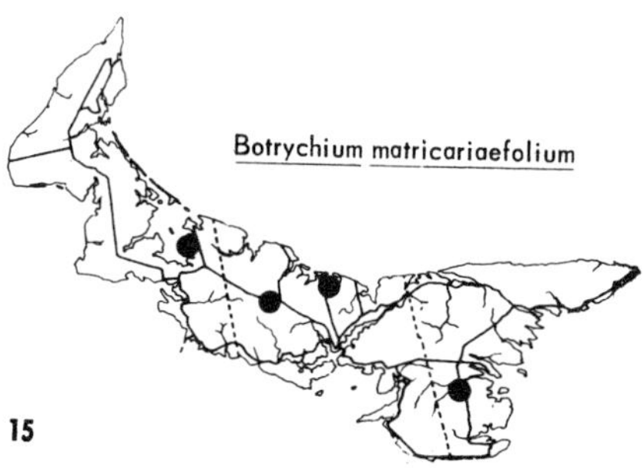

15

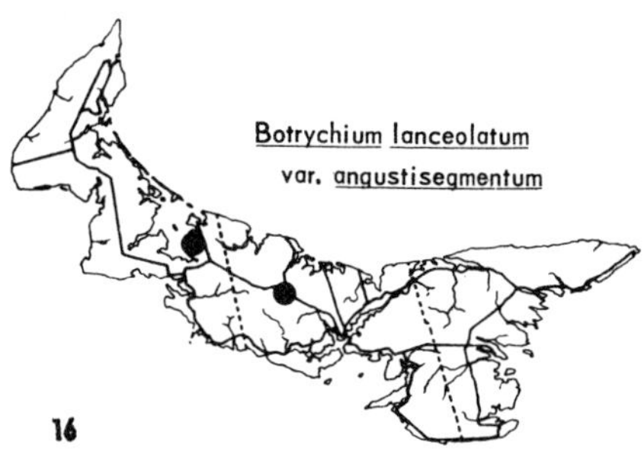

16

Botrychium multifidum (Gmel.) Rupr. Map 17. Rare: in coarse sandy soil, Cape Aylesbury (P), Brackley point, North River (Q), Morell, North Lake, Kilmuir (K). Scattered solitary plants in acid turf, among blueberry bushes. (*B.* ***lunarioides***, MacSwain and Bain; *B.* ***ternatum*** Bain, Burgess in Macoun, MacSwain, Hurst).

Botrychium virginianum (L.) Sw. Map 18. Scattered from east to west. In rather dense shade: deciduous woods or swampy but not boggy coniferous woods; in leaf mould or moss. (*B.* ***simplex***, Bain); MacSwain, Hurst.

Ophioglossum vulgatum L. var. ***pseudopodum*** (Blake) Farwell Map 19. Brackley Point, Dalvay House, in damp shady hollows behind dunes. Burgess in Macoun, Hurst.

OSMUNDACEAE

Osmunda regalis L. var. ***spectabilis*** (Willd.) Gray Map 20 Scattered, rather scarce. In swamps, especially at ditches and along river margins. MacSwain and Bain, MacSwain, Hurst.

Osmunda claytoniana L. Map 21. Common everywhere except, apparently, in the southeast. Damp, partly shaded sites, at edge of woods or swamps. MacSwain and Bain, MacSwain, Hurst.

Osmunda cinnamomea L. Map 22. Common throughout, acid soils, among heath bushes or in open woods. MacSwain and Bain, MacSwain, Hurst.

PTERIDACEAE

(Copeland's classification is followed)

Dennstaedtia punctilobula (Michx.) Moore Map 23. Distribution parallels that of the Alberry and Charlottetown fine sandy loams. Common in open deciduous or mixed woods and old fields or banks in hilly well-drained country. (***Dicksonia punctilobula*** (Michx.) Gray, MacSwain and Bain, Hurst; ***Dicksonia pilosiuscula*** Willd., Burgess in Macoun, MacSwain).

Pteridium aquilinum (L.) Kuhn var. ***latiusculum*** (Desv.) Underw. Map 24. Common throughout. Open sites, dry or damp, very abundant in acid bushy old pastures. (***Pteris aquilina***, MacSwain and Bain, MacSwain, Hurst).

MacSwain's record of ***Pteris aquilina*** var. ***lanuginosa*** Bong. from Rustico would not refer to the western var. ***pubescens*** Underw., but to an unusual pubescent specimen of the common variety.

[***Cryptogramma stelleri*** (Gmel.) Prantl Holman's report of it in 1935 from the cliffs of the Dunk River has been repeated by Adams and by Hurst, but no substantiating specimens were retained. Not in Roland.]

[***Adiantum pedatum*** L. was listed by MacSwain and Bain, and even earlier in Stewart's history (1806). No suitable habitats for it exist.]

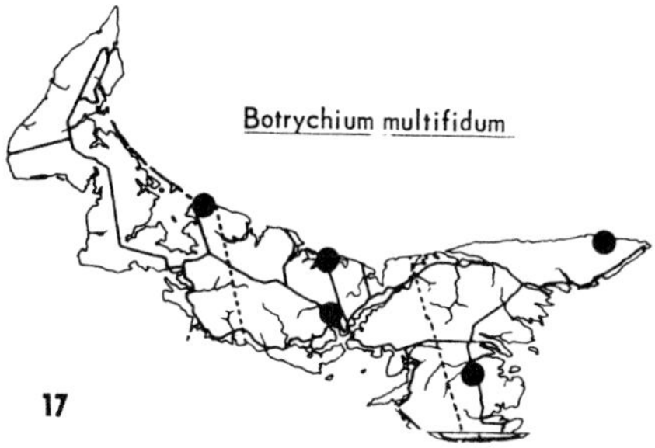

17

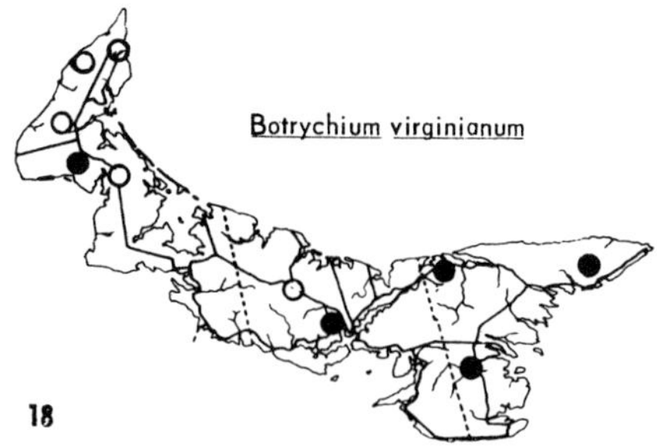

18

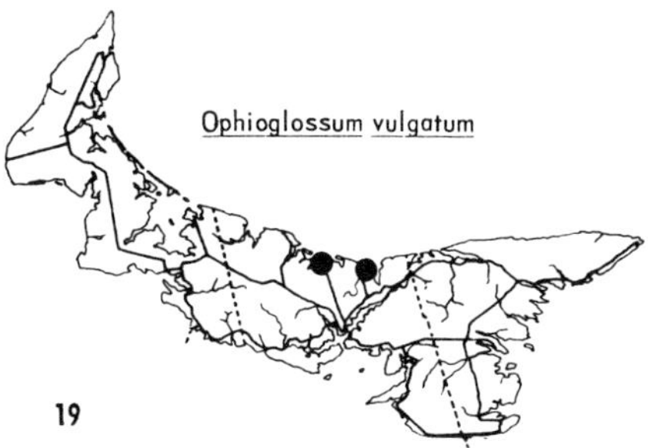

19

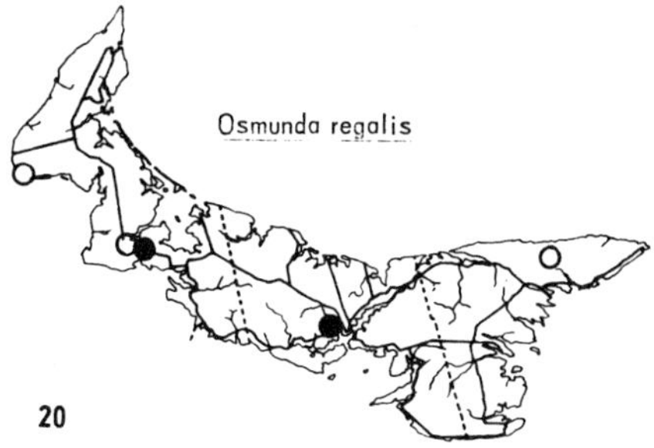

20

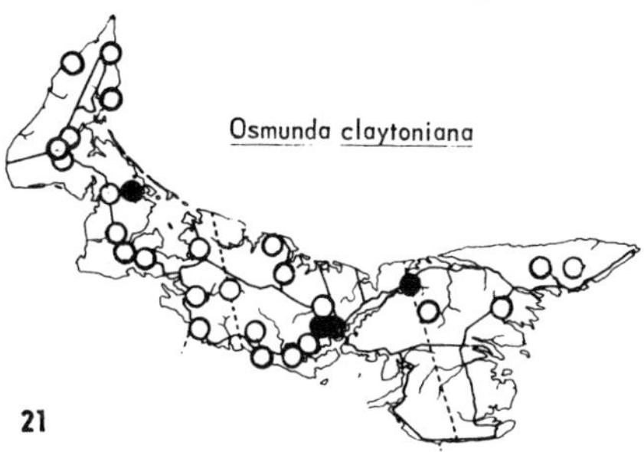

21

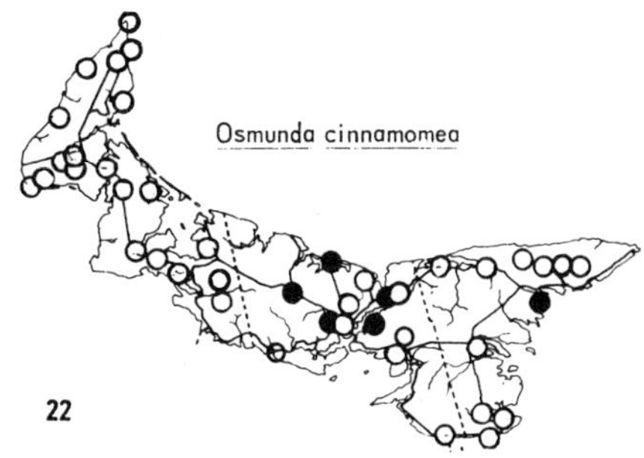

22

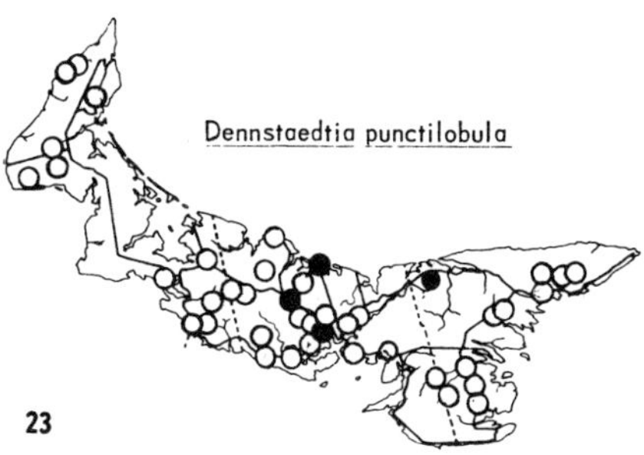

23

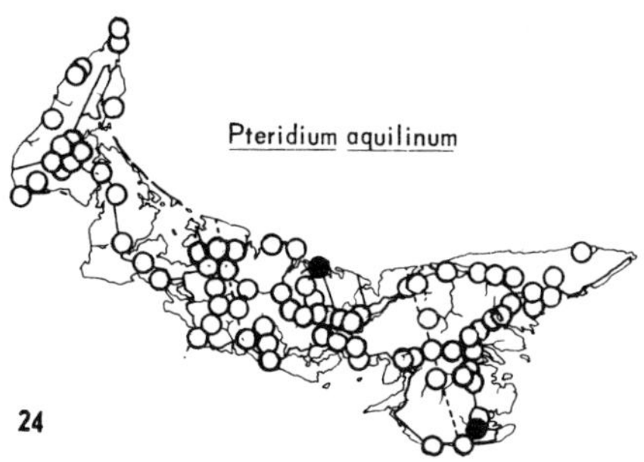

24

ASPIDIACEAE

Matteuccia pensylvanica (Willd.) Raymond (Nat. Can. 77: 55. 1950) Map 25. Common in the creek-bottom alder thickets of the Prince County clay regions, rarer eastward on the flood deposits of the larger streams. It should be noted that this, rather than ***Osmunda cinnamomea***, provides the commercial "fiddleheads" of New Brunswick. (***Struthiopteris germanica***, MacSwain and Bain; ***Onoclea Struthiopteris***, MacSwain, Hurst; ***Pteretis pensylvanica*** (Willd.) Fern., Campbell).

Onoclea sensibilis L. Map 26. Throughout. Common in damp sunlit sites: meadows and ditches. When vegetative fronds are cut, as often in haying operations, the unrolling fertile fronds may revert to a partially vegetative function with intermediate form: the "var. ***obtusilobata*** (Schkuhr) Torr." reported by Bain 1890). MacSwain and Bain, MacSwain, Hurst.

Polystichum acrostichoides (Michx.) Schott Map 27. Rare: Pleasant View (P); North River (possibly); scattered around Souris. Deciduous woods. (***Aspidium acrostichoides*** (Michx.) Sw., Burgess in Macoun, MacSwain and Bain, MacSwain, Hurst).

[***Dryopteris fragans*** (L.) Schott Bain's (1892) record of this species, as ***Aspidium fragans***, is quite improbable.]

[***Dryopteris marginale*** (L.) Gray All records of this species from P.E.I. can be traced to Francis Bain. Until collected, it can be excluded from the flora. (***Aspidium marginale*** (L.) Sw., Burgess in Macoun, Britton and Brown, Hurst).]

Dryopteris cristata (L.) Gray Map 28. Common in western Prince County, scattered eastward. Damp shady sites; very characteristic of cedar swamps; unusually, in a dry beech grove at the Experimental Farm. (***Aspidium cristatum*** (L.) Sw., Burgess in Macoun, MacSwain and Bain, MacSwain, Hurst).

[X ***Dryopteris boottii*** (Tuckerm.) Underw., a hybrid of ***D. cristata*** and ***D. spinulosa*** var. ***intermedia***, should, at least for the present, be excluded from the flora. MacSwain and Bain recorded ***Aspidium spinulosum*** var. ***Boottii*** as common, through an evident misidentification of ***D. spinulosa*** var. ***intermedia***. (***Aspidium Boottii*** Tuckerm., MacSwain, Hurst); Campbell.]

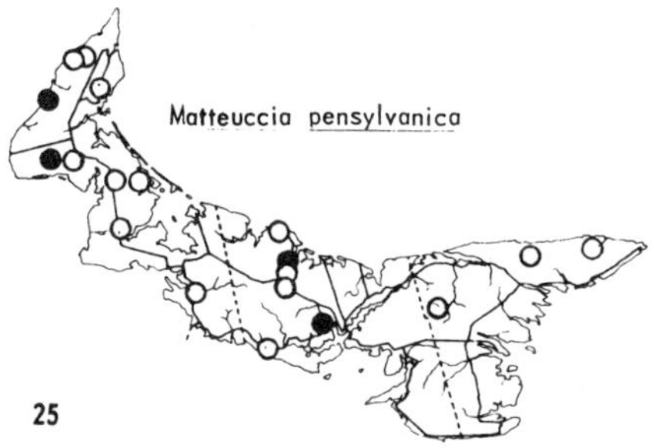

25

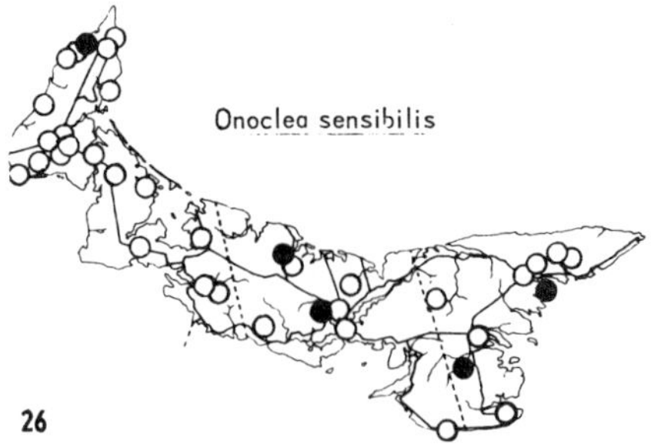

26

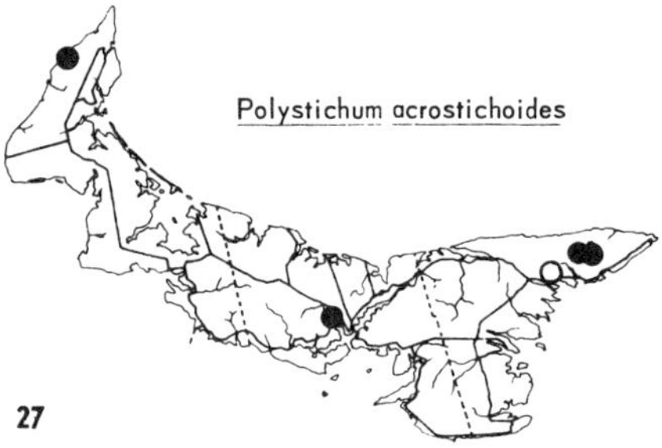

27

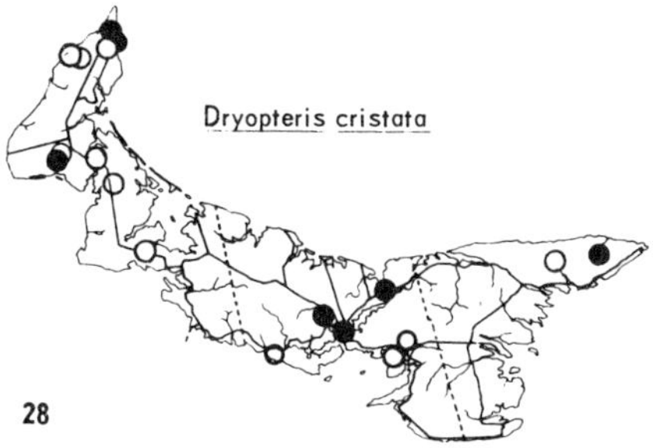

28

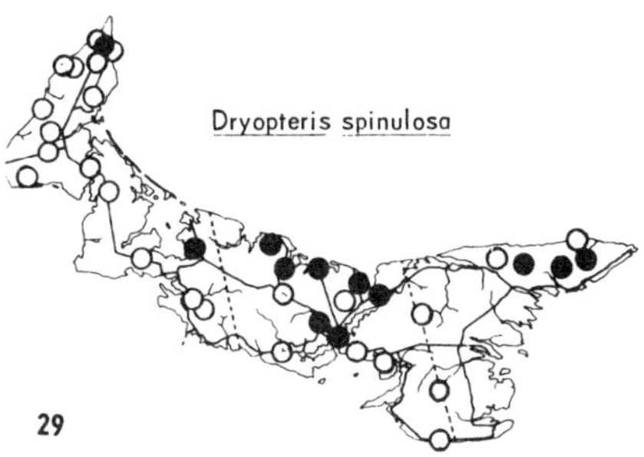

29

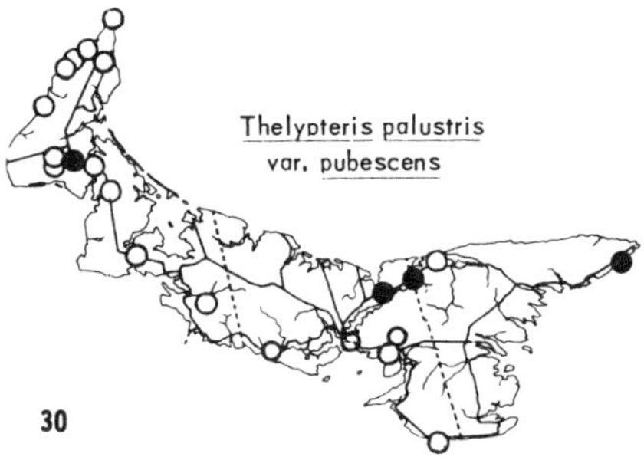

30

Dryopteris spinulosa (O.F. Müller) Watt, vars. Map 29. Common throughout, in woods. Specimens from central Queens County have been identified by Fernald as var. ***spinulosa.*** The widespread form is var. ***intermedia*** (Muhl.) Underw., sometimes regarded as a distinct species, and intermediate between them is the var. ***fructuosa*** (Gilbert) Trudell, represented by a Macoun collection from Brackley Point Road. (***Aspidium spinulosum*** (O.F. Müller) Sw., Burgess in Macoun, and var. ***Boottii*** MacSwain and Bain).

Var. ***americana*** (Fisch.) Fern. Scattered in deciduous woods near the North Shore from Cavendish eastward, particularly well developed on the ridge north of Souris.

Thelypteris palustris (Salisb.) Schott var. ***pubescens*** (Lawson) Fern. Map 30. Common in Prince County, scattered eastward. Mucky alder thickets and marshy runs; often in the damp hollows behind sand dunes, damp sites with considerable sunlight. (***Aspidium Thelypteris,*** MacSwain and Bain, MacSwain, Hurst; ***Dryopteris Thelypteris*** (L.) Gray var. ***pubescens*** (Lawson) Nakai, Campbell). The genus ***Thelypteris*** has been separated from ***Dryopteris*** by Copeland on morphological evidence and by Manton on chromosome numbers. The name ***Lastrea*** used by Copeland is antedated by ***Thelypteris.***

Thelypteris simulata (Davenp.) Nieuwl. is reported from the Island by Roland (1947). (***Dryopteris*** Davenp.).

Thelypteris noveboracensis (L.) Nieuwl. Map 31. Rather scarce. In shady fairly dry sites: sandy loam under hardwood trees or in dry thickets along streams. (***Aspidium noveboracensis*** (L.) Sw., MacSwain and Bain, MacSwain, Hurst; ***Dryopteris noveboracensis*** (L.) Gray, Campbell).

Thelypteris phegopteris (L.) Slosson Map 32. Common between Charlottetown and Summerside, scattered in Prince County, apparently rare in Kings County. Open, usually coniferous woods on well-drained banks, more rarely along damp runs. (***Phegopteris polypodioides*** Fée, Macoun, MacSwain and Bain, MacSwain, Hurst; including ***P. hexagonoptera*** Bain 1892; ***Dryopteris Phegopteris*** (L.) C. Chr., Campbell).

Thelypteris dryopteris (L.) Slosson Map 33. Throughout. Partially shaded sites; stream-bottom thickets and their banks, or dry hardwoods. (***Phegopteris Dryopteris*** (L.) Fée, MacSwain and Bain, MacSwain, Hurst; including ***P. calcaria,*** Bain 1892; ***Dryopteris disjuncta*** (Ledeb.) C.V. Morton, Campbell). In this case the species should possibly be segregated from Thelypteris, but the generic name cannot be ***Gymnocarpium*** as proposed by Morton in the New Britton and Brown, that name referring (Copeland says) primarily to ***T. phegopteris.***

[***Cystopteris fragilis*** (L.) Bernh. Holman's record of it from cliffs of the Dunk River in 1935 requires substantiation. Adams, Hurst.]

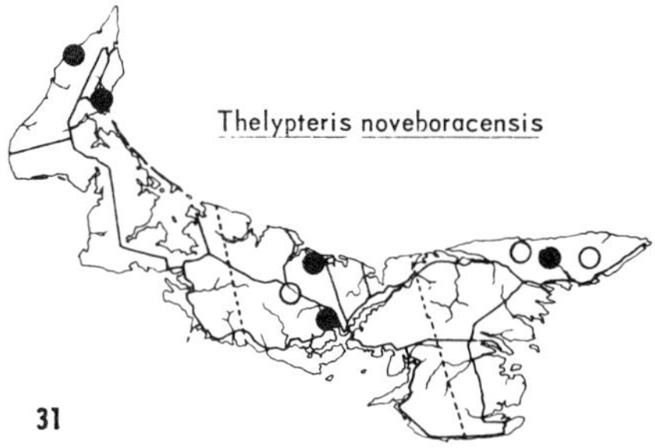

31

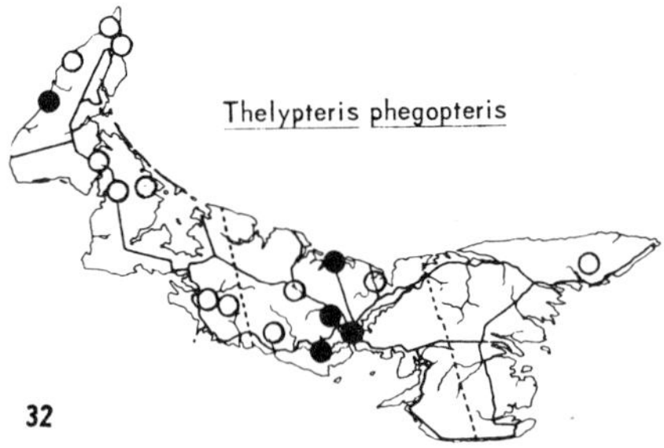

32

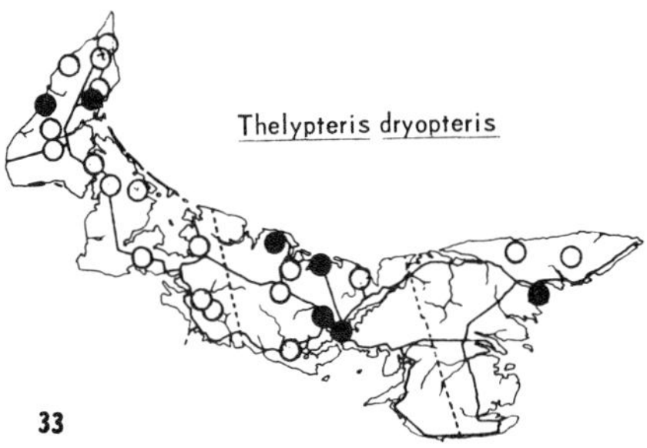

33

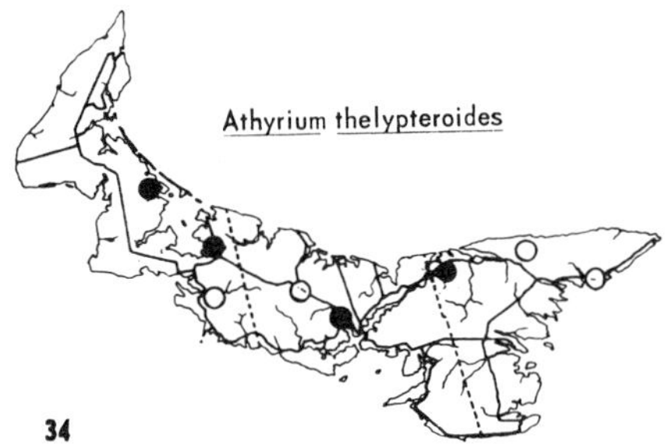

34

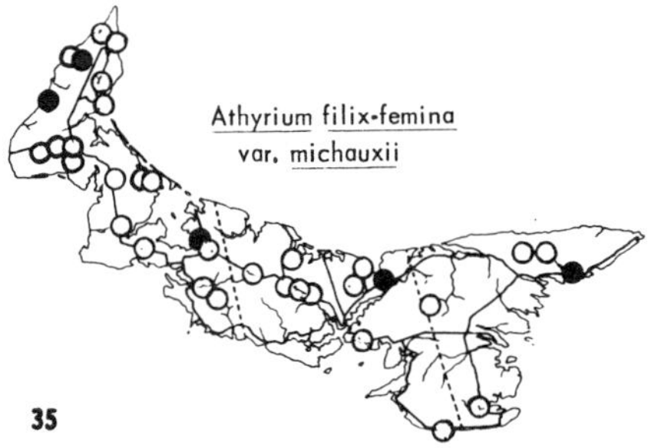

35

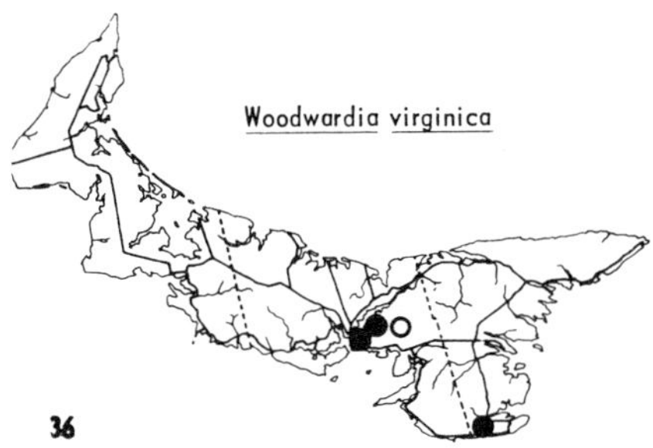

36

Athyrium thelypteroides (Michx.) Desv. Map 34. Scarce; stream banks, in shade and mould of mixed woods. (***Asplenium thelypteroides*** Michx., MacSwain and Bain, MacSwain; ***Asplenium acrostichoides*** Sw., Hurst).

Athyrium filix-femina (L.) Roth var. ***michauxii*** (Spreng.) Farwell Map 35. Common throughout. Ecological tolerance broad: dry wood-edge banks to damp shady runs. In dense shade the f. ***rubellum*** (Gilbert) Farwell is produced. (***Asplenium filix-femina*** (L.) Bernh., MacSwain & Bain, MacSwain, Hurst).

BLECHNACEAE

Woodwardia virginica (L.) Sm. Map 36. Pond south of Murray River; pond at Village Green; Mermaid Lake, Southport (discovered by A. Smallwood; MacSwain's only station, Erskine in Rhodora 53: 265. 1951). Floating mat of bog ponds.

POLYPODIACEAE

Polypodium virginianum L. Map 37. Dunk River above Central Bedeque, moss on shady sandstone ledges. Only station; known since 1903. (***P. vulgare***, MacSwain, Hurst), Campbell.

TAXACEAE

Taxus canadensis Marsh. Map 38. From east to west on clay banks or along streams in ravines and woodland; absent from the steep dry central hills and eastern acid swamps. Undershrub of the best mixed forest land; more common on good hardwood land (Stewart). "GROUND SPRUCE" (Stewart); in Nova Scotia "Ground Hemlock". (***T. baccata***, MacSwain and Bain); MacSwain, Hurst.

Abies balsamea (L.) Mill. BALSAM FIR Map 39. Throughout, but most abundant in West Prince, seldom reaching any great height and on headlands low or even depressed. Possibly favored south of the true Boreal Forest (as also in Cape Breton) by clay soils. Characteristic of mixed forest on the best soils (Stewart). Nodes may be swollen by attacks of Balsam Woolly Aphis. Timber poor, used, if straight, for fence rails and laths; Canada Balsam used as internal and external medicament (Stewart). Now in demand as a Christmas tree. At least in West Prince represented by var. ***phanerolepis*** Fern. (***Pinus balsamea*** L., Stewart); Macoun, MacSwain and Bain, MacSwain, Hurst.

Tsuga canadensis (L.) Carr. HEMLOCK Map 40. Scattered, in woods on well-drained land: ravines, knolls. The durable wood used for wharves and pilings (Stewart), too brittle for beams but boards used for siding, the bark excellent for tanning, the shoots for a spruce beer (Stewart). (***Pinus Abies***, Stewart; ***Abies canadensis*** (L.) Michx., MacSwain and Bain); MacSwain, Hurst.

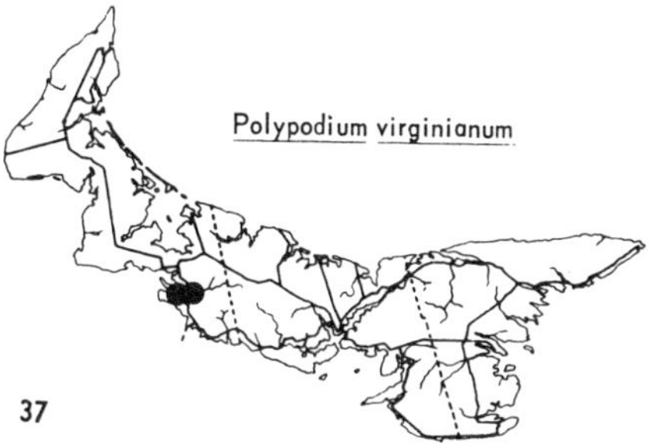

37

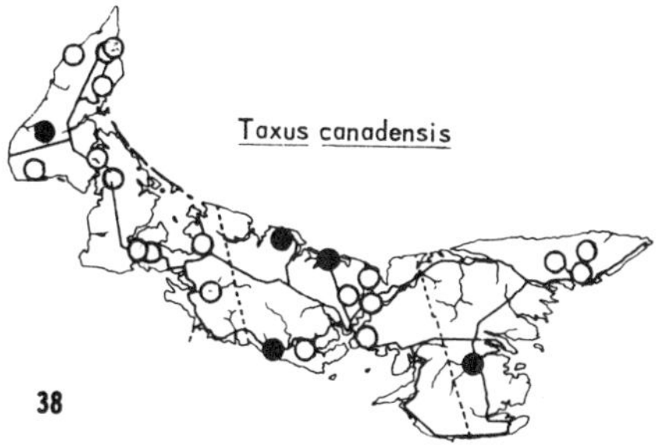

38

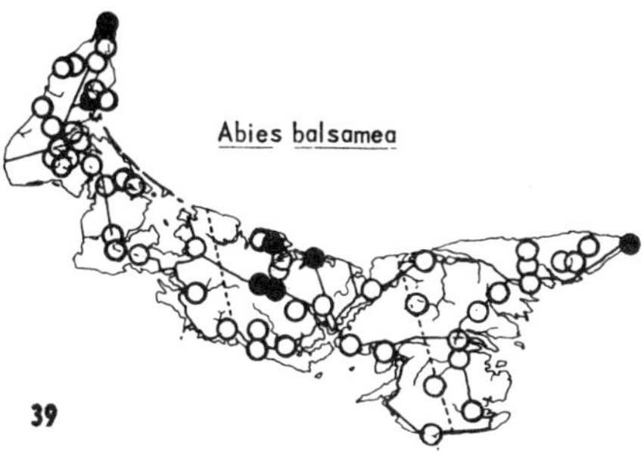

39

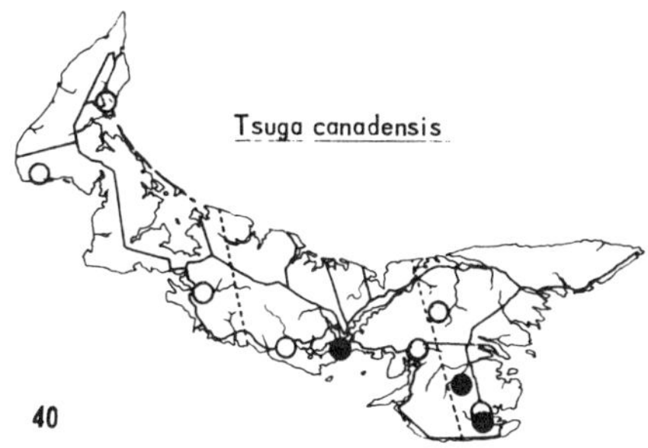

40

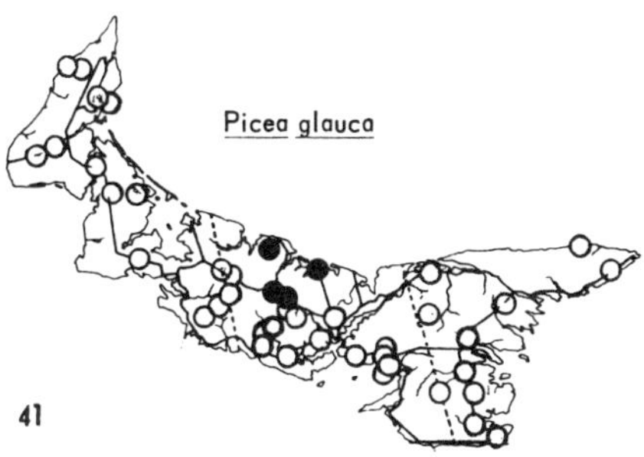

41

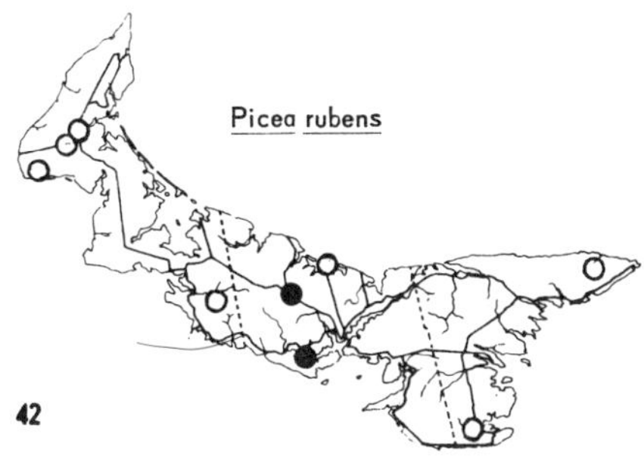

42

Picea glauca (Moench) Voss WHITE SPRUCE Map 41. Throughout, a pioneer of open land establishing itself only in sunlit sites: in woodland only on the old dunes where no trees precede it; nowadays very common on abandoned farms and pastures of the lighter soils, and a major source of pulpwood; there growing so densely as to shade out all green vegetation.

Growing with ***Betula populifolia***, and indicator of the poorest land for farming, The light wood useful for spars and rafters (Stewart). (***Pinus canadensis***, white type, Stewart; ***Abies alba*** (Ait.) Link, Macoun, MacSwain; ***P. canadensis***, Hurst), Campbell.

Picea rubens Sarg. RED SPRUCE Map 42. Scattered, in well-drained 'hardwood forest' sites only: notably in ravines of Queens, and on knolls of West Prince. Forming beautiful groves, superior to the white spruce (Stewart). Stewart, however, calls it a species of clearings and old fields, which would seem likely to mean *P. glauca*. An element of the Northern Hardwood Forest but closely related to *P. mariana* of the Boreal Forest. Macoun said of them "distinct enough in P.E.I. to be regarded as species and variety." (***Pinus canadensis***, red type, Stewart; ***Picea nigra*** var. ***rubra*** (DuRoi) Engelm., Macoun; ***P. rubra*** (DuRoi) Link, not Dietr., Hurst), Campbell.

Picea mariana (Mill.) BSP. BLACK SPRUCE Map 43. Common in Price and Kings; in Queens along the North Shore. Poorly-drained or exposed sites: swamps, bogs; on the dunes and headlands, depressed. Wood useful for masts and spars; shoots for spruce beer (Stewart). (***Pinus canadensis***, black type, Stewart; ***Abies nigra*** (Ait.) Michx., Bain; ***Picea nigra*** (Ait.) Link, Macoun; MacSwain and Bain, MacSwain), Hurst.

Larix laricina (Du Roi) K. Koch "JUNIPER" Map 44. Common throughout, with ***Picea mariana***, characteristic of peat bogs, in swamps, even with *Thuja;* in hilly regions, forming pure stands where invading poorly drained pastures. Bushy-topped in exposed sites (e.g., the Black Marsh at North Point). The tough durable wood was the best in the Temperate Zone for ship's knees and trunnels; the turpentine used medicinally (Stewart). (***Pinus Larix***, Stewart; ***L. microcarpa*** Forbes, Bain; ***L. Americana*** Michx., MacSwain and Bain, MacSwain); Hurst. The common name "Juniper" is in local usage, as also in parts of New Brunswick; known in Cape Breton as "Cedar", in Nova Scoria as "Hackamatack" and in Ontario as "Tamarack."

Pinus strobus L. PINE Map 45. More common in sandy eastern Kings and west of Summerside than in Queens. Large trees remain uncut along steep banks (e.g. near Montague River or at Milton) and old stumps indicate the former widespread occurrence in the climax forest, the "White Pine-Northern Hardwoods" regional type of E. Lucy Braun (1950), reported by early settlers.

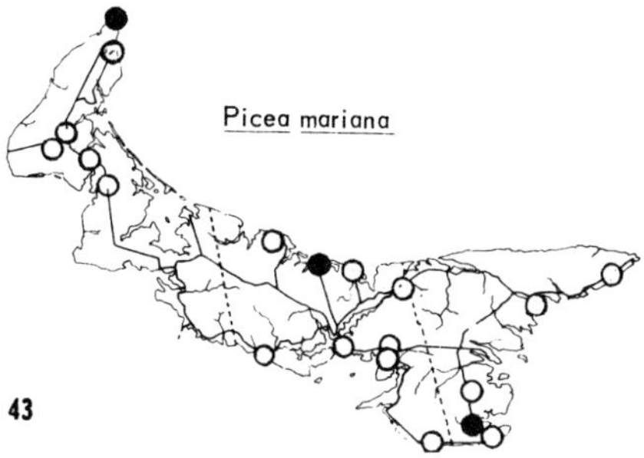

43

44

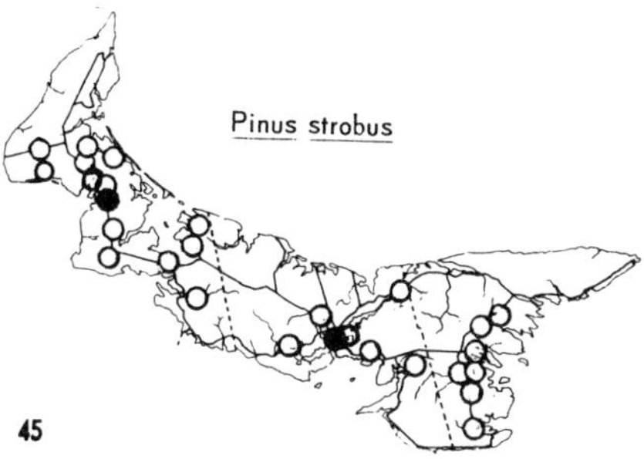

45

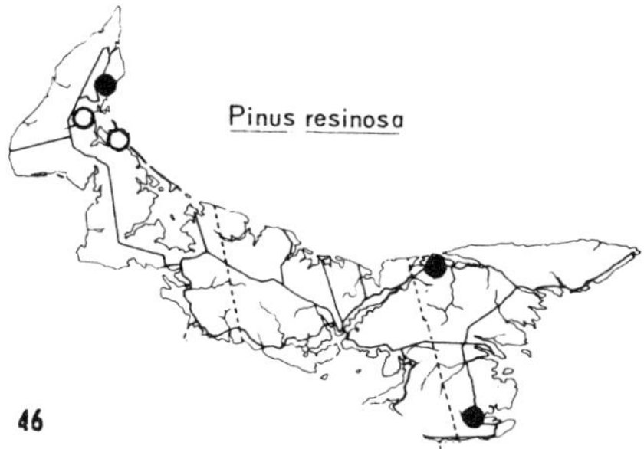

46

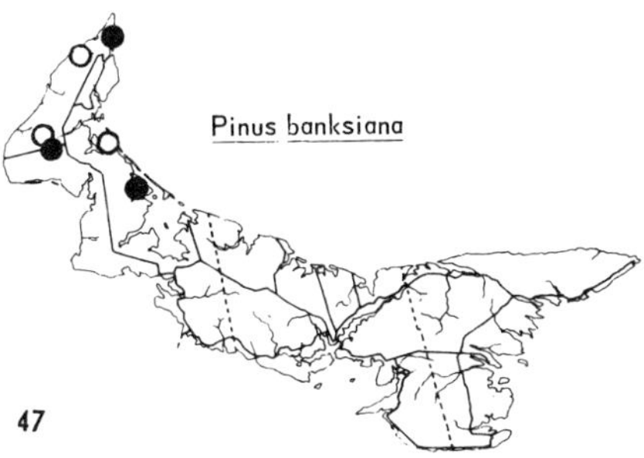

47

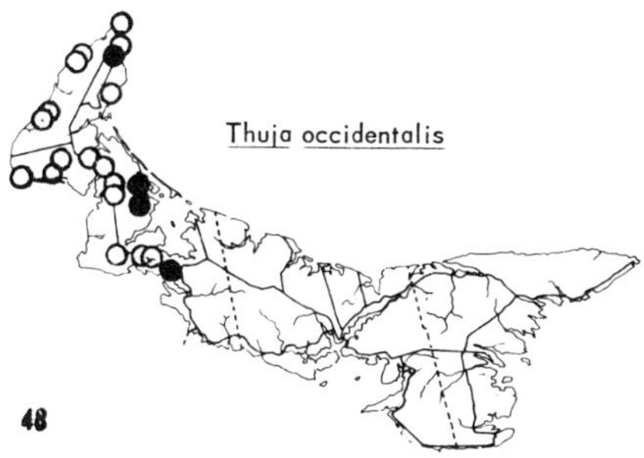

48

Under British rule in 1763, the pine was reserved to the navy for masts (one trunk of three to five feet in diameter could serve as mainmast of a 64-gun ship) and cutting by the settlers was forbidden. Stewart declares that it was not abundant and never found in pure stands but was characteristic of the mixed forest on the best (sandy loam) land. Groves of young trees are rare (e.g., at McNeill's Mills), and there is little replacement. MacSwain and Bain, MacSwain, Hurst.

Pinus resinosa Ait. Map 46. Very local: common around Cascumpeque Bay, e.g., at Mill River (Bain); locally through Kings County. Dry sandy soil. Small groves or scattered trees amid scrub; old trees only at Murray River. (*P. Taeda*, Stewart; *P. rigida*, Bain); MacSwain and Bain, MacSwain, Hurst.

Pinus banksiana Lamb. Map 47. West Prince. "Jack Pine" is said by MacLeod (1947) to occur in Kings. Sandy soils, after fire; boggy meadow and barrens. Heat frees seed from the cones. (*P. Pinea*, Stewart); Hurst 1941.

Jack Pine *(Pinus banksiana)* on the fire-barren on sand of a post-glacial beach at East Bideford in western Prince County, where it is quite localized.

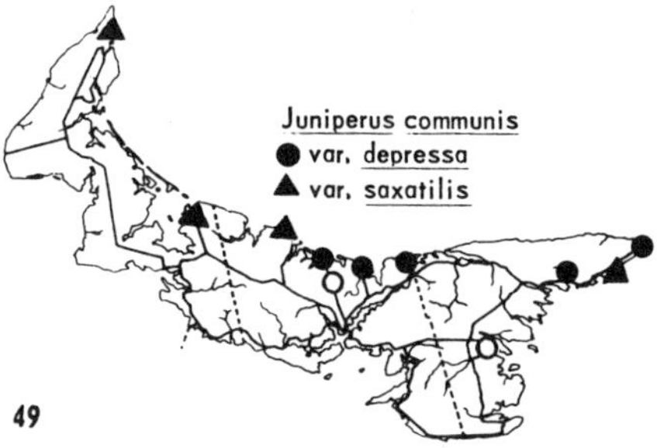

49

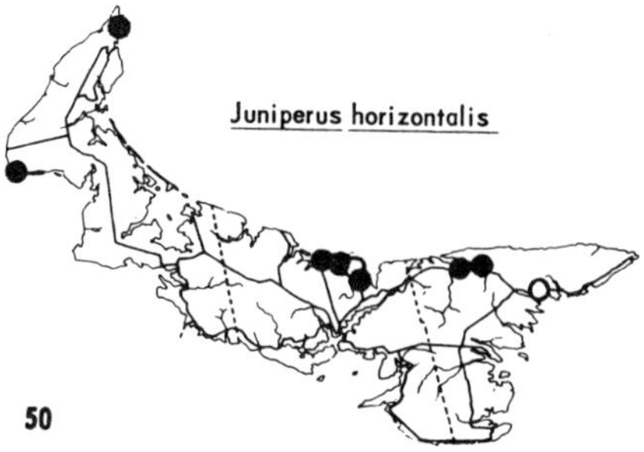

50

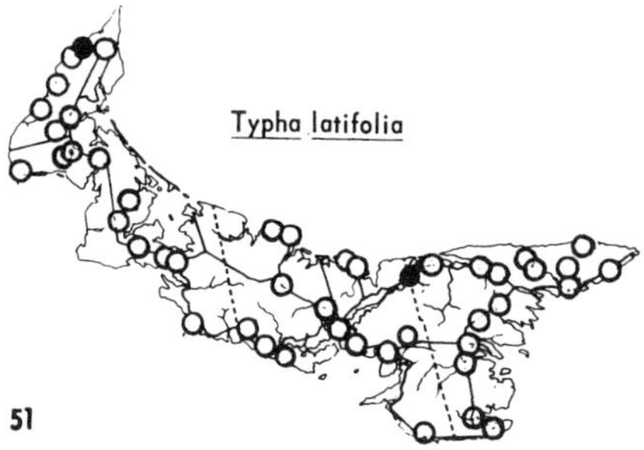

51

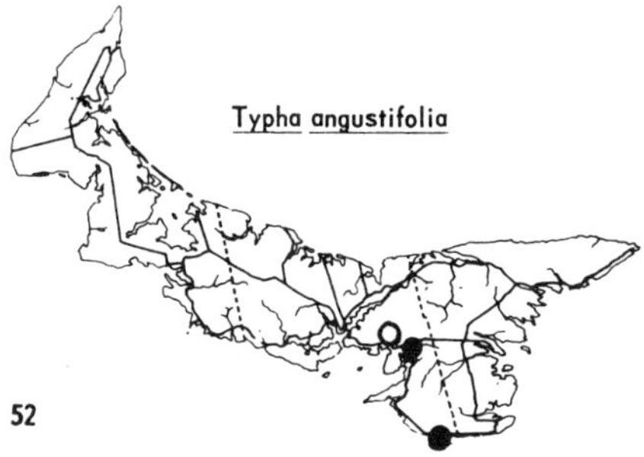

52

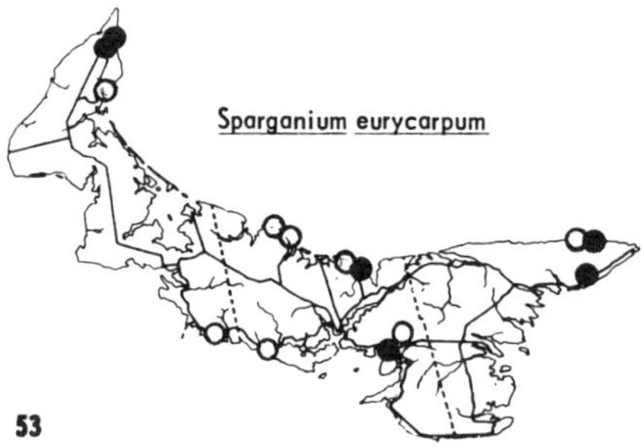

53

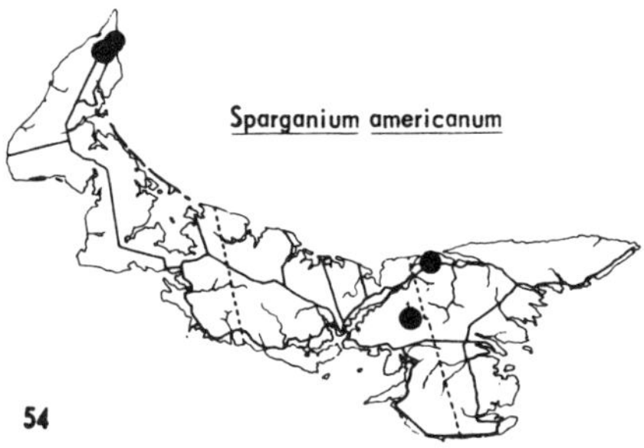

54

CUPRESSACEAE

Thuja occidentalis L. CEDAR Map 48. Prince County west of Summerside, its easternmost limit the Miscouche Woods. Swamps, mainly on clay; a little in the west on poorly drained sandy loam, where it may invade pastures. Indicator of a calcareous subsoil. Source of fence rails. (***Cupressus thyoides,*** Bain); MacSwain and Bain, MacSwain, Hurst.

Juniperus communis L., vars. Map 49. Along the North Shore and around to Montague. Dry open land of headlands and beaches.

Var. ***saxatilis*** Pallas Exposed sites: headlands and in the path of wind over sand dunes, along the North Shore. (Var. ***alpina*** Gaudin, MacSwain and Bain, MacSwain; var. ***montana*** Ait., Hurst; var. ***canadensis,*** Churchill), Campbell.

Var. ***depressa*** Pursh Open dry sites: openings in spruce woods, pasture weed near North Shore and eastern headlands. Poisonous; 1 Lot (Groh). Hurst.

Juniperus horizontalis Moench Map 50 Coastal, from West Point around the North Shore to Bay Fortune, on stabilized dunes and exposed sites nearby. Poisonous (Groh). (*J.* ***prostrata,*** Bain; *J.* ***Sabina*** var. ***procumbens*** Pursh, MacSwain and Bain, Churchill, MacSwain); Hurst.

TYPHACEAE

Typha latifolia L. "BULRUSH" Map 51. Throughout. Mucky shallows of fresh-water creeks or ponds, its rotting dead leaves and peduncles affording a habitat for many subaquatic species. MacSwain and Bain, MacSwain, Hurst.

****Typha angustifolia*** L. Map 52. Wood Island, brackish pool among sandy flats; an unconfirmed report from Village Green; Vernon (*J.S.* ***Erskine***).

SPARGANIACEAE

(included in Typhaceae by MacSwain and Bain, MacSwain)

Sparganium eurycarpum Engelm. Map 53. Margins of barrier-beach ponds and creeks along the North Shore; on the South, from a quaking marsh at Cherry Valley; scattered west to Summerside. A common element in the Typha stage of aquatic succession. MacSwain and Bain, MacSwain, Hurst.

Sparganium americanum Nutt. Map 54. Lower Sea Cow Pond; Tignish; Watervale (Q); Bristol, on muddy banks and at margin of slow streams. Hurst.

Sparganium chlorocarpum Rydb. Map 55. Common in east-central lowland, scattered elsewhere. Along streams, and shores of temporary ponds. (*S.* ***simplex,*** Macoun, Bain). Commonly represented by var. ***acaule*** (Beeby) Fern. (*S.* ***simplex,*** var. ***acaule*** Beeby, Macoun; *S.* ***diversifolium*** var. ***acaule*** (Beeby) Fern. & Eames, Hurst); Campbell.

****Sparganium fluctuans*** (Morong) Rob. Map 56. Pond between Stanhope Beach and Dalvay House, P.E.I. National Park.

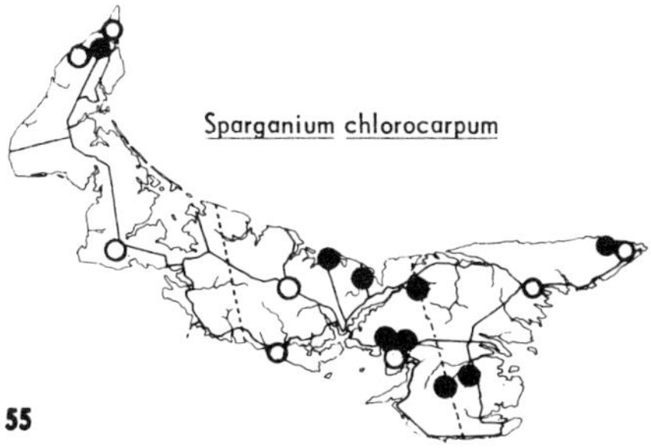

55

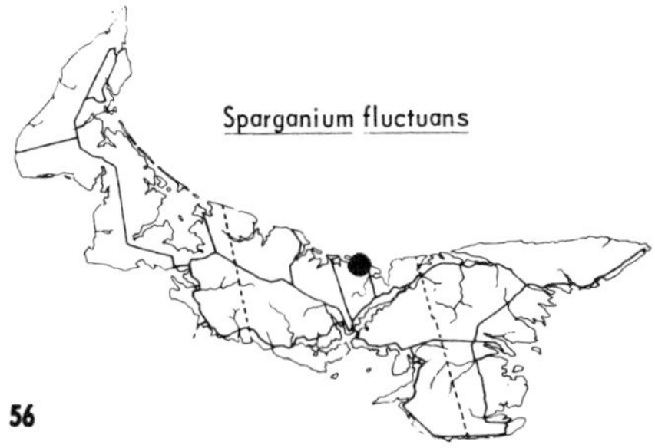

56

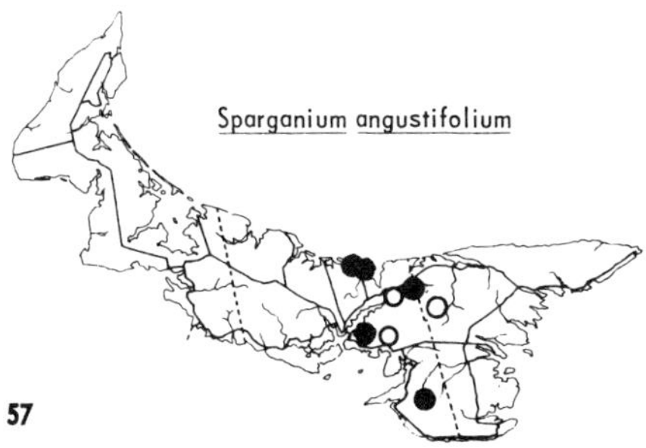

57

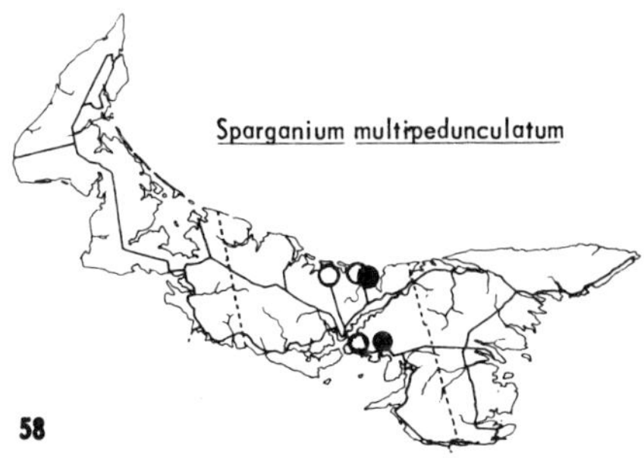

58

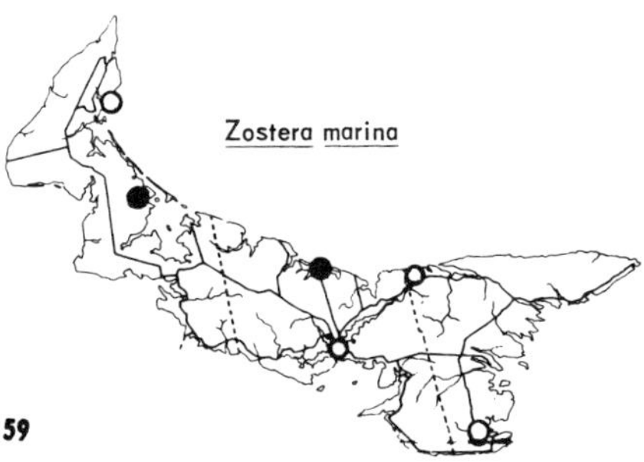

59

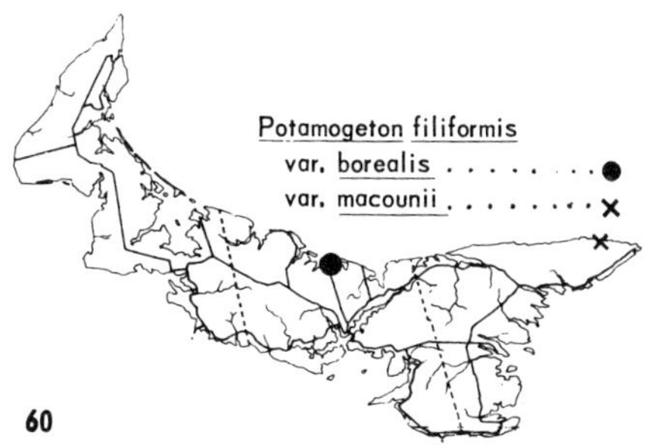

60

Sparganium angustifolium Michx. Map 57. Common in eastern Queens and adjacent parts of Kings, in lakes, standing or flowing water; streams, seldom flowering. (*S. simplex* var. *Nuttallii*, Macoun; var. *angustifolium* (Michx.) Engelm., MacSwain); Hurst.

Sparganium multipedunculatum (Morong) Rydb. Map 58. North Shore ponds from Brackley Beach to Grand Tracadie; Lake Verde; Mermaid Lake.

[*Sparganium simplex* Huds., a European species, is to be excluded. As used by older authors, it referred to all species other than *S. eurycarpum* (Bain, 1892), to *S. chlorocarpum* especially (Macoun), and to *S. multipedunculatum* in the seventh edition of Gray's Manual.]

Sparganium minimum Fries Tignish, muddy border of old millpond (*Fernald et al.*).

ZOSTERACEAE

(NAJADACEAE, MacSwain and Bain, MacSwain, Hurst)

Zostera marina L. var. *stenophylla* Asch. & Graebn. EELGRASS Map 59. Scattered around the coast: e.g., Alberton, Bideford, Brackley Beach, Savage Harbour, Murray River, Charlottetown harbor. Most completely marine species of the P.E.I. flora: salt-water estuaries and beaches, on moderately sandy or rocky bottom. Now nearly recovered from the wasting disease of 1933 and later years. Locally used as fertilizer; gathered in heaps on the beach for composting. ("SEAWEED") Hurst.

The salt-water eelgrass *(Zostera marina)*, before its mysterious decline in the 1930's was an economic source of land fertilizer and house insulation as well as a food source for brant. It is now returning to the coastal waters and here at Brackley Beach it has been gathered up for use on the land. A dense carpet of silverweed *(Potentilla anserina)* is spreading over it in the foreground.

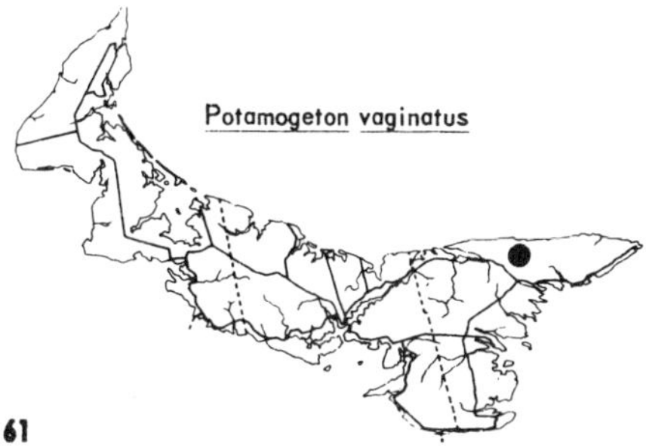

61

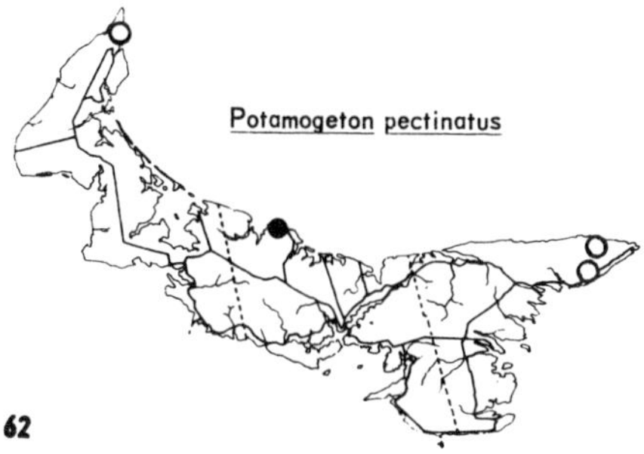

62

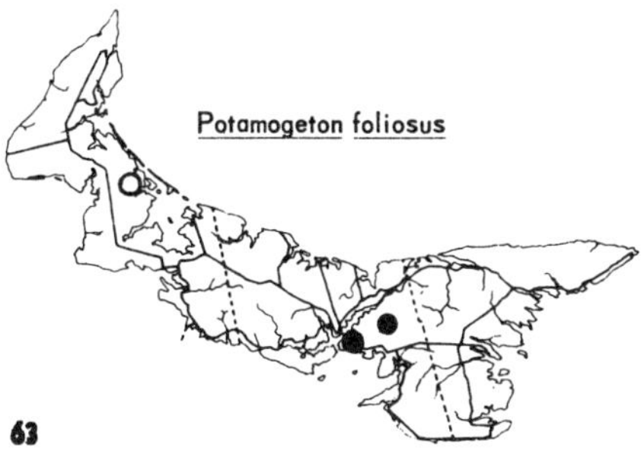

63

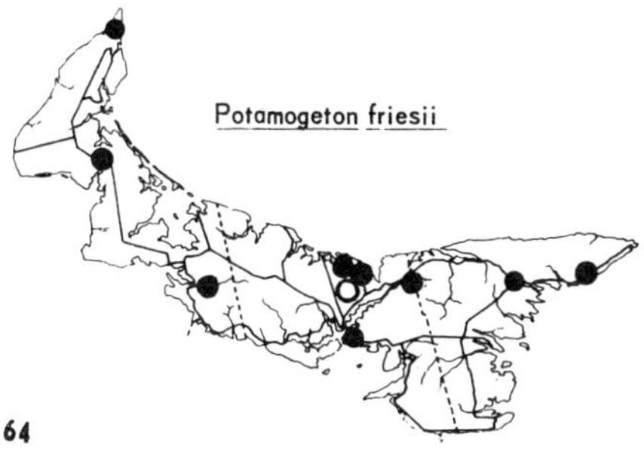

64

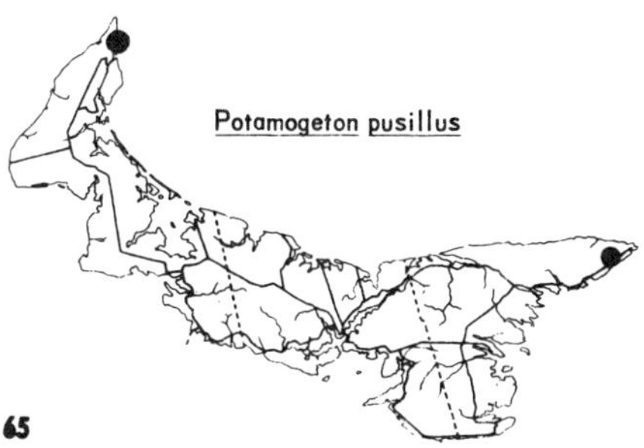

65

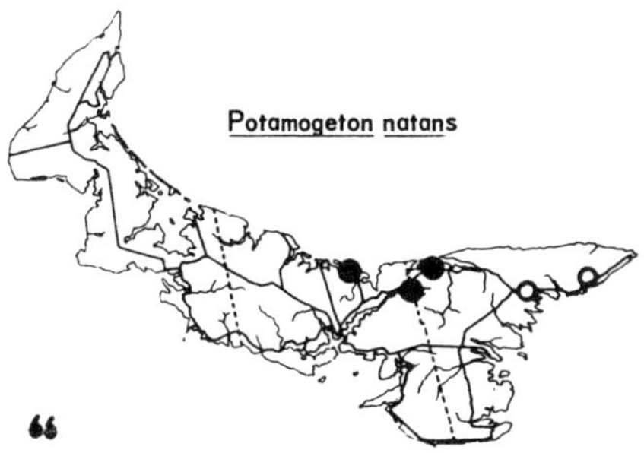

66

Potamogeton filiformis Pers., vars. Map 60. North Shore ponds.

Var. *borealis* (Raf.) St. John, reported from pond at Brackley Beach by St. John, Rhodora 18: 135. 1916. M.W. Smith's report (1946) from Clark's Pond is based on a specimen of *P. pectinatus*, however. Campbell.

Var. *macounii* Morong North Lake (*Macoun* 4373), variously reported as *P. flabellatus* in Macoun's Catalogue, *P. interior* in Gray's Manual (1908) and as *P. juncifolius* by A.R. Bennett (Jour. Bot. 46: 162. 1908). Campbell.

Potamogeton vaginatus Turcz. Map 61. "Selkirk, abundant in clear brook, above tidewater" (*Fernald* and *St. John* 10,897), referred by D.H. Webster (unpublished thesis, Acadia University) to *P. filiformis* var. *macounii*.

Potamogeton pectinatus L. Map 62. North Shore ponds: North Lake and Black Pond, Kings; Clark's Pond at Cavendish, Queens; Lower Sea Cow Pond. Brackish waters. MacSwain and Bain, MacSwain, Hurst, also St. John (Rhodora 18: 125. 1916).

**Potamogeton foliosus* Raf. var. *macellus* Fern. Map 63. Southport, Ellerslie, Watervale. Fresh-water ponds.

Potamogeton friesii Rupr. Map 64. Rather general: the common linear-leaved pond-weed of fresh-water ponds and streams, often found in mill-dams. (*P. mucronatus* Schrad., Macoun); Hurst.

Potamogeton pusillus L. Map 65. Black Pond (K); Sea Cow Pond near Tignish; brackish waters. Macoun's collection recorded as var. *capitatus* in Gray's Manual (1908). (*P. panormitanus* Biv., Fernald).

**Potamogeton epihydrus* Raf. var. *nuttallii* (C. & S.) Fern. Muddy Creek (St. Nicholas), stagnant pond near shore.

Potamogeton natans L. Map 66. Local: St. Peters Lake (Britain Pond), ponds at Stanhope, Mount Stewart, Canavoy, Vernon River. Fresh water. MacSwain, Hurst.

**Potamogeton alpinus* Balbis var. *tenuifolius* (Raf.) Ogden Map 67. Pleasant View (P), Watervale (Q). Muddy bottom of creeks.

**Potamogeton gramineus* L. Map 68. Cherry Valley. Stagnant water in marsh..

Potamogeton perfoliatus L. var. *bupleuroides* (Fern.) Farwell Map 69. General: less common by far than *P. praelongus*, and more tolerant of brackish waters. (*P. bupleuroides* Fern., Hurst 1940); Campbell.

Potamogeton praelongus Wulf. Map 70. The common broad-leaved pondweed, forming beds in streams and milldams. MacSwain, Hurst.

Ruppia maritima L. vars. Map 71. Around the coasts, not common. The varieties show no ecological segregation, var. *rostrata* Agardh, the most common, growing with var. *subcapitata* Fern. & Wieg. at Bunbury and transitional to it at Tignish (Rhodora 16: 127. 1914), and with var. *obliqua* (Schur) Asch. & Graebn. at Brackley Beach. Adams, Hurst 1940.

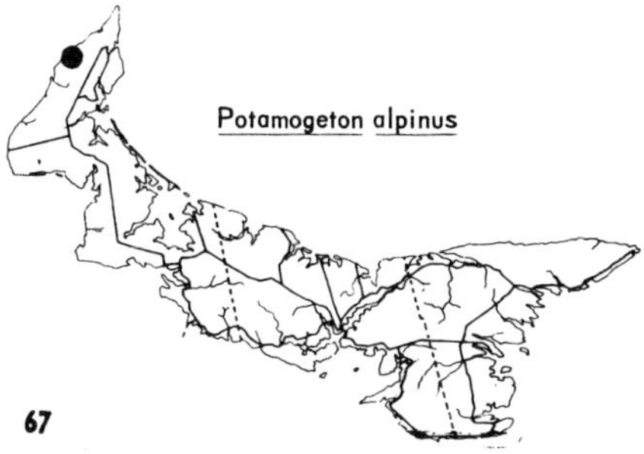

67

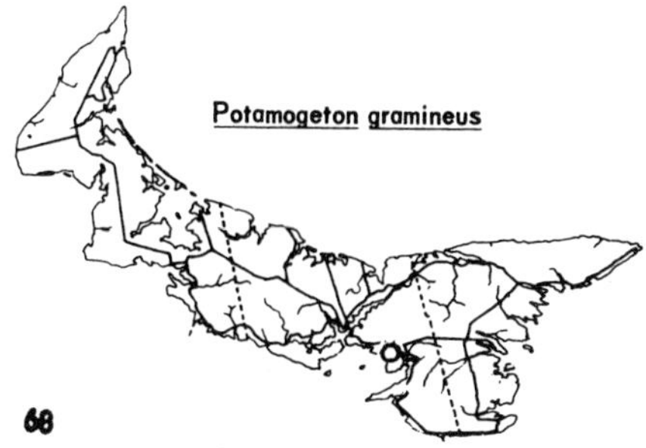

68

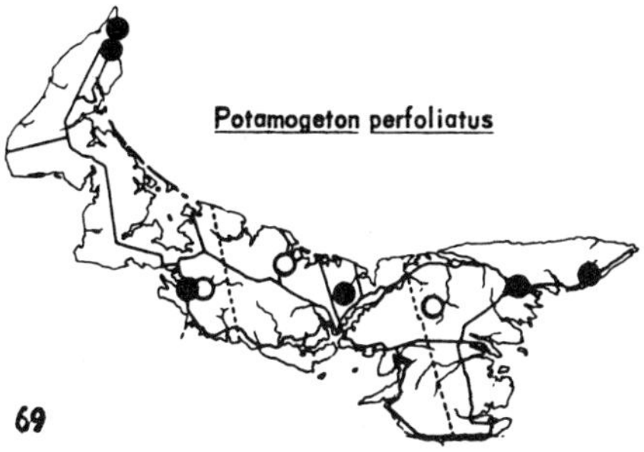

69

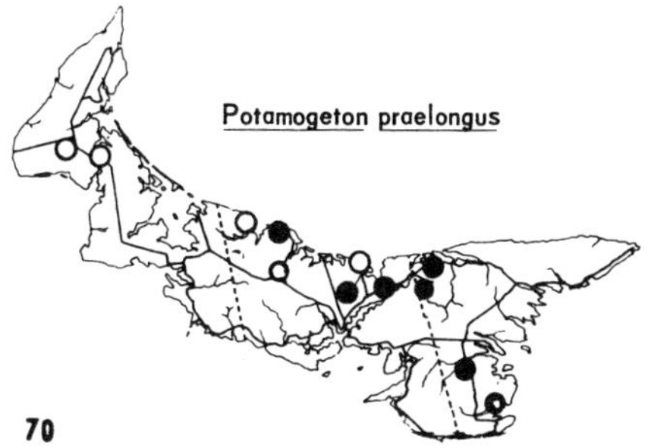

70

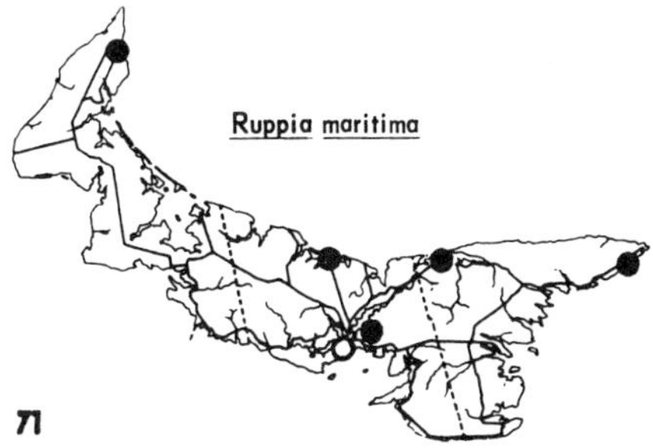

71

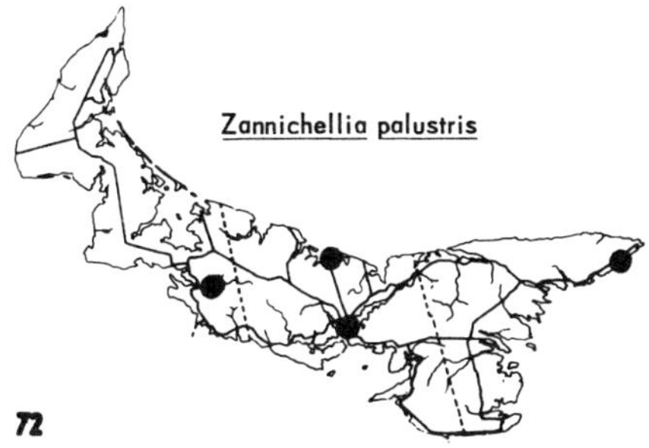

72

Zannichellia palustris L. var. ***major*** (Boenn.) Koch Map 72. Dunk River, in fresh water near sea level: Charlottetown, in harbor at North River estuary; Bothwell, a salty lagoon; Brackley Point, brackish waters. Not common. Brackish or salty waters, ascending larger streams to fresh water.

JUNCAGINACEAE

(included in Alismataceae by MacSwain and Bain, in Najadaceae by MacSwain)

Triglochin elata Nutt., Löve in Nat. Can. 85:162. 1958. Map 73. Common along coasts, on salt marshes or sandy shores once mixed with organic matter. Rarely in fresh marsh as at Mount Stewart. Two forms: 1. tall stout plants with erect, usually dense spikes of flowers (***T. elata sensu stricto***), and 2. small slender plants with upcurving loose spikes of few flowers (***T. concinna sensu*** Fernald 1950)., ***T. maritima***, MacSwain and Bain, MacSwain, Hurst 1940 omitted by printer from Hurst 1933).

Triglochin palustris L. Map 74. Grand Tracadie and Mount Stewart, Cape Wolfe. Less typical of salt marshes than is *T. maritima*; brackish and fresh marshes along coasts.

ALISMATACEAE

Sagittaria latifolia Willd. Map 75. Scattered: marshy borders of ponds, thus often coastal. Plants with narrow leaf lobes are called forma ***gracilis*** (Pursh) Rob. (*S.* ***variabilis*** Engelm., MacSwain and Bain, MacSwain); Hurst.

Sagittaria graminea Michx. Map 76. Keefe's Lake (*Groh & Malte*).

Alisma triviale Pursh Map 77. Widely scattered but rather uncommon, fresh pond-margins or marshes, often behind barrier beaches. (***A. Plantago***, MacSwain; ***A. Plantago-aquatica***, Hurst); Campbell.

BUTOMACEAE

Butomus umbellatus L. Map 78. Introduced European aquatic, naturalized on the St. Lawrence, whence introduced in 1935 (Groh, Can. Weed Surv. 2: 8. 1944) to the pond at Charlottetown Experimental Farm (extinct there by 1945), and from there to Hurst's (Smelt Creek) at Brackley Point in 1941, where he noted it persisting in 1945, though it is now extinct there. Also introduced to Reddin's Pond in Southport and still flourishing. Hurst 1940.

HYDROCHARITACEAE

[***Vallisneria americana*** Michx. Aquatic of rivers and lakes in northern N.S. has not been found. (***V. spiralis***, MacSwain and Bain).]

GRAMINEAE

[***Bromus racemosus*** L. Map 79. Montague railway station, 1945 (*Dore & Gorham*). European adventive.]

Bromus inermis Leyss. Map 80. Introduced as a fodder grass, becoming widespread on roadsides. Charlottetown (1921), well established by the Experimental Farm; since 1940, elsewhere. Campbell.

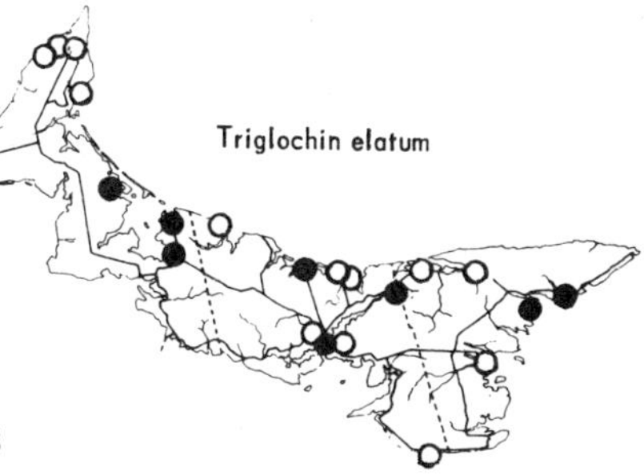

73

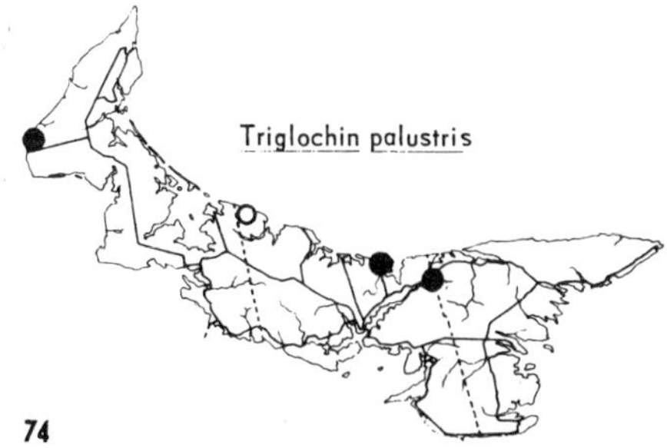

74

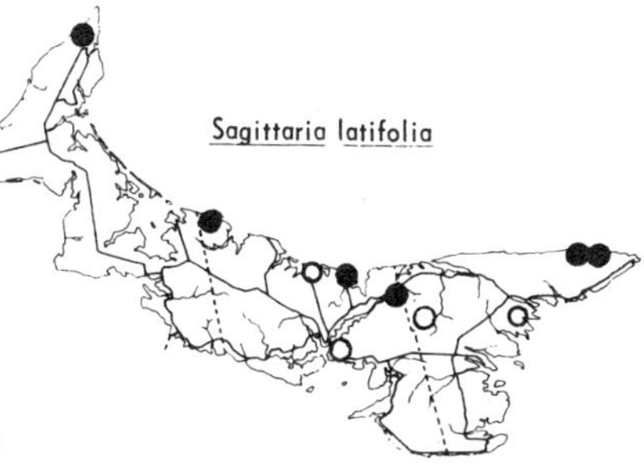

75

76

77

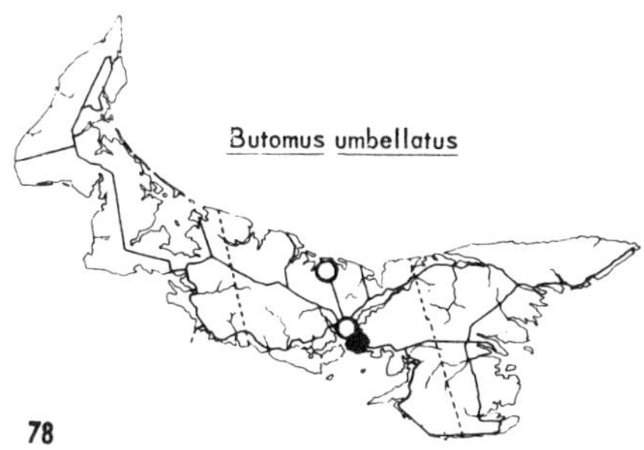

78

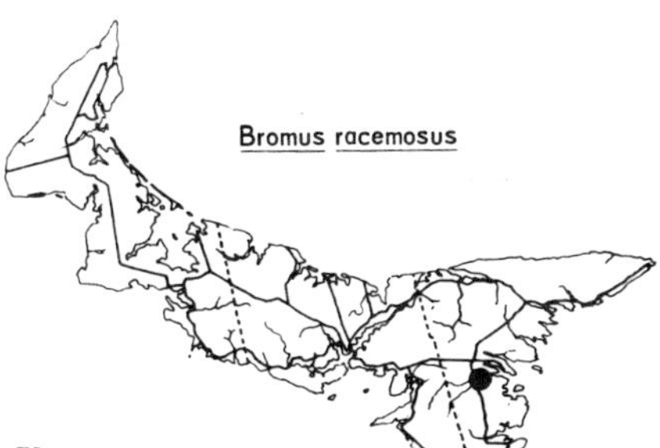

79

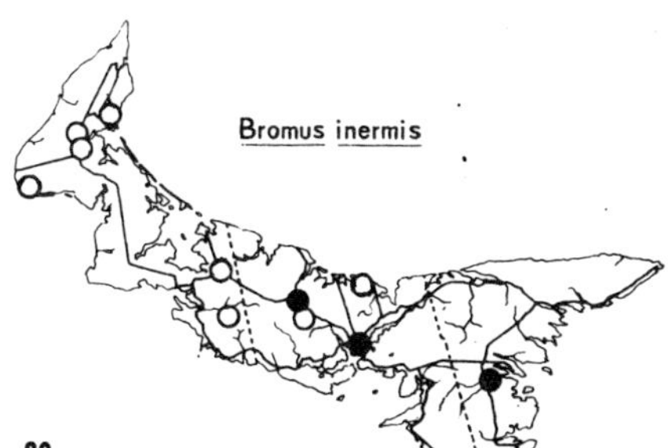

80

Bromus ciliatus L. Map 81. Locally throughout. Muddy banks usually along streams, often in partial shade. The original description of *B. dudleyi* Fern. (Rhodora 32: 64. 1930), a weak segregate of this species, included specimens from Indian River. Campbell.

Schizachne purpurascens (Torr.) Swallen Map 82. Indian River, St. Charles; pockets of spruce-needle litter in half-shade of woods on bank. The St. Charles (K) specimens belong to the forma ***albicans*** Fern. lacking anthocyanin throughout.

Festuca capillata Lam. Map 83. Locally, throughout. Established since before 1912 at Charlottetown. Open ground: old fields, roadsides, saltmarsh margin in disturbed soil. Forming large pale tufts.

* ***Festuca ovina*** L. Map 84. North Rustico, sea face of dunes at harbor, associated with the forma ***hispidula*** (Hack.) Holmb. Older records (MacSwain and Bain; Churchill; Macoun's identification of a Watson specimen from Governor's Island; Groh) may have referred to *F. rubra* (see Dore and Roland, 1942).

Festuca rubra L. Map 85. Common throughout. Sandy soil, forming extensive patches in poor pasture and on headlands or above beaches. 2 Lots (Groh).

Var. ***juncea*** (Hackel) Richter Brackley Point; Borden; Wood Islands - sand dunes; High Bank.

Festuca elatior L. Map 86. The few records mostly from towns. Introduced from Europe as a fodder grass. Roadsides; Bain's "tall fescue" of intervales probably not this species. (MacSwain and Bain, including *F. pratensis* Huds., MacSwain and Bain).

Glyceria borealis (Nash) Batchelder Map 87. Local: Tignish area, Sea View, Upton, St. Peters Lake, Selkirk. Fresh waters: streams or pond margins.

****Glyceria canadensis*** (Michx.) Trin. Map 88. Rare: Portage; Bristol. Muck at borders of lakes & swamps.

Glyceria laxa Scribn. Map 89. Bloomfield, springy railroad ditch (*Fernald, Long & St. John* 6905). This species appears as if a hybrid of *C. canadensis* with *G. striata*. (*G. striata* var. *laxa* (Scribn.) Hitchc., Dove and Roland).

Glyceria striata (Lam.) Hitchc. Map 90. Throughout. Damp places: in fields, ditches, woods, usually along runs. (*Poa nervata* Willd., Bain; *nervata* (Willd.) Trin., Hurst), Campbell.

Glyceria grandis S. Wats. Map 91. Throughout. Wet ground: streams, ditches, swales. (*Poa aquatica*, Bain); Adams, Hurst 1940.

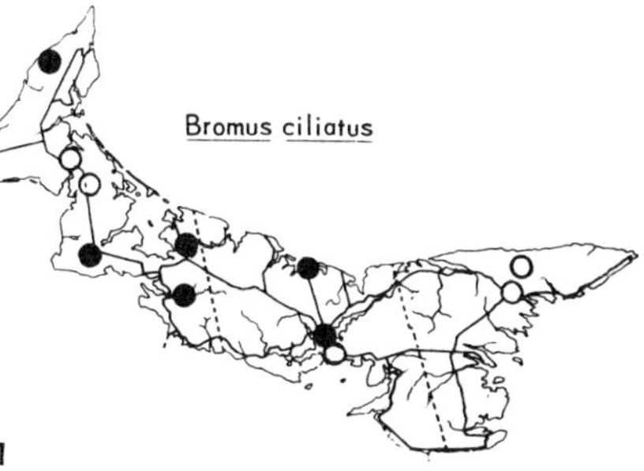

81

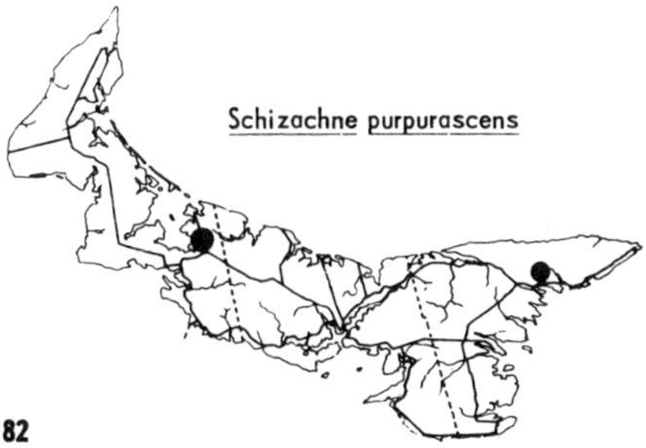

82

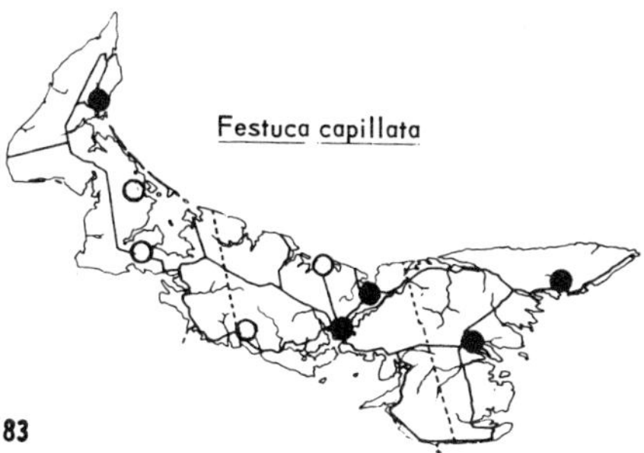

83

84

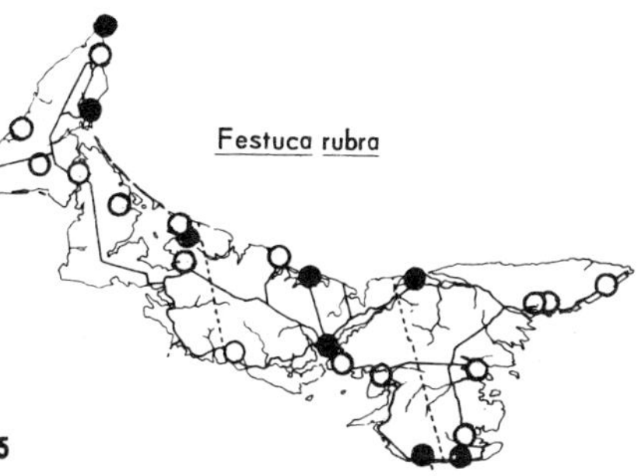

85

86

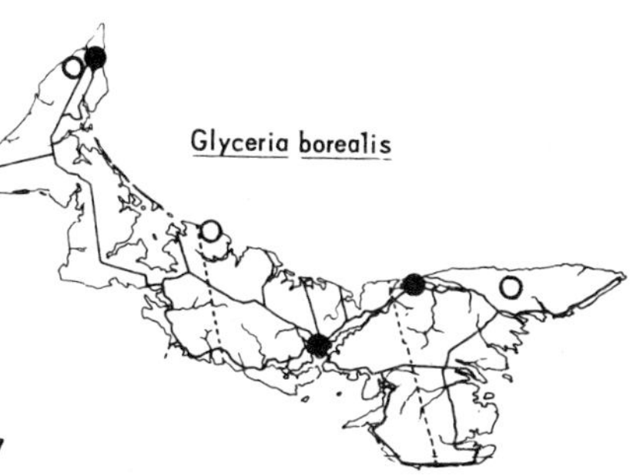

87

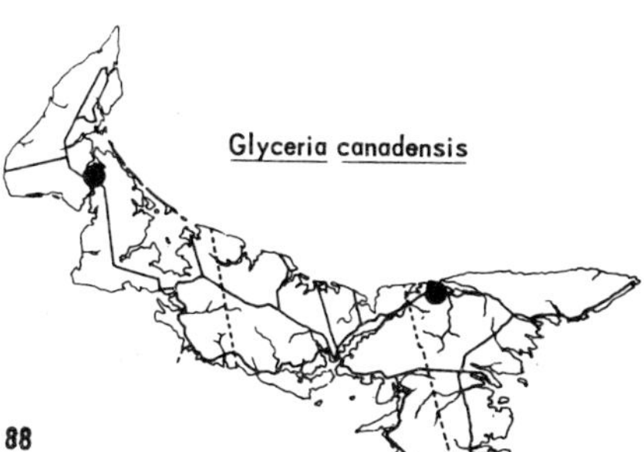

88

Torreyochloa fernaldii (Hitchc.) Church, Am. Jour. Bot. 36:164. 1949. Map 92. Local. Marshy pond margins, wet ditches and slow creeks. (***Glyceria fernaldii*** (Hitchc.) St. John, Dore and Roland).

Puccinellia americana Sørensen, Medd. om Grönl. 136(3): 67. 1953. Map 93. Salt marshes at Bunbury, Southport. Near its northeastern limit, the Bay of Chaleur. (*P. maritima*, Fernald 1950, Dore and Roland).

[****Puccinellia nuttalliana*** (Schult.) Hitchc. Map 94. Charlottetown, one clump in railway yards, 1953. Adventive, western North American.]

[****Puccinellia fasciculata*** (Torr.) Bickn. Map 95. Charlottetown, 1953, a single clump by Grafton Street near Canada Packers'. Adventive, though considered native in Nova Scotia.]

Puccinellia pumila (Vasey) Hitchc. Map 96. Coastal; common, both on salt marshes and sands. Fernald and Weatherby (Rhodora 18: 20 1916), treating this species as *P. paupercula* var. *alaskana*, distinguished var. *longiglumis* from Bunbury by the longer floral bracts. Collected subsequently only at Gaspé and not relocated at Bunbury. This elongation may have been, as is so usual in *Agrostis*, due to nematode infection.

Puccinellia ambigua Sørensen, Medd. om Grönl. 136(3): 64. 1953. Alberton, salt marsh (*Fernald*), type locality.

Poa annua L. Map 97. Throughout, common. European, naturalized. Bare, usually moist ground; roadsides and trampled pastures. Weed, 21 Lots (Groh). MacSwain and Bain, Hurst.

Poa compressa L. Map 98. Throughout. European, naturalized Abundant in dry sites: roadsides, banks (even in woodland) dunes and headlands. No less common on clay soil. Large areas thinly colonized by means of stolons. 27 Lots (Groh). Bain, MacSwain and Bain, Hurst.

Poa pratensis L. Map 99. Locally "JUNE GRASS." Common throughout. European. Fields, pastures, meadows and roadsides; introduced as a forage grass (Kentucky bluegrass) and now abundant. 11 Lots (Groh). Bain, MacSwain and Bain, Hurst. Nematode infestation, noted at several stations, induces elongation of spikelets, a form reported (Rhodora 23: 133. 1921) as *P. costata*.

Var. ***alpigena*** Fries Coastal, more rarely inland (Bloomfield, Marshfield (Q), Harmony (K), the stoloniferous group, including *P. subcaerulea* Sm. and *P. alpigena* (Fries) Lindm. f. of Gray's Manual (1950). Transitions to *P. pratensis* are suggested by Fernald's collections of both types at Rocky Point, and by the ecological contrast at North Rustico where plants like var. *alpigena* grew in the tall *Juncus balticus* of the sand flat while shorter less tufted types grew in the open turf.

Poa palustris L. Map 100. Throughout; introduced European. Damp open grounds; creek margins, dune slacks, ditches. 1 Lot (Groh). (*P. serotina* Ehrh., MacSwain and Bain; *P. triflora* Gilib., Groh, Hurst); Campbell.

Poa trivialis L. Map 101. Queens and Kings Counties; rare in Prince. European, introduced. Muddy margins of slow creeks, or ditch. MacSwain and Bain.

89

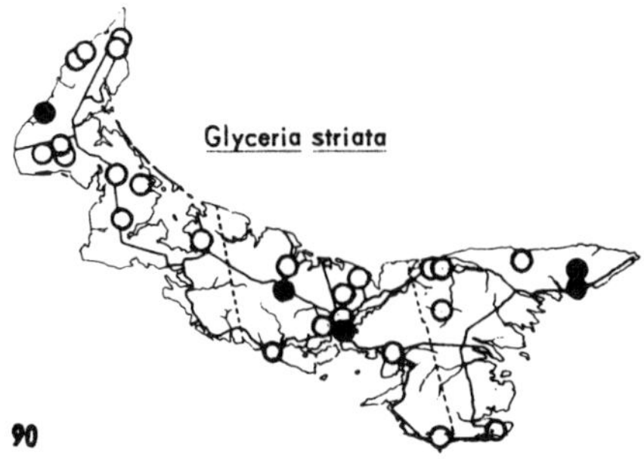

90

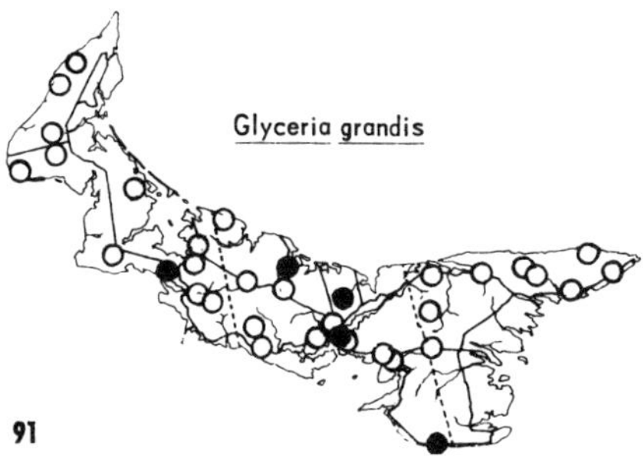

91

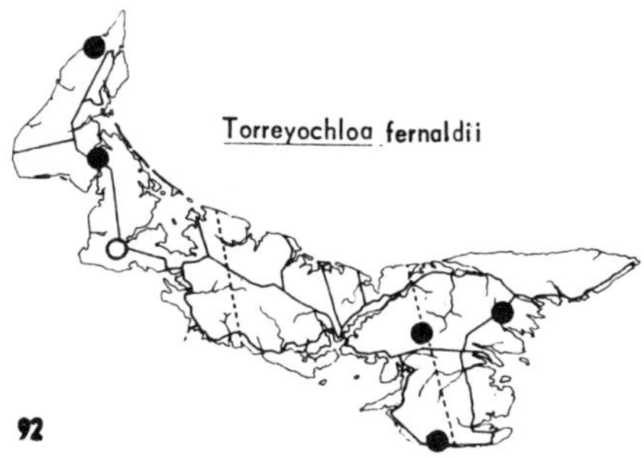

92

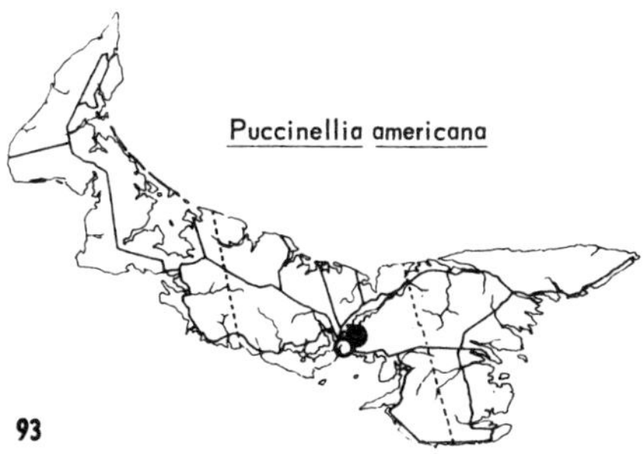

93

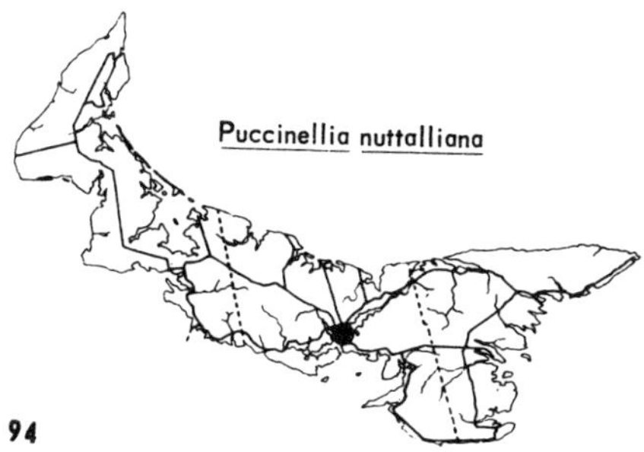

94

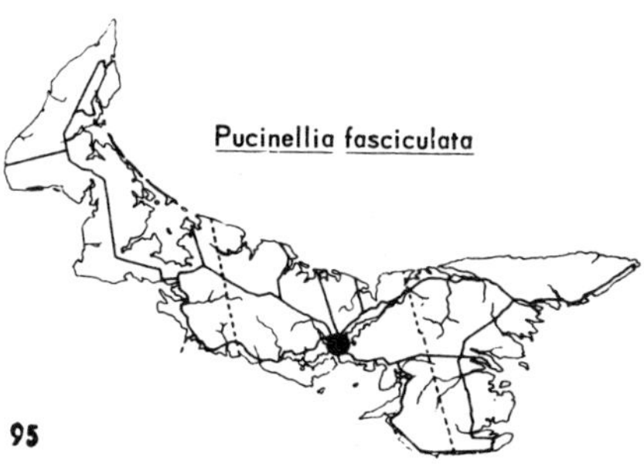

95

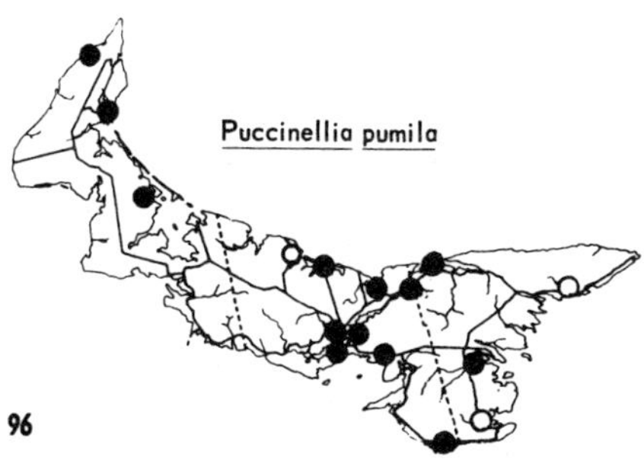

96

Poa saltuensis Fern & Wieg. Map 102. From west to east. Indigenous. Open, dry mixed woods; thus, common around Tignish and Alberton, and northeast of Souris. The size range of spikelets is that of var. *microlepis* Fern. & Wieg. in collections from the east and west; Fernald's specimens from St. Dunstan's are cited in the original description of *P. saltuensis* (typical).

Dactylis glomerata L. Map 103. Established from Summerside east; not abundant. European, introduced. Roadsides and waste lots; tolerant of shade (called "orchard grass"), seldom planted, but very persistent. 7 Lots (Groh). Bain, MacSwain and Bain, Groh, Hurst.

****Eragrostis poaeoides*** Beauv. Map 104. Charlottetown railway yards (1953). European; adventive from Canadian mainland.

Catabrosa aquatica (L.) Beauv. var. ***laurentiana*** Fern. Map 105. Vicinities of Launching, Charlottetown and Tignish. In perennial springs and streams.

Distichlis spicata (L.) Greene Map 106. Charlottetown harbor and its estuaries. Near head of salt marshes. A species of warm shores here at its northeastern limit.

Phragmites communis Trin. var. ***berlandieri*** (Fourn.) Fern. Map 107. Dunk River estuary, borders of salt marsh. Only known station for seventy years. (*P. sylvestris*, MacSwain and Bain), MacSwain and Bain.

Lolium perenne L. Map 108. Charlottetown, frequently since 1941. European; introduced in lawn seed, never long persistent. Hurst 1941.

Agropyron trachycaulum (Link) Malte, vars. Map 109. Very local; open woods or swamps, never weedy.

Var. ***majus*** (Vasey) Fern. Indian River, dry clearing.

Var. ***novae-angliae*** (Scribn.) Fern. Bloomfield (P) and Dundee, larch swamps.

Agropyron repens (L.) Beauv. Acadian "CHIENDENT" Map 110. Throughout, common. Native to the head of beaches and salt marshes (a tension zone); an ineradicable weed of cultivation. 42 Lots (Groh). Fernald separates the salt marsh form as var. ***subulatum*** (Schreb.) Reichenb., which has been collected at Brackley Point by Macoun. The var. ***repens*** is said to be more common on sandy shores and more aggressive as a weed. Each variety is divided into a series of forms depending on the combination of presence or absence of awns and pubescence on the rachis. (*Triticum repens* L., MacSwain and Bain); Groh, Hurst.

[***Triticum aestivum*** L. WHEAT Occasionally seed fallen from trucks or trains may germinate.]

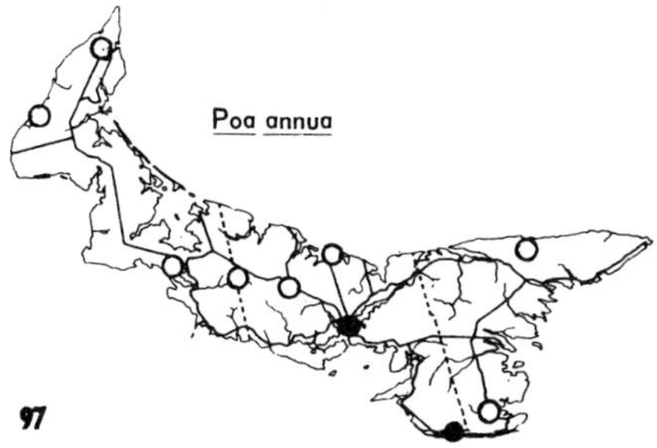

97

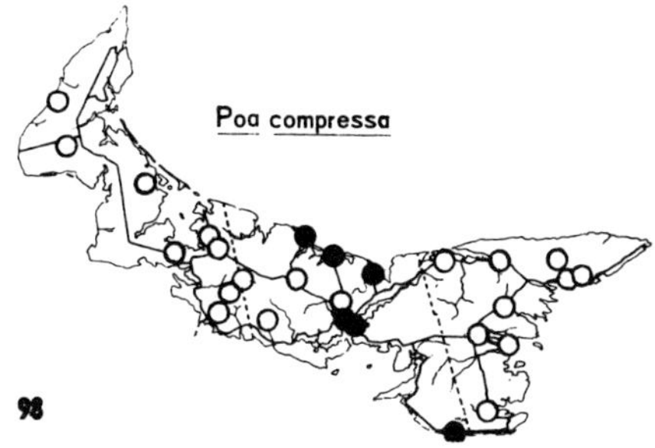

98

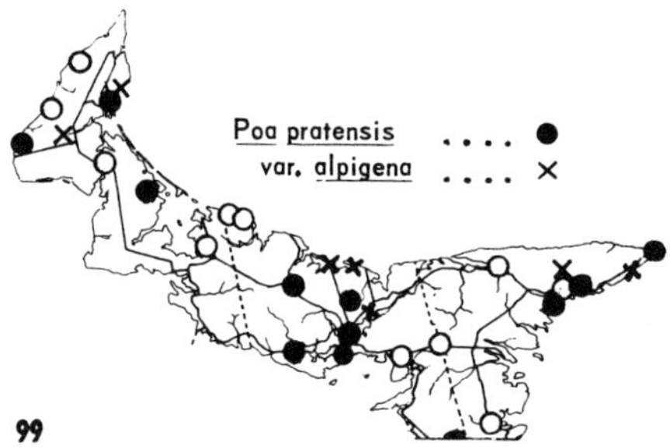

99

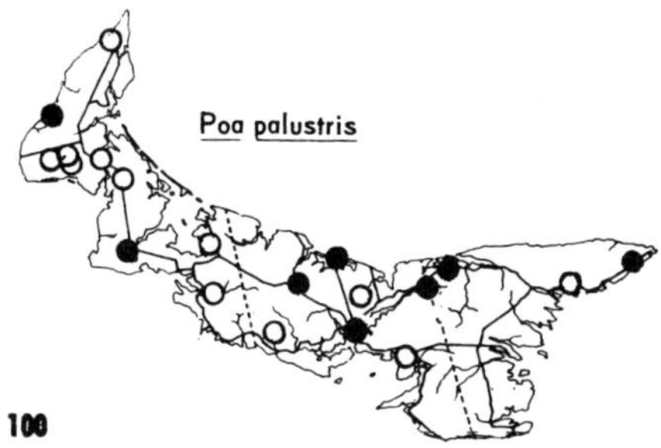

100

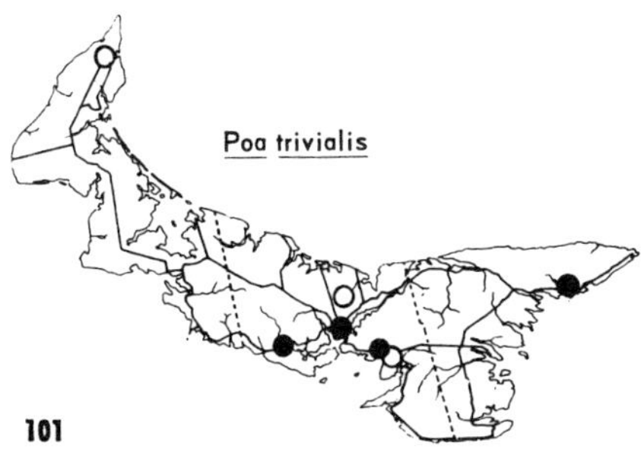

101

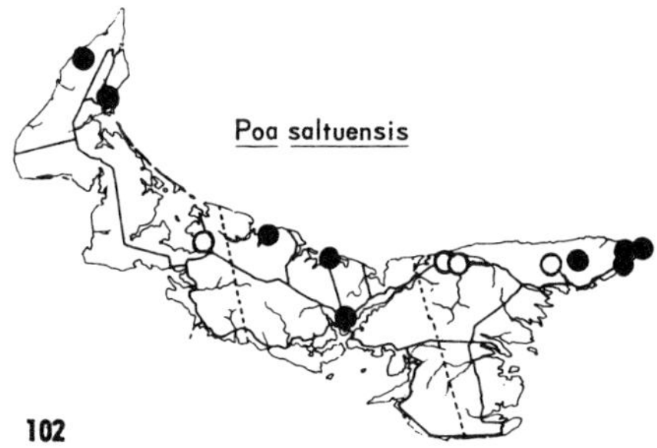

102

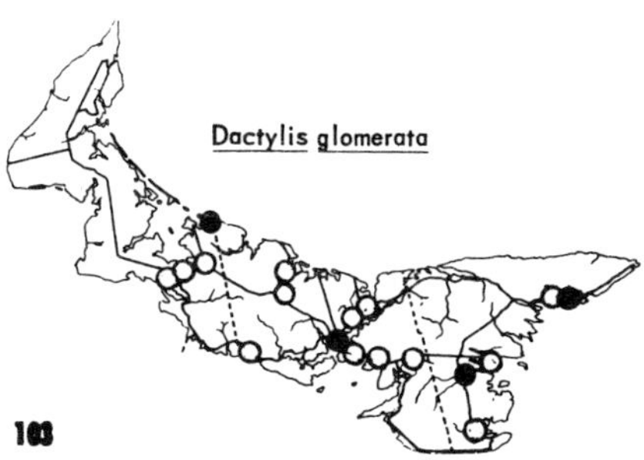

103

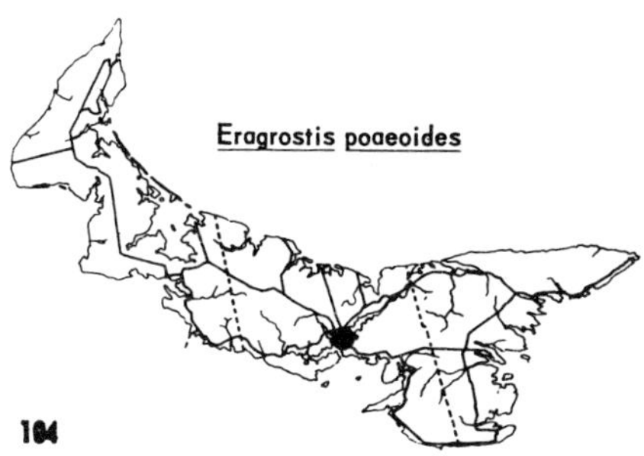

104

Elymus mollis Trin. Map 111. Locally about the coasts. Strands, not salt marshes or dunes. (*E. arenarius* var. *villosus* E. Mey.).

****Elymus virginicus*** L. Map 112. Selkirk, flood-plain woods of the Naufrage River.

Hordeum jubatum L. Acadian "FUBETTE" Map 113. Fairly common on the Northumberland Strait coast. Tension zone of salt marsh; spreading inland along railways; common on the sidings in Prince County clay. First collection, 1912. 8 Lots (Groh). MacSwain and Bain, Groh, Hurst.

[***Hordeum vulgare*** L. (*H. hexastichon*) BARLEY Widely planted but seldom spontaneous.]

[***Secale cereale*** L. RYE Cultivated occasionally. Quite often spontaneous in waste places. 1 Lot (Groh).]

****Sphenopholis intermedia*** Rydb. Map 114. Dunk River, sandstone cliff at high-flood level.

Avena sativa L. OATS Map 115. The commonly planted grain; as with the climate and Scottish settlement, one would expect. Spontaneous along roadsides and railway.

****Avena fatua*** L. var. ***vilis*** (Wallr.) Haussk. (det. D.R. Lindsay) Charlottetown, (1953), Kensington (1953); railway sidings. European; naturalized in western Canada; possibly introduced in western feed-grain (never found along roadsides, but usually associated with *A. sativa*).

Deschampsia flexuosa (L.) Trin. Map 117. Near the North Shore from Indian River to East Point; Victoria Park in Charlottetown. Dry ground: sandy spruce woods, exposed headlands and banks. Adams, Hurst 1940.

****Deschampsia cespitosa*** (L.) Beauv. var. ***glauca*** (Hartm.) Lindm. f. Map 118. Indian River, muddy banks of river and marshy ground at head of estuary.

Danthonia spicata (L.) Beauv. Map 119. Common throughout. Clearings and open woods; old fields, (apparently light-loving). Abundant during establishment of *Picea glauca*. Very variable. On burns much larger plants develop, corresponding to *D. alleni* Austin. *D. alleni* has been recorded from Alberton (dry clearings) and Bristol (recent burn). Weed, 48 Lots (Groh). MacSwain and Bain; Groh; Hurst.

Danthonia compressa Austin Map 120. Elmira, East Point (*Malte*). The woodland counterpart of *D. spicata*, which in woodland may similarly have spreading spikelets.

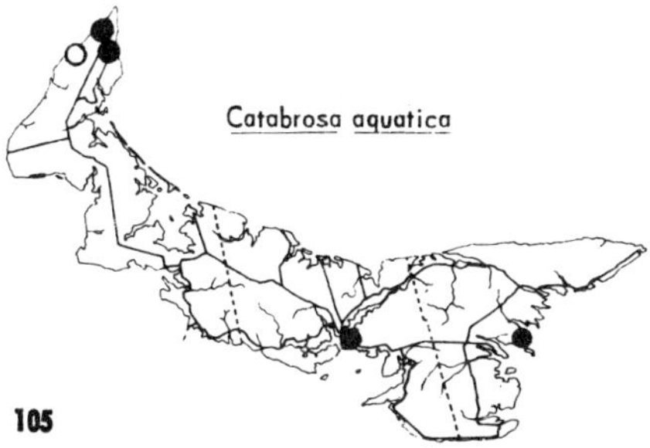

Catabrosa aquatica

105

Distichlis spicata

106

Phragmites communis

107

Lolium perenne

108

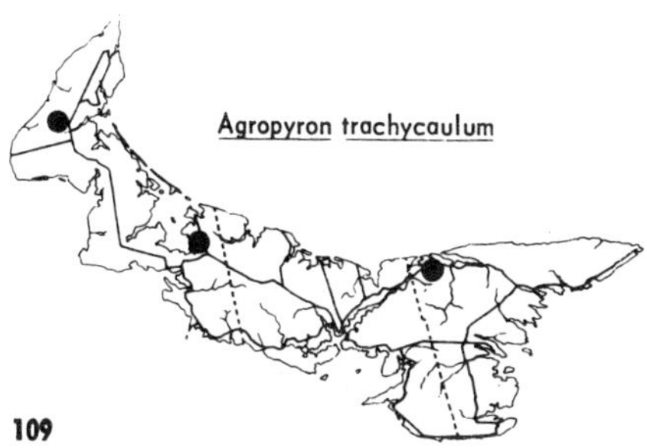

Agropyron trachycaulum

109

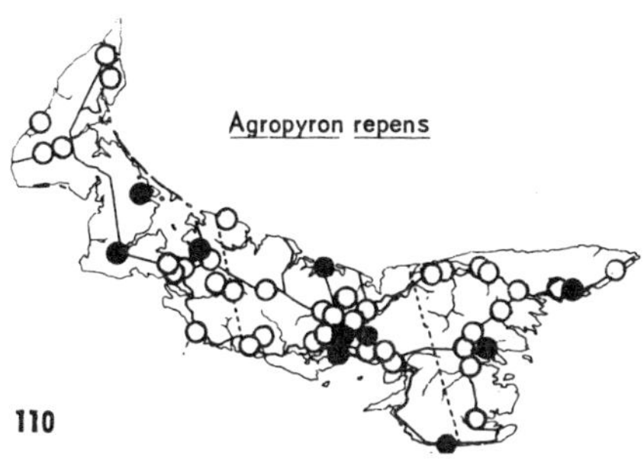

Agropyron repens

110

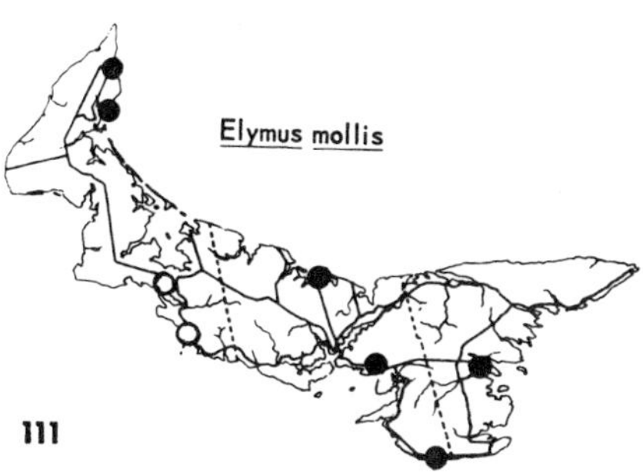

Elymus mollis

111

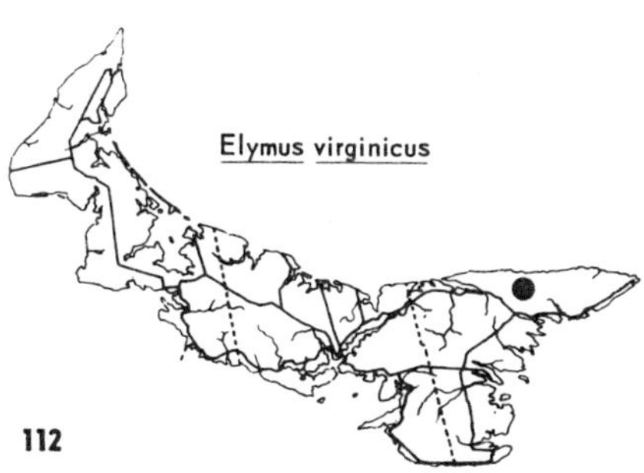

Elymus virginicus

112

Calamagrostis canadensis (Michx.) Nutt. Map 121. Throughout; particularly common in Prince County. Damp or wet places in open bushland and swamp, or along streams and ditches, forming extensive pure stands. Bain, MacSwain and Bain, Groh, Hurst.

Var. ***macouniana*** (Vasey) Stebbins, marked off by the shorter, denser inflorescence, is cited from Mount Stewart, by Stebbins (Rhodora 32: 41. 1930).

*Var. ***robusta*** Vasey Specimens from North Point attain the spikelet measurements of this variety. The species seems to include a cline from var. *macouniana* by var. ***typica*** to var. ***robusta*** and on to the more northern var. ***scabra***, the spikelet and its parts becoming longer, the glumes becoming longer relative to the lemma, more long-scabrous on the keel, more scabrous-hirtellous generally from the nearly glabrous var. ***macouniana***.

Calamagrostis neglecta (Ehrh.) Gaertn., Mey. & Scherb. Map 122. Local: Mount Stewart, Brackley Beach, Summerside. Fresh to brackish marshes along the coast.

Ammophila breviligulata Fern. Map 123. Around the coast. Drifting sand, very abundant. Usually the first sand-binder, stabilizing the dunes and persisting until the establishment of a forest of white spruce. (***Psamma arenaria***, Bain, MacSwain and Bain; *A. **arenaria***, Churchill, Hurst).

Dune grass *(Ammophila breviligulata)* is the characteristic, and often sole, species of plant on the coastal dunes of the North Shore.

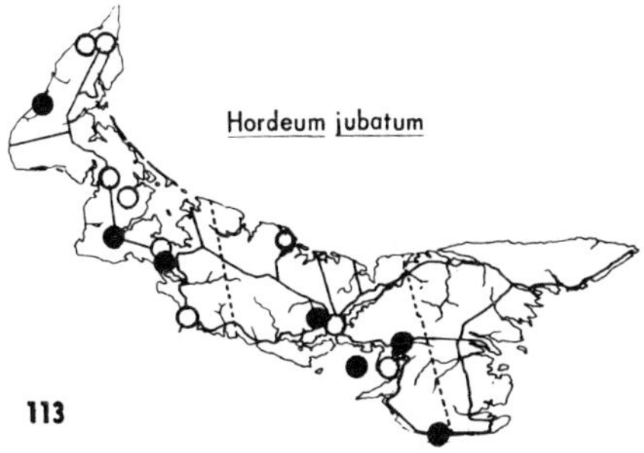

113

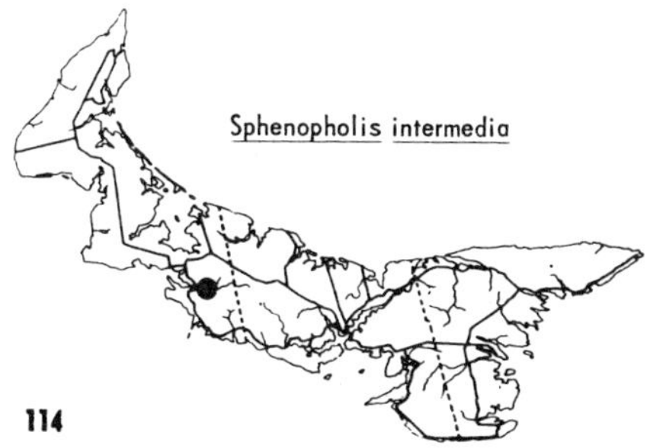

114

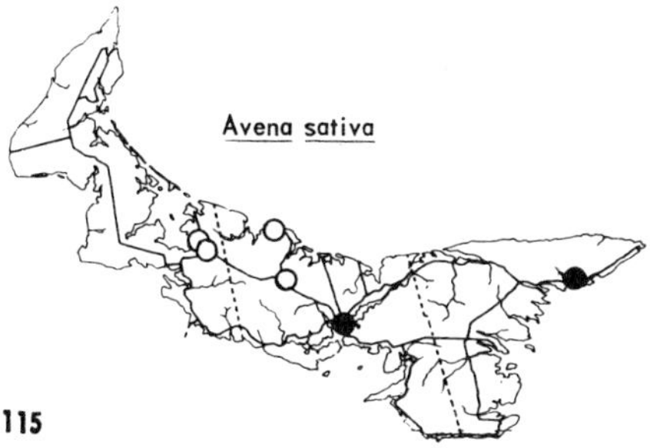

115

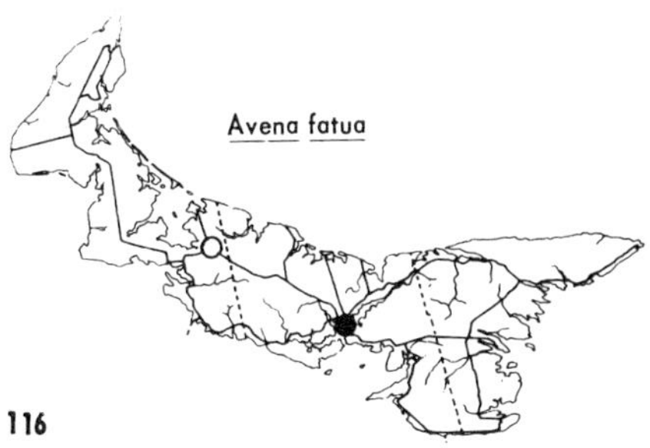

116

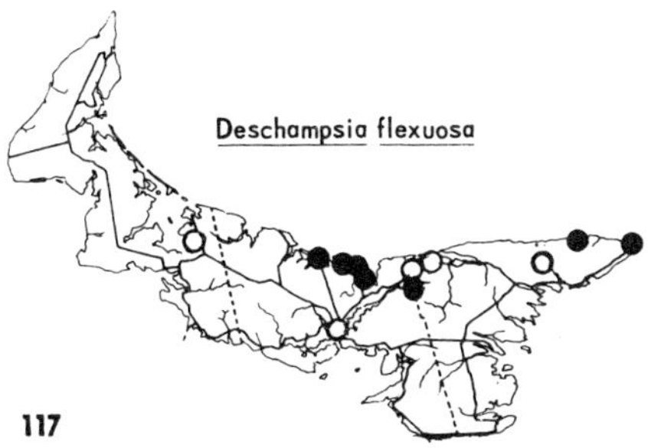

117

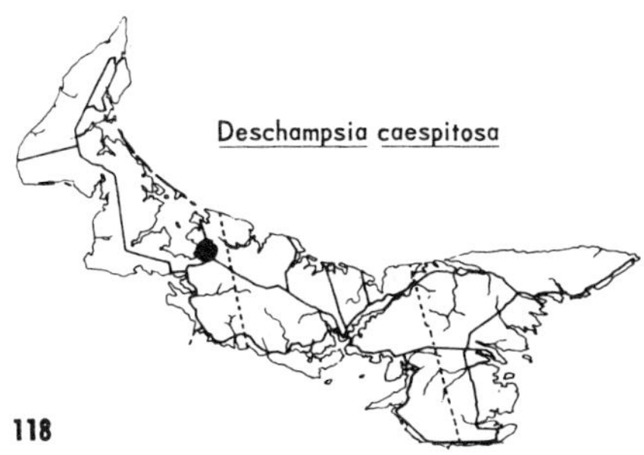

118

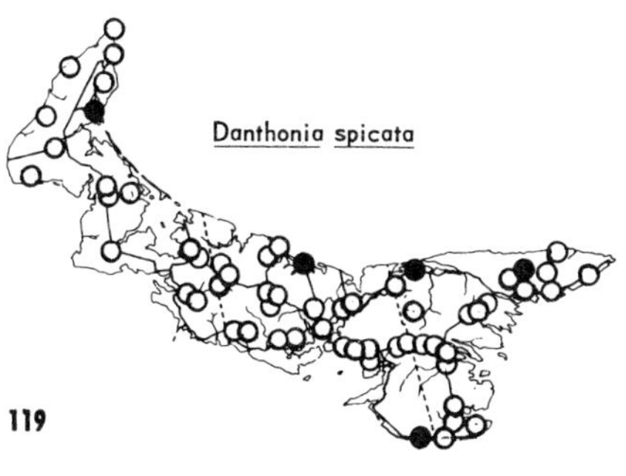

119

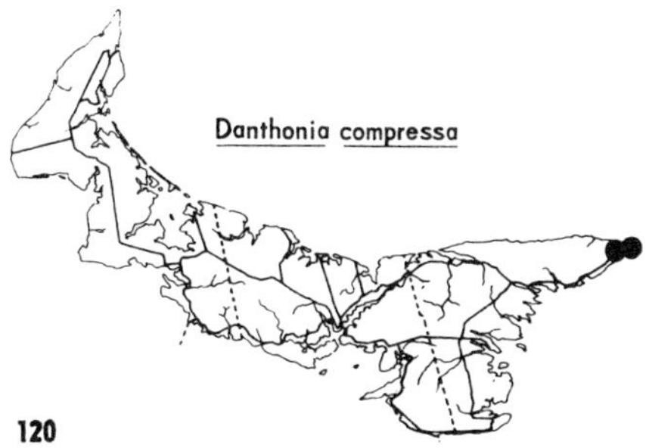

120

Agrostis gigantea Roth var. ***dispar*** (Willd.) Philipson Map 124. A frequent roadside weed naturalized from Europe, often planted or weedy in hayfields in the uplands, tending to form clumps; a lax form in springs or streams. Weedy, 5 Lots (Groh). (*A. alba,* Groh; *A. stolonifera* var. *major* (Gaud.) Farwell, Dore and Roland.

Agrostis palustris Huds. Map 125. Scattered throughout, presumed indigenous. Marshy sites, salt or fresh: damp fields along streams, and ditches. A weed of waterlogged sites, more common inland on clays. Purple color lacking in specimens found at Wellington. "Introduced" weed, 25 Lots (Groh). (*A. stolonifera,* Groh, and var. *compacta* Hartm., Malte; *A. maritima* Lam., Adams; *A. alba* var. *maritima* (Lam.) G.F. W. Mey., Hurst; and var. *palustris* (Huds.) Pers., Campbell).

Agrostis tenuis Sibth. Map 126. Throughout, European, presumed introduced. Formerly grown for seed at Souris and Georgetown to be used in growing turf for golf greens, and sold as 'Prince Edward Island bent' or 'brown top' (Malte, Ann. Rep. Nat. Mus. Can. 50: 105-126. 1928). Fields and roadsides, on light soil, forming a good sod. Best developing in areas with the cool summers of a maritime climate. Nematode infestation and distortion of flowering parts common. Awned forms (f. *aristata* (Parnell) Wieg.) at Borden (*Malte*), Hunter River. 49 Lots (Groh). (*A. vulgaris* With., Bain, MacSwain and Bain); Hurst.

Agrostis scabra Willd. Map 127. Scattered throughout. Open dry (or even damp) sites, clearings in coniferous woods, becoming very abundant on new roads and burns, dikeland, sandy seashores even marshes. Weedy. 10 Lots (Groh). (*Trichodium laxiflorum,* MacSwain and Bain; *Agrostis hyemalis,* Hurst).

Agrostis perennans (Walt.) Tuckerm. Map 128. Ellerslie Brook, Lake Verde, Selkirk (Armedale). Damp, shady sites. MacSwain and Bain.

Agrostis canina L. Map 129. European, introduced at Georgetown as 'Rhode Island bent.' Small clump in a damp grassy lot at Alberton; seed for planting golf greens formerly cultivated at Georgetown (Malte). Malte determined as hybrids with *A. tenuis* collections from Borden, Georgetown and Wood Islands. 1 Lot (Groh). Hurst.

Cinna latifolia (Trev.) Griseb. Map 130 Locally throughout. Damp woods, often along streams.

[*Cinna arundinacea* L. Not known from the Maritimes. Reported by MacSwain and Bain; doubted by Hurst.]

Phleum pratense L. TIMOTHY Map 131. Throughout. Introduced, European. Commonly planted in clover hay mixtures; also grown for seed. Persistent in fields; weed of roadsides. 42 Lots (Groh). Bain, MacSwain and Bain, Hurst.

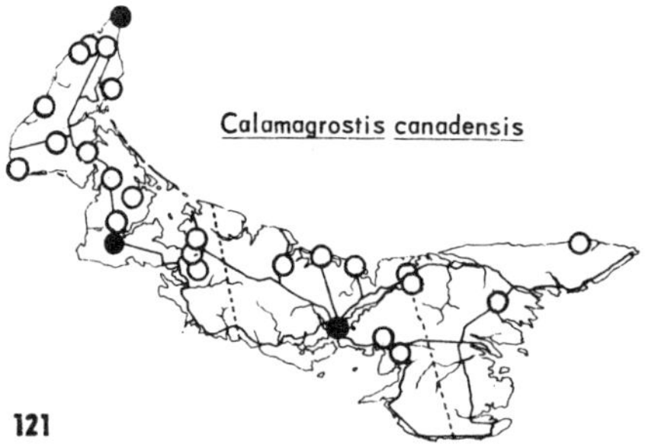

121

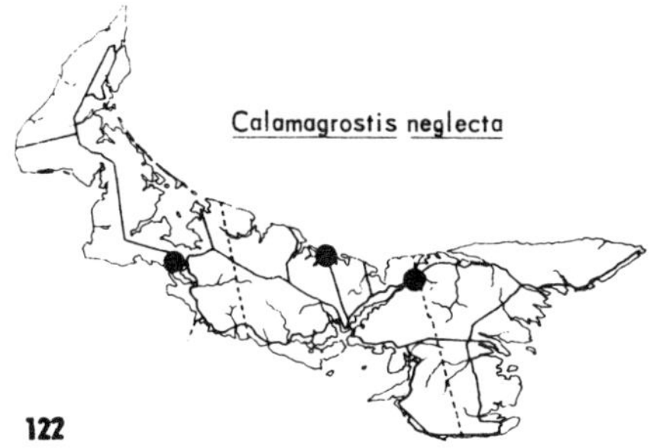

122

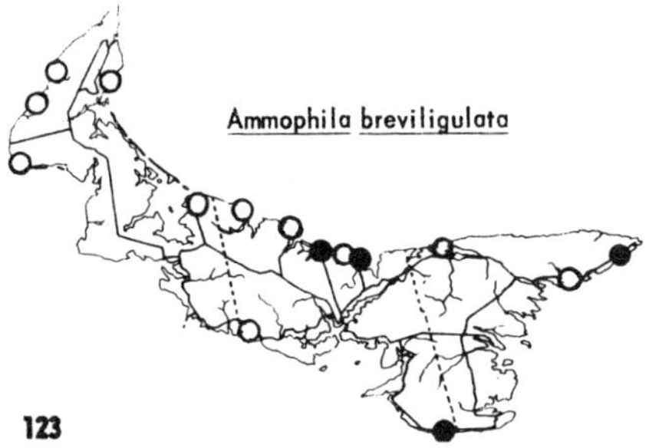

123

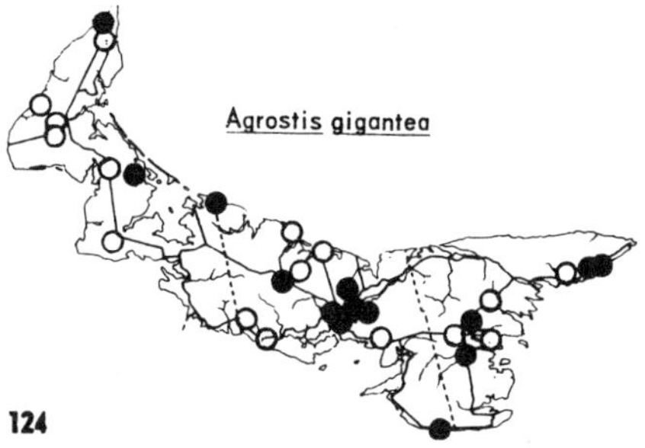

124

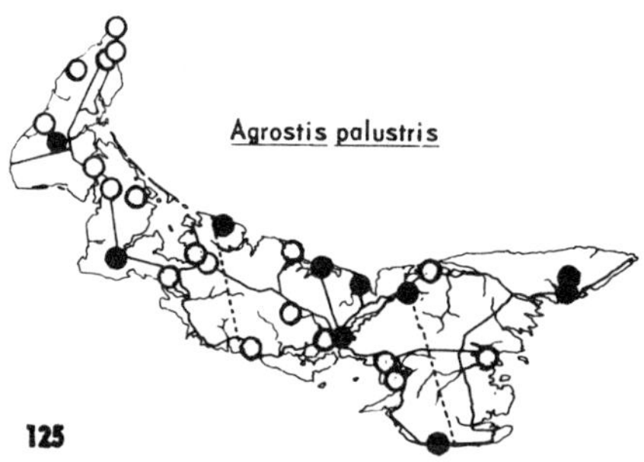

125

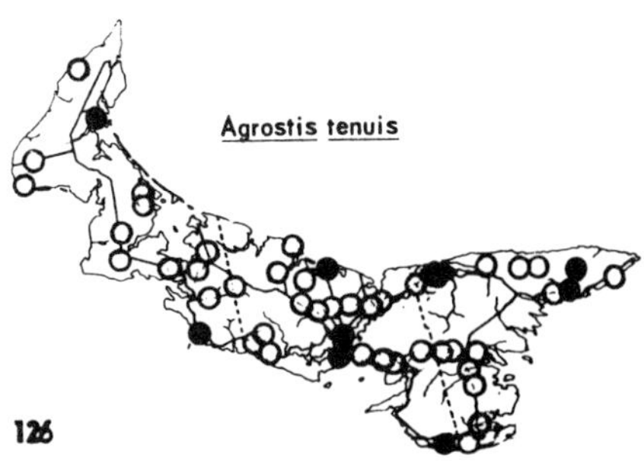

126

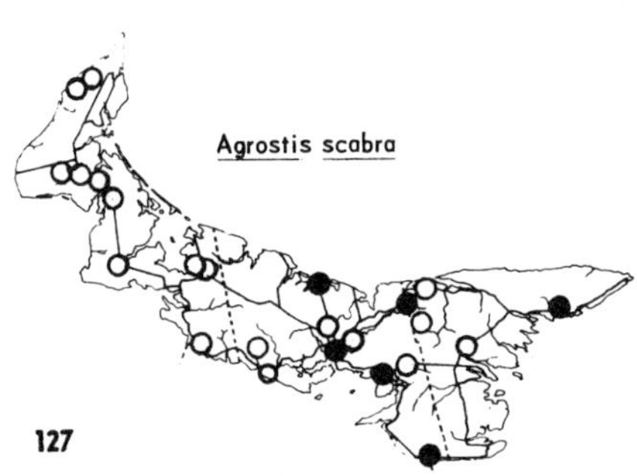

127

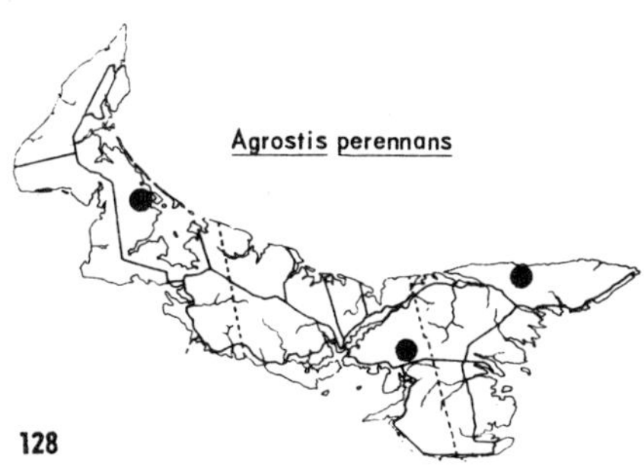

128

Alopecurus pratensis L. Map 132. Rare; Alberton, O'Leary, Charlottetown. European, introduced. Locally naturalized weed of damp roadsides and fields.

Alopecurus aequalis Sobol. Map 133. Rare: Wellington, Southport. Springs and creek banks.

Alopecurus geniculatus L. Map 134. Charlottetown, brook at Spring Park. European, established since before 1912.

Muhlenbergia glomerata (Willd.) Trin. var. ***cinnoides*** (Link) F. J. Hermann Map 135. Indian River: damp thicket (***Fernald, Long & St. John***), Hillsborough River marshes above Mount Stewart. (***M. racemosa,*** Dore and Roland, not (Michx.) BSP.)

Oryzopsis canadensis (Poir.) Torr. Map 136. Bristol: dry heathy barren north of Lot 40 station (***Fernald & St. John***).

****Brachyelytrum erectum*** (Schreb.) Beauv. var. ***septentrionale*** Babel Map 137. McNeills Mills corner on Western Road, dry leaf mold in alder thicket along stream.

Spartina pectinata Link Map 138. Fairly common. Brackish to fresh sites: head of salt marsh or sandy shores, along estuaries and fresh water creeks forming extensive stands. (*S.* ***cynosuroides,*** MacSwain and Bain, including *S.* ***polystachya,*** Bain; *S.* ***Michauxiana*** Hitchc.); Adams, Hurst 1940. M.W. Smith's report of *S.* ***Michauxiana*** from the barrier dunes of the National Park ponds apparently refers to ***Ammophila breviligulata.***

Spartina patens (Ait.) Muhl. Map 139. Almost limited to salt marsh; at Wood Islands and Charlottetown on extensive brackish sand flats. A dominant species of salt marsh, forming an uninterrupted zone above the *S.* ***alterniflora.*** "It is cut for fodder where it is abundant, as on the broad marshes of the East River, above Mount Stewart" (Bain). Where cut each year, it is replaced as a zone former by the interspersed ***Limonium*** and ***Puccinellia.*** (*S.* ***juncea,*** Bain, MacSwain and Bain); Hurst.

* X ***Spartina caespitosa*** A.A. Eaton Map 140. Bunbury; Charlottetown. Salt marsh tension zone, brackish flats. Combining the characteristics of *S.* ***pectinata*** and *S.* ***patens,*** which grow with it at each station.

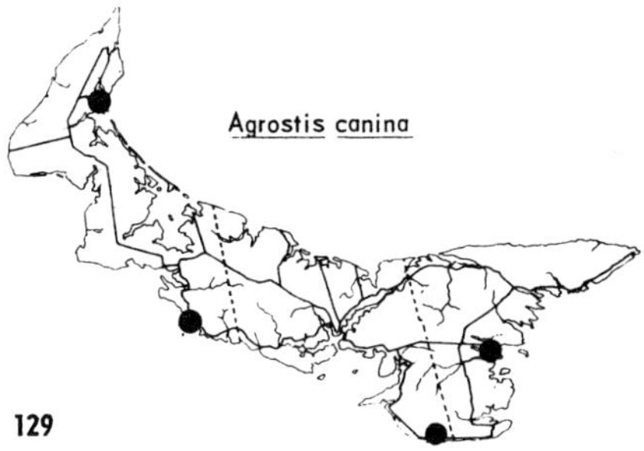

129

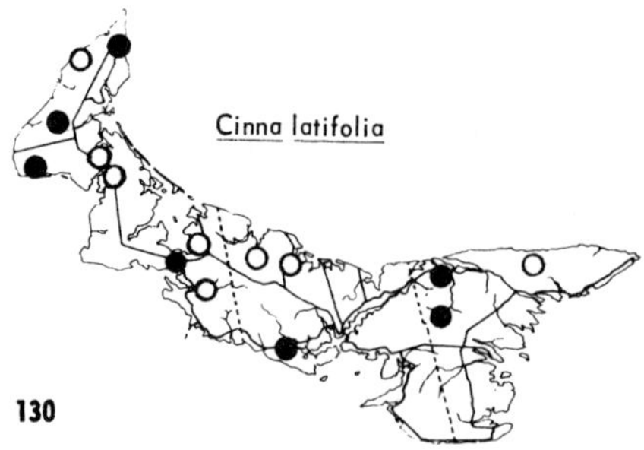

130

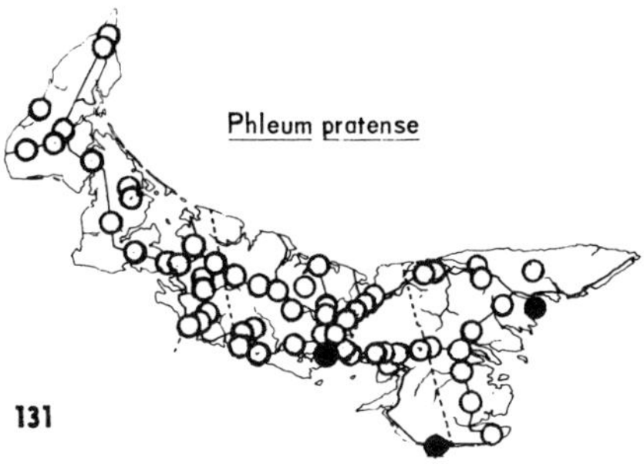

131

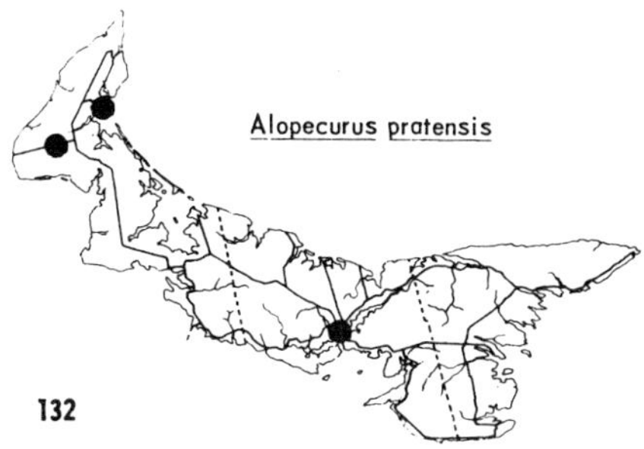

132

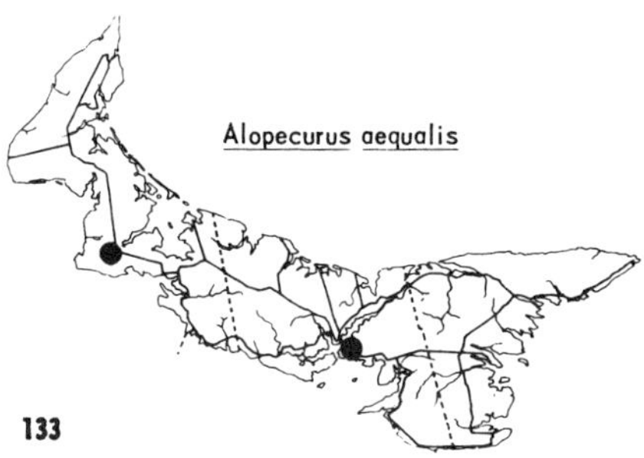

133

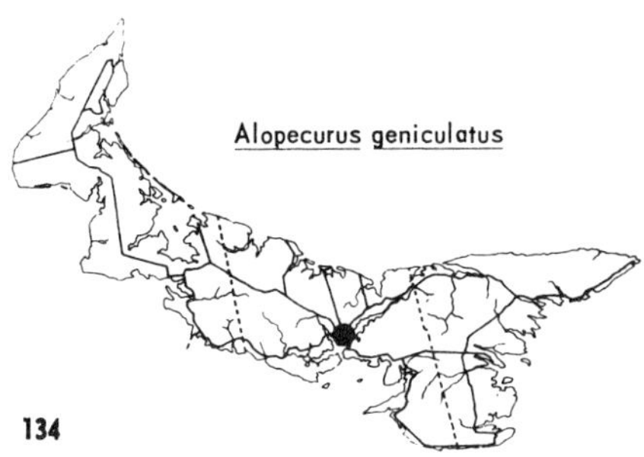

134

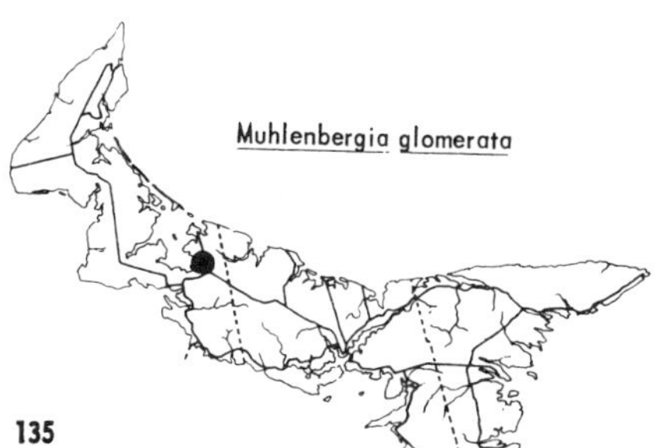

135

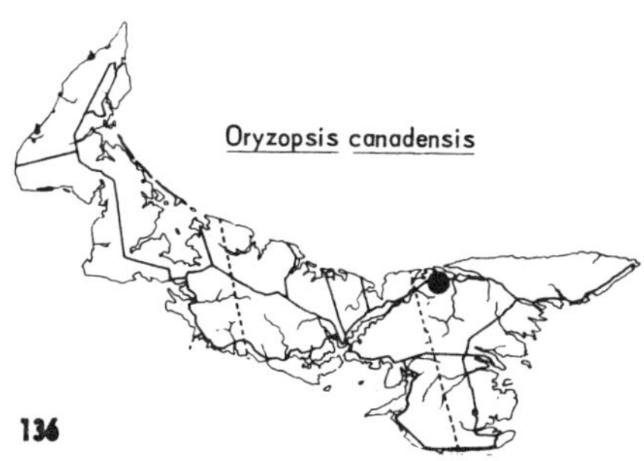

136

Spartina alterniflora Loisel. var. ***alterniflora*** Map 141. Around the coast on salt marshes. A dominant, forming salt marsh and often acting as a mud binder along creeks. (*S. glabra*, MacSwain and Bain, and var. ***alterniflora*** (Loisel.) Merr., Hurst), Campbell.

[****Beckmannia syzigachne*** (Steud.) Fern. Map 142. Western Canadian, adventive: Charlottetown, black mud of pool drying on cinders by Rocky Point ferry wharf, 1953.]

[****Phalaris canariensis*** L. Map 143. European, introduced as birdseed and adventive: Summerside, dump at edge of salt marsh, 1953. Hurst's 1940 record erroneous, based on Bain's ''canary grass''.]

Phalaris arundinacea L., forms. Forma ***arundinacea*** Map 144. Rare. West Point; Charlottetown; Vernon River. European, probably introduced. Forming dense clumps in standing water or damp waste places. Colonies lacking anthocyanin also grow in Charlottetown. The forma ***variegata*** (Parnell) Druce, established at Pleasant Valley, Bonshaw and Charlottetown is introduced as a garden plant. 6 Lots (Groh). (var. *picta* L., Groh, Hurst; ''canary grass'', Bain; *P. Americana*, MacSwain and Bain).

Anthoxanthum odoratum L. Map 145. European, introduced. Rare: O'Leary, Bay View; around Charlottetown; Montague to Georgetown. ''Not much grown'' -- Bain 1890. Light dry soil, in turf. 4 Lots (Groh).

Hierochloë odorata (L.) Wahl. Map 146. Local, around the coasts. Upper border of salt marsh; in and below tension line. Leaves formerly used by the Micmacs of Lennox Island in basketry. (*H. borealis* R. & S., Bain 1892); Groh, Hurst.

Leersia oryzoides (L.) Sw. Map 147. Local; more frequent in Prince County. Muddy or mucky fresh-water shores, forming stands.

Zizania aquatica L., vars. WILD RICE Hurst 1940. Map 148.
Var. ***angustifolia*** Hitchc. Pisquid Pond; Mermaid Lake; forming belts around each in shallow water and deep organic muck. Reputedly introduced as duck food, (if so, a success), native in the St. John River of New Brunswick.

[Var. ***interior*** Fassett Reported by W. Reddin as introduced at Southport, where it has not persisted. Wisconsin seed source indicated this variety.]

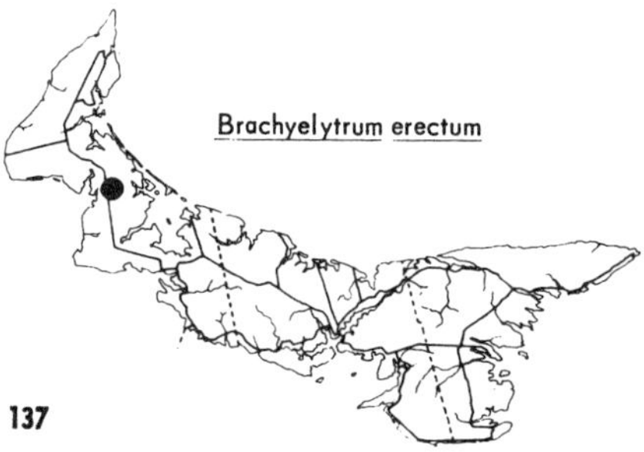

137

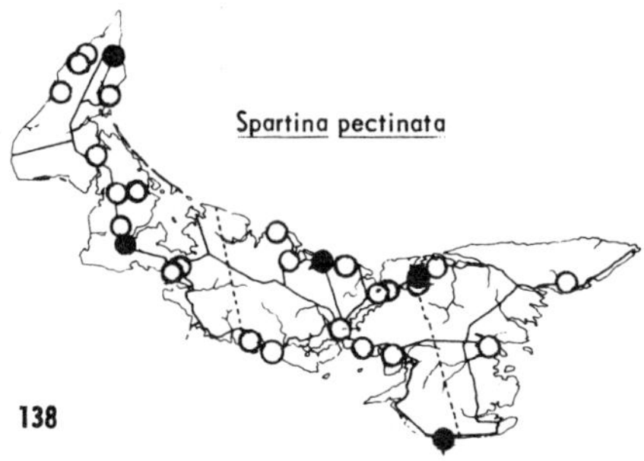

138

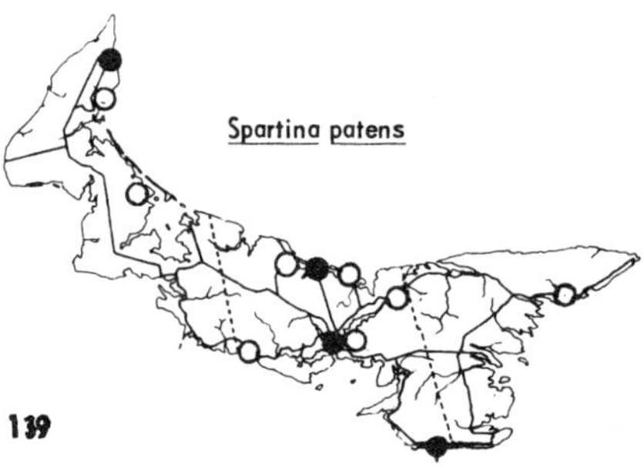

139

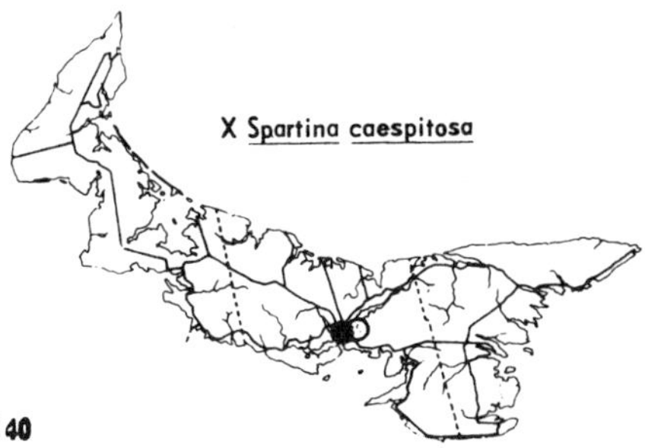

140

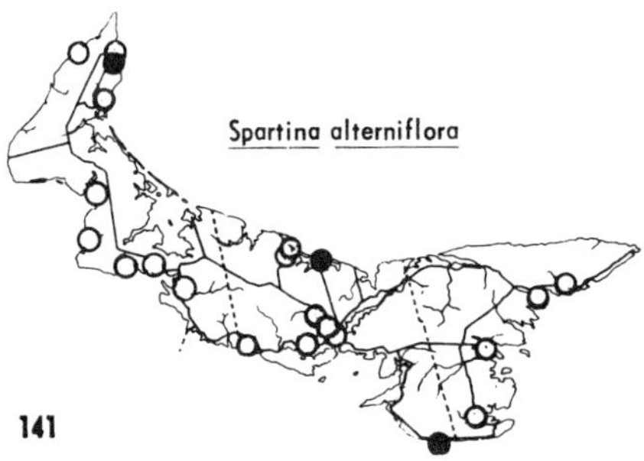

141

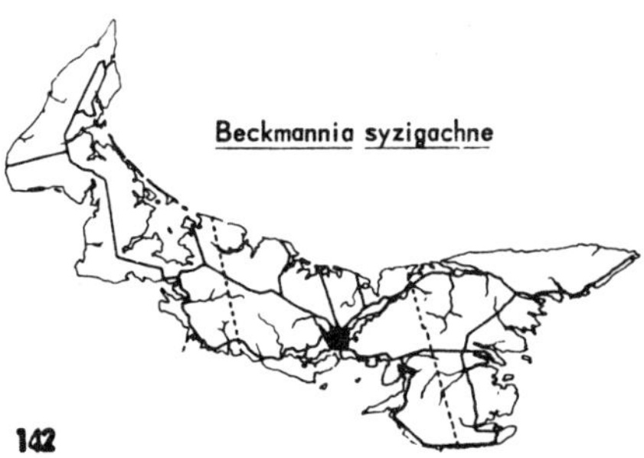

142

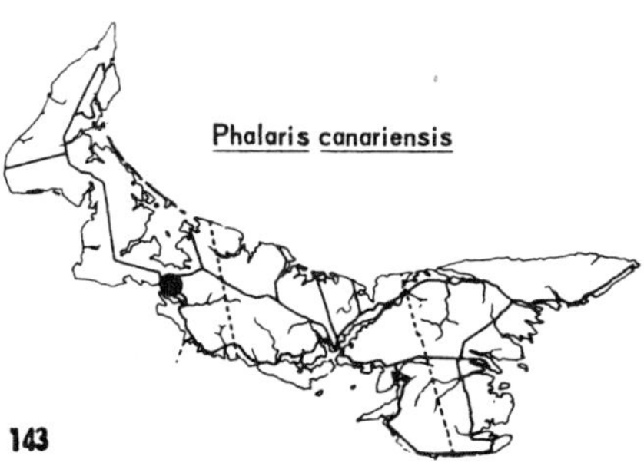

143

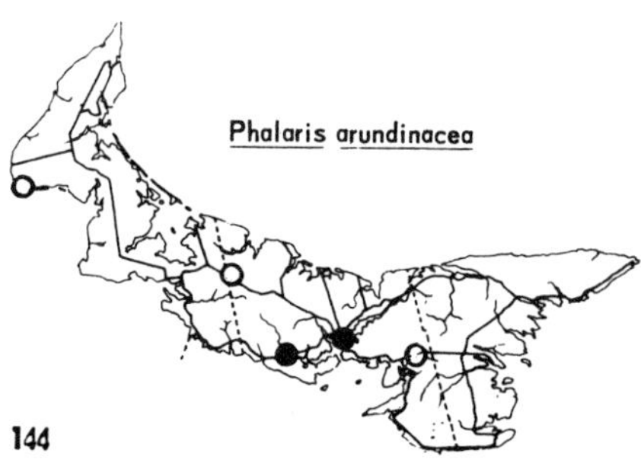

144

Digitaria ischaemum (Schreb.) Muhl. Map 149. European, naturalized from Wellington to Wood Islands, C.N.R. and North Shore. Railway cinders and dry sandy sites near coast, a weed since before 1912. 1 Lot (Groh). (***D. humifusa*** Pers., Groh, Hurst), Campbell.

[*Digitaria sanguinalis* (L.) Scop. Map 150. European, adventive. Charlottetown railway yard, 1912.]

Panicum capillare L. Map 151. Common; adventive from Canada by C.N.R., perhaps naturalized. Scattered along the railways; also at Brackley Beach. Adams, Hurst 1940.

Panicum depauperatum Muhl. var. *psilophyllum* Fern. Map 152. Pisquid West, Morell, Wood Islands Station, sandy to gravelly cut banks. A colonizer of dry sites.

[*Panicum miliaceum* L. Map 153. Brackley Beach (1930); Summerside (1953). European, casual adventive; possibly introduced from the West in chicken feed. Hurst.]

Panicum boreale Nash Map 154. Commonly in central P.E.I.; Campbellton (P). Sunny sites, dry to damp: burns, clearings, old fields, pond borders on sand flats. Rapidly shaded out. (***P. capillare*** var. ***boreale*** (a lapse), Hurst 1940); Campbell.

Panicum subvillosum Ashe Map 155. Scattered throughout: considered by Fernald the common species of section Lanuginosa. Open dry sites: old fields, barrens and maritime sand.

Panicum lanuginosum Ell. var. *implicatum* (Scribn.) Fern. Common in central Queens and at least to Wood Islands. Old hayfields or sandy places, weedy and never found in turf. (***P. lanuginosum***, Adams, Hurst).

Echinochloa crus-galli (L.) Beauv. Map 156. Frequent only in Queens County; Central Bedeque (1905). European, naturalized by 1888. Weed of broken cultivated ground, thriving in muddy or well-manured soil. 8 Lots (Groh), Hurst.

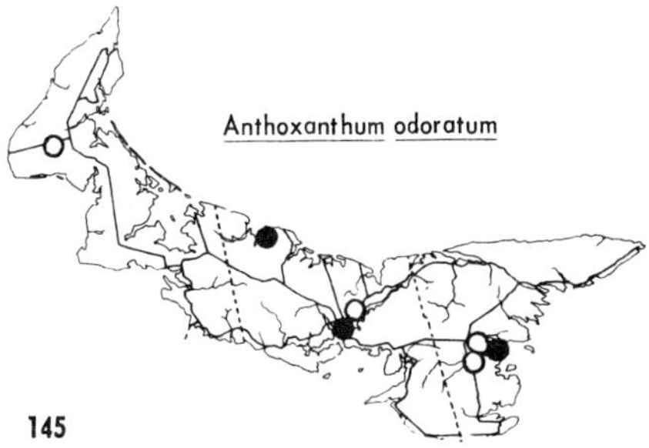

145

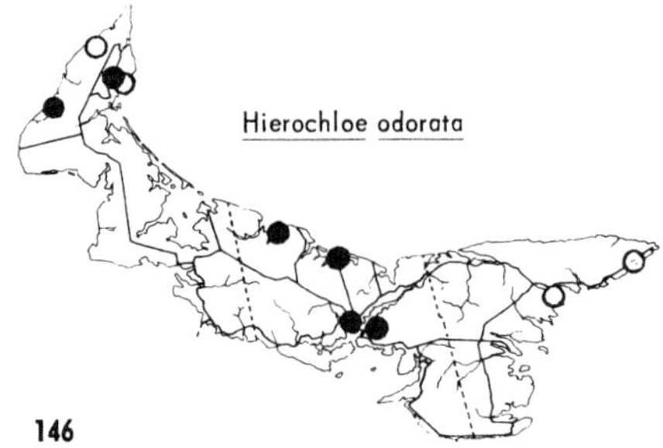

146

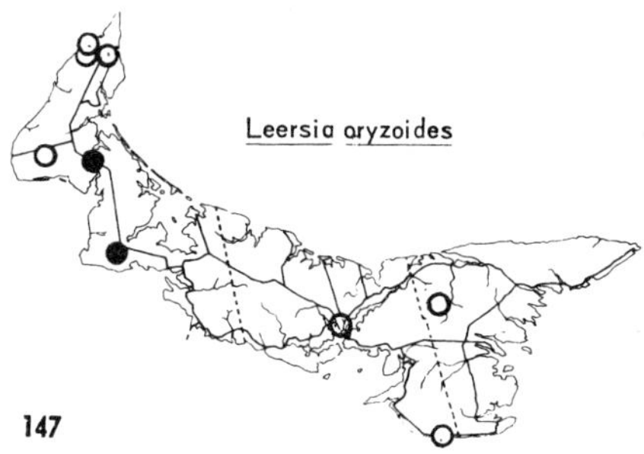

147

148

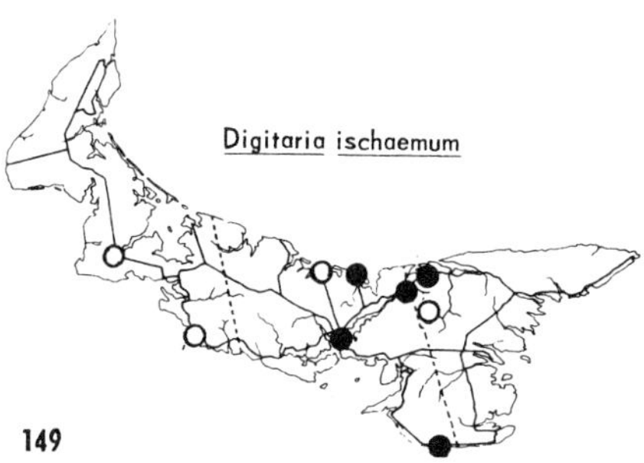

149

150

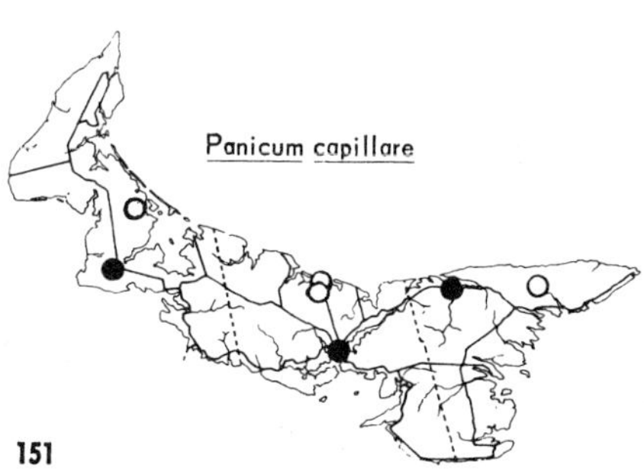

151

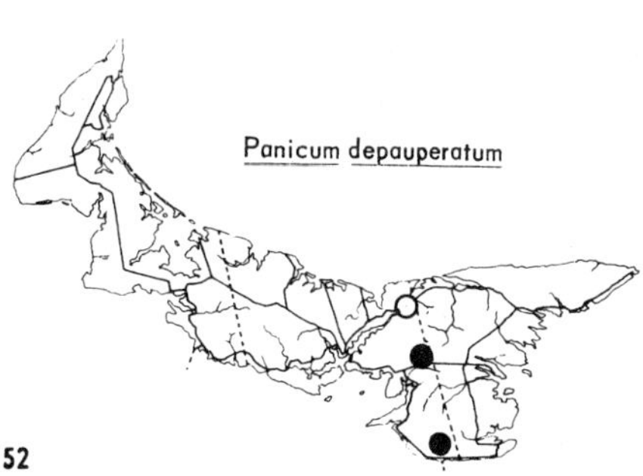

152

Setaria glauca (L.) Beauv. Map 157. Rare: Charlottetown; only on the Georgetown railway (Groh). European, naturalized. Weed of bare or disturbed ground. Groh, Hurst.

Setaria viridis (L.) Beauv. Map 158. Scattered, most abundant in towns and villages. European; in Charlottetown, naturalized before 1953. Weed of railways and gardens. Var. ***weinmannii*** (R. & S.) Brand Often present to the exclusion of var. ***viridis:*** Lot 6 Point; Montague. 8 Lots (Groh). *(S. **verticillata,*** MacSwain and Bain); Hurst. (*S.* ***italica,*** Groh & Frankton, Can. Weed Surv. 5: 4. 1948).

[*Setaria* ***italica*** (L.) Beauv. Mrs. Heath Haviland's collection from Charlottetown was of *S.* ***viridis,*** as was that of Dore & Gorham from Montague.]

CYPERACEAE

Dulichium arundinaceum (L.) Britt. Map 159. Rare: Portage; Lake Verde, Watervale. Mucky shores of fresh water lakes.

Eleocharis parvula (R. & S.) Link Map 160. Local; Souris, Mount Stewart, Brackley Beach, Andrews Mills, Ellerslie, Skinners Pond. Brackish mud at the mouth of creeks, forming carpets.

****Eleocharis pauciflora*** (Lightf.) Link var. ***fernaldii*** Svenson Map 161. Nail Pond, damp sand of dune slack. (*E.* ***fernaldii*** (Svenson) Löve).

****Eleocharis ovata*** (Roth) R. & S. Map 162. Wellington, in clumps on muddy bottom of old millpond. Specimens fairly typical, but occupying ecological place of the related and similar *E.* ***obtusa.***

Eleocharis obtusa (Willd.) Schult. Map 163. Tignish; rather common in the southeast, replaced by *E.* ***ovata*** at Wellington. Muddy shores. Bain (1892).

Eleocharis palustris (L.) R. & S. var. ***major*** Sonder Map 164. Scattered, shallow margins of freshwater ponds. (*E.* ***palustris*** var. ***vigens*** Bailey, Macoun; *E.* ***palustris,*** Hurst). (*E.* ***Smallii*** Britton by Svenson's (Rhodora 49: 63-5. 1947) simplified treatment).

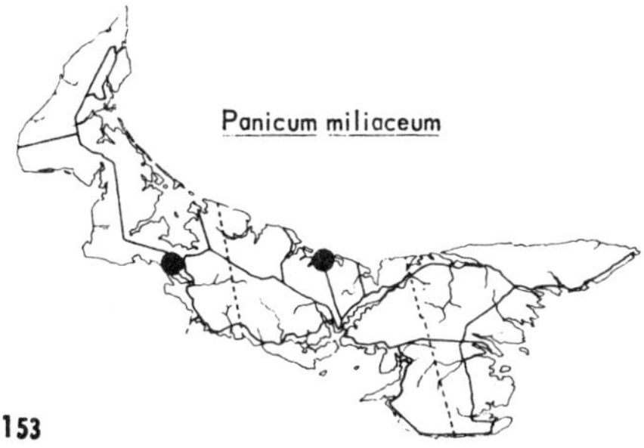

153

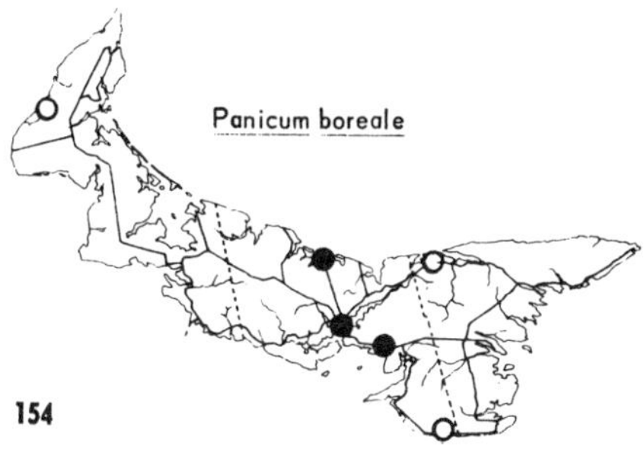

154

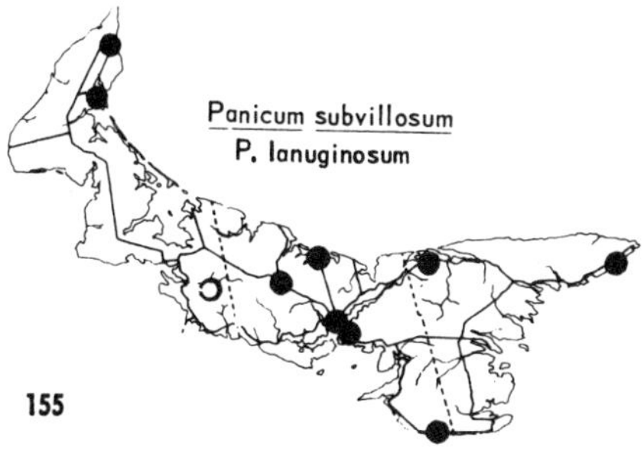

155

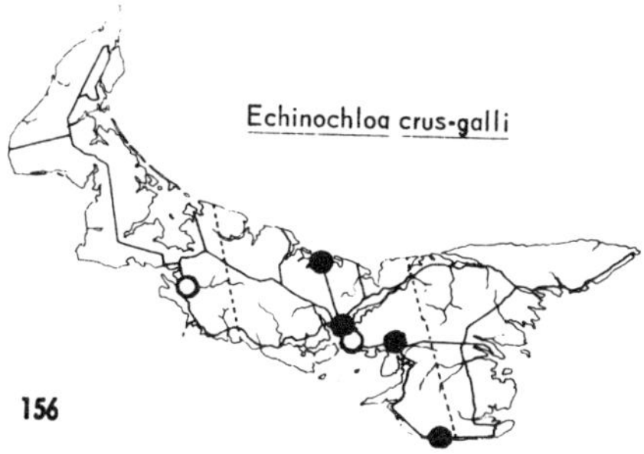

156

157

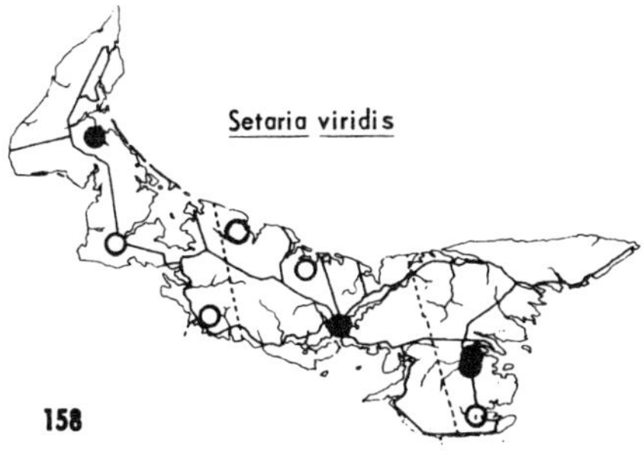

158

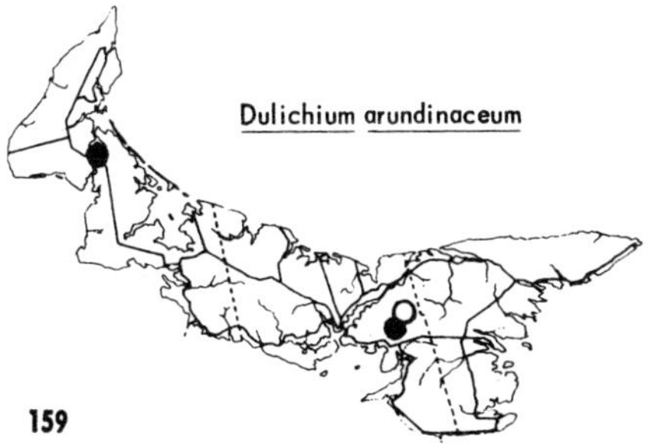

159

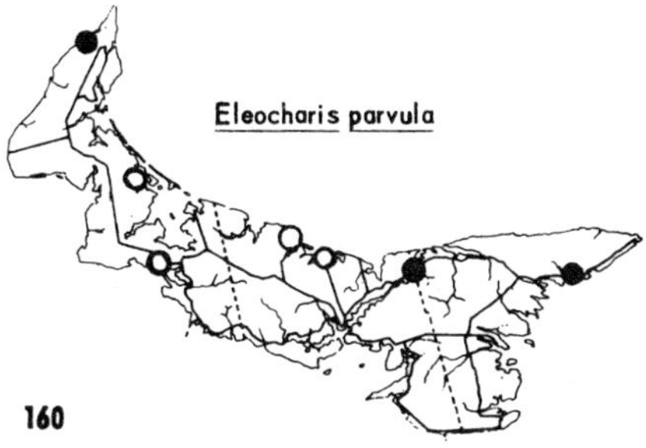

160

Eleocharis halophila Fern. & Brackett Map 165. Maritime; not observed in Kings County. Coastal marshes: dune slacks, salt to brackish marsh of shore, even in quite fresh marsh. At Portage plants growing around a bog lake show characteristics of ***E. calva*** Torr. (det. W.B. Schofield). (*E. palustris* var. ***Watsoni,*** Macoun).

Eleocharis acicularis (L.) R. & S. Rare: Reddin's Pond, Southport, muddy shore; shallows of the Eastern Queens lakes.

Eleocharis tenuis (Willd.) Schult. Bideford (P): marshy field; Long Swamp, Bunbury.

Scirpus cespitosus L. var. ***callosus*** Bigel. Map 166. Western Prince County peat bogs; Village Green, floating mat of bog-pond.

Scirpus subterminalis Torr. Mermaid Lake, extensive beds in shallow water on sandy northern shore.

Scirpus rufus (Huds.) Schrad. var. ***neogaeus*** Fern. Map 167. Summerside; North Rustico; Brackley Beach. Seashores; salt marsh and dune slacks.

Scirpus americanus Pers. Map 168. Coastal, almost absent from Northumberland Strait shore. Salt marsh, sand flats (often dwarf, forming pure stands) and dune slacks; seldom in fresh marsh: Mount Stewart, in larch swamp; Nail Pond, along lower course of brook, thus no evidence of salt-water influence; Clark's Pond, P.E.I. National Park. (*S.* ***pungens*** Vahl, Churchill); Hurst.

Scirpus validus Vahl var. ***creber*** Fern. Map 169. Throughout, more or less coastal; fresh to brackish creeks, and marshes, peat bog (alkaline ?) at Bristol. Restricted largely to firm soils, thus dependent on clay for substratum, and abundant in Dalvay Pond while rare in other barrier beach ponds with mucky shores (M.W. Smith 1946). (*S.* ***lacustris,*** MacSwain and Bain, in part); Campbell.

161

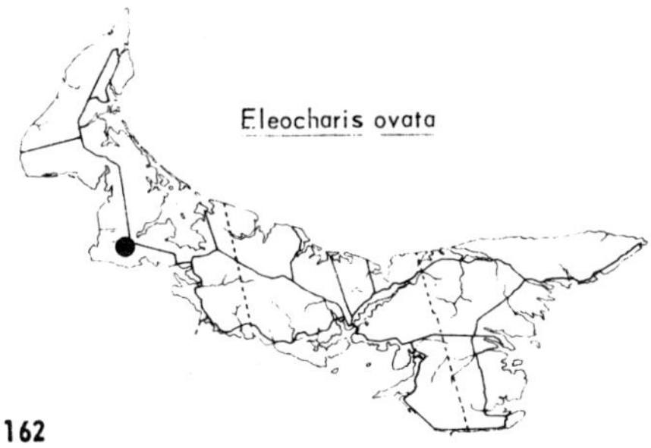

162

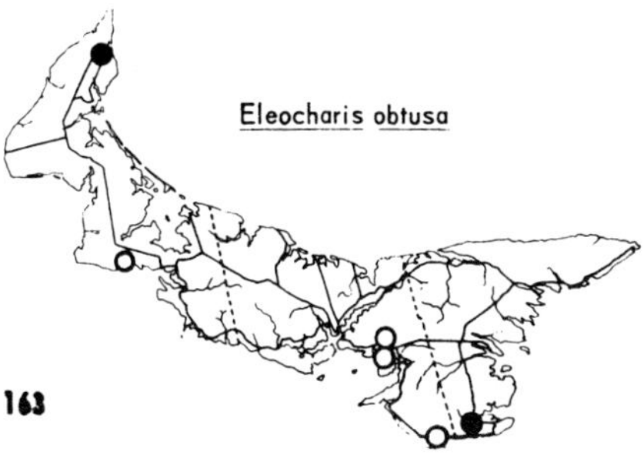

163

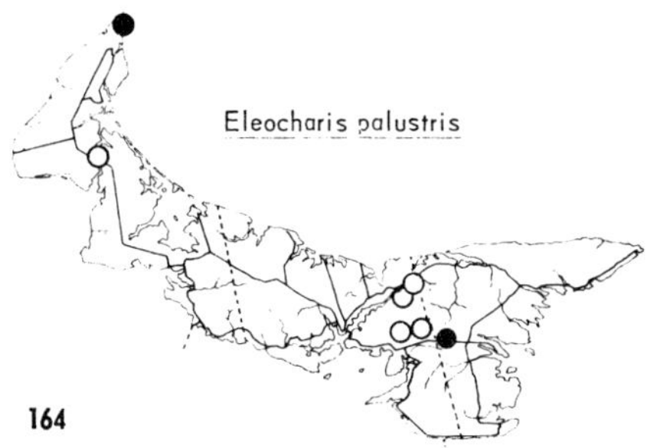

164

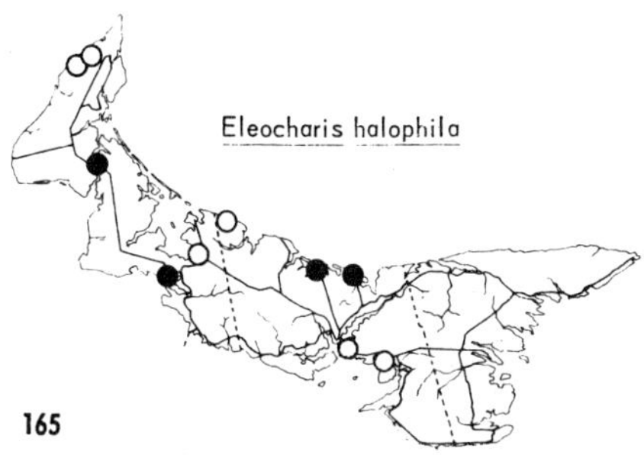

165

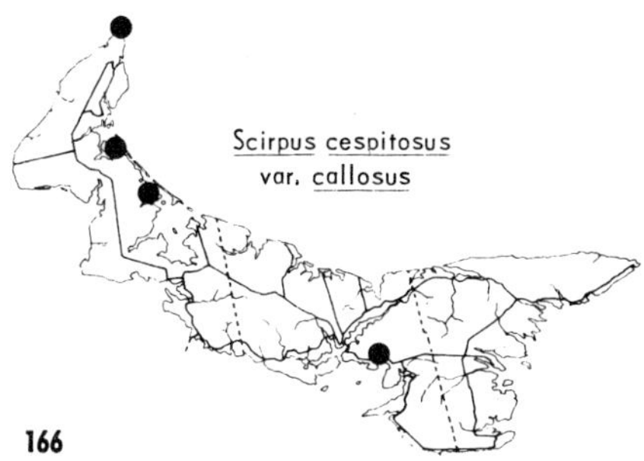

166

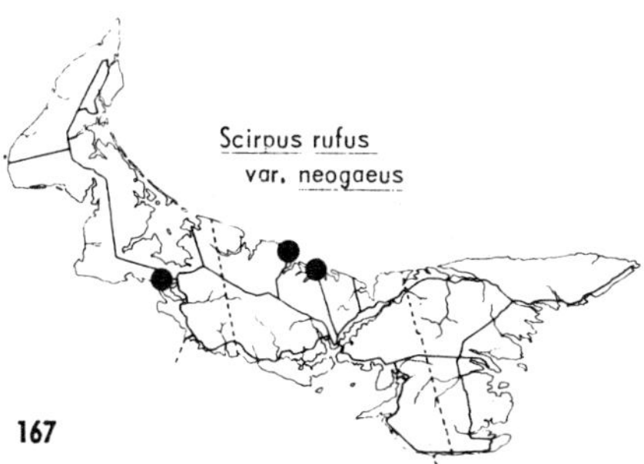

167

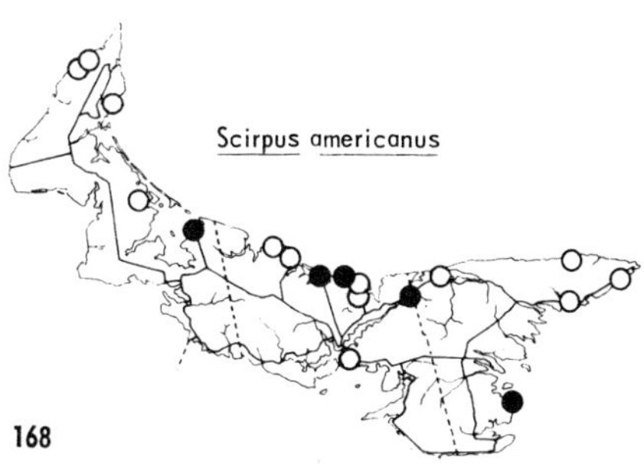

168

Scirpus acutus Muhl. Map 170. Local, coastal. Forming stands. Transition from salt to fresh marsh, especially at creek mouths. Not distinguished from *S.* ***validus*** by Roland. (*S.* ***lacustris,*** MacSwain and Bain in part).

Scirpus paludosus A. Nelson var. ***atlanticus*** Fern. Map 171. Local, on the warmer coast, the type collection from Bunbury. Salt marsh or sand flats. Campbell.

Scirpus maritimus L. var. ***fernaldii*** (Bickn.) Beetle Map 172. Local, coastal. Salt marshes, but more tolerant of brackish conditions than *S.* ***paludosus;*** dune slacks; of barrier-beach ponds in the Park, only in Clark's Pond, the most saline (!), (M.W. Smith 1946). Forma ***agonus*** Fern. forms colonies with forma ***fernaldii*** at Mount Stewart, Bunbury and Brackley Beach. (*S.* ***macrostachys,*** MacSwain and Bain).

****Scirpus atrovirens*** Willd. var. ***georgianus*** (Harper) Fern., Map 173. Local in Prince County: Wellington, McNeills Mills corner. Damp fields. (*S.* ***atrovirens,*** MacSwain and Bain, may have referred to the next).

Scirpus rubrotinctus Fern. Map 174. Throughout; absent from the southeast. Wet open sites: along ditches and streams. (*S.* ***atrovirens,*** MacSwain and Bain (?)).

Scirpus cyperinus (L.) Kunth var. ***pelius*** Fern. Map 175. Common throughout. Damp sites: swamps and marshy places. The forma ***condensatus*** (Fern.) Blake has persisted at the same station in Southport for forty years. (*S.* ***eriophorum*** Michx., MacSwain and Bain in part; *S.* ***cyperinus,*** Hurst 1940).

Scirpus atrocinctus Fern. Map 176. Throughout, in similar habitats to *S.* ***cyperinus:*** swamps and bog meadows; a little more shade tolerant. Fruiting a week or so earlier than *S.* ***cyperinus.*** (*S.* ***eriophorum*** Michx., MacSwain and Bain in part); Hurst 1940.

Eriophorum russeolum Fries Map 177. (Raymond, Svensk. Bot. Tidskr. 48 (1): 70-2. 1954). Rare: Black Marsh at North Point, Cape Wolfe, Brackley Point, vicinity of Charlottetown. Wet peat. (*E.* ***Chamissonis,*** Fernald 1950).

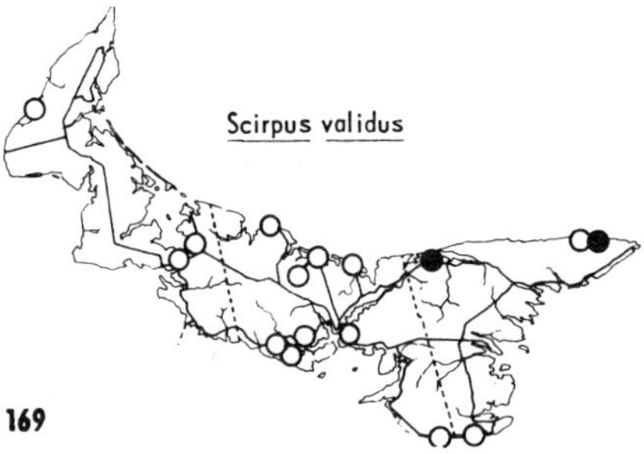

169

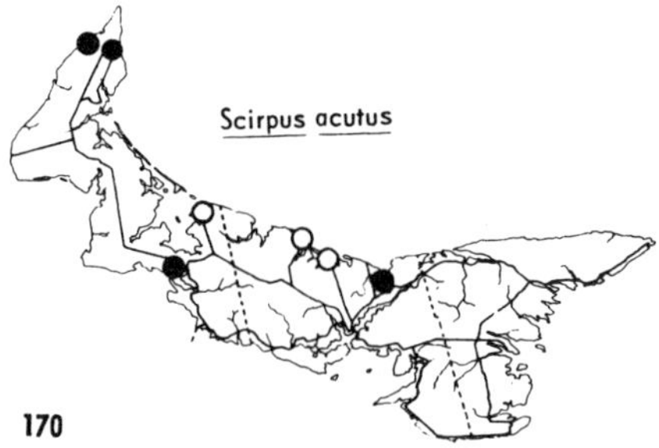

170

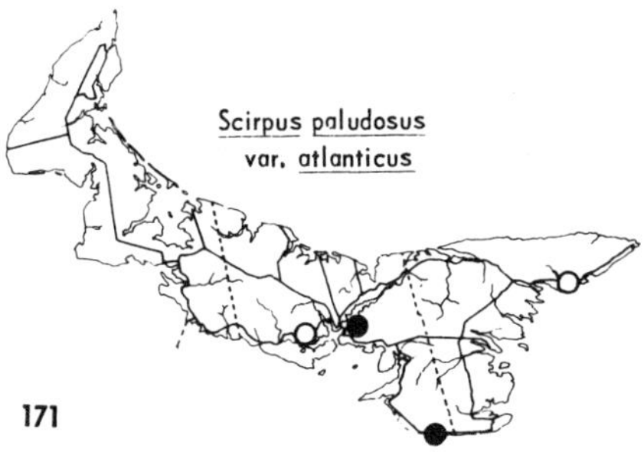

171

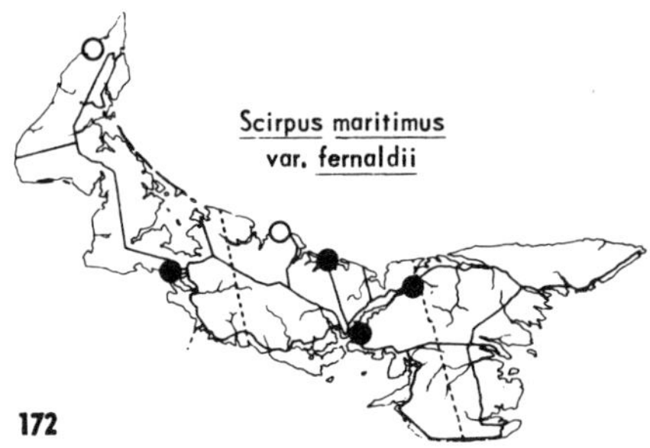

172

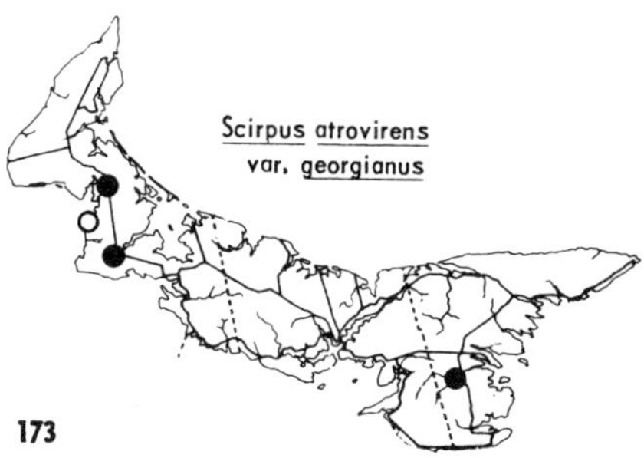

173

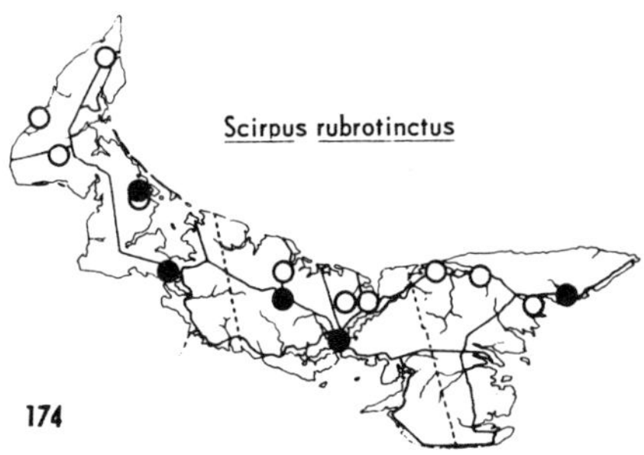

174

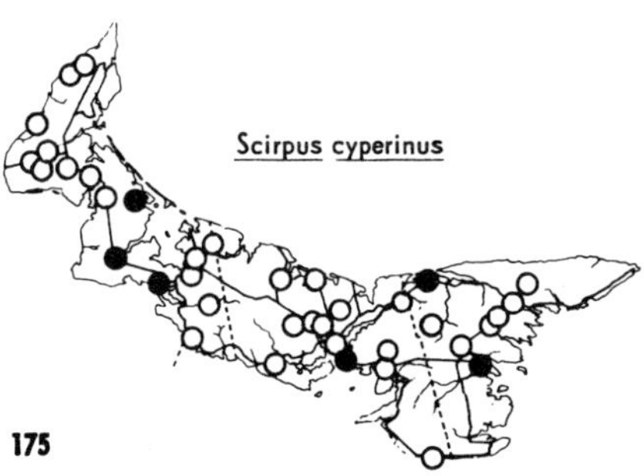

175

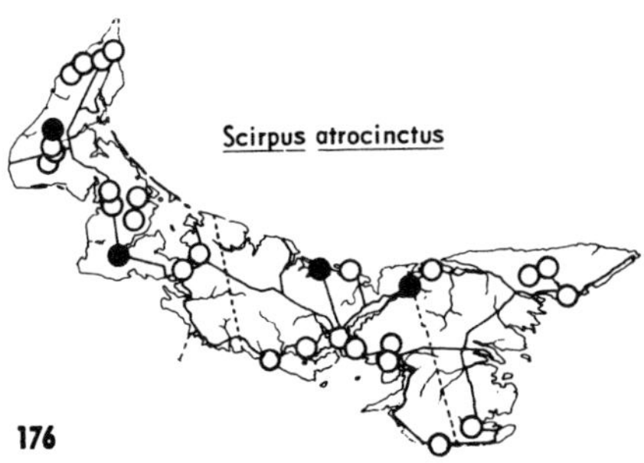

176

Eriophorum spissum Fern. Map 178. Local throughout. Acid peat bogs; a weed where stripped of peat. (*E. vaginatum*, MacSwain and Bain; *E. callitrix*, Hurst).

* ***Eriophorum gracile*** Koch Map 179. Watervale; Bristol (K), floating mats of *Typha* marsh.

Eriophorum tenellum Nutt. Map 180. Locally throughout. Wet peat. Later flowering than the several preceding species.

Eriophorum angustifolium Honck. Map 181. Scattered throughout: except in the northeast the most common cotton-grass. Wet acid sites; bogs, swamps and marshy fields, seldom in peat. The larger var. ***majus*** Schult. from Charlottetown (*Fernald et al.*). (*E. polystachyon* L., in part, MacSwain and Bain); Hurst.

Eriophorum viridi-carinatum (Engelm.) Fern. Map 182. Rare: Grand Tracadie, Bristol. Wet peat bogs.

Eriophorum virginicum L. Map 183. The common cotton-grass in the east; local also in West Prince. Last of the species to flower. Acid peat: wet or dry bogs and swamps. Hurst 1940.

Rhynchospora alba (L.) Vahl Map 184. Quite local; throughout. Peat bogs.

Carex vulpinoidea Michx. Map 185. Local in Prince County; Bunbury; reported from Brackley Beach by Hurst. Damp sites, in open places or disturbed soil, roadside gutters and old fields. (*C. setacea*, MacSwain and Bain (?)); Adams, Hurst 1940.

Carex diandra Schrank Map 186. Scattered. Shallow brown water; especially around bogs and ponds, where the tussocks build up a marsh. Confined to low sites.

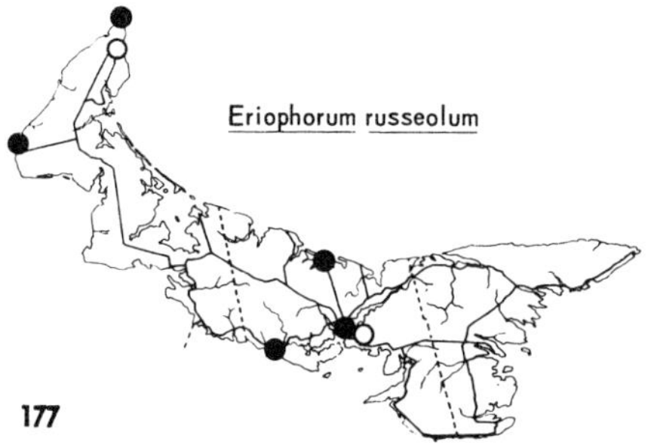

177

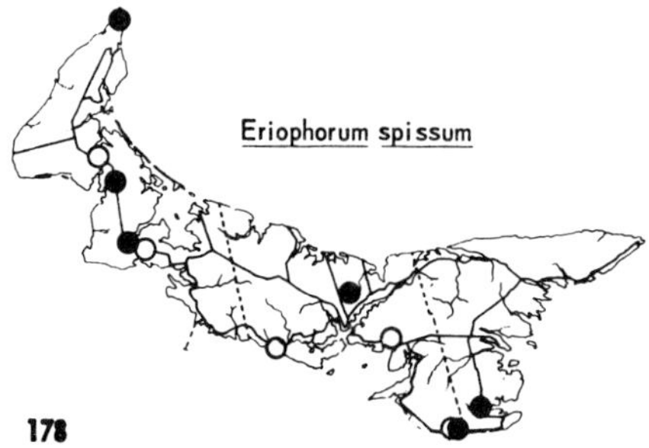

178

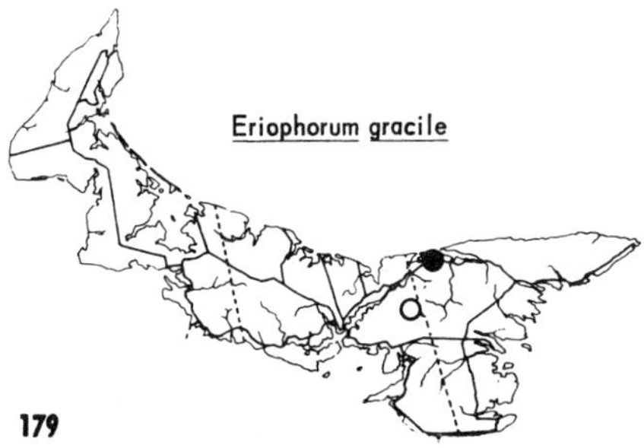

179

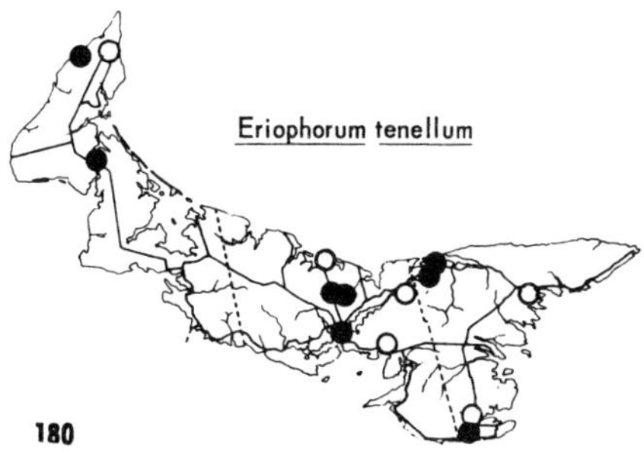

180

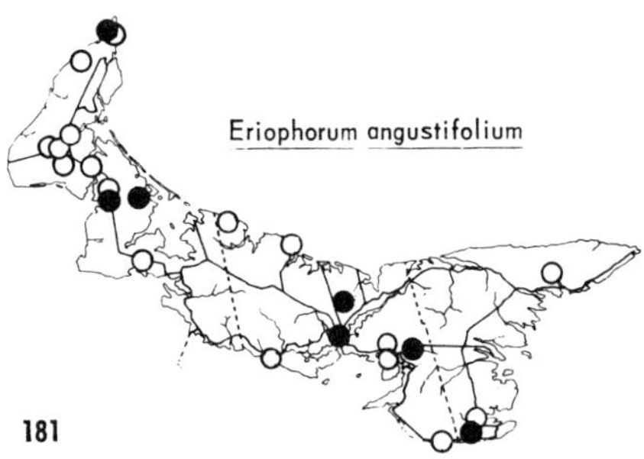

181

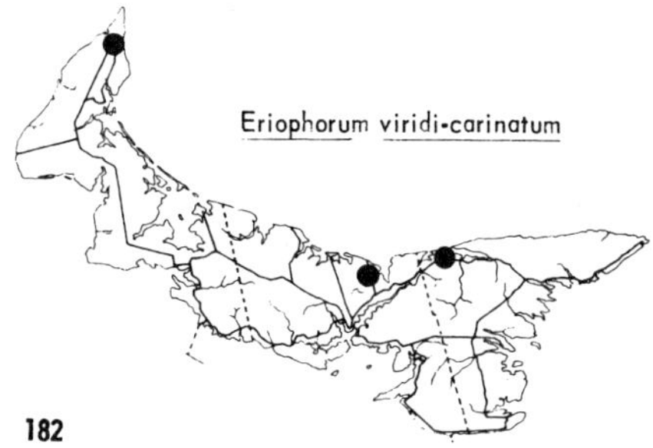

182

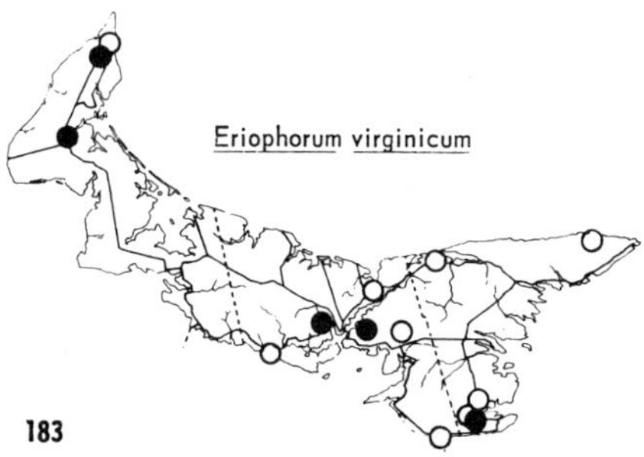

183

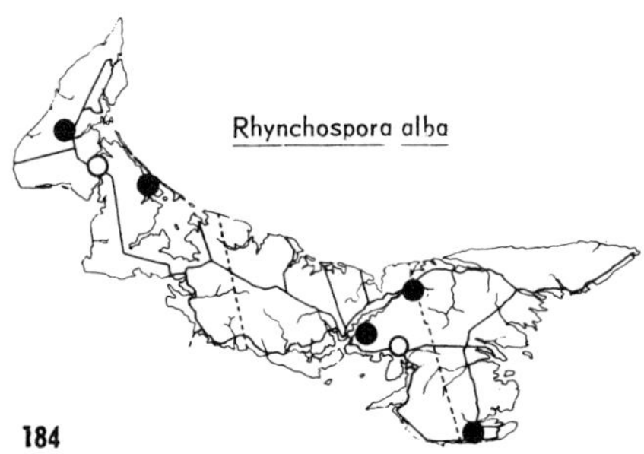

184

Carex stipata Muhl. Map 187. Common throughout; distinctive and late-flowering. Damp meadows, stream banks in fields and woods (*C. Mulenbergii*, MacSwain and Bain); Adams, Hurst 1940.

Carex disperma Dewey Map 188. Local: not observed in Kings. Shady sites; damp pockets in woods. Adams, Hurst 1940.

Carex trisperma Dewey Map 189. Throughout, perhaps most common in the southeast. Damp, mossy or boggy woods; in open peat represented by var. ***billingsii*** Knight with setaceous leaves. Adams, Hurst 1940.

Carex mackenziei Krecz. Map 190. Rocky Point; Green's Shore at Summerside, head of salt marshes. (*C. **norvegica*** Willd. (not Retz.), Adams, Hurst 1940); Campbell.

Carex brunnescens (Pers.) Poir. var. ***sphaerostachya*** (Tuckerm.) Kük. Map 191. Scattered throughout. Damp or coniferous woods, rarely in dry woods as in Quebec Province (Victorin). Adams, Hurst 1940.

Carex canescens L. var. ***disjuncta*** Fern. Map 192. Throughout in low grounds. Mucky swamps, bogs and at edges of marshy pools. Fruit often smutted. Bain 1892, Adams, Hurst 1940.

Carex interior Bailey Map 193. Frequent in east-central lowlands (Southport, Bristol); McNeills Mills (P), O'Leary; in boggy swamps, often under black spruce. Adams, Hurst 1940.

****Carex atlantica*** Bailey Map 194. Murray River, Bristol; swamps.

Carex cephalantha (Bailey) Bickn. Map 195. Common in Prince County; local in the northeast open marshes and damp fields, only rarely in spruce swamps, the usual habitat of *C. **interior***. Adams, Hurst 1940.

185

186

187

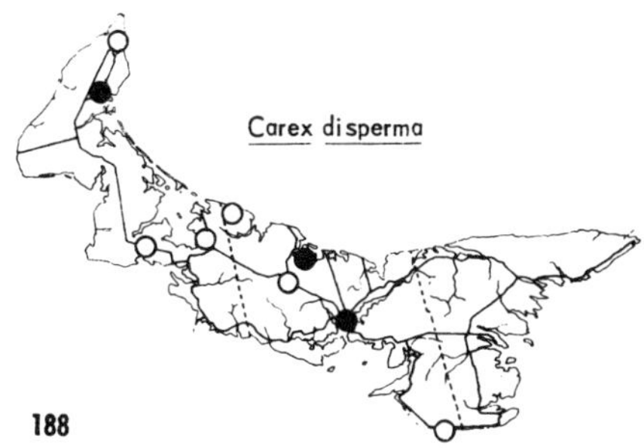

188

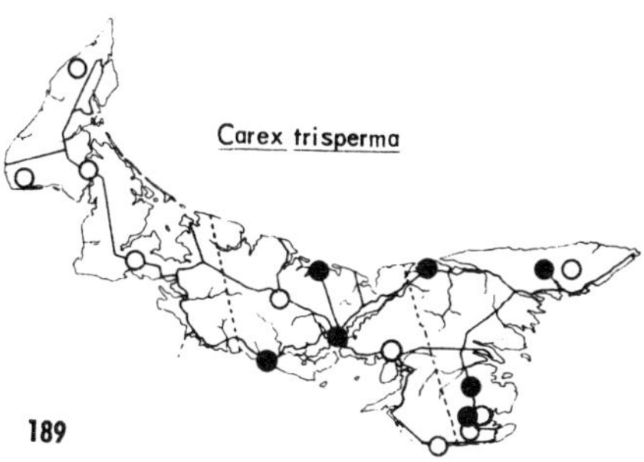

189

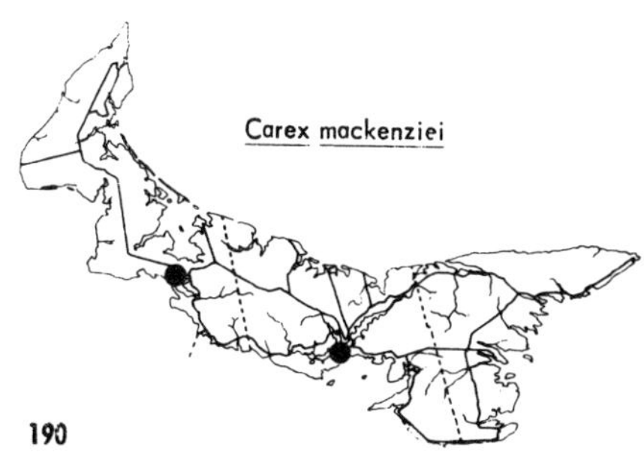

190

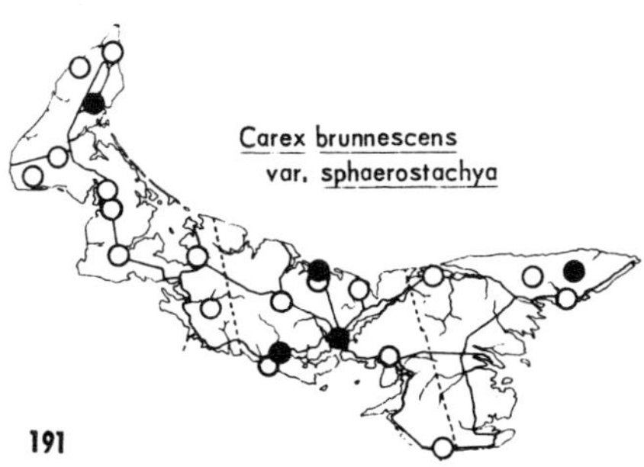

191

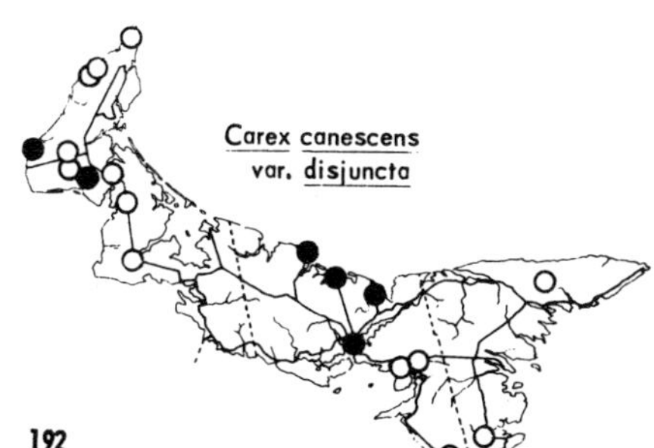

192

Carex angustior Mackenzie   Map 196.   Scattered in all counties. Streams and shores.

Carex deweyana Schwein.   Map 197.   Western and central areas. Well-drained sites: second growth or open woodland, beneath red maple, red or white spruce. Bain 1892, Adams, Hurst 1940.

Carex leporina L.   Map 198.   Summerside (*Fernald & St. John* in 1912); Charlottetown (*Malte* in 1913). Naturalized from Europe. Pastureland. Adams, Hurst 1940.

Carex scoparia Schkuhr   Map 199.   Common throughout. Open damp sites; meadows and stream banks, abundant and variable in old fields. In the Fernald & St. John collections from Charlottetown are represented: *f. **moniliformis*** (Tuckerm.) Kük., of which only a single clump; an approach to var. ***tesselata*** Fern. & Wieg.; and a hybrid with *C. **projecta***. Bain 1892, Adams, Hurst 1940.

[*Carex **tribuloides*** Wahl. Reported by Hurst from Brackley Beach (Adams, Hurst 1940). No specimens seen.]

Carex projecta Mackenzie   Map 200.   Throughout rather common. Damp sites: thickets and stream banks. Adams, Hurst 1940.

Carex crawfordii Fern.   Map 201.   Fairly common in Prince and Queens, absent from the northeast. Open, dry or damp sites, thickets, grassy runs, colonizing old fields. 8 Lots (Groh), probably including *C. **scoparia*** sight records. Hurst 1940.

Carex bebbii Olney   Map 202.   Local: Alberton, O'Leary, Wellington, Charlottetown, Southport, Cherry Valley, Murray River. Damp open sites; meadows, marshes and mucky ditches. Adams, Hurst 1940.

Carex tincta Fern.   Map 203.   Hunter River, abundant loose clumps in dry cindery soil along railway track.

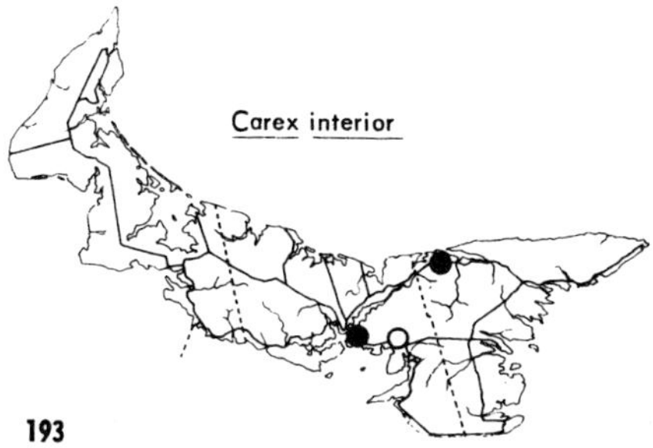

193

194

195

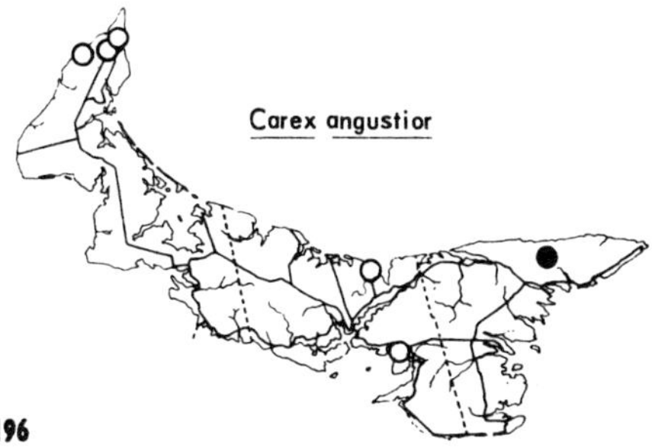

196

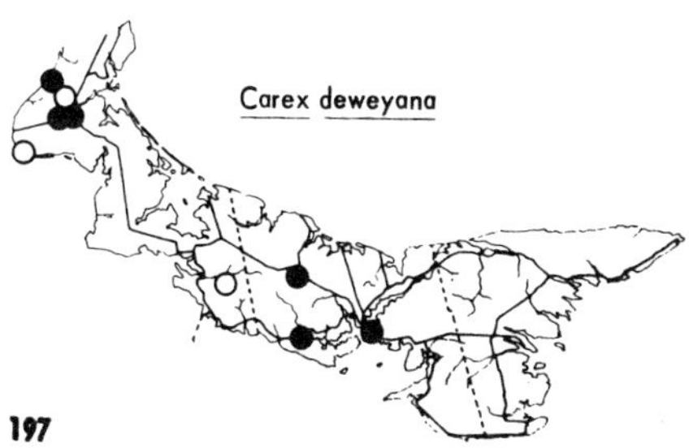

197

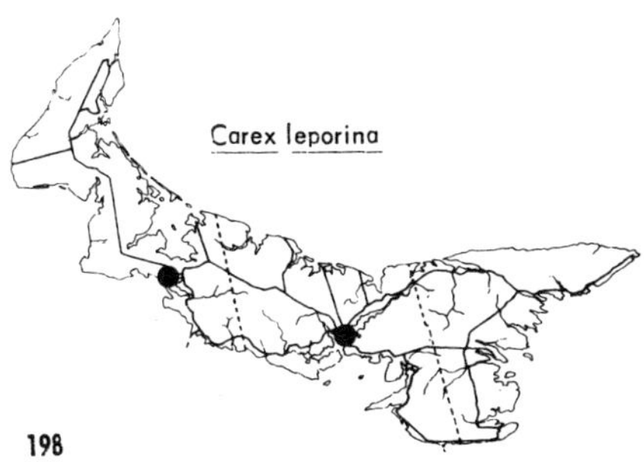

198

Carex cumulata (Bailey) Mackenzie Map 204. Bristol, burnt over blueberry barren by Lot 40 Station (*Fernald & St. John* in 1914, still present).

Carex silicea Olney Map 205. North Shore round to Souris; Wood Islands, West Point. Abundant on coastal sand dunes. Churchill 1901; Adams, Hurst 1940. (*C. foenea*, Adams, in part).

Carex hormathodes Fern. Map 206. Maritime or low-lying places-apparently absent from the southeast. Inland (e.g., at Lake Verde, Charlottetown, Bristol) in meadows along streams, thus preserving as its habitat the open damp sites equally provided around barrier ponds and at the head of salt marsh or spring-moistened sand beaches. (*C. straminea*, MacSwain and Bain (perhaps referring also to *C. silicea* and other Ovales); Adams, including Adams *C. foenea* in part, Hurst 1940).

Carex aenea Fern. Map 207. Mount Stewart to Fortune Bridge. Sandy banks in open woods, now growing tall and weedy along road shoulders. Adams, Hurst 1940.

* **Carex argyrantha** Tuckerm. Map 208. Alliston (K), blueberry barrens; Indian River, roadside cut into acid soil or spruce woods. Adams records of *C. foenea* belong to *C. hormathodes* and *C. silicea.*

Carex leptalea Wahl. Map 209. Local; more common in West Prince County. Damp, shady thickets; pockets in coniferous woods. Adams; Hurst 1940.

Carex emmonsii Dewey Map 210. Confined to Alberry-Charlottetown soil series. Dry hardwoods and successional phases: colonizing moss pats of cleared or burnt hardwood lands, and affording a foothold in turn for seedlings of erect species; more robust than in closed forest. (*C. albicans*, Adams, Hurst 1940, Roland; *C. Peckii*, Campbell).

Carex novae-angliae Schwein. Map 211. Scattered throughout. Well-drained sites supporting open woods. Macoun, MacSwain, Hurst.

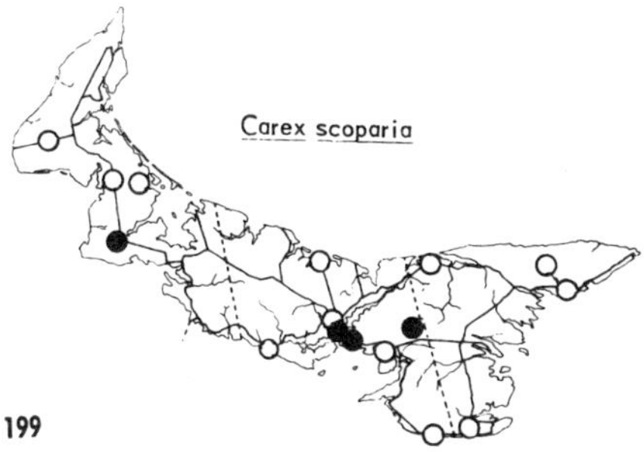

199

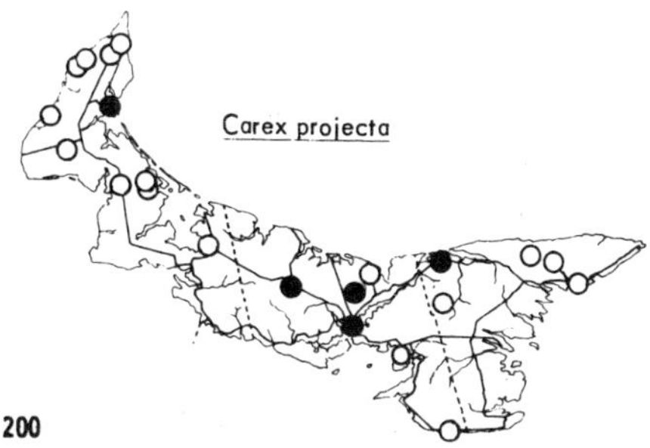

200

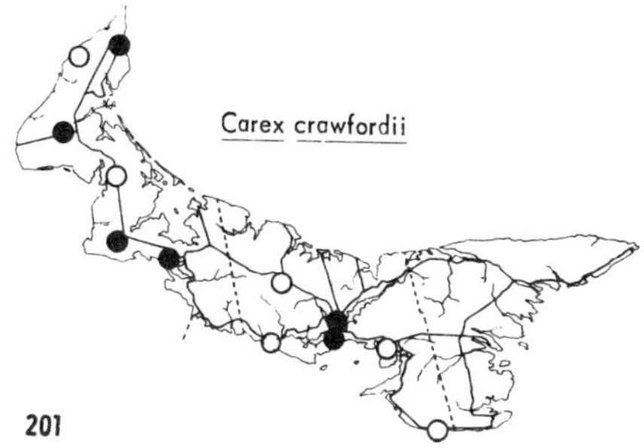

201

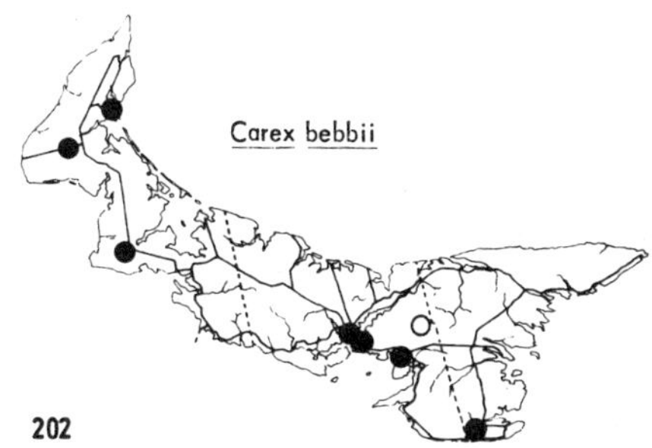

202

203

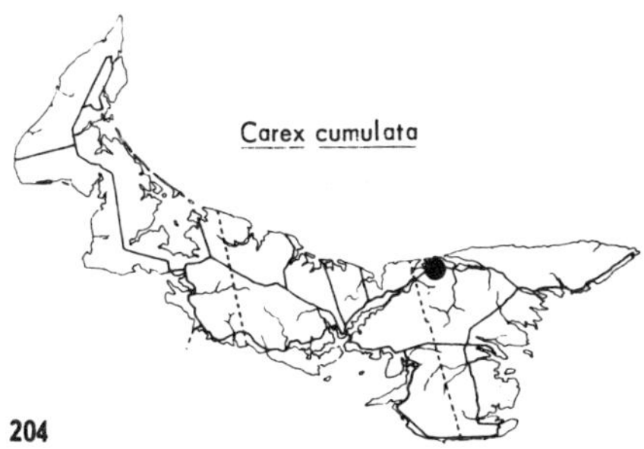

204

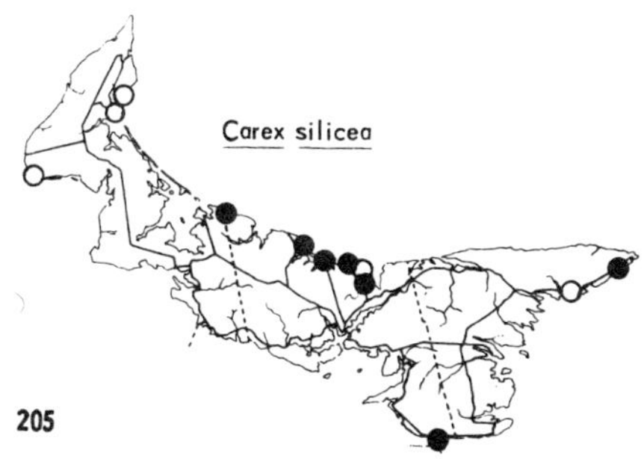

205

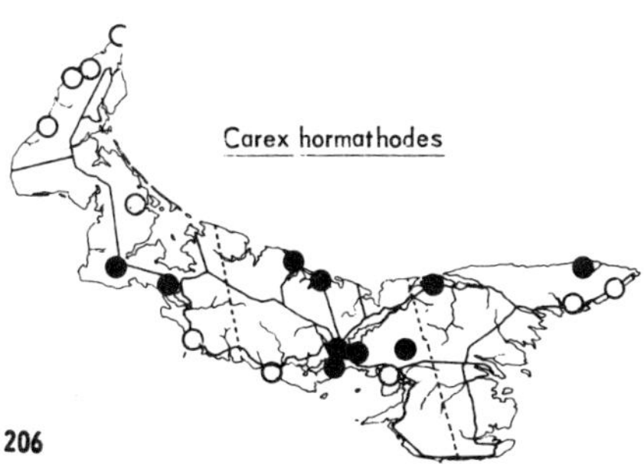

206

Carex communis Bailey   Map 212.   Confined to Alberry-Charlottetown soil series. Dry woodland. Adams, Hurst 1940.

Carex deflexa Hornem.   Map 213.   Rare: Charlottetown, Brackley Point (***Macoun***), Glencorradale corner near Souris. Well-drained woods. Adams, Hurst 1940.

Carex umbellata Schkuhr.   Map 214.   Morell; Dundee; Pisquid West; Brackley Beach, stony sandy roadsides. Intergrading with *C.* ***tonsa*** on the inner wooded dunes. Open sandy sites; more shade-tolerant than *C.* ***tonsa.*** Its forma ***vicina*** (Dewey) Wieg. is paralleled in each species of the ***umbellata*** complex. (*C.* ***deflexa*** var. ***Deanii,*** Macoun; *C.* ***rugosperma*** Mackenzie, Adams); Campbell.

Carex tonsa (Fern.) Bickn.   Map 215.   North shore: Tignish, Brackley Beach, St. Peters, and Bothwell. Open, sandy sites (e.g. fixed dunes, old beaches). At Brackley Beach, collections from the dunes are extreme *C.* ***tonsa,*** from the roadside good **C. *umbellata,*** and from the inner dunes intermediate, with long but ***tonsa***-like leaves and short pubescence on the dark green ***tonsa***-like perigynia. Adams, Hurst 1940.

[***Carex abdita*** Bickn. (*C.* ***umbellata*** var. ***brevirostris*** Boott) recorded for "P.E.I." by the seventh edition of Gray's Manual, is not supported by specimens. *C.* ***umbellata*** sensu Mackenzie, Hurst 1940.]

****Carex aurea*** Nutt.   Map 216.   Local: from Prince County east to Bristol. Damp open sites; dune-slacks (often), clay fields, pond-shores.

Carex paleacea Wahl.   Map 217.   Coastal; frequent. Salt marshes. At Summerside and Bideford, apparently extending inland to fresh or brackish marsh. Adams, Hurst 1940.

Carex salina Wahl. var. ***kattegatensis*** (Fries) Almq.   Map 218. Coastal, rare: McWilliams Cove, North Rustico, North Lake, Souris, Head of salt marsh, but in less saline habitats than *C.* ***paleacea,*** never descending creeks to salt water; in fresh marsh around barrier ponds. (*C.* ***recta*** Boott).

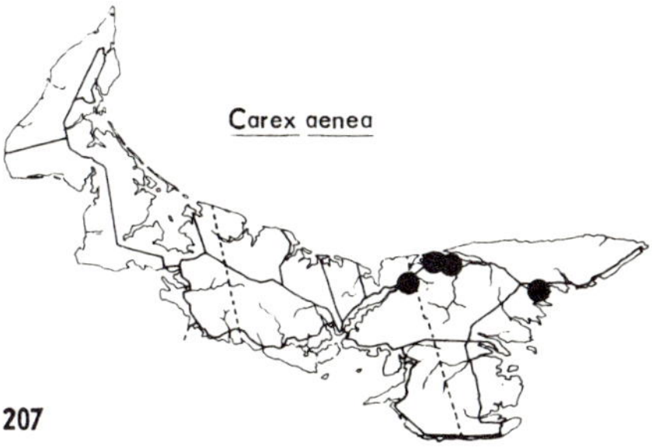

207

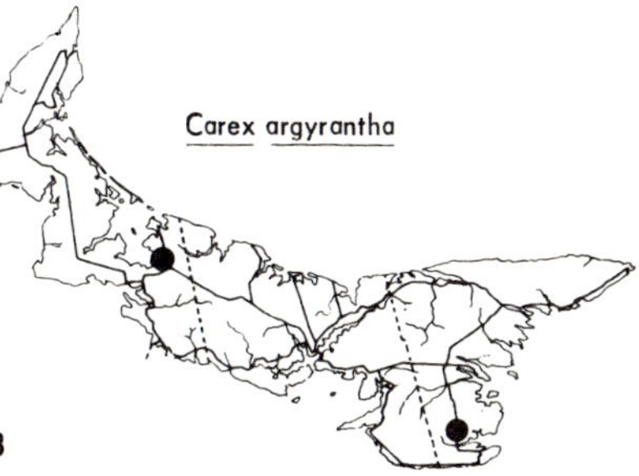

208

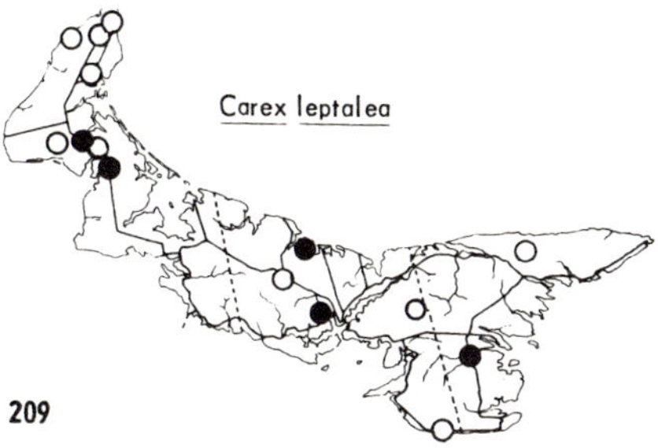

209

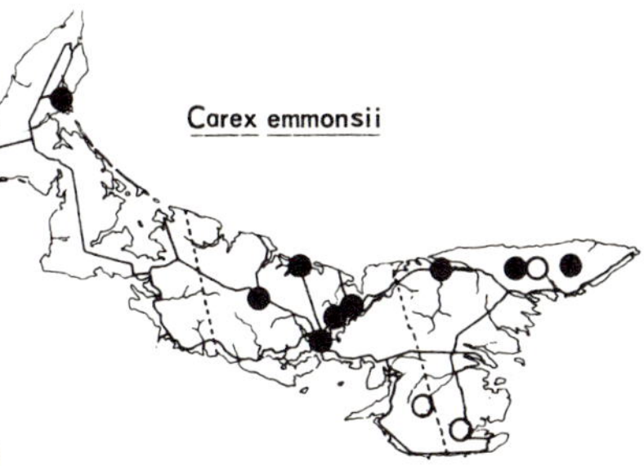

210

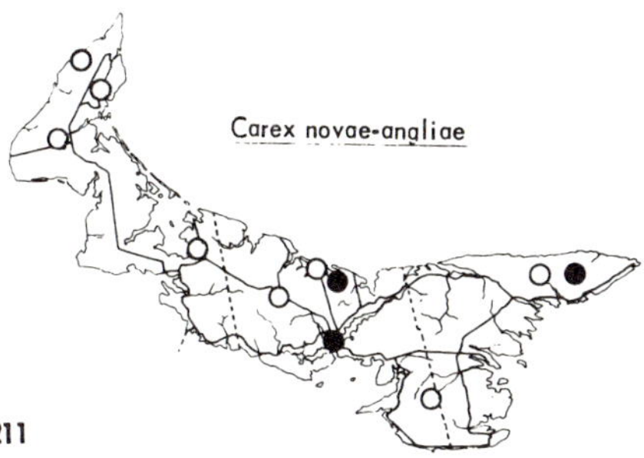

211

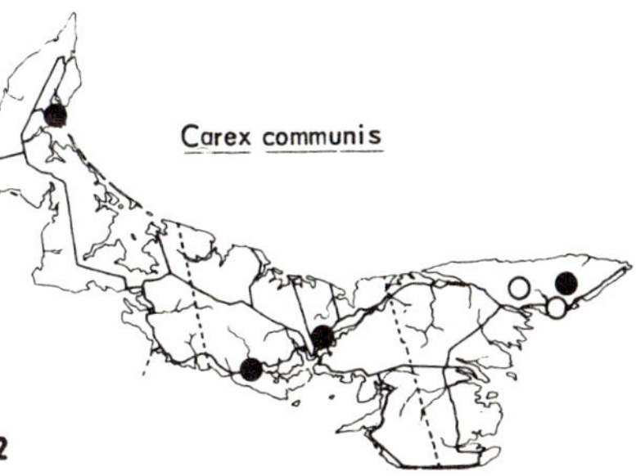

212

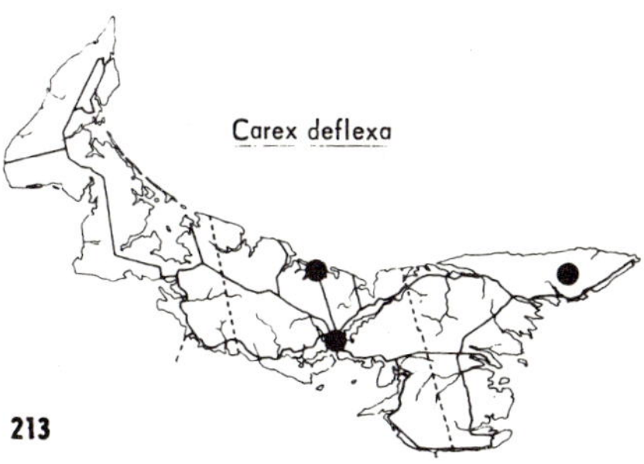

213

214

Carex crinita Lam., vars. Map 219. Open wet sites: along streams, ditches and in marshy thickets. Common; absent from the northeast.

Var. **_crinita_** Common in Prince County, locally east to Dalvay. MacSwain and Bain, Adams, Hurst 1940.

Var. **_gynandra_** (Schwein.) Schwein. & Torr. Scattered, from Prince County to Murray River. (*C. **gynandra*** Schwein., Adams, Hurst 1940), Campbell.

Carex aquatilis Wahl. var. **_altior_** (Rydb.) Fern. Map 220. Portage, dense stands in damp hollow along railway; Lake Verde, shallow water at sandy lake margin.

Carex nigra (L.) Reich. Map 221. Scattered and locally abundant, absent from the east. Open places: sandy beach heads, damp or clayey fields, often forming sod; in damp hollows often forming clumps. Often appearing weedy in clay land. In waste land at Spring Park near Charlottetown, forming dense stands over a foot tall, with long leaves equaling the culms, quite unlike the form growing in turf. In temporary grassy pools, at Wellington and North Point, a form with short, narrow and quite distant sessile spikes, the male spike long-peduncled, though as stoloniferous as the species. These differ enough to be set off as var. **_strictiformis_** (Bailey) Fern. (*C. **acuta*** L., in part, Adams, Hurst 1940); Campbell.

Carex limosa L. Map 222. Rare: Bloomfield, East Bideford, Prince County; Stanhope. Wet peat of open bogs; floating mats, even around barrier beach ponds. Adams, Hurst 1940.

Carex paupercula Michx. var. **_irrigua_** (Wahl.) Fern. Map 223. West Prince; east-central region (Queens and adjacent Kings). Peat-bogs and mossy woods, more shade-tolerant than *C. limosa*. Plants with pale-green scales collected by Fernald and St. John at Cape Wolfe, seem to approach var. *pallens* Fern., but may be a shade form. Adams, Hurst 1940.

Carex scabrata Schwein. Map 224. Hills of western Queens: DeSable, Bonshaw, Hunter River; and Kings: Harmony, Montague. On sandy and gravelly alluvial reaches or banks along ravine brooks in mixed woods; very specific for this habitat. Adams, Hurst 1940.

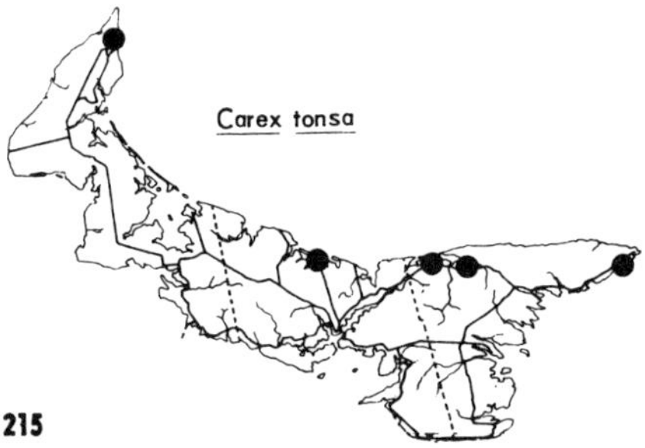

215

216

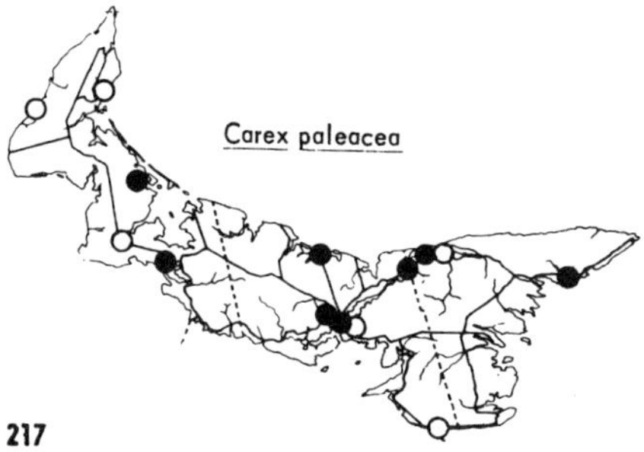

217

218

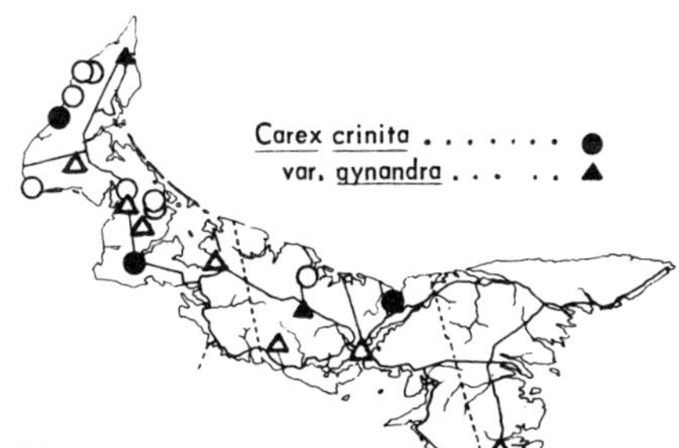

219

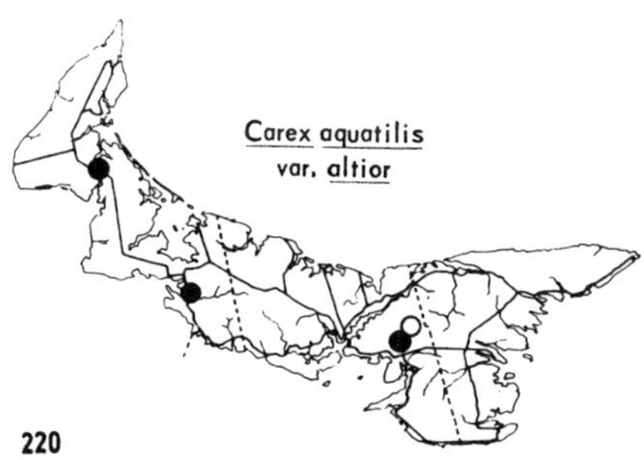

220

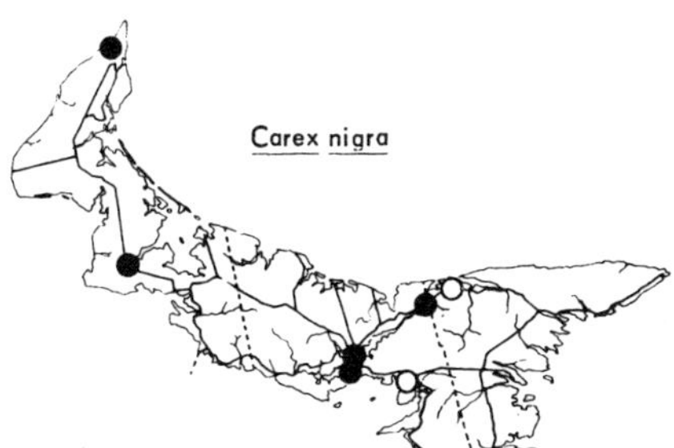

221

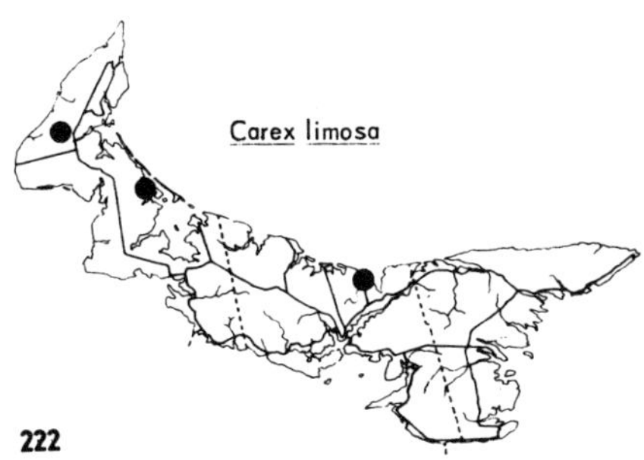

222

Carex hirta L. Spring Park, Charlottetown; a naturalized European species. Damp, pastured fields (*Fernald & St. John* in 1912); waste ground, tallest on sandy banks (1953). Adams, Hurst 1940.

* ***Carex pallescens*** L. var. ***neogaea*** Fern. Map 225. Local: O'Leary, Charlottetown, Bear River, Georgetown. In a natural habitat at Bear River: damp, grassy bank around pool in open woods; elsewhere in old fields.

[*Carex vestita* Willd., reported by MacSwain and Bain, unsubstantiated, and unaccounted for.]

****Carex gracillima*** Schwein. Map 226. Western Prince, east to Ellerslie. Mucky stream-banks, usually in partial shade.

Carex arctata Boott Map 227. Throughout; more common in the hill districts. Slopes and stream banks in woods, never in moist sites. Adams, Hurst 1940.

Carex debilis Michx. var. ***rudgei*** Bailey Map 228. Throughout, scattered. In damp woods, dark green, lax plants with few fruiting culms occur; in sunny clearings, yellowish green, stiffer plants with many fruiting culms. (*C. flexuosa* Muhl., Adams); Hurst 1940.

* ***Carex livida*** (Wahl.) Willd. var. ***grayana*** (Dewey) Fern. Bristol, floating mat of *Typha* bog.

Carex leptonervia Fern. Map 229. From west to east, in Prince County mesophytic, along streams in low woods; in Queens and Kings, dry mesophytic, along streams and in woods of the well-drained hill regions. Adams, Hurst 1940.

* ***Carex flava*** L. Map 230. Rare: Portage, Cherry Valley, Bay Fortune. Open swamps, cattail marsh.

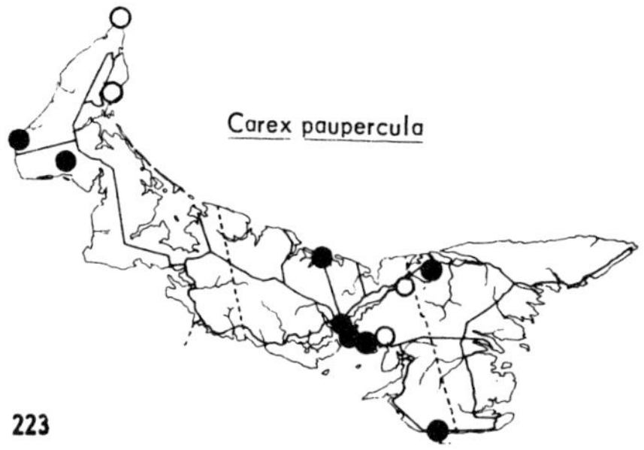

223

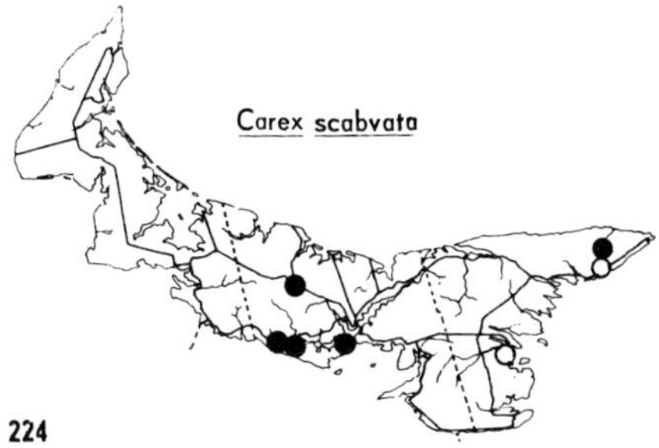

224

225

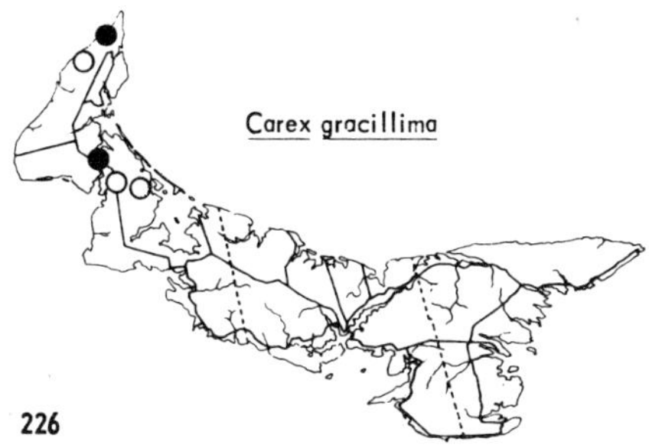

226

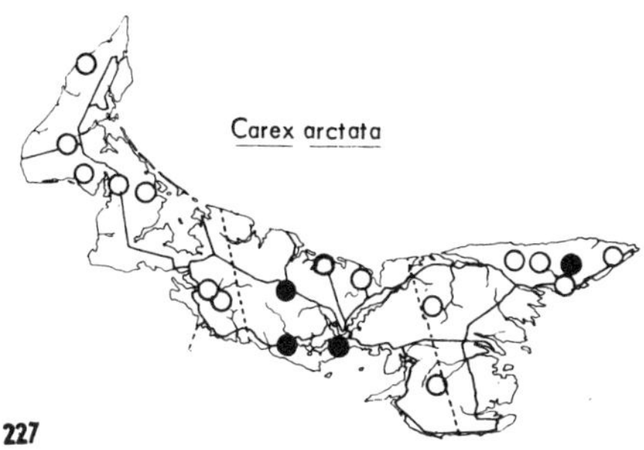

227

228

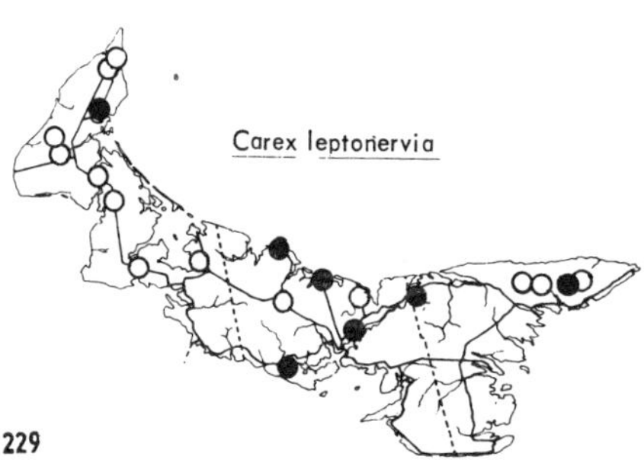

229

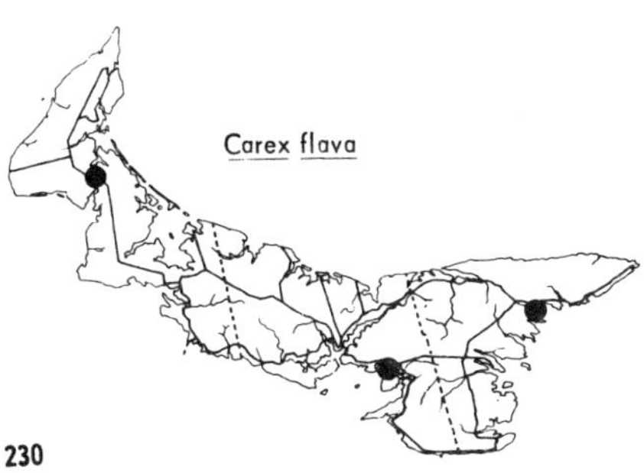

230

Carex viridula Michx. Map 231. Local: Prince County and eastward along the North Shore. Open, damp or wet sites. Often calciphilous (Fernald 1950); usually in dune slacks, and at Wellington, in a hardwater–region, in clay fields; at Portage in floating mat around pond. Producing fresh fruiting culms all summer long. Adams, Hurst 1940.

Carex lacustris Willd. Map 232. Rare: Alberton and O'Leary, Prince County, reported found at New Glasgow by Sterling Campbell. Forming a zone in shallow water around ponds. Hurst 1940.

Carex pseudo-cyperus L. Map 233. Throughout; lowland and coastal, never upland in sandy soils. Muddy deposits of bogs, swamps and pond margins. MacSwain and Bain, Hurst 1933.

Carex hystricina Muhl. Map 234. West Prince: Bloomfield, Nail Pond; also Murray River in Kings. Damp sites: peat bog, and stream bank. The yellow inflated perigynia and the complete infertility of clumps from shores of the Howlan millpond suggest a hybrid with ***C. vesicaria***. (***C. lurida*** var. ***gracilis***, Roland, not (Boott) Bailey).

Carex lurida Wahl. Mapped by Roland (1947) from Kings; no collection seen. Grassy stream banks. (***C. tentaculata*** Muhl., MacSwain and Bain).

* ***Carex folliculata*** L. Map 235. North Point, the Black Marsh; swampy margins of ***Picea mariana*** bog.

Carex intumescens Rudge var. ***fernaldii*** Bailey Map 236. Common in Prince County; rare eastward: Marshfield, St. Charles. Damp woods, stream bank thickets. Adams, Hurst 1940.

Carex retrorsa Schwein. Map 237. Prince County; rare eastward, Southport (***Macoun***) and Peakes Station. Mucky stream banks; forming clumps. Adams, Hurst 1940.

Carex rostrata Stokes var. ***rostrata*** Map 238. St. Charles (K), in boggy railroad ditch in coniferous woods.

231

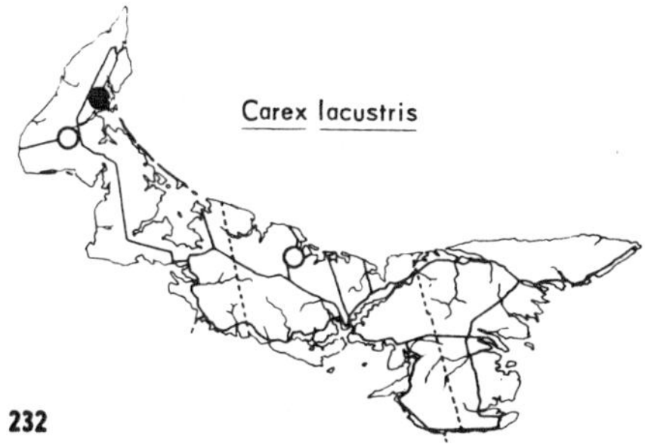

232

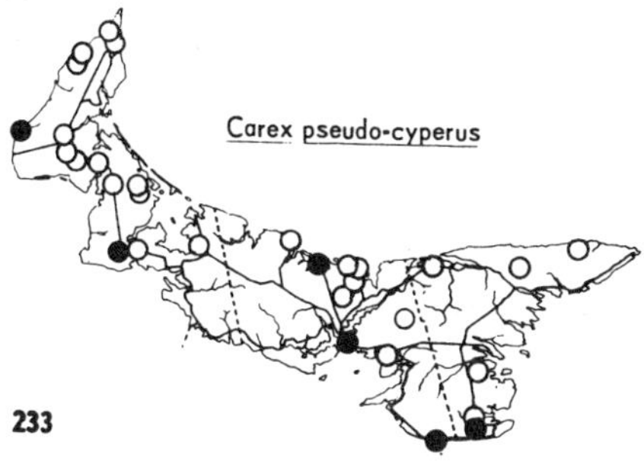

233

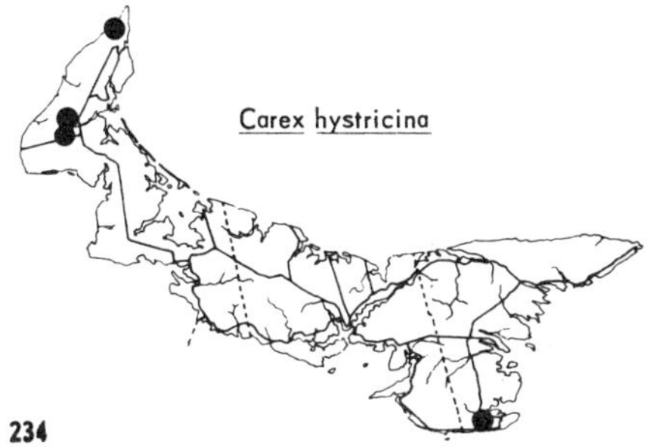

234

235

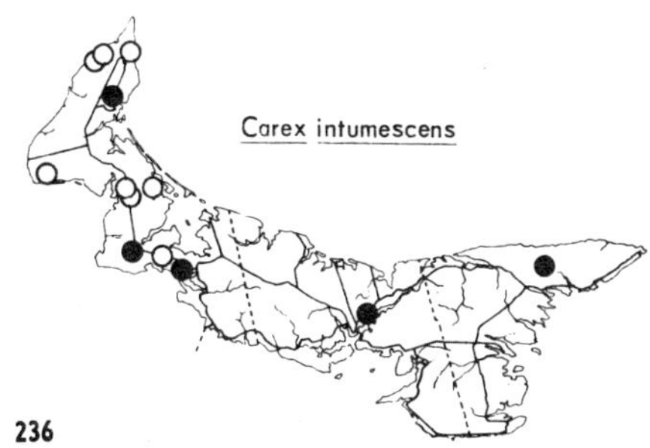

236

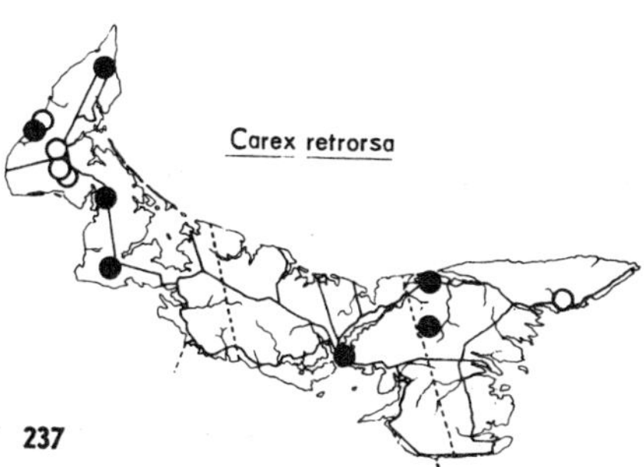

237

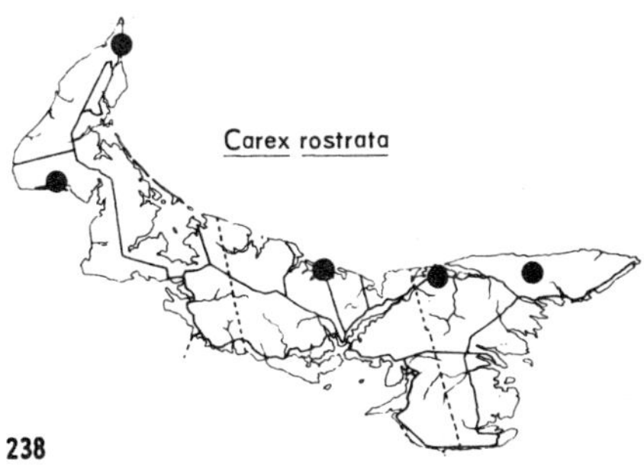

238

Var. ***utriculata*** (Boott) Bailey Coastwise, scattered from West Point to Bristol. Boggy pond margins. Adams, Hurst 1940.

Carex vesicaria L. var. ***jejuna*** Fern. Map 239. Prince County, rare. Damp sites: shores, thickets, old fields.

ARACEAE

Arisaema stewardsonii Britton Map 249. Common in Prince County, scattered eastward on northern streams. Alder thickets, muck along streams; "rich swamps" - Bain. (*A. triphyllum*, MacSwain and Bain, MacSwain, Hurst, Campbell).

Calla palustris L. Map 241. West Prince; east-central lowlands. Shallow water on mucky bottom: ponds, creeks and alder swamps. MacSwain and Bain, MacSwain, Hurst.

Acorus calamus L. Map 242. Local, coastal: Sea Cow Pond near Tignish; Wellington; Long Pond and Dalvay in the National Park; North Lake (K). Shallow standing water, especially where well-manured; "wet bogs" - Bain. MacSwain and Bain, MacSwain, Hurst.

LEMNACEAE

Spirodela polyrhiza (L.) Schleid. Map 243. Grand Tracadie, large pond (*Fernald, Long & St. John*); Pisquid Pond; Watervale pond.

Lenma triscula L. Map 244. West Prince; northeast Kings. Fresh-water ponds.

Lemna minor L. Map 245. Lowlands throughout, but rare in Prince County. Slow streams and ponds. Perhaps more abundant in streams draining cultivated and hence fertilized regions, as it depends (Eaton in Rhodora 49: 165. 1947) on the nitrate and phosphate content of the water. MacSwain and Bain, MacSwain, Hurst.

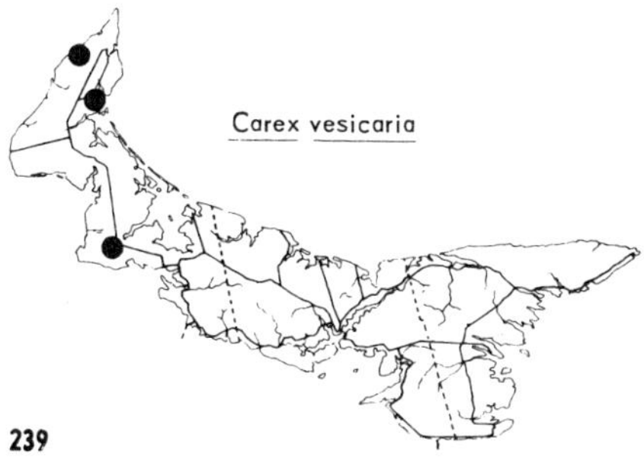

239

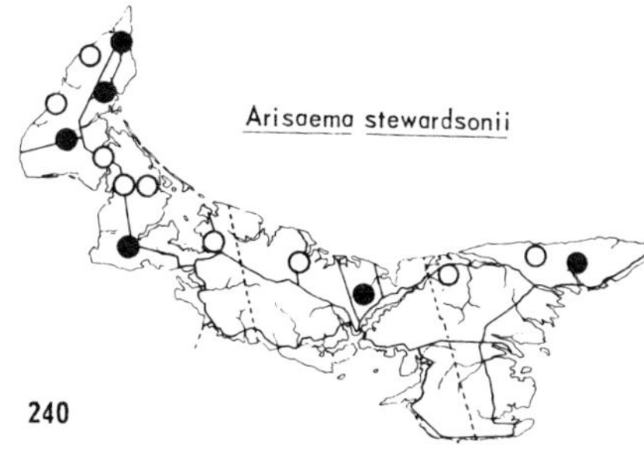

240

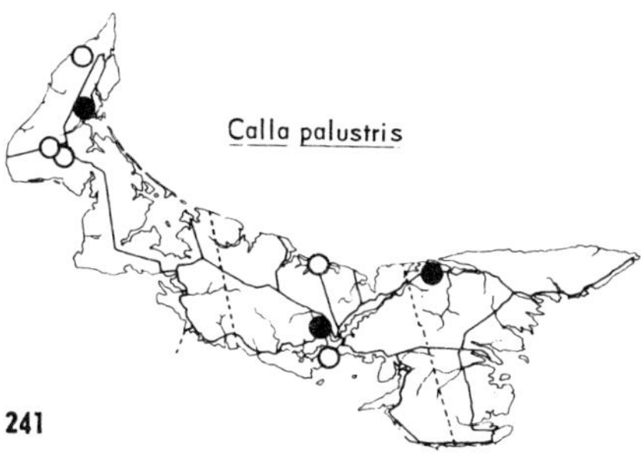

241

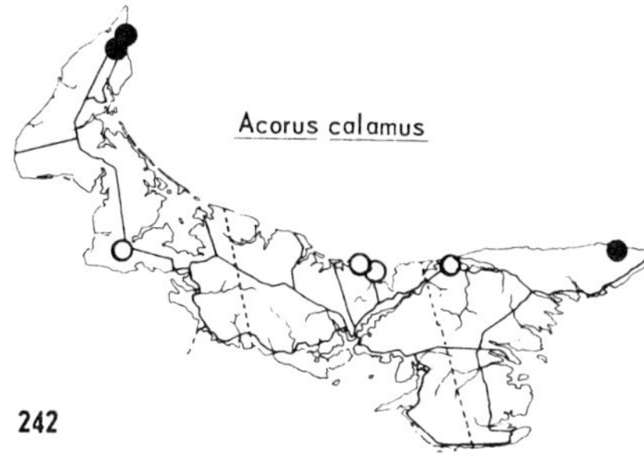

242

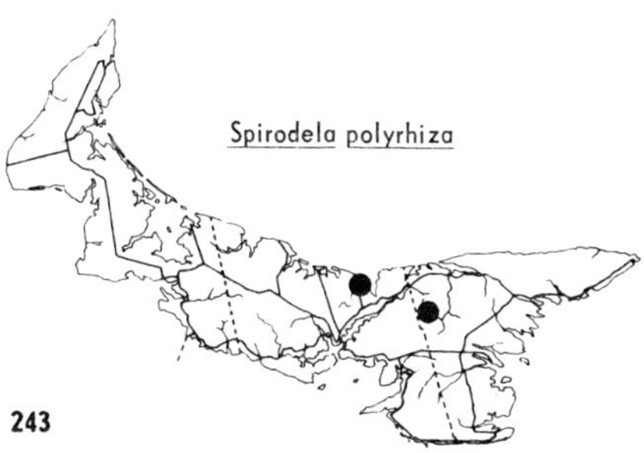

243

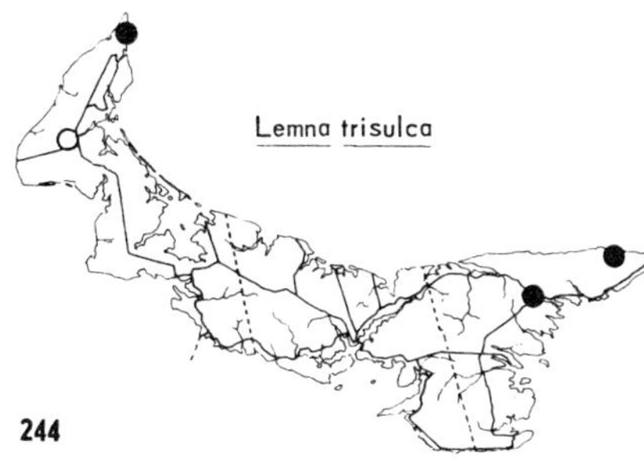

244

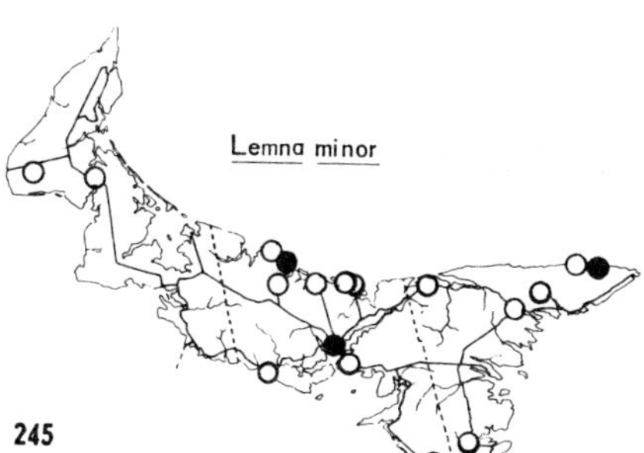

245

246

ERIOCAULACEAE

Eriocaulon septangulare With. Map 246. Lake Verde and Keefe's Lake. Submerged rosettes in shallows of lake shores. (*E. pellucidum* Michx.).

PONTEDERIACEAE

Pontederia cordata L. Map 247. Lake Verde, mucky shores in shallow water; forming beds, and represented mainly by *f. angustifolia* (Torr.) Solms. Collected by Fernald and St. John. Listed by Hurst as an addition to the flora by Walter Reddin, who successfully introduced it to a pond at Southport in 1930. Hurst.

JUNCACEAE

Juncus bufonius L. Map 248. Common throughout. Abundant weed on bare damp ground: roadside gutters, cart tracks, footpaths, damp coastal sands. An annual unable to stand competition, it seems out of place in "spring fed marsh, Southport," as reported by Fernald & St. John. 10 Lots (Groh). Hurst.

Juncus compressus Jacq. Salt marsh, Charlottetown (*Fernald et al.*) (Fernald 1950).

Juncus gerardii Loisel. Map 249. Local, around the coast. On salt marsh at some places (notably Souris and North Rustico) forming an upper zone. Rarely in nonsaline sites: in O'Leary, clay at roadside.

Juncus tenuis Willd. Map 250. Common throughout. Bare damp ground liable to drying, as with *J. bufonius*, but, as a perennial, persistent in damp fields and native perhaps in marshy spots. 7 Lots (Groh). MacSwain and Bain, MacSwain, Hurst.

Juncus dudleyi Wieg. Map 251. "Scattered, eastern P.E.I." – Roland (1947); rare in Prince County: Skinners Pond and Wellington. Damp clay fields.

* ***Juncus greenei*** Oakes & Tuckerm. Wellington, drier parts of a runout hayfield.

Juncus filiformis L. Map 252. Rare: Mount Pleasant, Wellington, Charlottetown Experimental Farm, Connaught Station, Wood Islands (?). Damp sand, often in ditches, sometimes in thin peat.

Juncus effusus L. vars. Map 253. Common throughout. Damp open ground: swamps, stream marshes, wet meadows or fields trodden by cattle in which often the only species left ungrazed.

Var. ***effusus*** Lot 40 (Bristol), black spruce swamp, apparently a shade form. Common, dikeland and meadows.

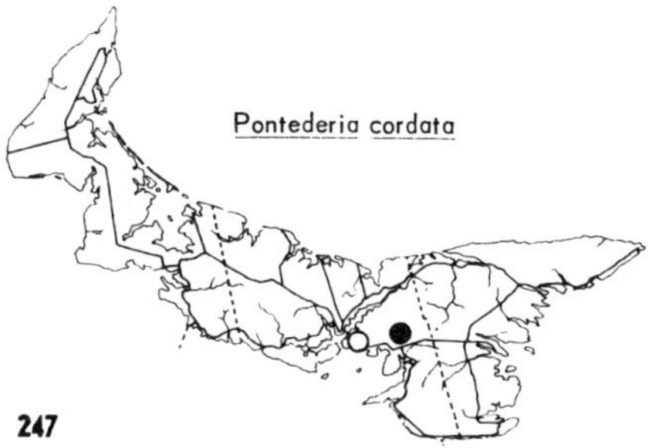

247

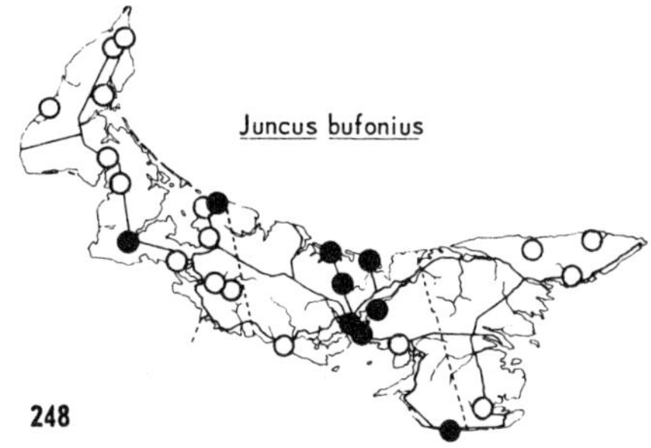

248

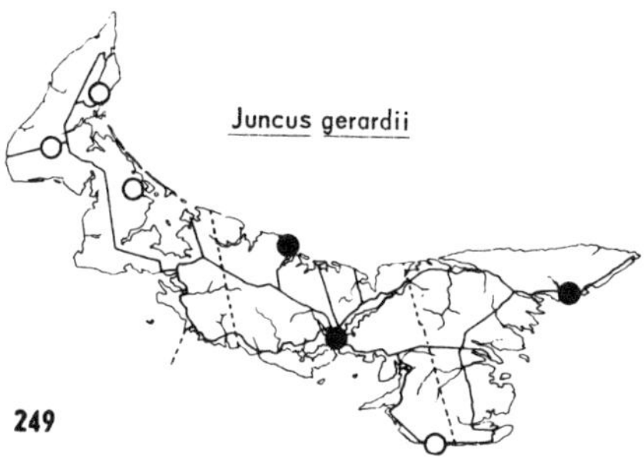

249

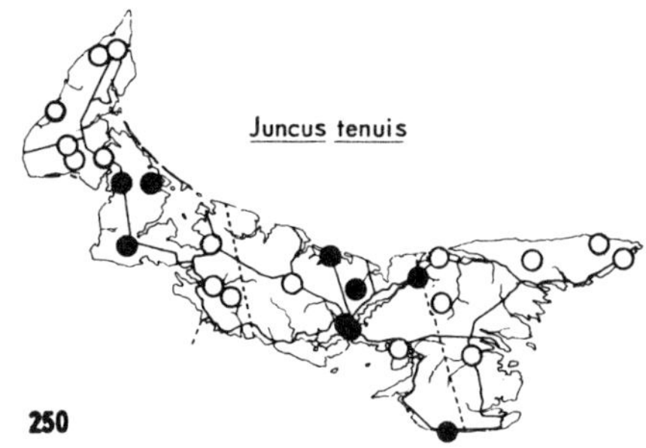

250

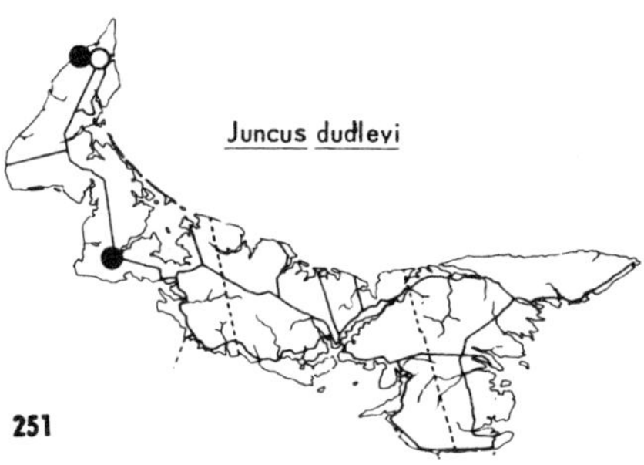

251

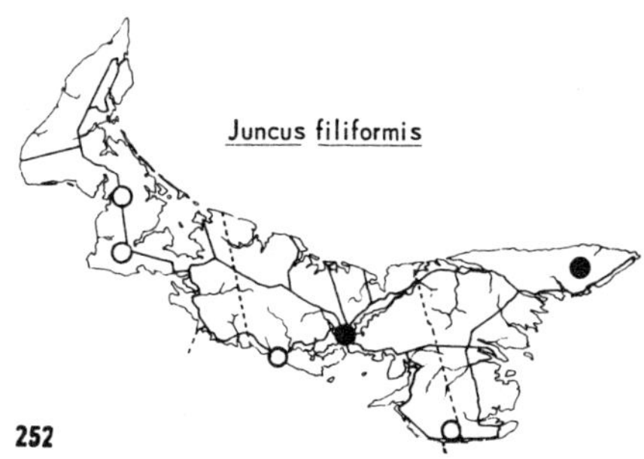

252

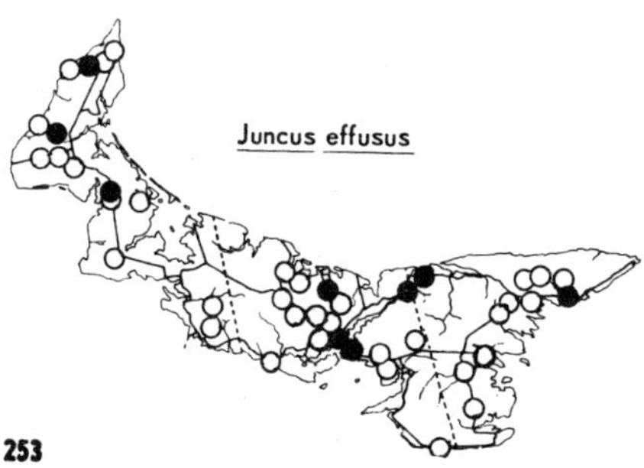

253

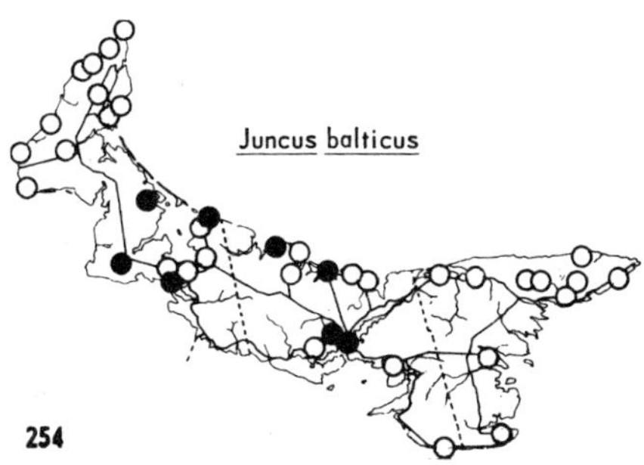

254

Var. **compactus** Lej. & Court. Common; appears to tolerate wetter habitats than the associated var. *solutus*. However, even the collections indicate that they are found together at all stations.

Var. **pylaiei** (Laharpe) Fern. & Wieg. More local: Bloomfield, Charlottetown, Mount Stewart. Slender type most typical of swamp sites.

Var. **costulatus** Fern. Mount Stewart (*Fernald et al.*), rather like the last.

Var. **solutus** Fern. & Wieg. Most common and widespread variety. A lax form of thickets and open swamps. 10 Lots (Groh), including records of *J. balticus*. MacSwain and Bain, MacSwain, Hurst.

Juncus balticus Willd. var. ***littoralis*** Engelm. Map 254. All coasts; in Prince County and in the northeast inland up to about 125 feet above sea level. Damp sites: very characteristic of the uppermost zone of salt marsh in contact with fresh water marsh or seepage; at Souris farther out on salt marsh following fresh-water springs. Around dune slacks, ponds, marshes, ditches and fields. (*Juncus Roemerianus*, MacSwain and Bain, Bain 1892 (as new!); Churchill, Hurst 1933.

[*Juncus marginatus* Rostk. Not seen; not recorded east of Annapolis County, Nova Scotia. Bain 1892.]

Juncus nodosus L. Map 255. Local: more common in Prince County, scattered in southeast and on North Shore. Damp sand (shores and ditches), marshes (on clay or organic mats).

Juncus canadensis J. Gay Map 256. Rare: Bloomfield, Bristol, Wood Islands. Thin peat, or, in muddy areas denuded of it, growing more thickly.

Juncus brevicaudatus (Engelm.) Fern. Map 257. Throughout: absent from the uplands. Muddy shores, swamps or marshes; now an abundant weed of muddy roadsides and ditches, and of overgrazed damp pastures.

[*Juncus acuminatus* Michx. Not seen; not recorded north of southern Nova Scotia. Hurst's record (1940) based on a specimen of *J. articulatus*.]

X ***Juncus fulvescens*** Fern., an abundant sterile hybrid of *J. brevicaudatus* with *J. articulatus* of similar habitats has been collected at Tignish.

Juncus articulatus L. Map 258. Common throughout. Fresh to brackish marshes; muddy stream banks. Most abundant as a weed of ditches, damp roadsides and fields. On brackish marshes and sand flats as at Wood Islands, var. *obtusatus* Engelm. may replace var. *articulatus*. Hybridizes with *J. brevicaudatus*, from which it is quite easily distinguishable in growth form. Each dense clump is surrounded underground by a circle of pale, conical buds which will become new short rhizomes. In *J. brevicaudatus*, each clump is a tuft, producing the new shoots centrally. (*J. acuminatus*, Hurst 1940); Hurst 1941.

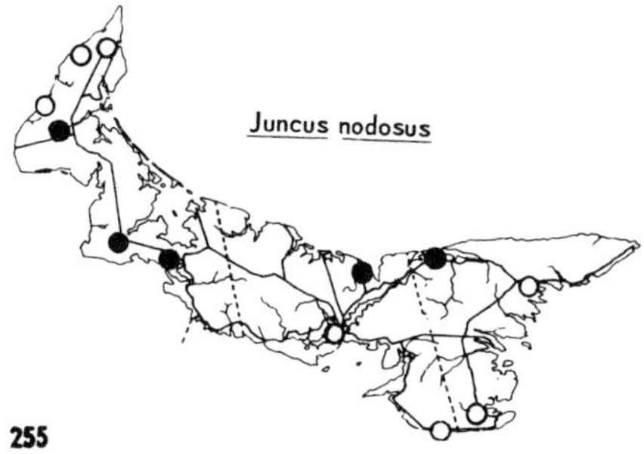

255

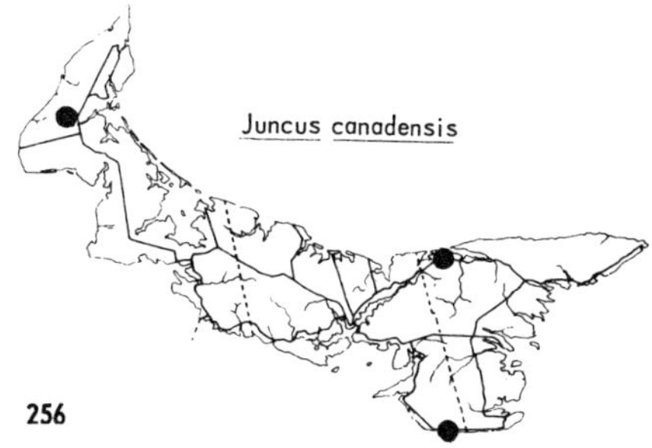

256

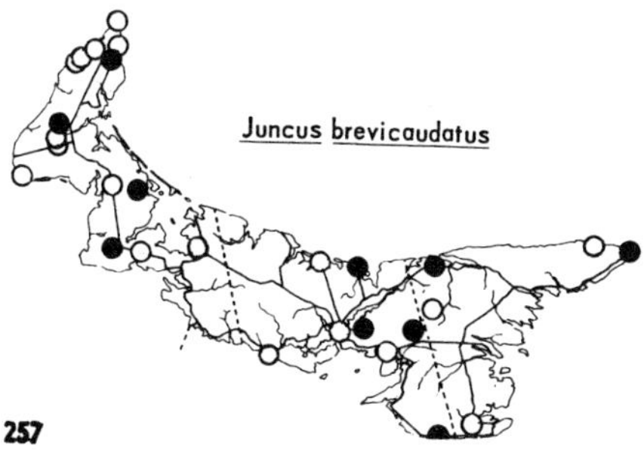

257

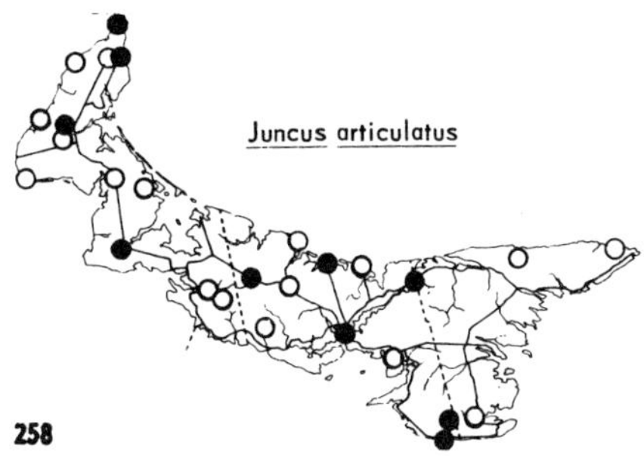

258

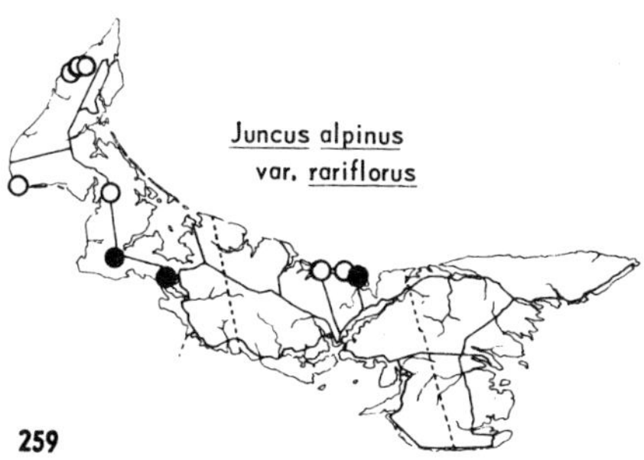

259

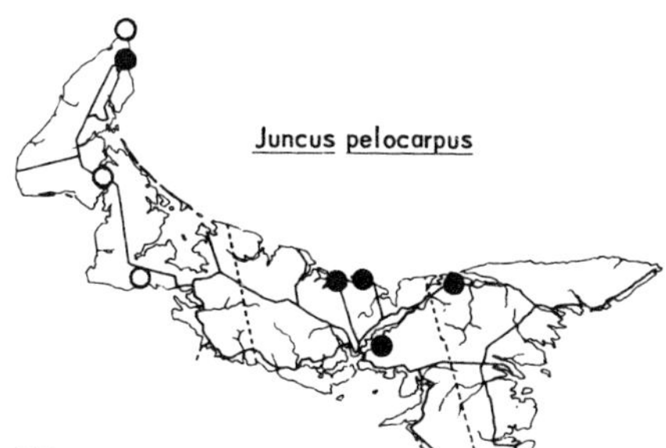

260

Juncus alpinus Vill. var. ***rariflorus*** Hartm. Map 259. Prince County and North Shore east to Dalvay. Damp open sites: dune slacks; fresh marsh, often where bordering on brackish marsh.

Juncus pelocarpus Mey. Map 260. West Prince; east-central lowlands. Marshy or mucky sites, in the dune slacks on the North Shore. Hurst 1940.

Luzula acuminata Raf. Map 261. East Point to Morell along the hardwood ridge; one station five miles southwest of Tignish. Dry mesophytic, shady sites: hardwood or old-age spruce forest, thickets in sandy soil. (*L. pilosa*, MacSwain and Bain).

[***Luzula spicata*** (L.) DC., an Arctic montane species, not occurring south of Gaspé, is an impossible record, but may have referred to *L. multiflora*. (Includes var. *racemosa*, MacSwain and Bain); with habitat "fields", MacSwain, Hurst).]

[***Luzula arcuata*** Sw., a species of northern Europe; recorded by MacSwain and Bain, dropped by Hurst.]

Luzula multiflora (Retz.) Lej., vars. Map 262. Throughout; less common in West Prince. Light soil; open sites: abundant as a weed of old fields. 19 Lots (Groh).

Var. ***multiflora*** Open fields; the common variety. (*L. campestris*, MacSwain and Bain, MacSwain; *L. campestris* var. *multiflora* (Retz.) Celak., Groh, Hurst; *L. campestris* var. *frigida*, Groh (!), Hurst 1940?); Campbell.

Var. ***acadiensis*** Fern. Eastward; also Miscouche. More often in partial shade: in dry thickets; persisting from old field to pasture spruce woods. The type collection from St. Peters Bay South.

Var. ***fusconigra*** Čelak. Similar or slightly damper habitats; often along roadsides or in clearings. (*L. campestris* var. *frigida*, Hurst 1940?); Campbell.

LILIACEAE
(including SMILACACEAE, Bain)

[***Helonias dioica*** L. MacSwain and Bain, is an impossible record; dropped by MacSwain.]

[****Hemerocallis fulva*** L. Map 263. ORANGE DAY-LILY Gardens, commonly introduced as a self-perpetuating ornamental and sometimes outlasting the house; occasionally on garden dumps.]

[***Asparagus officinalis*** L. var. ***altilis*** L. Map 264. Gardens, introduced as an ornamental; escaping to sandy roadside banks, Queens County (Groh). Hurst.]

261

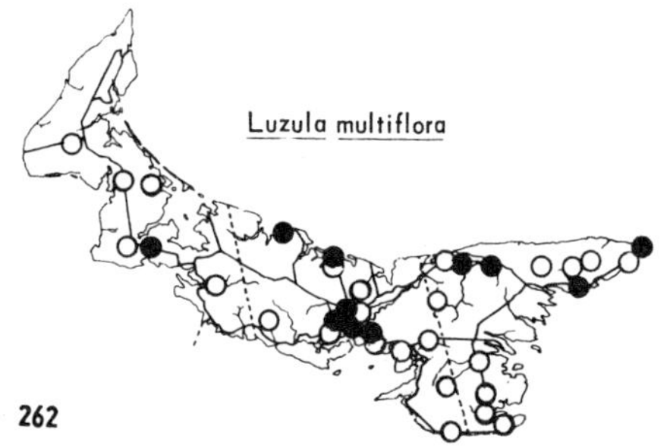

262

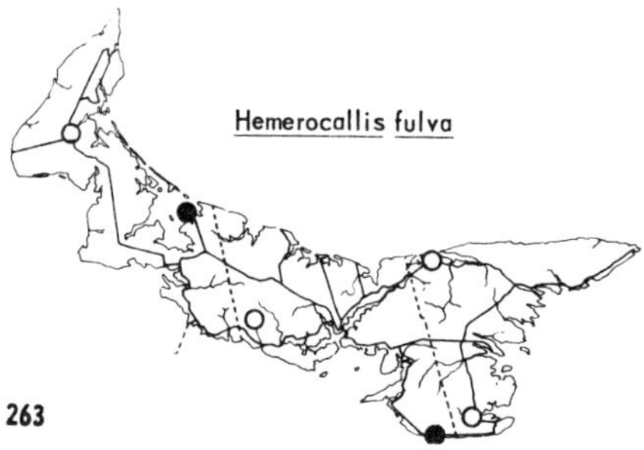

263

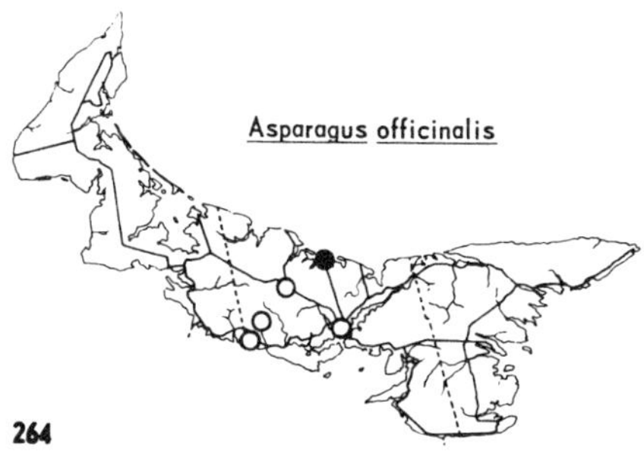

264

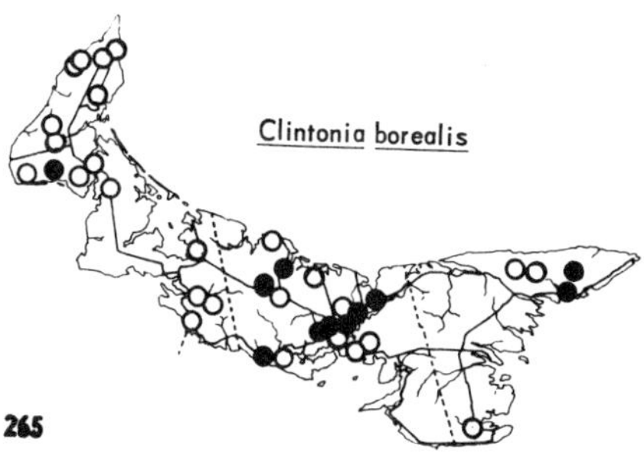

265

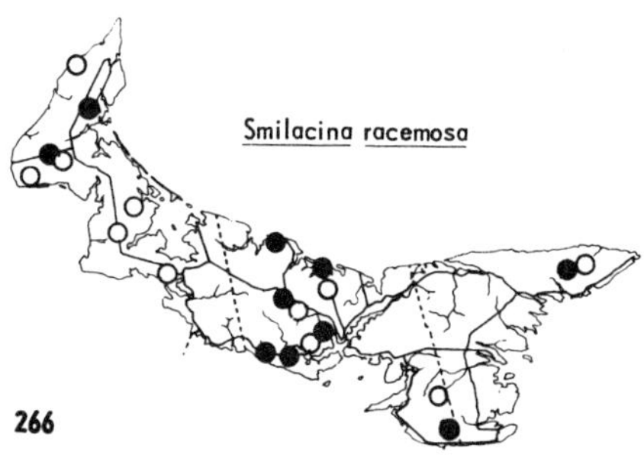

266

Clintonia borealis (Ait.) Raf. Map 265. Throughout; absent from lowlands and rare in the southeast. Fresh to dry woodland: forming beds in mixed and hardwood forest. The flowers are often borne in a single umbel but often with a smaller secondary umbel below the primary. (Including *C. umbellulata,* MacSwain and Bain, referring to the form with secondary umbels); MacSwain and Bain, MacSwain, Hurst.

Smilacina racemosa (L.) Desf. Map 266. Throughout; in the east restricted to uplands. Well-drained woods: mixed or hardwood forest and ravines. MacSwain and Bain, MacSwain, Hurst.

Smilacina stellata (L.) Desf. Map 267. Rare, coastal. Low inner dunes, or at Cavendish, along a drain. MacSwain and Bain, Churchill, MacSwain, Hurst.

Smilacina trifolia (L.) Desf. Map 268. Scattered: rare in Prince County, more frequent in eastern lowlands. Wet swamps and bogs. MacSwain and Bain, MacSwain, Hurst.

Maianthemum canadense Desf. Map 269. Common throughout. Characteristic of mossy floor under spruce forest in well-drained sites. Three-leaved specimens occur. (*Smilacina canadensis* (Desf.) Pursh, MacSwain and Bain); MacSwain, Hurst.

[***Convallaria majalis*** L. Map 270. Gardens, introduced as a self-perpetuating ornamental, flourishing in shade. Often spreading beyond garden; Summerside, by railway (dumped?). MacSwain and Bain.]

Streptopus amplexifolius (L.) DC. var. ***americanus*** Schult. Map 271. Frequent in Prince County and eastward in hilly regions. Mossy woods: ravines and cedar swamps. Hurst.

Streptopus roseus Michx. var. ***perspectus*** Fassett Map 272. Common, Prince and Western Queens; also in the northeast. Hardwood hills and mossy mixed woods. Bain 1892, MacSwain, Hurst.

Medeola virginiana L. Common, uplands of Kings and Queens; local in West Prince. Hardwood hills and mixed woods, in leaf mold. "Dry woods" (Bain). Bain unaccountably referred *Trillium* to the Smilacaceae, but left *Medeola* in the Liliaceae. (*M. virginica,* MacSwain and Bain); MacSwain, Hurst.

Trillium cernuum L. Map 274. Scattered; apparently absent from the southeast. Damp woods, along brooks or in thick leaf mold. MacSwain and Bain, MacSwain, Hurst. Used as a tonic (Bain).

[***Trillium erectum*** L. "Rare". Not known in the Island, nor from nearer than Baie Chaleur, New Brunswick. Introduced in a garden at Charlottetown. MacSwain, Hurst.]

267

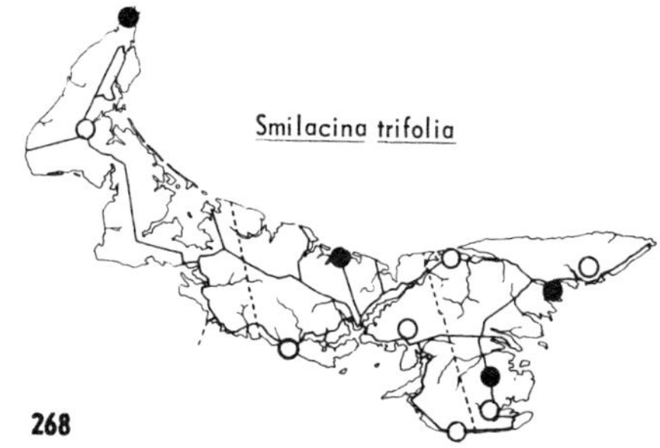

268

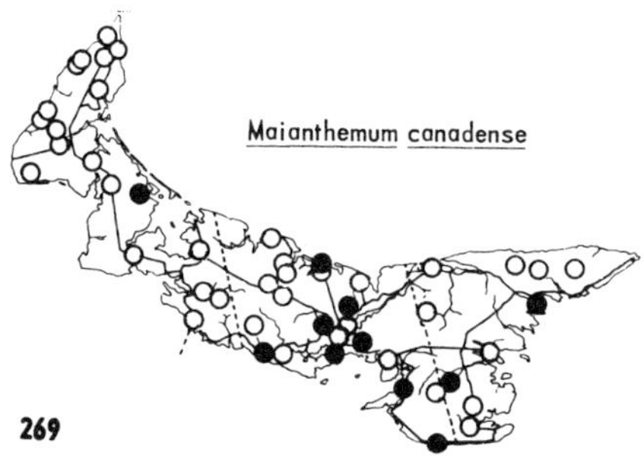

269

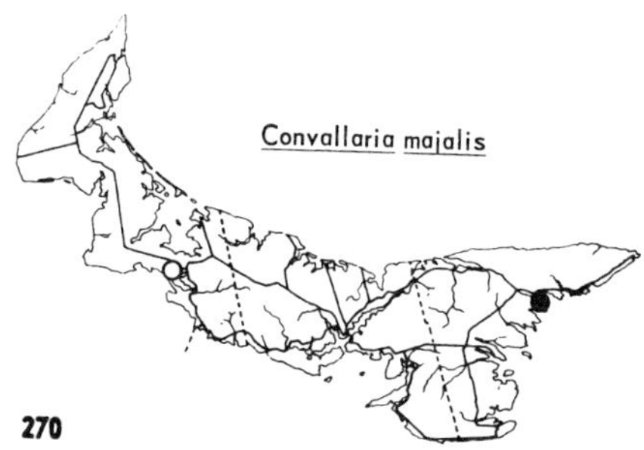

270

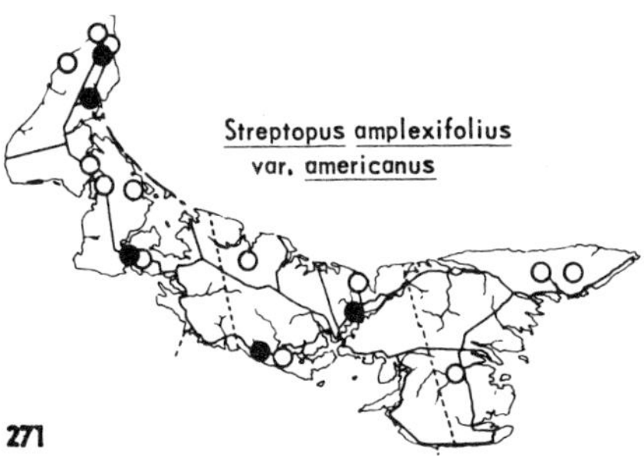

271

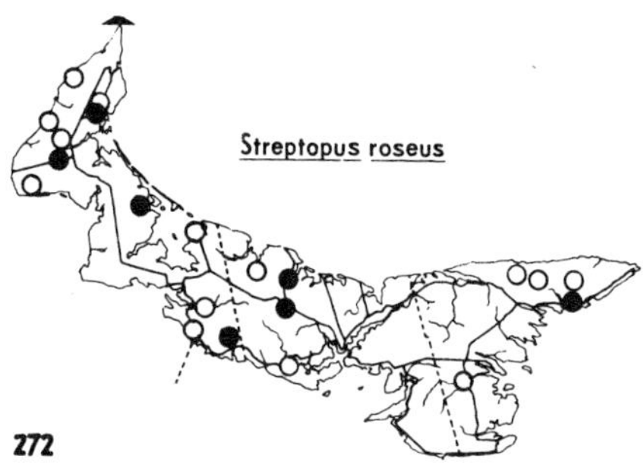

272

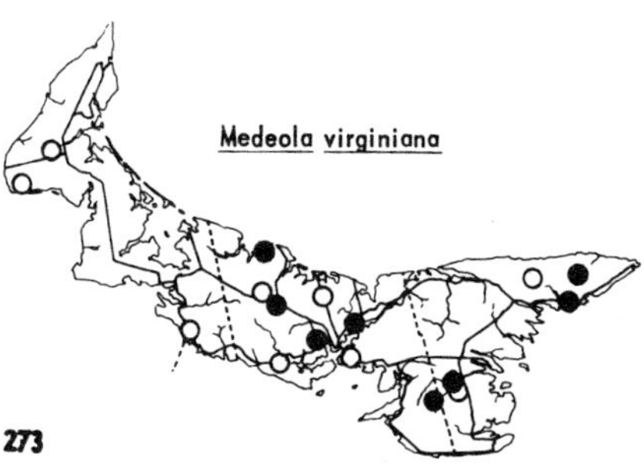

273

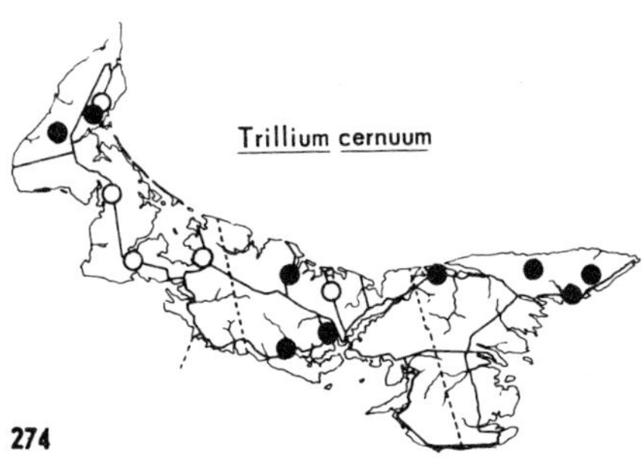

274

Trillium undulatum Willd. Map 275. Upland regions: common in the central and southeastern hills, rare in Prince and the northeast. Hardwood hills; dry deciduous woods, on light friable soil, which contrasts with its habitat in Nova Scotia on acid soil in spruce or pine woods. (*T.* ***erythrocarpum*** Michx., MacSwain and Bain, MacSwain); Hurst.

IRIDACEAE

Sisyrinchium montanum Greene var. **crebrum** Fern. Map 276. Rare in Prince; apparently absent from the southeast; common in Queens. Damp open sites: dune slacks; a weed of old fields and pastures. (*S.* ***Bermudiana*** var. ***anceps***, MacSwain and Bain; *S.* ***angustifolium***, MacSwain, Hurst, Campbell).

Iris hookeri Penny Map 277. North Shore: Nail Pond; National Park, Brackley Beach to Dalvay; Bothwell. Dunes and dune slacks. (*I.* ***Setosa*** var. ***canadensis*** Foster, Roland).

Iris versicolor L. Map 278. Throughout; more common in lowlands. Damp open sites: marshes, meadows pond margins, "fresh-water ponds" (Bain). Rootstocks used in medicine (Bain), poisonous but perhaps purgative.

F. ***murrayana*** Fern., the albino, from millpond south of O'Leary. MacSwain and Bain, MacSwain, Hurst.

Iris pseudacorus L. Map 279. Campbellton; Brackley Point (introduced by Hurst in 1939); near Suffolk (Q). Introduced and naturalized in streams and ditches.

[*Iris foetidissima* L., according to Hurst (Charlottetown *Guardian*, Mar. 3, 1939), shows a tendency to become naturalized at Brackley Point.]

ORCHIDACEAE

Cypripedium calceolus L. var. ***parviflorum*** (Salisb.) Fern. Map 280. YELLOW LADY'S SLIPPER Rare: cedar swamp along road from O'Leary to Milo (discovered by the Webbs); swampy woods southwest of Fortune Bridge (discovered by E.S. Blanchard), at both stations associated with *C.* ***reginae***. (*C.* ***parviflorum*** Salisb., Hurst 1940), Campbell.

Cypripedium reginae Walt. Map 281. SHOWY LADY'S SLIPPER West Prince, several stations; rare eastward: Dundee, Fortune Bridge. Mucky woods and thickets, often in larch-cedar swamp. Provincial emblem; first proposed by Hurst. (*C.* ***spectabile*** Salisb., MacSwain and Bain, MacSwain; *C.* ***hirsutum***, Hurst); Campbell.

Cypripedium acaule Ait. Map 282. Throughout. Acid plant litter: floor of dry or damp woods, swamps, and bog hummocks. "Dry beech woods" (Bain), but equally at home beneath spruce or pine. MacSwain and Bain, MacSwain, Hurst.

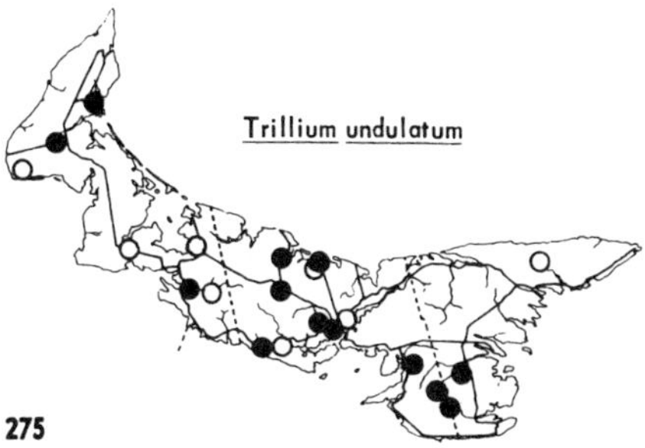

275

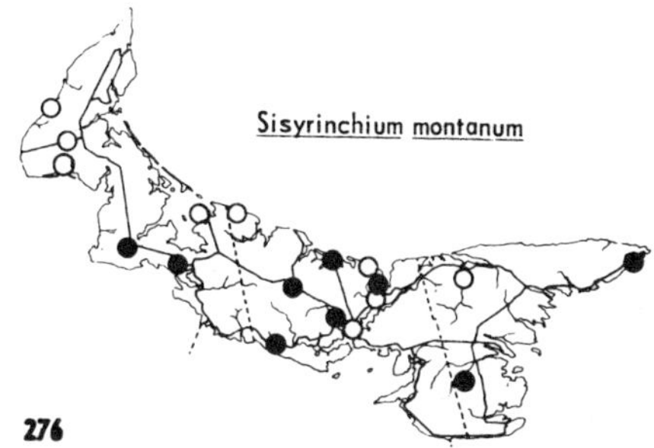

276

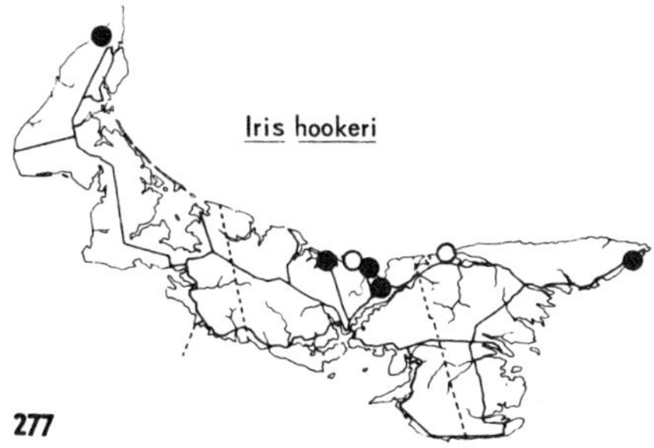

277

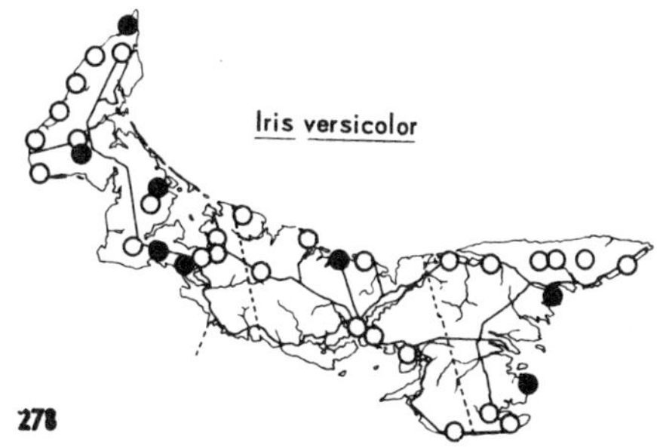

278

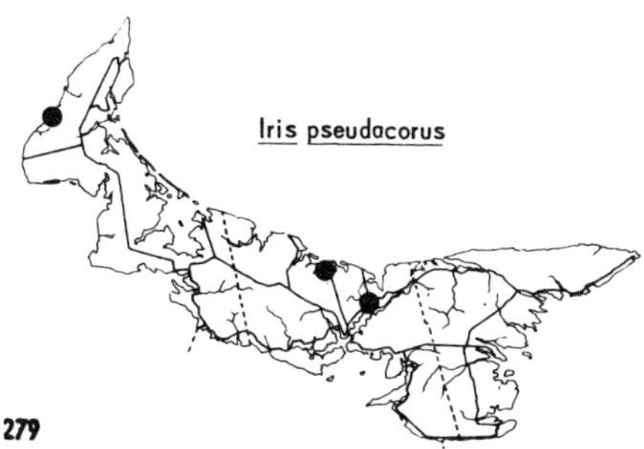

279

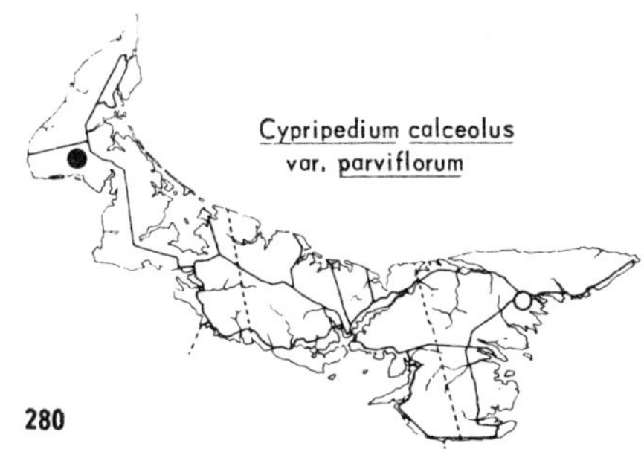

280

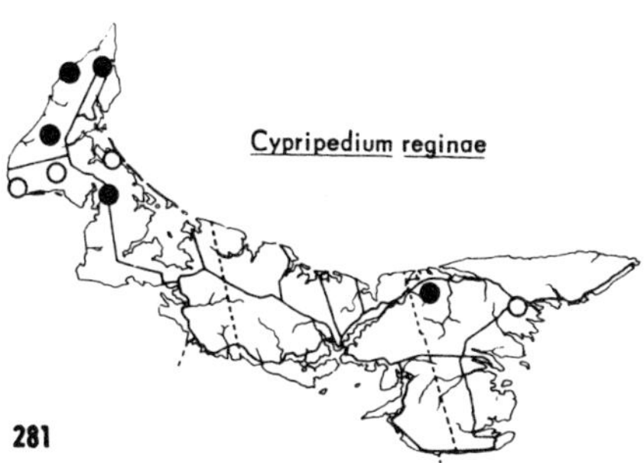

281

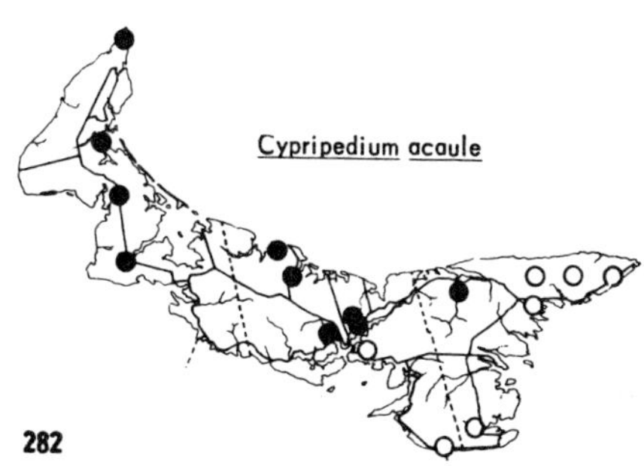

282

Habenaria clavellata (Michx.) Spreng. Map 283. Local: Sea Cow Pond near Tignish; mainly in southeast and east. Damp, open sites: thickets, runs; damp or boggy meadows. Represented by var. ***ophioglossoides*** Fern., with rounded leaf bases. (*Gymnadeniopsis* Rydb.; *Orchis tridentata* Pursh, Bain; *H. tridentata* (Pursh) Hook., MacSwain and Bain, MacSwain); Hurst.

Habenaria hyperborea (L.) R. Br. Map 284. Common in West Prince, scattered eastward: Miscouche Woods, Grand Tracadie, Bristol, Fortune Bridge. Marshy streambanks, bogs. Macoun labeled his collection from North Pond as "green form of *H. dilatata*", but the lip is scarcely dilated. Fernald calls our material var. ***huronensis*** (Nutt.) Farwell, as it is less dwarfed than the original *hyperborea*. (*Platanthera* Lindl.); Hurst 1940.

Habenaria dilatata (Pursh) Gray Map 285. Scattered in Prince County and eastward to the vicinity of Souris. Swampy woods; persisting if cleared. Hurst considered material from Brackley Point as var. *media* (Rydb.) Ames (X *H. media* (Rydb.) Niles), intermediate with *H. hyperborea*. (*Platanthera* Lindl.); Hurst 1933.

Habenaria hookeri Torr. Orwell (*J.W. Eastham*, 1912). Habitat in Nova Scotia, dry mixed woods. (*Platanthera* Lindl.); Bain 1892.

Habenaria orbiculata (Pursh) Torr. Map 286. Scattered in central uplands; also Bear River in northeast. Leaf mould in mixed woods. A specimen from Hunter River has the dimensions of *H. macrophylla* Goldie, a segregate of doubtful value. (*Platanthera* Lindl.); MacSwain, Hurst.

Habenaria obtusata (Pursh) Richards. Harmony, mixed forest floor (*Fernald & St. John*). (*Platanthera* Lindl.)

Habenaria blephariglottis (Willd.) Torr. Map 287. Rare: eastern lowlands; also reported from Black Banks. Peat bogs. (*Platanthera* Lindl.); MacSwain and Bain, MacSwain, Hurst.

Habenaria lacera (Michx.) Lodd. Map 288. Throughout, absent only from well-drained upland areas. Frequent in poorly drained meadows and runs, along woods, roads and ditches. (*H. psycoides*, Bain, MacSwain and Bain; *H. leucophaea*, MacSwain and Bain, MacSwain; *Platanthera lacera* (Michx.) G. Don); Hurst.

Habenaria psycodes (L.) Spreng. Map 289. Western Prince and from Charlottetown to Souris. Thickets along rivers, rarely elsewhere; meadows and marshes. Less restricted to open habitats than *H. lacera*. (*H. grandiflora*, Bain, MacSwain and Bain; *Platanthera psycodes* (L.) Lindl.); MacSwain, Hurst.

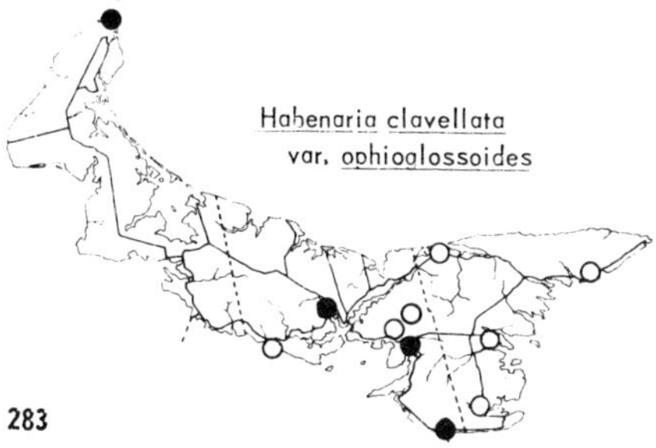

283

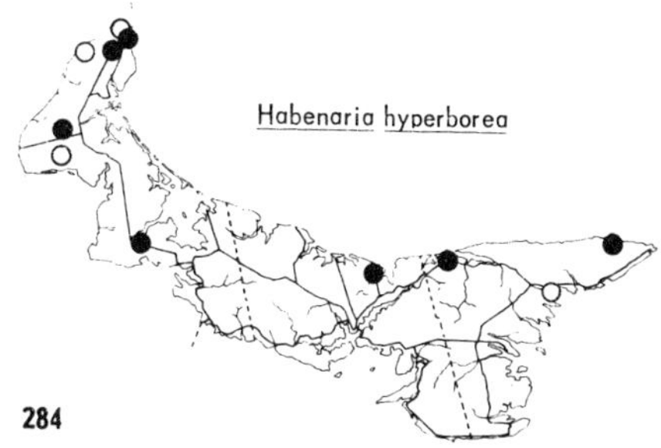

284

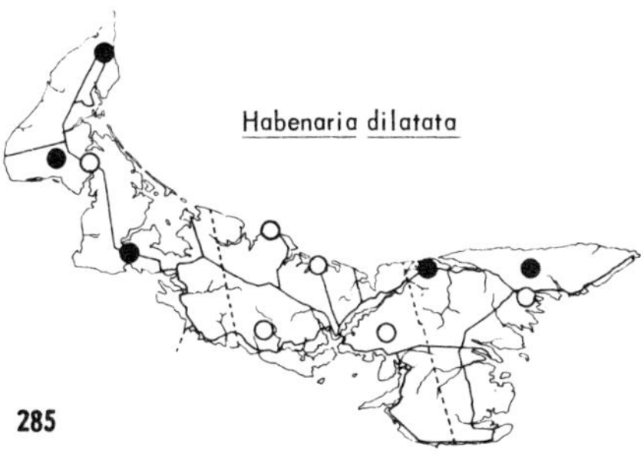

285

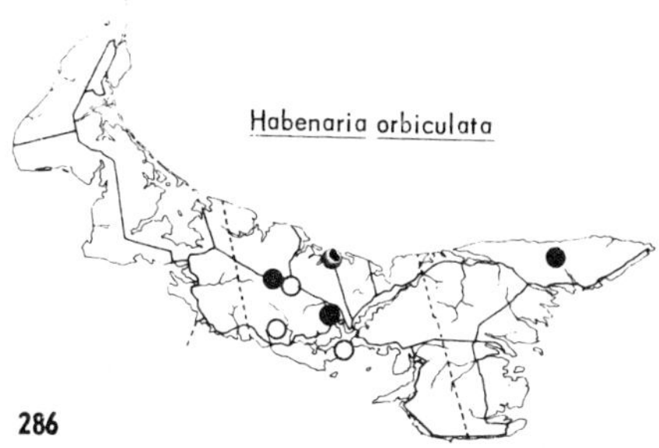

286

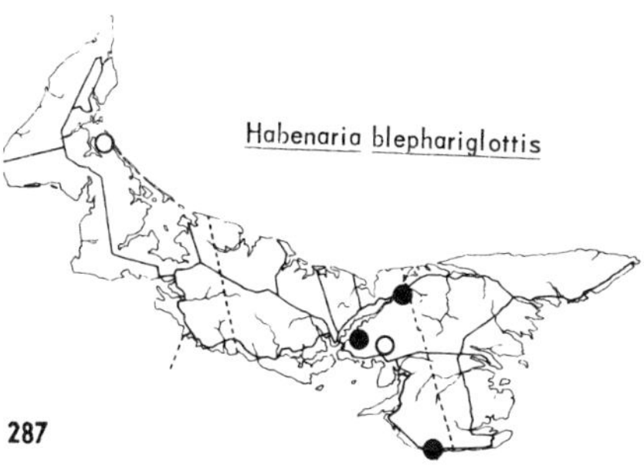

287

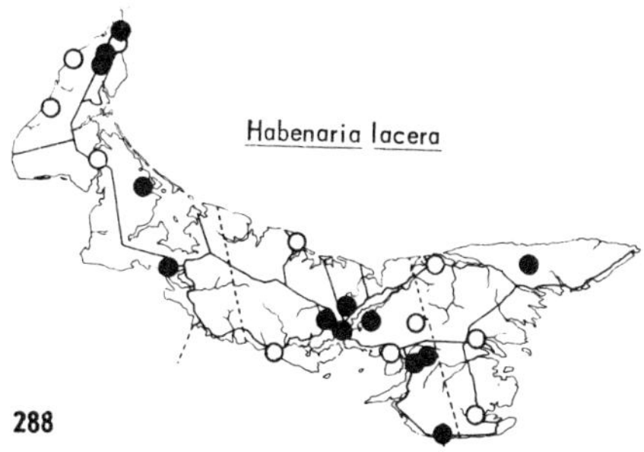

288

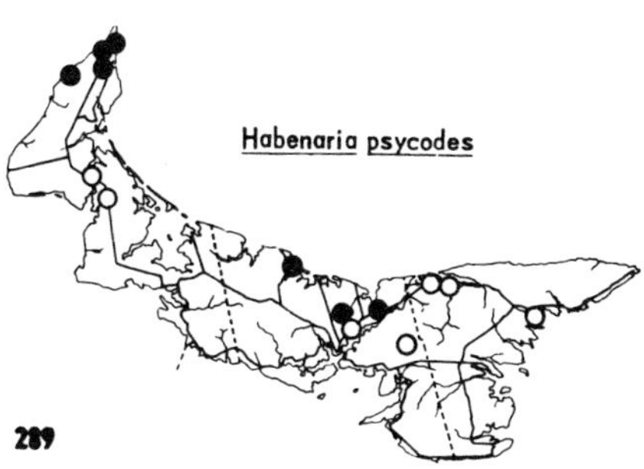

289

290

F. **albiflora** Rand & Redfield Black Pond (P). A few plants in *Calamagrostis* thicket along cliff; a few in woods along ravine above Pleasant View School. Pink-flowered (rather than magenta) forms along York River are basis of the X *H. Andrewsii* M. White reports of Hurst (1933).

[*Habenaria fimbriata* (Ait.) R. Br. *H. grandiflora* (Bigel). Torr. was listed by Hurst (1933) on the basis of Bain's record of *H. grandiflora,* which referred to *H. psycodes* (q.v.) Messervy's sight records from York and Bay Fortune refer to *H. psycodes,* the species present. Campbell's record from Campbell's Mills, New Glasgow, may refer to *H. fimbriata,* but needs confirmation.]

Pogonia ophioglossoides (L.) Ker Map 290. Western Prince County; Dundee; Bothwell. Peat bogs. Bain, MacSwain and Bain, MacSwain, Hurst.

Calopogon pulchellus (Salisb.) R. Br. Map 291. Western Prince; rare in the east; Mount Stewart, Dundee, Murray River. Swamps and peat bogs. Bain, MacSwain and Bain, MacSwain, Hurst.

Arethusa bulbosa L. Map 292. Rare in West Prince: Black Banks (*Messervy*); Waterford. Peat bogs. Hurst 1933, 1940.

Spiranthes lacera Raf. Map 293. Throughout, more common in lowlands. Open woods, and barrens; weedy in old fields. Along North Shore in banks kept grassy by exposure to sand and wind. (*S. gracilis,* MacSwain and Bain, MacSwain, Hurst, Campbell).

Spiranthes lucida (H.H. Eaton) Ames Heatherdale grayed alder thicket along Brown's Creek (*S. latifolia* Torr., MacSwain and Bain, *S. plantaginea* (Raf.) Torr., Roland).

Spiranthes romanzoffiana Cham. Map 294. Central Queens County; Alberry Plains; Bloomfield; Portage (P). Dune slacks, damp old fields and thickets. MacSwain and Bain, MacSwain, Hurst.

Goodyera repens (L.) R. Br. var. **ophioides** Fern. Map 295. Brackley Point; Wood Islands West. Damp coniferous woods. (*G. repens,* MacSwain, in part, Campbell; *Epipactis repens,* Hurst).

Goodyera tesselata Lodd. Brackley Point; Cornwall. Mixed or coniferous woods. (*G. repens,* MacSwain, in part).

Listera cordata (L.) R. Br. Map 296. Harmony, mossy spruce woods (*Fernald & St. John*). MacSwain and Bain.

Listera convallarioides (Sw.) Nutt. Map 297. Rare, central and northeastern uplands; Hunter River; Dundee, Harmony and Glencorradale. Humus, deciduous woods.

Corallorhiza trifida Chatelain var. **verna** (Nutt.) Fern. Map 298. Western Prince County; very local eastward. Dry to damp coniferous woods (as a saprophyte). (*C. innata,* Bain 1892; *C. trifida,* Hurst).

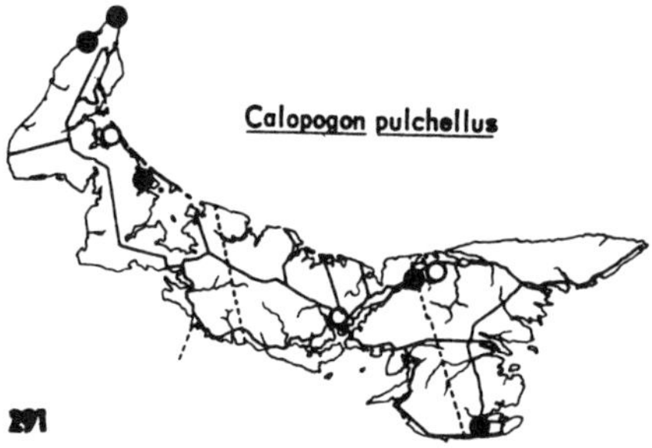

291

292

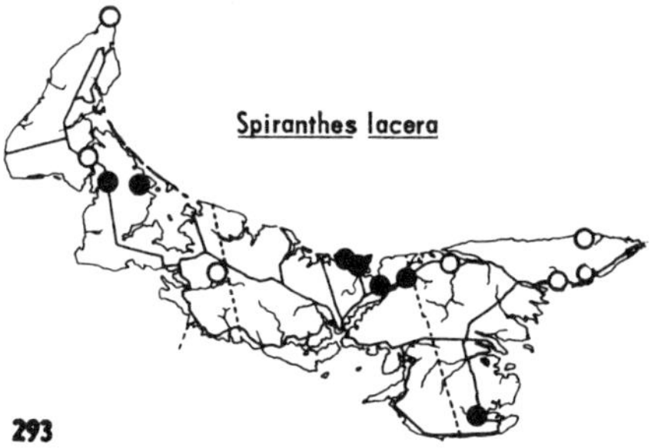

293

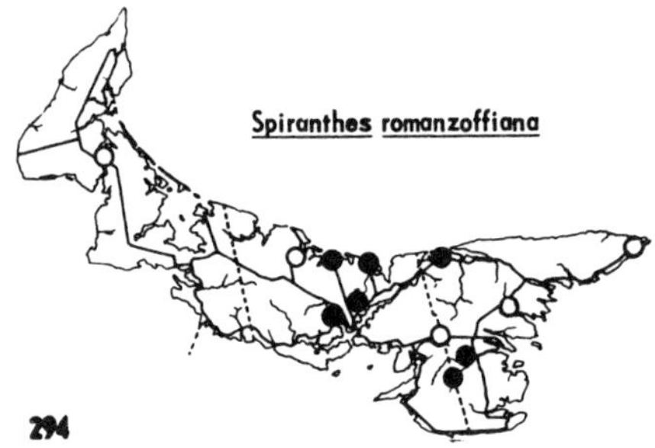

294

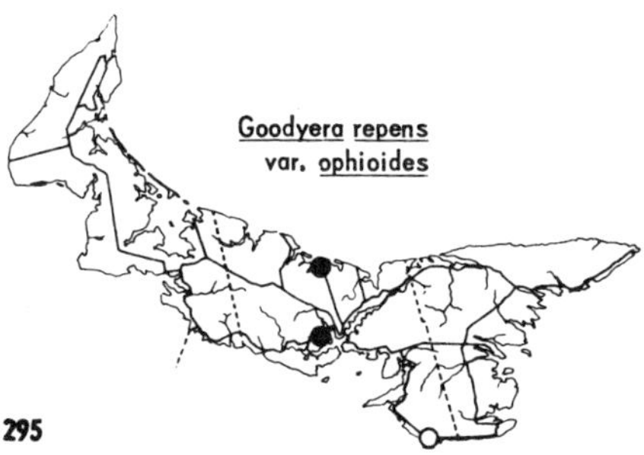

295

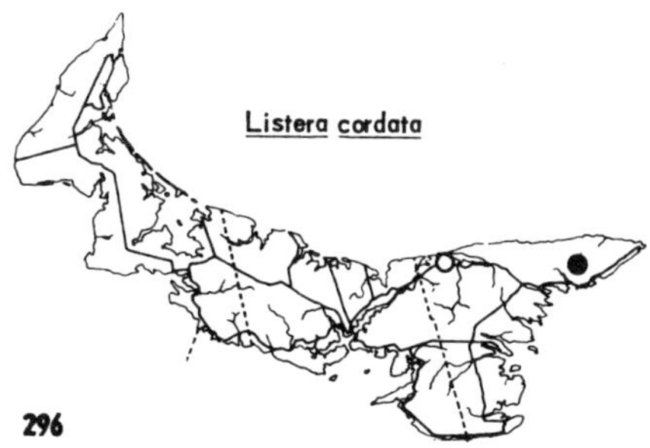

296

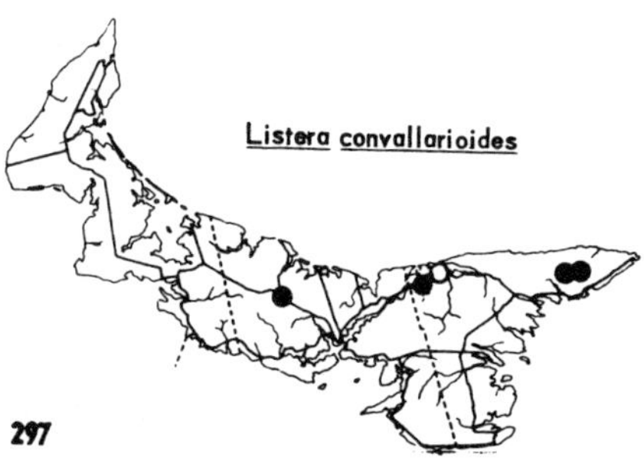

297

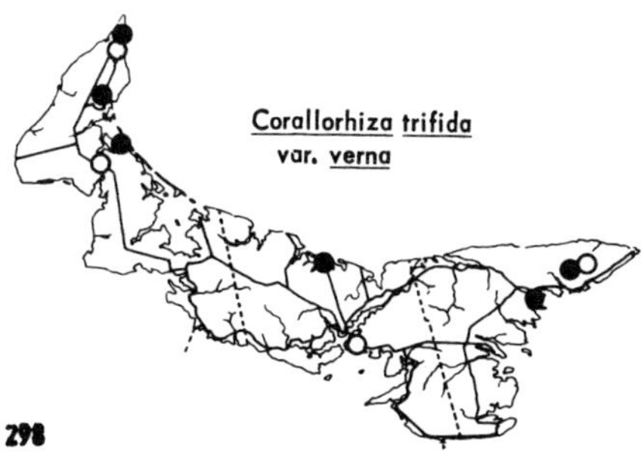

298

Corallorhiza maculata Raf. Map 299. Central and northeastern uplands; also Black Banks (***Messervy***). Mixed or deciduous woods (as a saprophyte). All specimens are f. ***punicea*** (Bartlett) Weath. & Adams. (***C. multiflora*** Nutt., Bain 1892, MacSwain); Hurst.

[*Corallorhiza odontorhiza* Nutt. Ranges northeast only to southwestern Maine. MacSwain and Bain listed it before they were aware (1892) that two native species were found, but MacSwain unfortunately listed it again in 1907.]

Microstylis unifolia (Michx.) BSP. Map 300. Local: West Prince; central Queens; Kilmuir, Bay Fortune. Damp, open places in spruce woods, thickets or bogs. (*M. ophioglossoides* Nutt., MacSwain and Bain, L.W. Watson 1899; *Malaxis unifolia* Michx., Campbell); Hurst (omitted by MacSwain).

Liparis loeselii (L.) Richard Map 301. Local: distributed throughout. Damp open places: boggy meadows or marshes; occasionally in dune slacks. MacSwain, Hurst.

SALICACEAE

Salix lucida Muhl. Map 302. Common in Prince County; Cove Head; Bear River. Open wet places: along streams, along roadsides in swampy woods where up to 15 feet tall and treelike.

Var. ***angustifolia*** Anderss. from low grounds, Cove Head (***Macoun***).

Salix pyrifolia Anderss. Lowlands throughout, most common in Prince County. Common in swamps and thickets. Hurst 1941; (***S. balsamifera*** Barratt, Hurst 1941).

Salix fragilis L. Map 304. Malpeque, Brackley Point, Southport, and probably elsewhere; an introduced tree. Twigs often strike root in damp roadside ditches. Hurst.

Salix alba L., vars. Map 305. Trees, introduced from Europe. Hurst.

Var. ***vitellina*** (L.) Stokes Roadside south of Uigg (Q), and probably elsewhere. Twigs strike root in damp roadside ditches, and form large bushes.

Var. ***calva*** G.F. W. Mey. North River Road, Charlottetown.

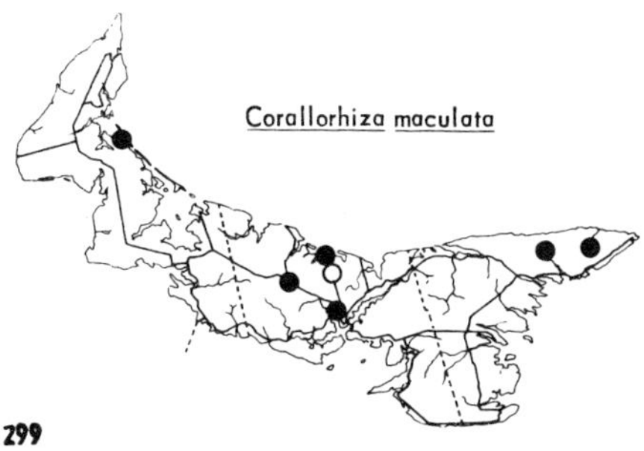

299

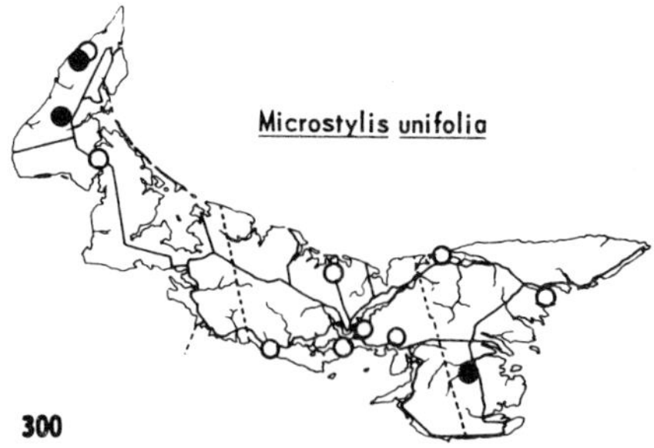

300

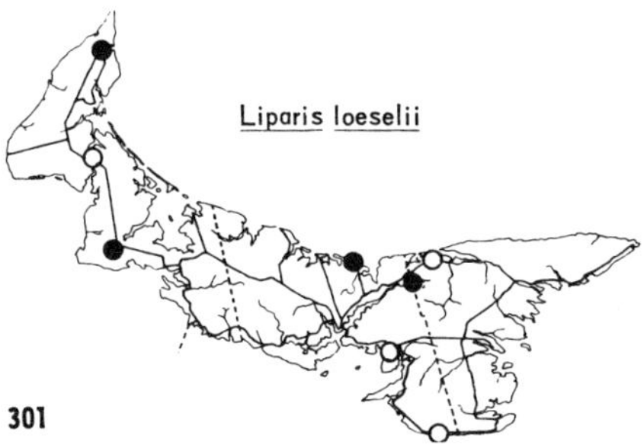

301

302

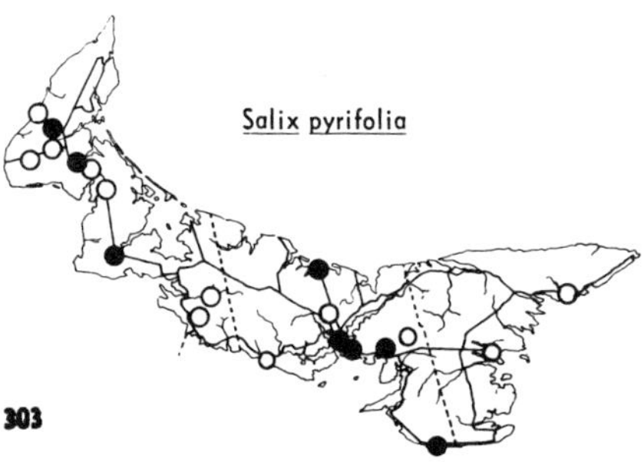

303

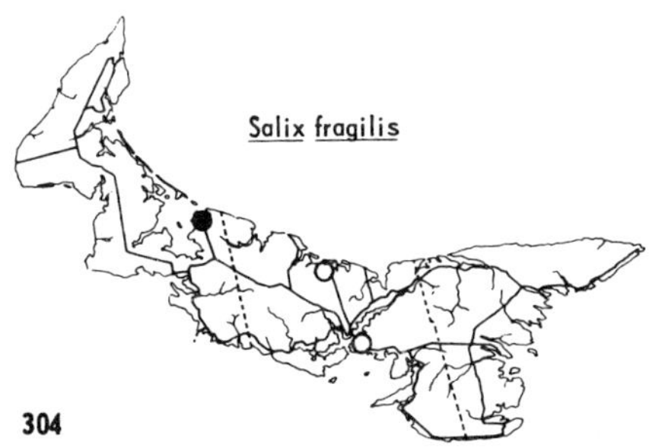

304

305

306

*__Salix rigida__ Muhl. Map 306. Common in Prince County; one station at Charlottetown (introduced ?). Marshy streambanks, ditches and thickets.

Salix bebbiana Sarg. Map 307. Very common throughout. Thickets, often near water but never in it.

Salix discolor Muhl. "PUSSY WILLOW" Map 308. Common throughout. Damp open sites: thickets, ditches, old fields. MacSwain and Bain, MacSwain, Hurst.

Var. **overi** C.R. Ball Specimen from Alberton seems to correspond to this glabrate form (Fernald 1950).

Salix humilis Marsh. Map 309. Common in Queens County and adjacent Kings; also O'Leary. Relatively dry and acid sandy soils, but less commonly in damp fields, thickets, and even bogs. Specimens from central Queens County referred by Fernald to his var. ***keweenawensis*** Farwell. (*S. **Muhlenbergiana***, MacSwain & Bain); MacSwain and Bain, MacSwain, Hurst.

[***Salix gracilis*** Anderss. var. ***textoris*** Fern. Charlottetown, 1945-55, cattail marsh below Experimental Farm, eliminated by reclamation.]

Salix viminalis L. Map 310. Scattered throughout: introduced European tree. Along streams, always on farms.

X ***Salix smithiana*** Willd. (stated to be *S. **viminalis*** X ***caprea***). Planted and occasionally spreading: introduced European tree. Near Charlottetown, dry thickets along St. Peters Road (*Fernald & St. John*).

Salix purpurea L. Charlottetown and along Brackley Road; introduced European shrub now seldom planted. Ditches along road and railroad. 1 Lot, or Royalty! (Groh). Hurst.

Populus tremuloides Michx. POPLAR Map 312. Very common throughout, the common native poplar (Bain). So-called "weed tree," quick-growing but short-lived, occupying old fields and clearings in any well-drained site and forming large uniform stands. Maybe cut as pulpwood, but fiber useful for paper only as an adulterant not to exceed 5%. (*P. **tremula***, Stewart); Bain, MacSwain and Bain, MacSwain, Hurst.

Populus grandidentata Michx. Map 313. Throughout, much less common than *P. **tremuloides***. "Weed tree" of clearings and burns, seldom forming pure stands; especially on sandy sites where often persisting in developing pine stands, and attaining a greater girth than in developing stands of fir on damp sites. MacSwain, Hurst.

Populus alba L. Map 314. Charlottetown vicinity; a European tree, planted as a windbreak on the Jones estate, Bunbury, and spreading by root suckers to form thickets. Introduced, 8 Lots (Groh). Hurst.

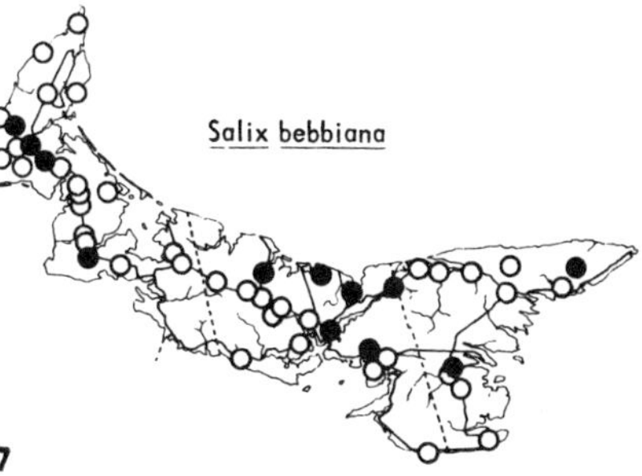

307

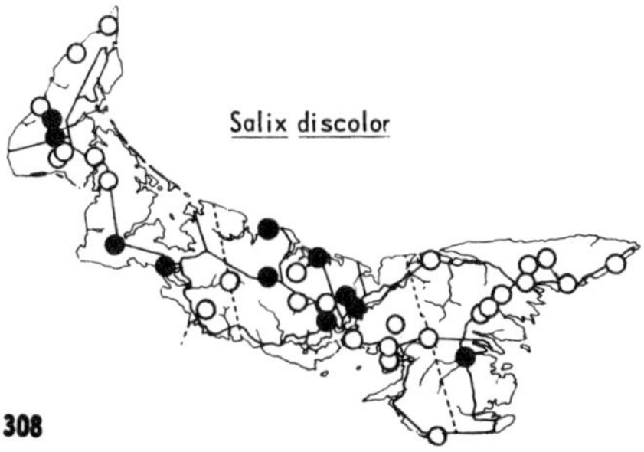

308

309

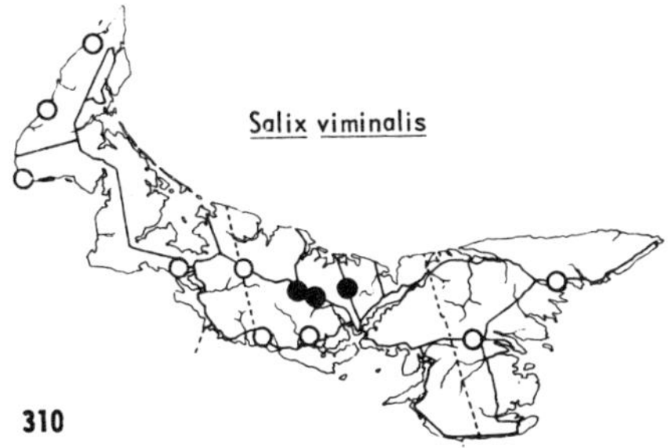

310

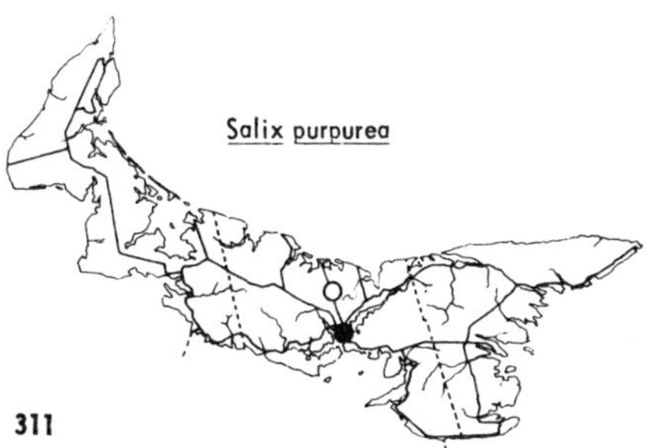

311

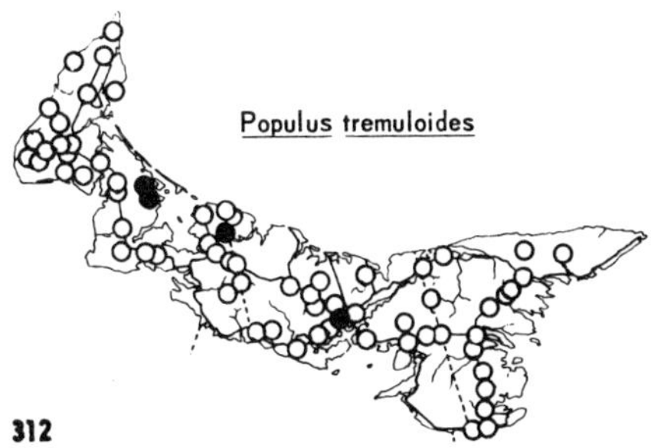

312

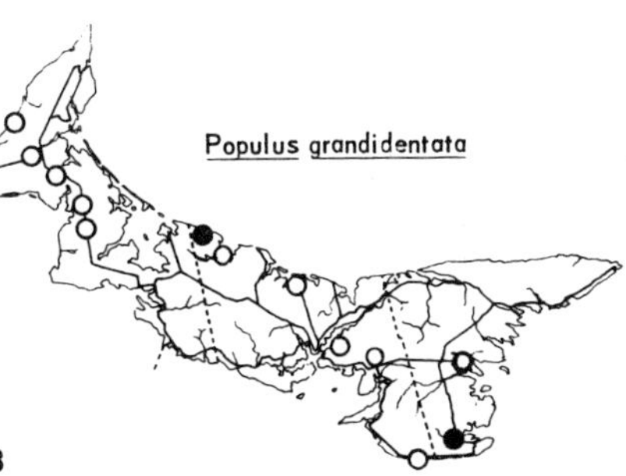

313

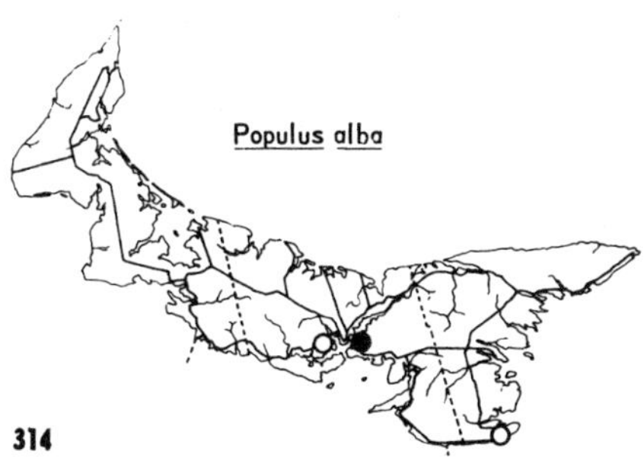

314

[*Populus nigra* L. European tree, the vegetatively reproduced fastigiate form, "var. *italica* Muench.", formerly introduced from Britain (locally known as "English Poplar") as an ornamental. At Eldon, and to some extent at Springfield West and Wellington, root suckers form spreading colonies. Hurst.]

Populus balsamifera L. Map 315. Common in Prince County; rare eastward: Bonshaw, St. Peters Road (Charlottetown) and Cardigan. Damp thickets along streams and in open swampy woods. Often planted as a shade tree (Bain) - around farmhouses in Prince, less commonly eastward; spreading from cultivation by suckering, as on St. Peters Road where of the more northerly var. ***subcordata*** Hylander, here associated with introduced willows. Bain, MacSwain and Bain, Hurst (omitted by MacSwain).

MYRICACEAE

Myrica gale L. Map 316. Throughout, most common near coast. Pond margins, where often forming a zone; bushy swamps. MacSwain and Bain, MacSwain, Campbell.

Var. ***subglabra*** (Chev.) Fern. West of St. Peters. Hurst.

Myrica pensylvanica Loisel. Map 317. Common near the coast, occasional on uplands. Open sites: dune slacks (where shrubs compact), borders of bogs, poor pastureland, persisting into weed tree woods as a straggly shrub. "Candleberry," when settlers melted down the wax from the fruits for candles; known from its aromatic leaves as "Myrtle," both names noted by Mrs. Heath Haviland in 1853-4. (***M. cerifera,*** Stewart, Macoun 1886, MacSwain and Bain, MacSwain; ***M. caroliniensis,*** Hurst); Campbell.

Comptonia peregrina (L.) Coult. Map 318. Near coast: common near North Shore from Malpeque Bay eastward; Murray River; Victoria (Q). Open sandy sites, near dunes or in pine woods; now in clearings and old fields. (***C. asplenifolia,*** MacSwain and Bain; ***Myrica asplenifolia,*** MacSwain, Hurst); Campbell.

BETULACEAE (CORYLACEAE)

Corylus cornuta Marsh. HAZEL Map 319. Throughout; common on uplands. Open well-drained sites, forming thickets on slopes and at wood edges. Nuts gather edand eaten; oil formerly extracted for use as a drier in paint (Bain). Included in CUPULIFERAE by MacSwain and Bain. (***C. Avellana,*** Stewart; ***C. Americana,*** Bain; ***C. rostrata*** Ait., MacSwain and Bain, MacSwain, Hurst); Campbell.

Betula lutea Michx. f. YELLOW BIRCH Map 320. Throughout; common on uplands, rare on coast. A dominant in the climax forest of the hardwood hills. On light friable loamy soil in well-drained sites. Old trees up to six feet thick (Bain), all dead of "birch dieback," but still cut for firewood. (***B. lenta,***

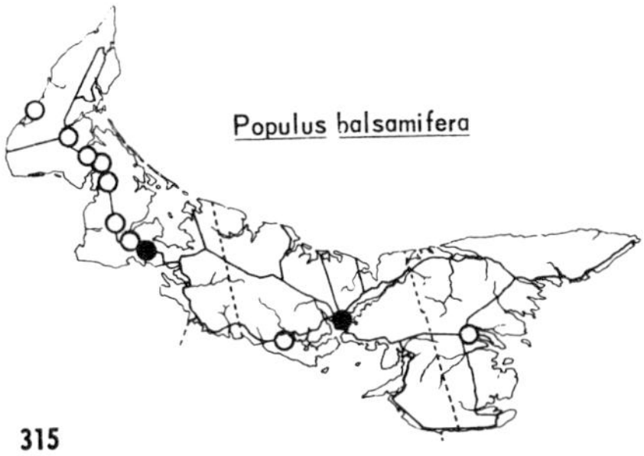

315

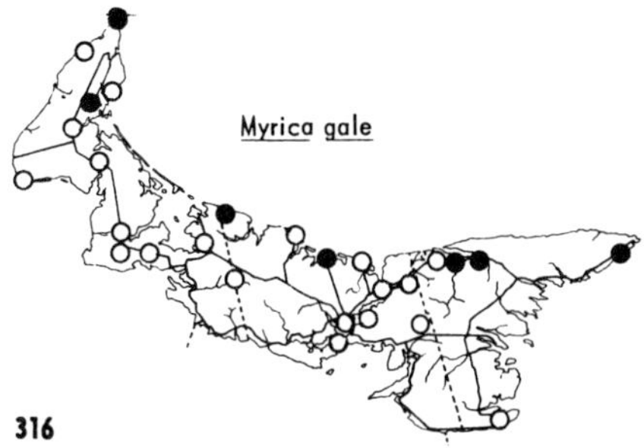

316

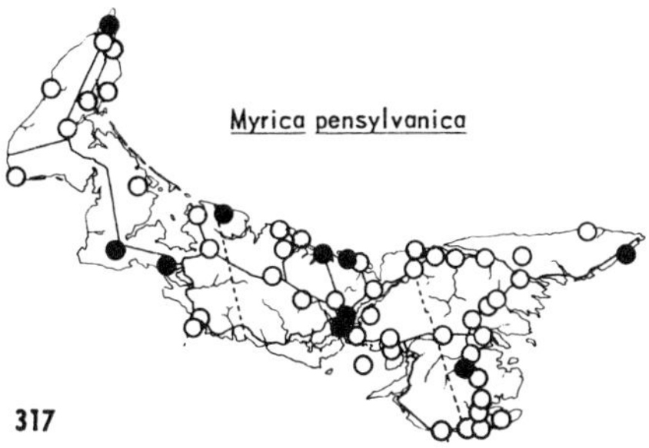

317

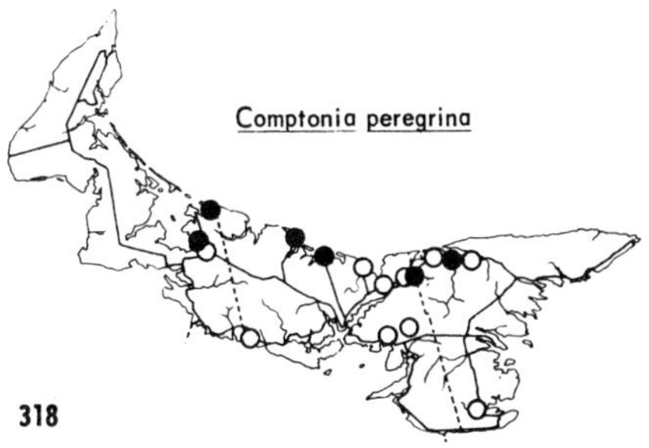

318

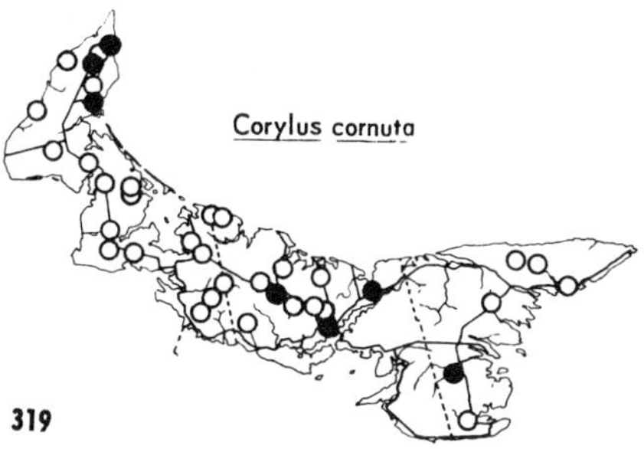

319

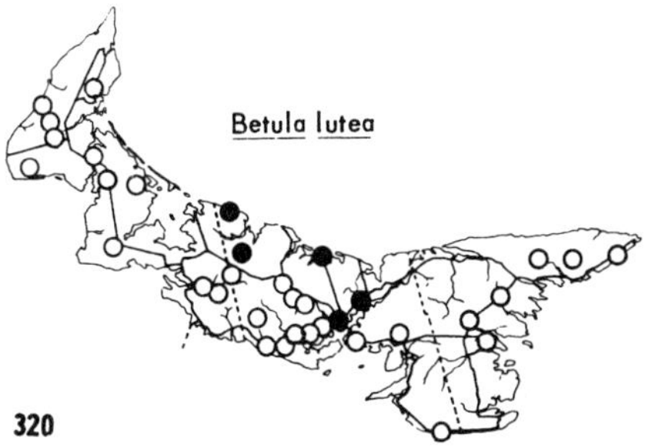

320

Stewart; *B. excelsa,* Bain, MacSwain and Bain); MacSwain and Bain, MacSwain, Hurst. *Betula lenta* L., "Black Birch," has always been distinguished from "Yellow Birch" on characters of the bark, but the names *B. nigra,* Stewart, and *B. lenta,* MacSwain, belong with *B. lutea.*

Betula populifolia Marsh. Map 321. Throughout; very common in the lowlands, less so on the uplands. Thin soil in sunny places, the characteristic weedy tree of sandy or boggy sites. (*B. alba,* first sort ("Grey Birch"), Stewart; *B. alba* var. *populifolia* (Marsh.) Spach, MacSwain and Bain); MacSwain, Hurst.

Betula coerulea-grandis Blanchard Map 322. Brackley Point Road, dry woods (*Fernald et al.*). Somewhat similar plants around lily pool, Experimental Farm, Charlottetown. Description suggests the hybrid origin, *B. populifolia* X *papyrifera.*

[***Betula alba*** L. Brackley Point, woods (*Macoun,* det. Fernald). European, cultivated but probably not naturalized.]

Betula papyrifera Marsh. WHITE BIRCH Map 323. Common throughout. Well-drained open sites: abundant even along streams in the hills, but in lowlands only on sandy soil, usually succeeded by the intermingling fir. White birches in fencerows between fields are characteristic of the uplands. Much planted in towns as ornamental shade tree. (*B. alba,* second sort ("White Birch"), Stewart; *B. papyracea* Ait., Bain, MacSwain and Bain; *B. alba* var. *papyrifera* (Marsh.) Spach, Hurst); MacSwain, Campbell.

Betula pumila L. var. ***renifolia*** Fern. Map 324. Rare: Miscouche Woods (P); St. Andrews (K) to Dundee. Open boggy ground, in barren, swamp or clearing.

Alnus crispa (Ait.) Pursh var. ***mollis*** Fern. Map 325. Scattered throughout, almost absent from uplands. Open, damp or heavy soil, on banks, never in swamps; often colonizing old fields. (*A. serrulata,* MacSwain and Bain (?); *A. viridis,* MacSwain; *A. crispa,* Hurst).

Alnus rugosa (DuRoi) Spreng. var. ***americana*** (Regel) Fern. Map 326. Common throughout. Open wet mucky sites: creek bottoms, pond holes, wet runs, forming characteristic thickets with dense undergrowth of marsh herbs. (*Betula Alnus,* Stewart; *Alnus incana,* MacSwain and Bain, MacSwain, Hurst); Campbell. Forma *malacophylla* Fern. from St. Charles, cited in Rhodora 47: 353. 1945.

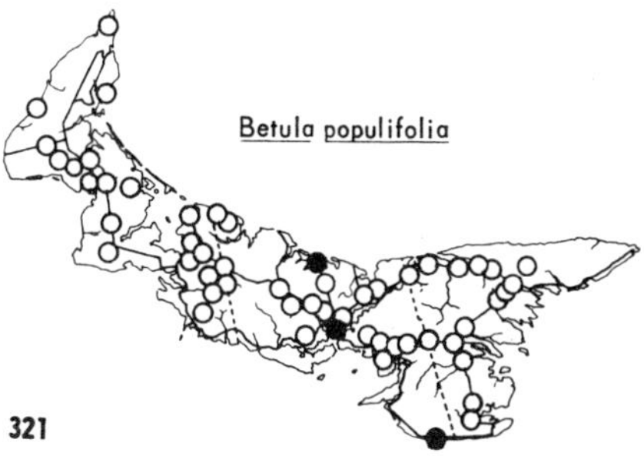

321

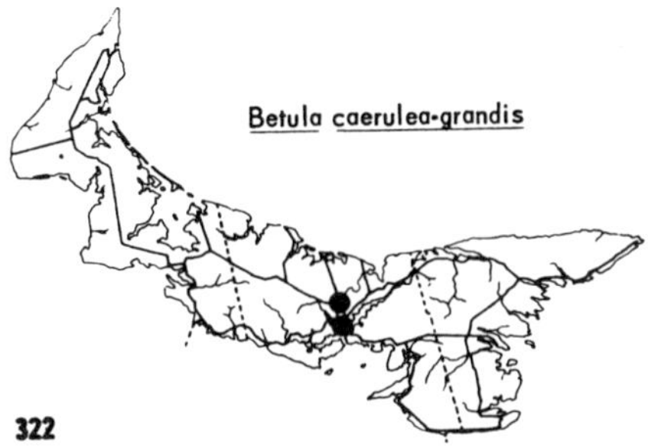

322

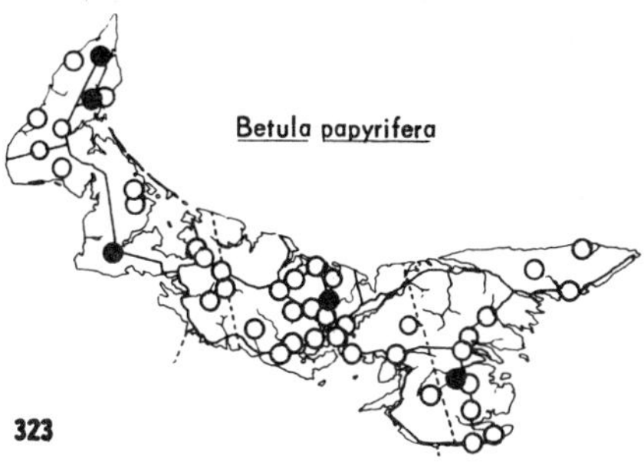

323

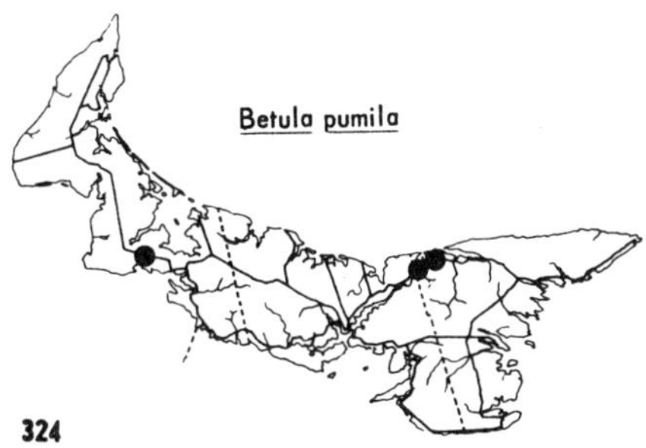

324

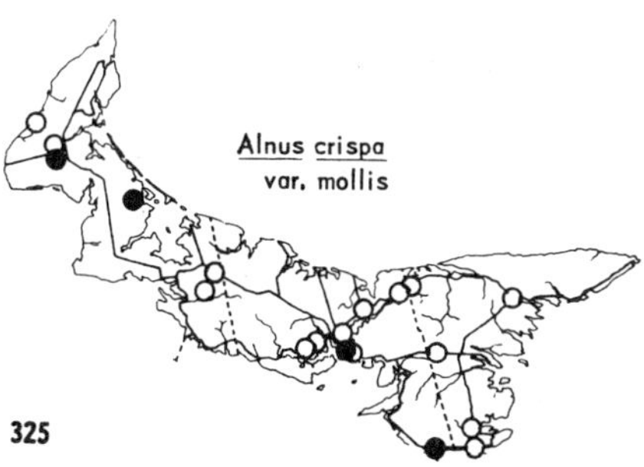

325

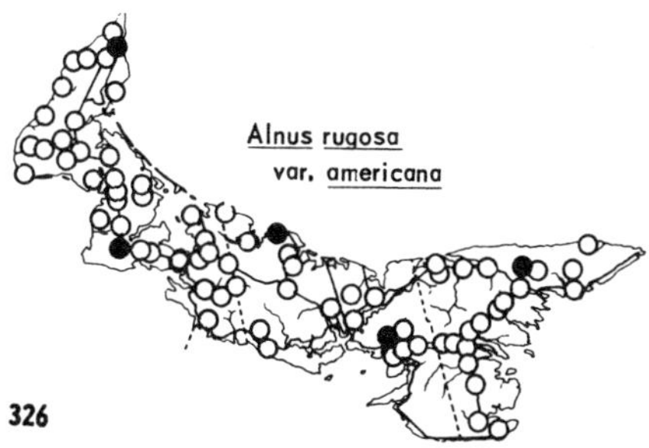

326

FAGACEAE (CUPULIFERAE, MacSwain and Bain)

Fagus grandifolia Ehrh. ssp. ***heterophylla*** Camp BEECH Map 327. Throughout, mainly in uplands. "Dry sandstone soils" (Bain), almost confined to the Alberry and Charlottetown soil series, on which codominant with ***Acer saccharophorum*** forming the deciduous climax forest with scanty undergrowth and springflowering herbs. Infected by beech canker (***Nectria coccinea***) since the 1930's, but with less extensive injury than in Nova Scotia. Wood formerly used by cabinetmakers and turners (Bain); good firewood. Fallen nuts (beech mast) served to fatten pigs (Bain). (***F. sylvatica,*** Stewart; ***F. ferruginea*** Ait., MacSwain and Bain, MacSwain); Hurst.

Quercus robur L. Map 328. Introduced from England as shade tree, fruiting and occasionally spreading to roadsides and wood edges, especially near Charlottetown and Georgetown. (***Q. alba,*** Hurst).

Quercus rubra L. var. ***borealis*** (Michx. f.) Farwell Map 329. Local: East Prince to central Kings. Dry, open soil, sandy or on river bluffs (as at Cardigan). Formerly more extensive? "Once a large oak forest" at Tracadie (Bain), recorded in French times (18th century). (***Q. rubra,*** Stewart, MacSwain and Bain, MacSwain, Hurst. ***Q. borealis*** Michx. f., Roland).

ULMACEAE

Ulmus americana L. ELM Map 330. It is frequent in Prince County; local in east-central lowland. Alluvial soil of creek bottoms, often in alder thickets in Prince. Wood nonsplintering, formerly used for wheel hubs, ship's keels and millwork. Often planted as a shade tree; less so in Charlottetown where ***U. glabra*** ssp. ***montana*** is the common elm. MacSwain and Bain, MacSwain, Hurst.

CANNABINACEAE

Humulus lupulus L. HOP Map 331. Formerly introduced, naturalized at scattered localities. Shady roadsides, old house sites; in alder thickets, as at Cousins Pond near Malpeque, possibly native. Abundant on site of old brewery, Mount Edward Road near Charlottetown; pistillate spikes in beer to impart bitterness. Hurst (wrongly implying listing by MacSwain).

URTICACEAE

Urtica gracilis Ait. Map 332. Scattered: Tignish; central and northeastern uplands. Shady, damp thickets. (***Laportea canadensis,*** MacSwain and Bain in part; ***Urtica gracilis,*** MacSwain and Hurst in part; ***U. procera,*** Campbell in part).

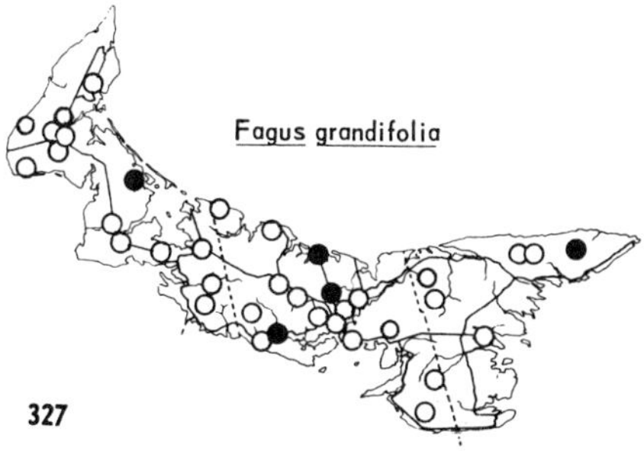

327

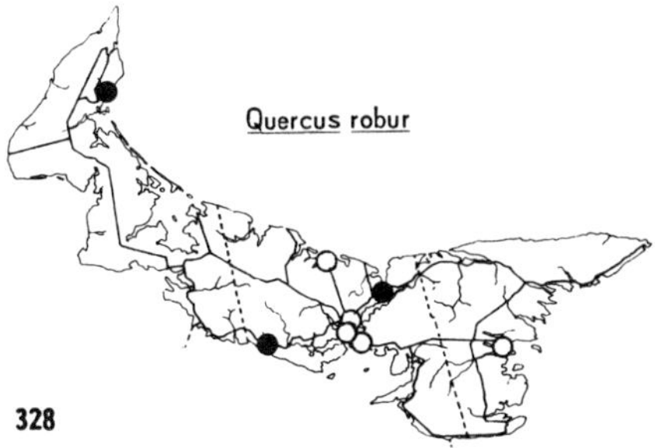

328

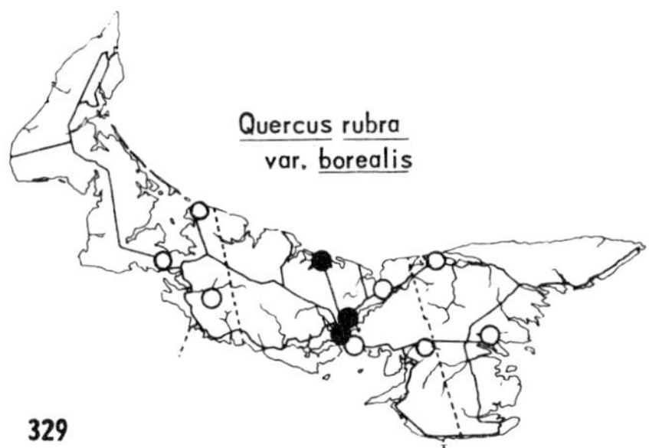

329

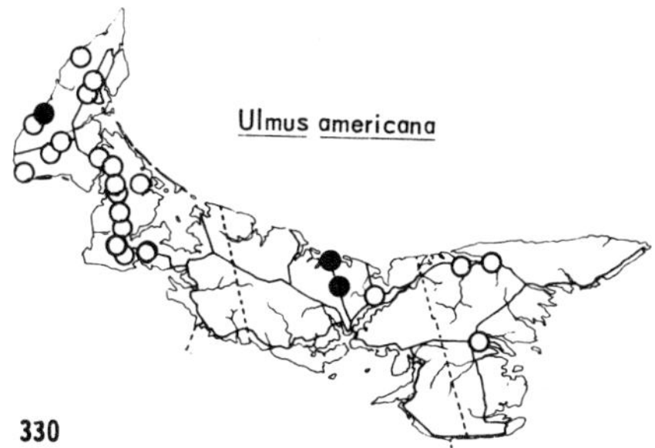

330

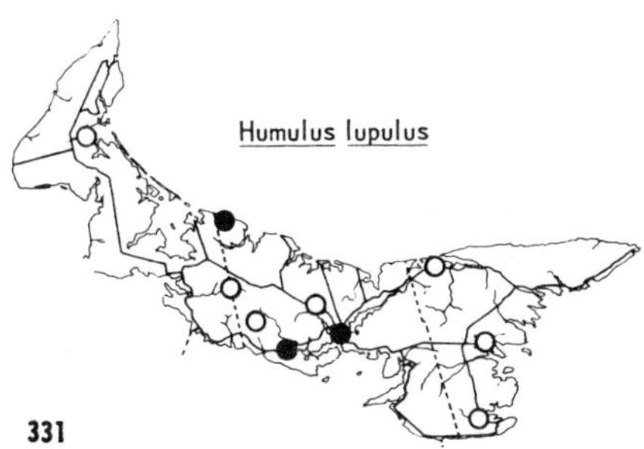

331

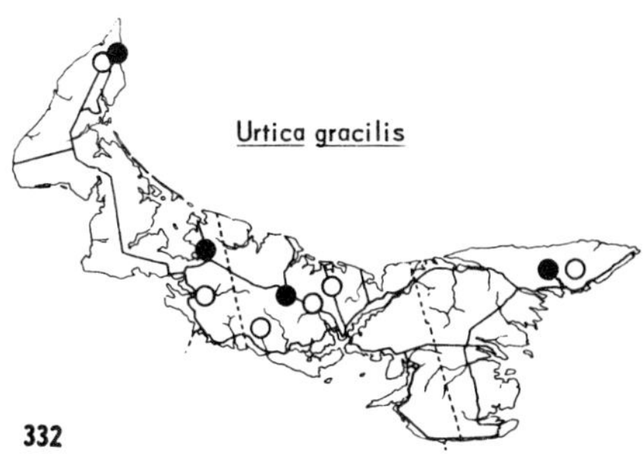

332

Urtica procera Muhl. Map 333. Rare: Hunter River, Brackley Point, York. Damp thickets; along streams, in sandy alluvium. (***Laportea canadensis,*** MacSwain and Bain, in part; ***U. gracilis,*** MacSwain, Hurst, in part; ***U. procera,*** Campbell in part).

Urtica dioica L. STINGING NETTLE Map 334. Borden, Kensington, Montague, Murray River (known for over forty years). European weed. Colonies along roadsides in villages. Campbell.

Urtica urens L. Map 335. Tignish (1888, 1926); Bideford (1953). European weed. Bare ground; farmyard, waste places. MacSwain, Hurst.

Pilea pumila (L.) Gray Map 336. Rare: Dundee, Mount Stewart; five miles southwest of Tignish. Damp shady places; larch swamps, alder thickets.

SANTALACEAE

Comandra richardsiana Fern. Map 337. North Shore, Cavendish around to Bothwell. In turf on sandy soil inland from dunes. (***C. umbellata,*** Hurst).

Arceuthobium pusillum Peck Map 338. Locally, Prince, North Shore, Kings. On ***Picea glauca*** and ***mariana*** in open habitats; a parasite inducing production of elongate fastigiate branches, known as witches'-brooms (Groh). Hurst.

POLYGONACEAE

****Rumex mexicanus*** Meisn. var. ***triangulivalvis*** (Danser) Lepage (in Nat. Can. 81: 66. 1954). Map 339. Recently adventive: Wellington (1952), Charlottetown (1953), a species of Boreal Forest range. Open, low, clay sites slightly alkaline (due to hard or brackish water). Closely related to the next.

Rumex pallidus Bigel. Plat River (on Malpeque Bay), erect or sprawling on sandy strands and dryish borders of salt marshes (***Fernald & St. John,*** det Rechinger).

Rumex domesticus Hartm. Map 340. St. Dunstans, Charlottetown (1914); Souris, waste land behind new hall (1952); a European weed.

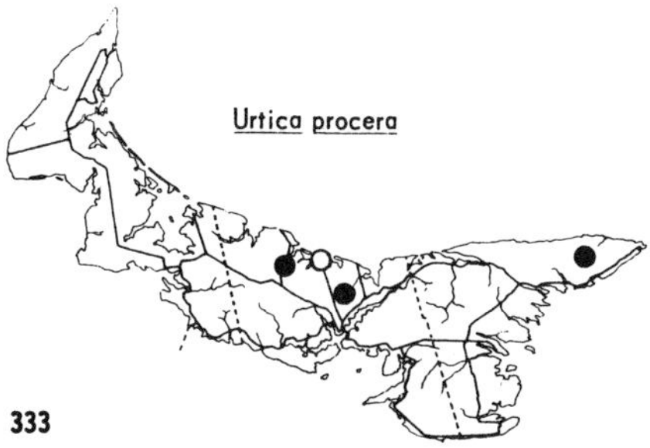

333

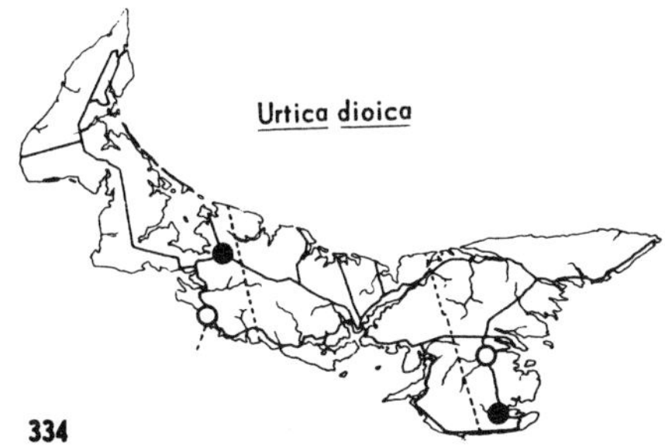

334

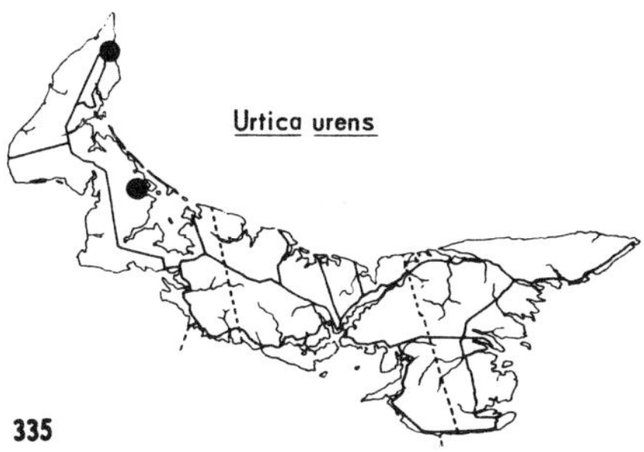

335

336

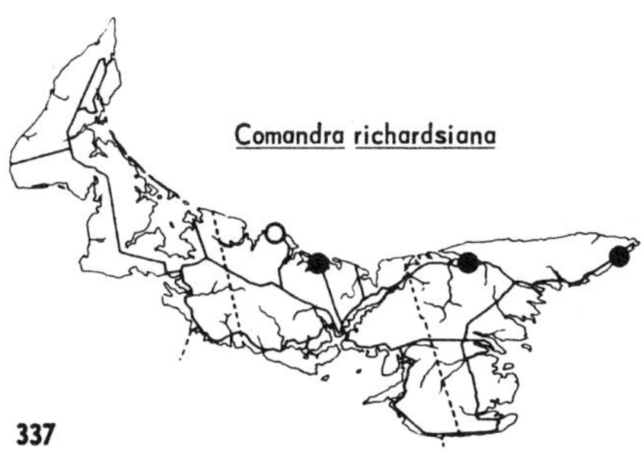

337

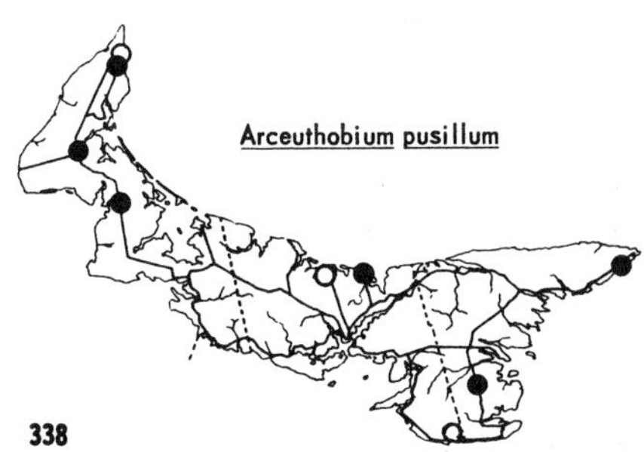

338

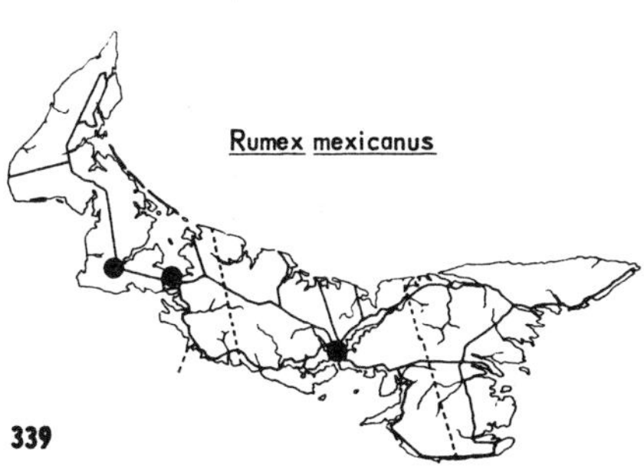

339

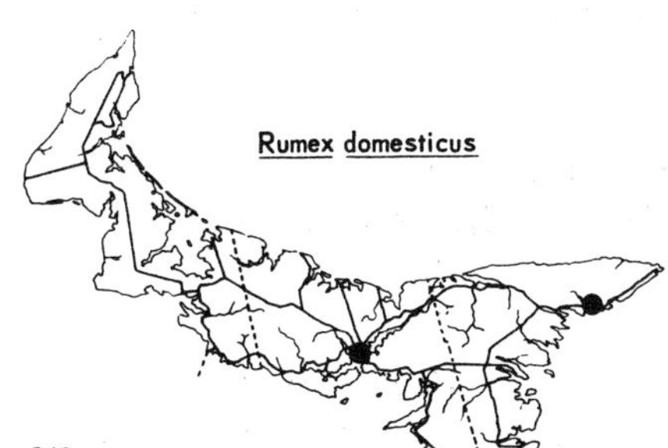

340

Rumex orbiculatus Gray Map 341. Scattered throughout. Lowland marshes; around ponds or creeks. (*R. Britannica,* Hurst); Campbell.

Rumex crispus L. Map 342. Common throughout; naturalized, European. Weed of cultivated or disturbed ground; 42 Lots (Groh). MacSwain and Bain, MacSwain, Hurst (including *R. elongatus* Guss., Hurst).

Rumex obtusifolius L. Map 343. Scattered throughout; naturalized, European. Weed of damp, shady or organically rich soil; 13 Lots (Groh). Var. *sylvestris* (Lam.) Koch Charlottetown (Fernald in Rhodora 23: 258. 1921). (*R. sanguineus,* Bain, MacSwain and Bain); MacSwain and Bain, MacSwain, Hurst.

Rumex persicarioides L. Map 344. Coastal, rare: Malpeque, Grand Tracadie. Border of salt marsh, damp sand flats.

Rumex maritimus L. var. ***fueginus*** (Philippi) Dusén Map 345. Coastal, rare: Brackley Beach, Bunbury. Sand flat or salt marsh. Closely related to the last. (*R. fueginus* Philippi, Roland).

Rumex acetosella L. "REDWEED", SORREL Map 346. Common, throughout. European, naturalized. In old fields and pastures; bare roadside shoulders; weed in 41 Lots (Groh). Bain, MacSwain and Bain, MacSwain, Hurst.

[*Tovara virginiana* (L.) Raf. Not known; quite improbable record. (*Polygonum Virginianum* L., MacSwain and Bain, MacSwain, Hurst); Campbell.]

Polygonum raii Bab. Map 347. North Shore: Brackley Beach, Grand Tracadie (?), North Lake. Sandy sea beaches (J.M. Macoun, Can. Record Sci., Jan. 1895: 208).

Polygonum allocarpum Blake Salt marshes of the Hillsborough estuary: Mount Stewart, Bunbury (det. J.F. Brenckle).

* ***Polygonum exsertum*** Small Map 348. Northumberland Strait salt marshes: Summerside, Bedeque, Charlottetown, Bunbury, Mount Stewart.

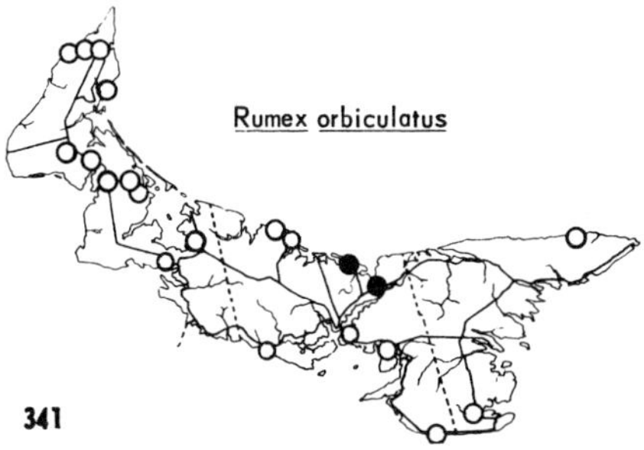

341

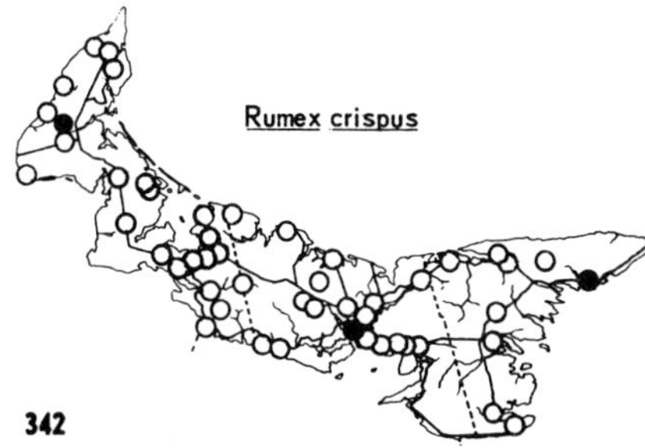

342

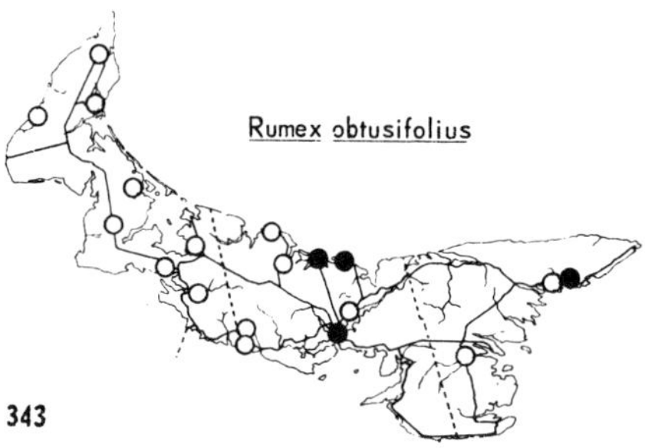

343

344

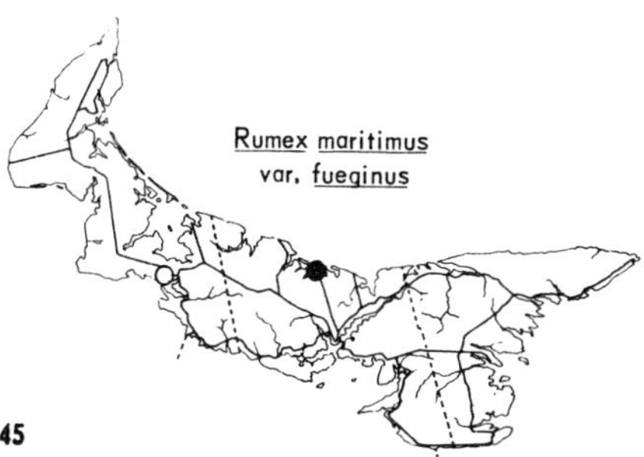

345

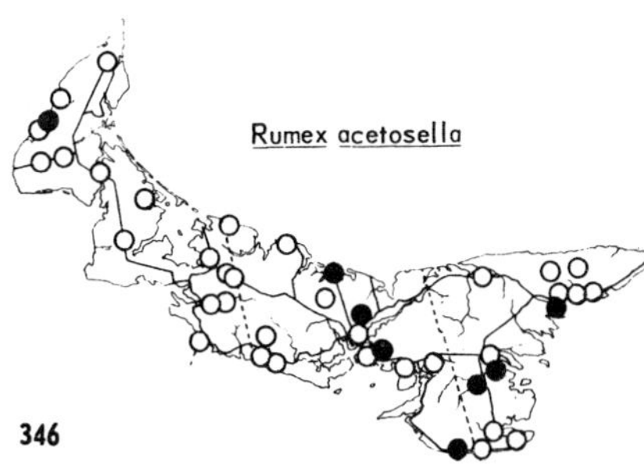

346

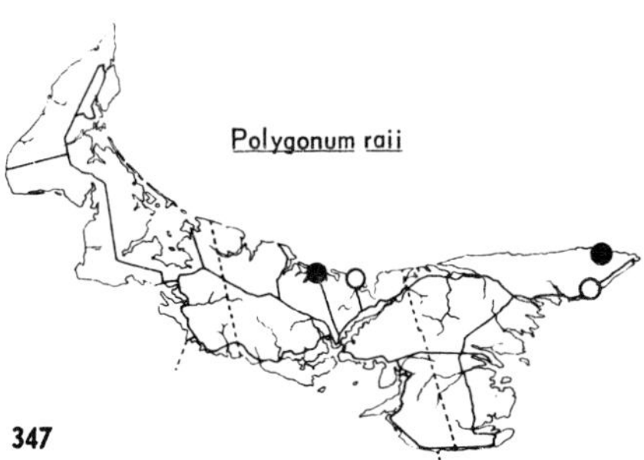

347

348

[***Polygonum erectum*** L. Not seen. The record from Charlottetown (Groh 1927) possibly refers to a member of the *P. aviculare* complex (e.g., *P. heterophyllum*).]

Polygonum aviculare L. Map 349. Scattered throughout. Naturalized European. Weed of beaten paths, waste places and the head of salt marsh. MacSwain and Bain, MacSwain, Hurst. 29 Lots (Groh).

Var. ***neglectum*** Bess., Plat River, head of salt marsh. Narrow-leaved.

In this complex species or group of species, some segregates recognized by J.F. Brenckle, authority on the group, appear. Specimens from Wood Islands have been identified as *P. heterophyllum* Lindm. f.; specimens from Charlottetown, Wood Islands, Mount Stewart and Bunbury as *P. pulchellum* Loisel. Without field study, these cannot usefully be separated here.

Polygonum amphibium L. var. ***stipulaceum*** (Coleman) Fern., formae: Map 350. Watervale (Q), North Shore ponds. Terrestrial plants, or plants developed in air; habitat: marshy borders of ponds are f. *stipulaceum*. Aquatic plants or plants developed in water; habitat: ponds are f. *fluitans* (Eaton) Fern. (*P. amphibium*, MacSwain and Bain, MacSwain, Hurst).

* ***Polygonum coccineum*** Muhl. Southport, choking muddy bed of stream above broken dam, Reddin's pond. May have come in with Reddin's introductions from Wisconsin (*Butomus, Zizania, Pontederia,* etc.) which it is now choking out.

[***Polygonum pensylvanicum*** L. Not seen, though not improbably as weed of organically rich soil. 3 Lots (Groh) who, however, overlooked *P. scabrum.* MacSwain; queried by Hurst.]

Polygonum scabrum Moench Map 351. Throughout; rather common. Soil rich in organic matter, weed of cultivation. (*P. lapathifolium*, Macoun, MacSwain, Hurst); Campbell.

Polygonum hydropiper L. Map 352. Common in lowlands. Damp, open ground; around ponds and streams, perhaps native, and a weed in damp gardens or waste places. (Including var. ***projectum*** Stanford); MacSwain and Bain, MacSwain, Hurst.

Polygonum persicaria L. Map 353. Scattered, central P.E.I. Locally abundant weed of cultivation and waste places. 24 Lots (Groh). MacSwain and Bain, MacSwain, Hurst.

Polygonum punctatum Ell. var. ***confertiflorum*** (Meisn.) Fassett, Brittonia 6: 377. 1948. Map 354. Rare: Mount Stewart; Tracadie (a form with short broad leaves). Marshy borders of bog or pond. (*P. acre,* MacSwain, Hurst; *P. punctatum* var. *leptostachyum* (Meisn.) Small, Fernald 1950).

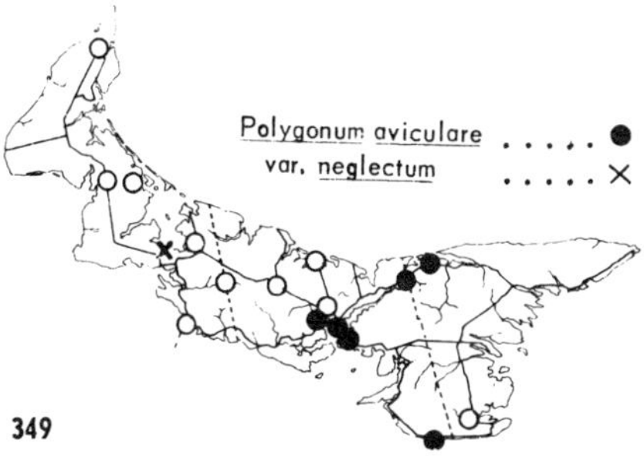

349

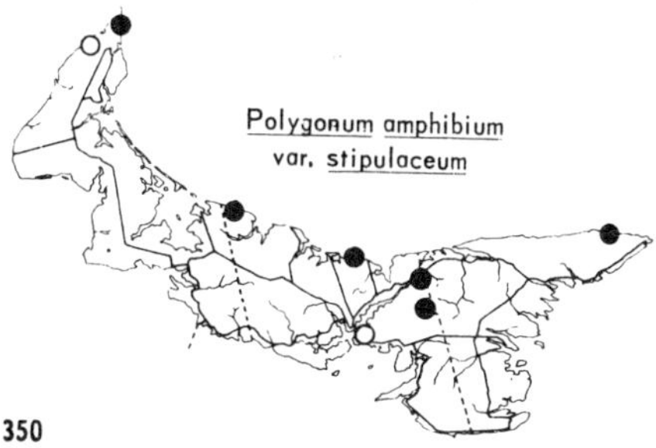

350

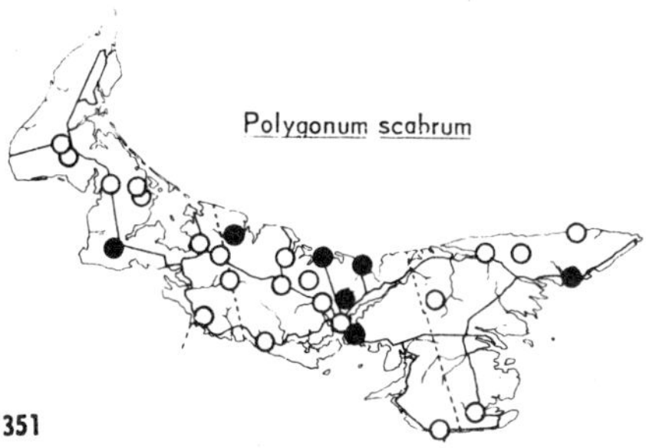

351

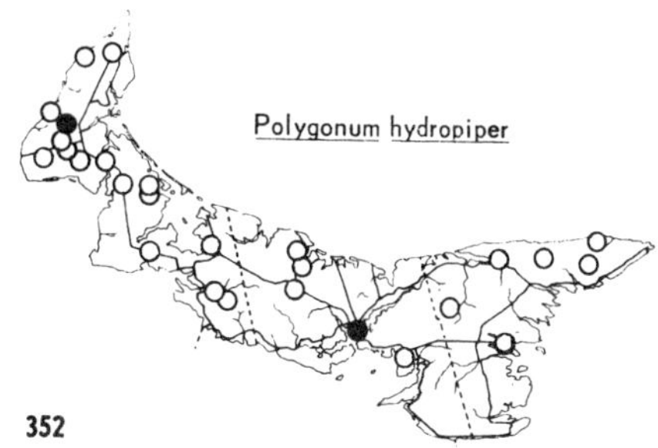

352

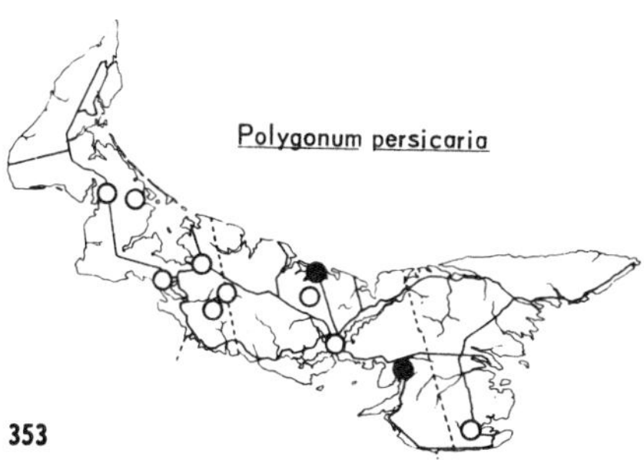

353

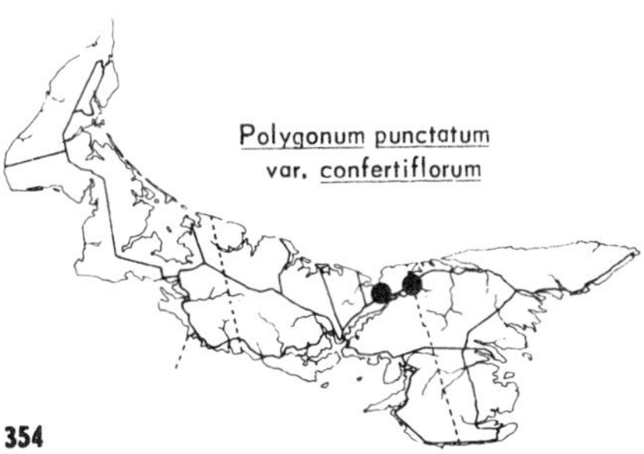

354

Polygonum sagittatum L. Map 355. Throughout; common in Prince and east-central lowlands. Open or partially shaded damp mucky sites. MacSwain and Bain, MacSwain, Hurst.

Polygonum arifolium L. var. **pubescens** (Keller) Fern. Map 356. Rare in the south and southeast: Beach Grove, Lake Verde, Watervale, Cherry Valley, Wood Islands. Marshy borders of thickets. (*P. arifolium* MacSwain and Bain, MacSwain, Hurst).

Polygonum cilinode Michx. Map 357. Rare mostly on northern slope: Indian River; Winsloe Road at Brackley Point (Messervy); Tracadie Beach; Avondale (Q); Glencorradale. Sandy soil: stream deposits, moraine. Hurst. (*Tiniaria* Small).

Polygonum convolvulus L. Map 358. Scattered, throughout. Weed of gardens waste and sandy places. European. 25 Lots (Groh). MacSwain and Bain, MacSwain, Hurst. (*Tiniaria* Webb. & Moq.).

***Polygonum scandens** L. Campbell's Mills near New Glasgow, cleared damp thickets along stream below mill. (*Tiniaria* Small).

[*Polygonum dumetorum* L., reported by MacSwain and Bain, may refer either to *P. scandens*, closely related, or to the habitally similar *P. cilinode*.]

Polygonum cuspidatum Sieb. & Zucc. Towns; most common in Charlottetown and Souris; Vernon River. Planted in gardens, where almost indestructible; introduced from Japan. Campbell. (*Tiniaria japonica* (Houtt.) Hedb.).

Polygonum sachalinense F. Schmidt Becoming established in Souris and district. Introduced to gardens from Japan, now spreading. (*Tiniaria* Janchen).

Fagopyrum esculentum Moench BUCKWHEAT Cultivated. Rarely spontaneous, in the fields or adjacent ditches. 11 Lots (Groh). (*F. sagittatum* Gilib., Campbell).

CHENOPODIACEAE

Chenopodium album L. Map 359. Only one of Bain's "Goosefeet" which is common throughout. European, naturalized. Open, disturbed soil; waste land and gardens. MacSwain and Bain, MacSwain, Hurst.

Chenopodium macrocalycium Aellen (Wahl, Bartonia 1952-3 (27): 45. 1954) Borden (*Bassett*). Gravelly beaches or head of beach. (*C. Bushianum*, Roland).

Chenopodium glaucum L. Map 360. European, adventive. Summerside (1912, 1926 (Groh), 1951-3), Tignish (1950). Railway and waste places. Hurst.

[*Chenopodium botrys* L. Not seen. Not recorded since MacSwain and Bain.]

[*Chenopodium urbicum* L. Not seen, record probably erroneous. MacSwain and Bain, MacSwain, Hurst.]

[*Chenopodium hybridum* L. Not seen. Record of MacSwain and Bain withdrawn by 1907.]

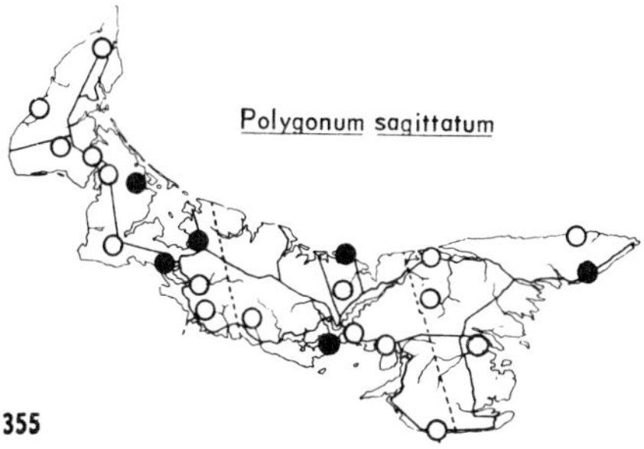

355

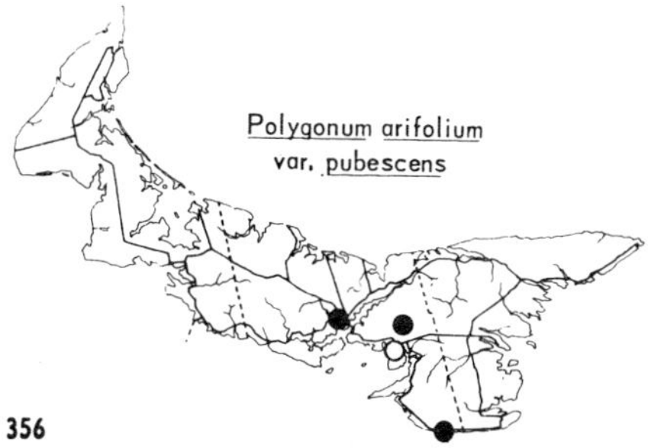

356

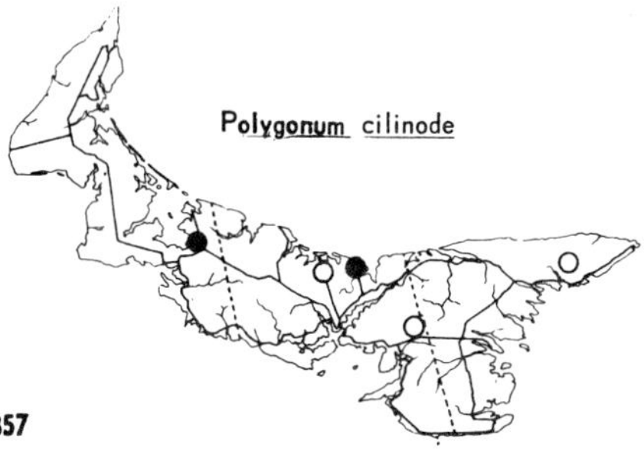

357

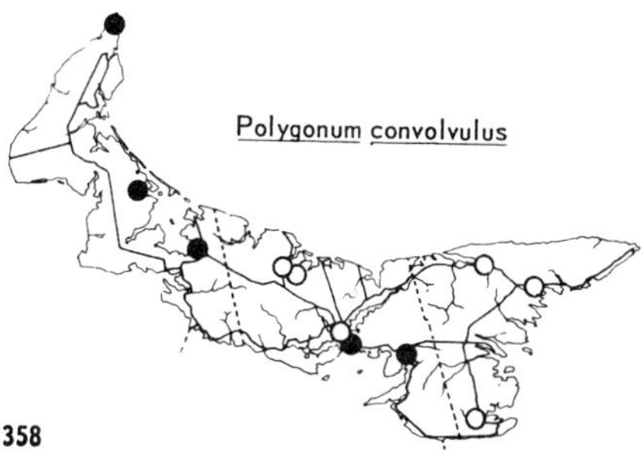

358

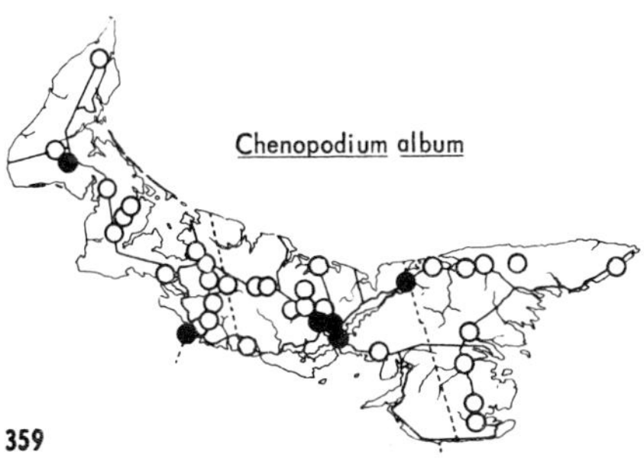

359

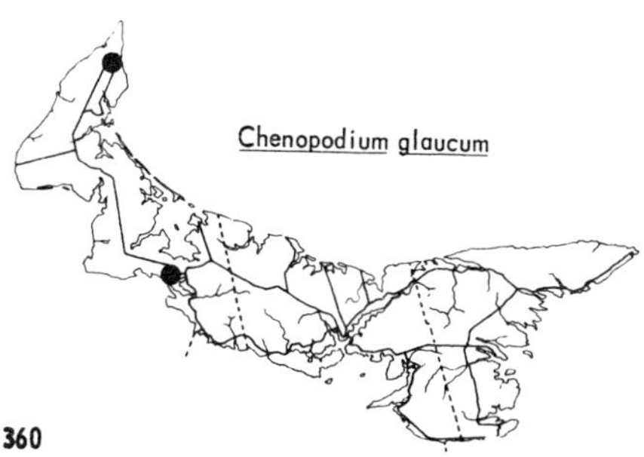

360

Chenopodium rubrum L. Map 361. C.N.R.: Summerside (1953), Emerald (1953), Charlottetown (1953). Adventive from mainland by rail. Cinders. MacSwain and Bain, MacSwain, Hurst.

[*Chenopodium bonus-henricus* L. Map 362. Brackley Point, roadside (*Macoun*). European, adventive. Macoun, Groh, Hurst.]

Atriplex patula L., vars. Widespread. Map 363.

* Var. *patula* Cultivated soils and waste places; seldom on beaches.

Var. *hastata* (L.) Gray Common, maritime; weed of gardens and waste places. 16 Lots (Groh). (*A. hastata* L., Groh); Hurst.

Var. *littoralis* (L.) Gray Map 364. Occasional, maritime (on sandy shores), rarely weedy. Strikingly different in erect habit, narrow fleshy leaves with shiny upper surface, from the other varieties. Macoun, MacSwain and Bain, MacSwain, Hurst.

Atriplex sabulosa Rouy Map 365. Brackley Beach, "salt marshes" (*Macoun*), first North American collection; Launching, Wood Islands, sandy beaches. (*A. maritima* E. Hall. Roland).

Salicornia europea L., vars. Map 366. Coastal salt marshes, especially as weed of bare patches in the *Spartina patens* zone. Turning red in fall, whence the name "sand fire" as corruption of samphire. (*S. herbacea* L., MacSwain and Bain, MacSwain); Hurst.

Var. *europea* Common.

Var. *simplex* (Pursh) Fern. Bunbury (*Fernald et al.*).

Var. *prostrata* (Pall.) Fern. Bunbury. A series collected by Fernald *et al.* to illustrate transition to var. *europea.*

Suaeda maritima (L.) Dumort. Map 367. Coastal, apparently rare in regions of sandy shores. Muddy salt marshes, abundant in denuded patches, scattered on middle and upper zones. MacSwain and Bain; (*S. linearis*, MacSwain; queries by Hurst 1933, Adams, Hurst 1940); Hurst 1940.

Suaeda americana (Pers.) Fern. Map 368. Cape Aylesbury, damp brackish sands (*Fernald et al.*).

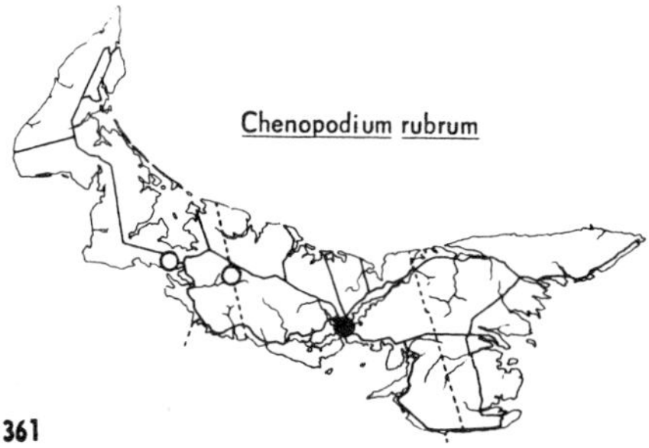

361

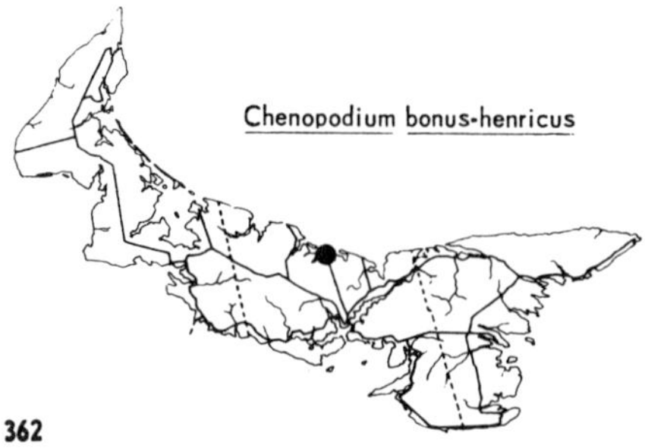

362

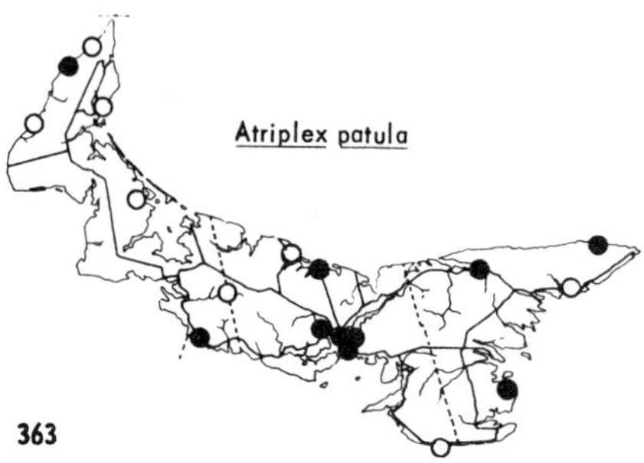

363

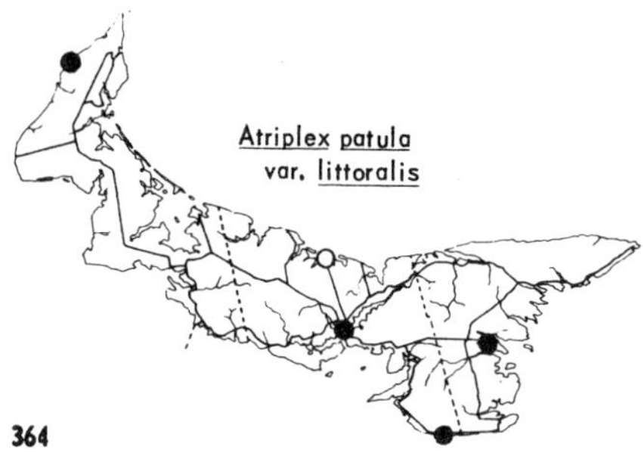

364

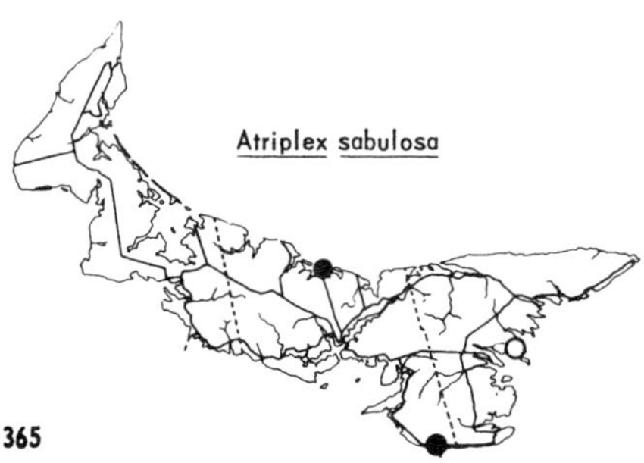

365

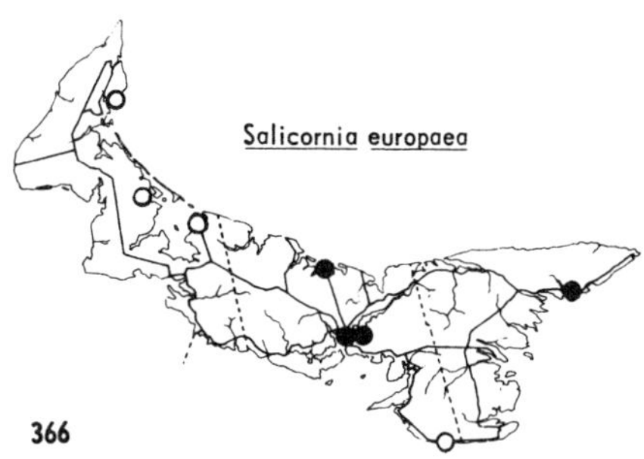

366

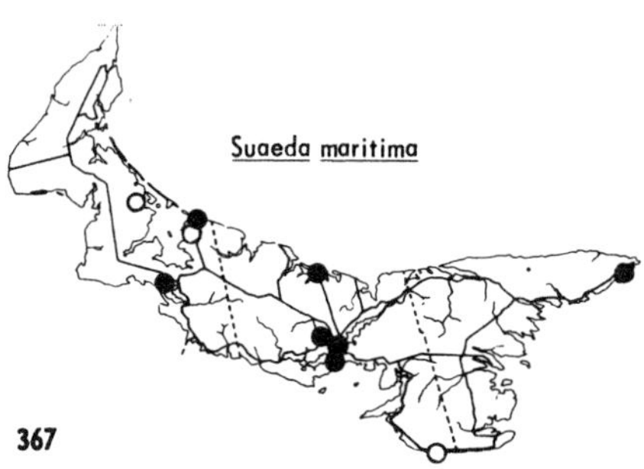

367

368

Salsola kali L., vars. Map 369.

Var. *kali* Coastal; sandy or gravelly beaches. Bain, MacSwain and Bain, MacSwain, Hurst.

Var. *tenuifolia* Tausch Railway weed: Emerald (1953), Charlottetown (1953), Brackley Station (1952), St. Peters (1950), from the west; native of Russia.

AMARANTHACEAE

Amaranthus retroflexus L. Map 370. Scattered: Ellerslie to Murray River, or farther, spread from towns to farms comparatively recently. Locally abundant in cultivation or along roadsides, in waste places. 4 Lots (Groh). MacSwain, Hurst.

**Amaranthus albus* L. Railway weed in Charlottetown; the record "Brackley Beach, cultivated ground, 1947-50" (B. Hurst), probably refers to *A. retroflexus*. (*A. gracizans*, Roland, not L.).

[*Acnida altissima* Riddell var. *subnuda* (S. Wats.) Fern. Specimen from L.W. Watson, determined by Fletcher, no longer extant. No other record east of Montreal. (*A. tuberculata* var. *subnuda* (S. Wats.) Uline & Bray, MacSwain, Hurst); Campbell.]

PORTULACACEAE

Portulaca oleracea L. Map 371. Common in Charlottetown; Union Corner (P). Naturalized from southwestern North America. Rapidly becoming an abundant weed of garden soils, flourishing in spite of hoeing since it does not wilt readily, and of mowing because prostrate. Lochhead's (1904) list of agricultural weeds, Campbell.

Montia fontana L. Map 372. (Walters, Watsonia 3: 1-6. 1953) Green's Shore at Summerside, about spring rills at border of a salt marsh, mixed with *Stellaria crassifolia* (*Fernald & St. John*). (*M. lamprosperma* Cham.).

Claytonia caroliniana Michx. Map 373. Local, central upland: Hunter River, Campbell's Mills, near New Glasgow, Wheatley River (MacSwain). Hardwood leaf mold; member of spring flora. MacSwain, Hurst.

CARYOPHYLLACEAE (including Illecebraceae, MacSwain, Gray's Manual (1908))

Scleranthus annuus L. Map 374. Western Prince; Hazel Grove (Q), Village Green; Morell and vicinity; Montague to Murray River. Sandy roadside shoulders; following stubble fields. Known to Roland (1947) evidently only from Dore's and his collections; listed 1891 and collected at Alberton in 1912 and Tignish in 1926. 1 Lot (Groh). MacSwain and Bain, MacSwain, Hurst.

369

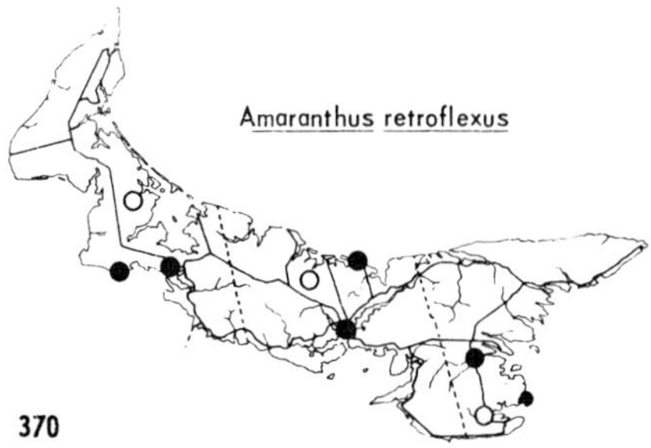

370

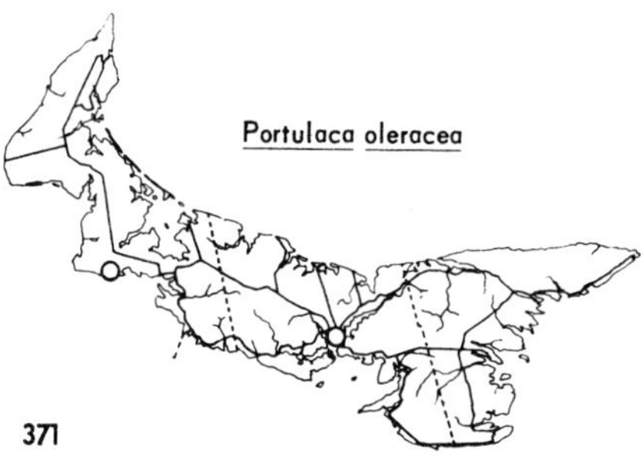

371

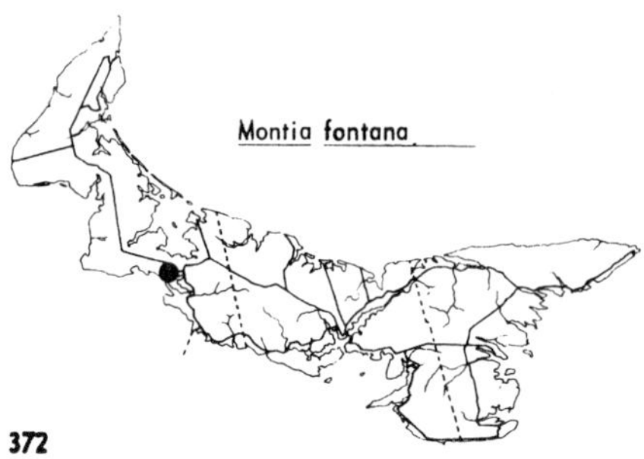

372

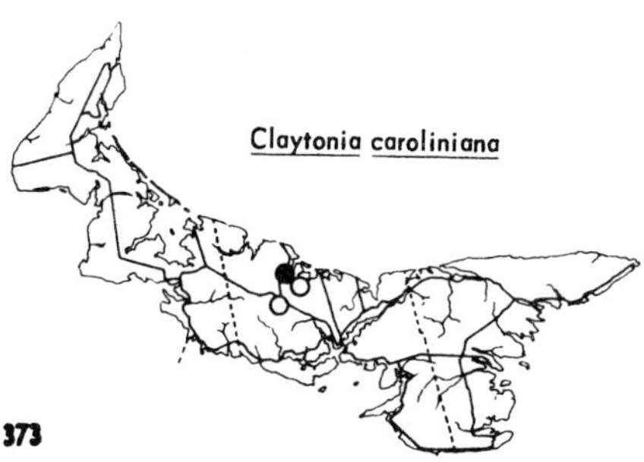

373

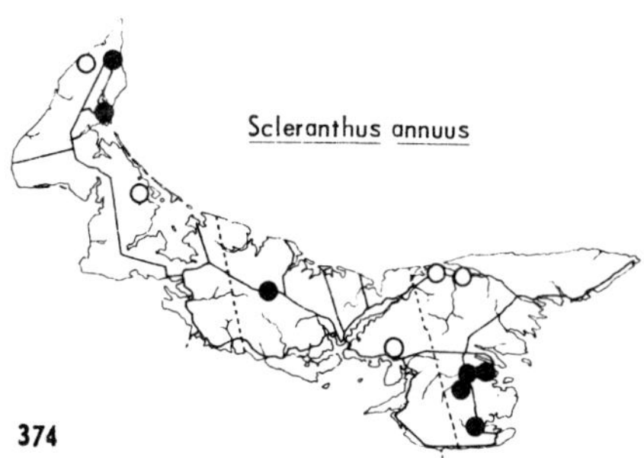

374

Spergularia rubra (L.) J. & C. Presl Map 375. West Prince; Kings and eastern Queens. Naturalized, European. Dry, open sandy soil: weed of roadsides and shores; 12 Lots (Groh). (*S. rubra* var. *campestris*, MacSwain and Bain; *Buda rubra* (L.) Dumort., MacSwain); Hurst.

Spergularia marina (L.) Griseb., vars. Map 376. Central Queens to West Prince; coastal. Open, usually saline sites: muddy patches in salt marsh, sand flats or nearby fields. (*Buda marina* (L.) Dumort., Macoun); Hurst.

Var. *marina* Alberton (cited by Rossbach in Rhodora 42: 127. 1940).

Var. *leiosperma* (Kindb.) Gürke Bunbury. Not sharply separated; transitional forms from Summerside and Charlottetown.

Spergularia canadensis (Pers.) G. Don Map 377. Coastal: scattered, Tignish to Souris. Locally abundant, salt marsh mud and damp brackish flats. (*Buda borealis* S. Wats., Macoun; *Spergularia borealis* (S. Wats.) Rob., Churchill); Hurst.

Spergula arvensis L. Map 378. Common; absent only from West Prince. European, naturalized. Weed of cultivation and scraped roadsides; 27 Lots (Groh), "serious" (Lochead). MacSwain and Bain, MacSwain, Hurst.

Sagina procumbens L. Map 379. Common in Queens; scattered in eastern Prince; Souris. Damp open ground: shores, springs, ditches, beaten paths, roadsides. Weedy. MacSwain, Hurst.

Sagina nodosa (L.) Fenzl Map 380. North Shore, around to Bothwell. Sand: damp, in dune slacks; dry, on lee of fixed dunes, becoming established in *Dicranum* moss tufts. MacSwain, Hurst.

Arenaria lateriflora L. Map 381. Throughout; seldom abundant. Thickets, mixed woods (in spruce litter), dune slacks and damp meadows. A peculiar, possibly diseased form, with congested inflorescence of greenish flowers, was collected at Charlottetown by Malte. (*Moehringia lateriflora* (L.) Fenzl); MacSwain and Bain, MacSwain, Hurst.

Arenaria serpyllifolia L. Map 382. Souris, large colony on waste sand behind beach dunes; perhaps formerly elsewhere. Adventive European weed. Not seen (Groh). MacSwain and Bain, MacSwain, Hurst.

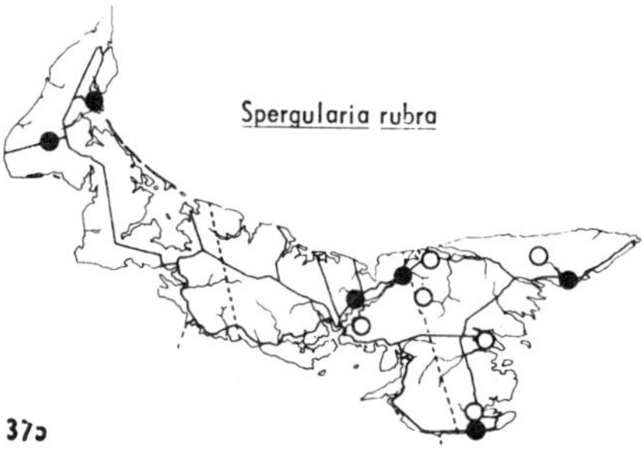

375

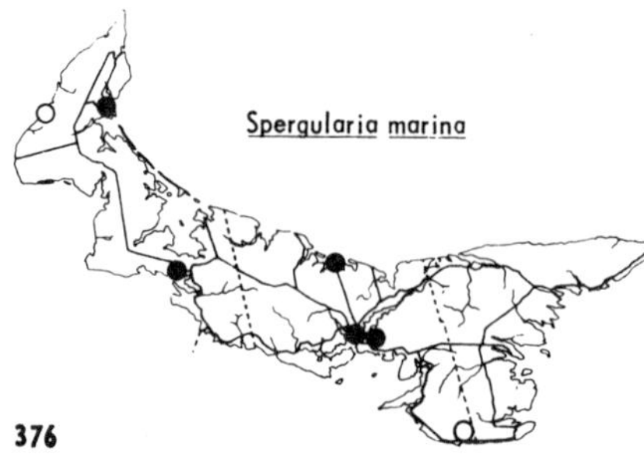

376

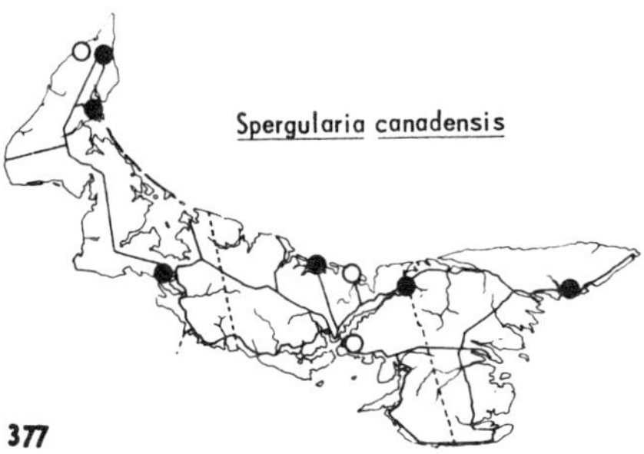

377

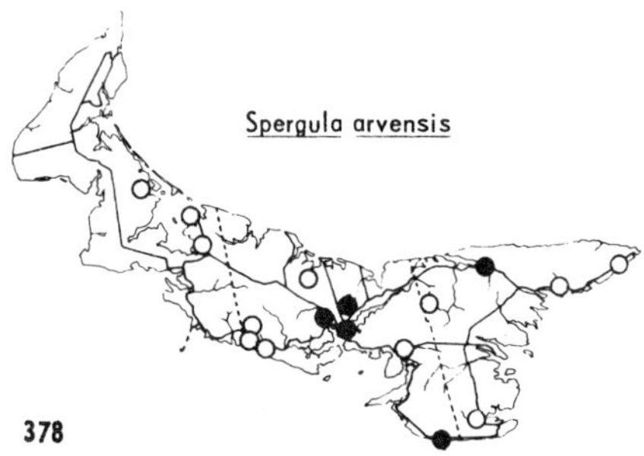

378

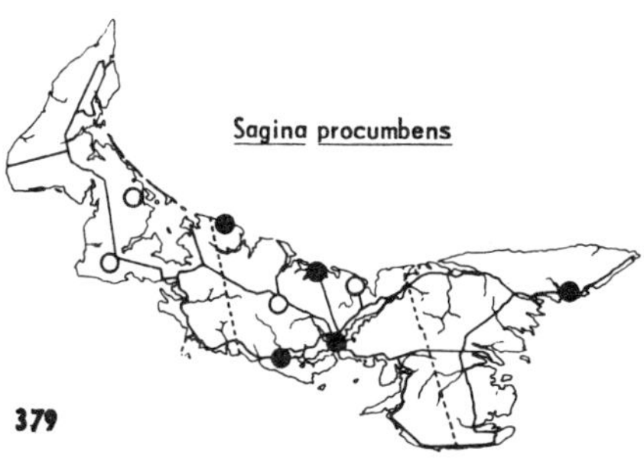

379

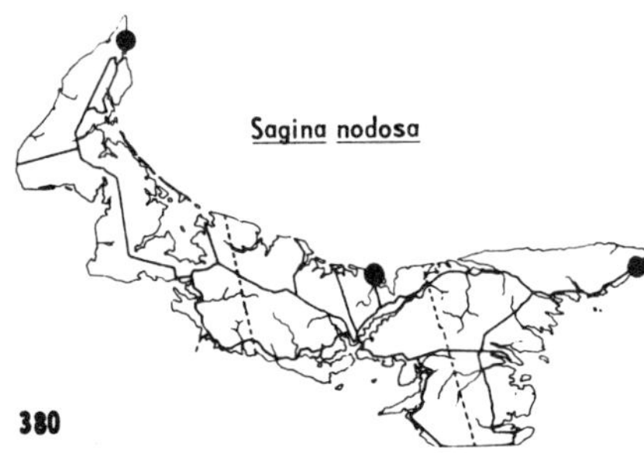

380

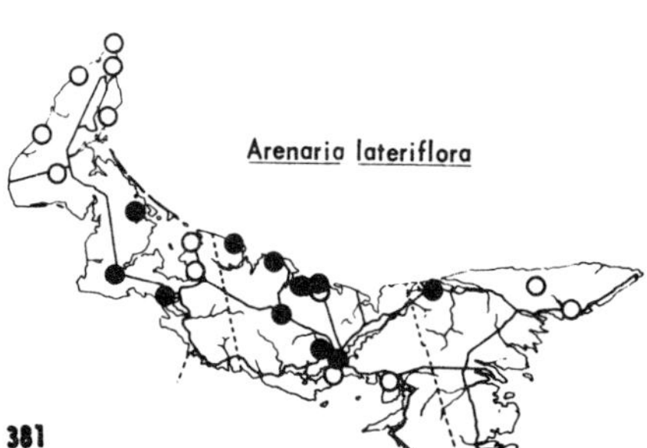

381

382

Arenaria peploides L. var. ***robusta*** Fern. Map 383. Coastal; rare. Sand beaches. MacSwain & Bain, MacSwain, Hurst.

[*Arenaria groenlandica* (Retz.) Spreng. Sterling Campbell's specimens from near New Glasgow were probably misidentified *Sagina*. Not otherwise known. MacSwain and Bain, dropped by MacSwain; Hurst 1940]

[*Arenaria stricta* Michx. MacSwain and Bain; dropped by MacSwain.]

Stellaria media (L.) Cyrillo Map 384. Scattered throughout. Naturalized, European. Weed of waste places and of cultivation; 22 Lots (Groh); too inconspicuous for recording by rapid survey methods, hence probably more common than indicated. MacSwain and Bain, MacSwain, Hurst.

Stellaria graminea L. Map 385. Common throughout. Naturalized, European. Weed of fields and grassy roadsides, 31 Lots (Groh). (*S. longipes*, MacSwain and Bain, MacSwain, Hurst); MacSwain, Hurst. *S. longifolia*, MacSwain and Bain.

Stellaria alsine Grimm Map 386. Common in the central upland: Charlottetown to Bedeque. Marshy runs and stream borders. (*S. uliginosa* Murr., J.M. Macoun, Can. Record Sci. Jan. 1895, 4), Campbell.

Stellaria humifusa Rottb. var. ***oblongifolia*** Fenzl Map 387. Rare, coastal; estuaries or brackish marsh: Souris, Rocky Point (where first noted by L.W. Watson), Alberton. Elsewhere on the east coast of North America known only from the Miramichi estuary of New Brunswick (Boivin, Ann. ACFAS 20: 98. 1954).

Stellaria crassifolia Ehrh. Map 387. Britain Pond near Bristol; Summerside, Green's Shore, about spring rills at border of a salt marsh, mixed with *Montia fontana* (*Fernald & St. John*). Fernald 1950.

Stellaria calycantha (Ledeb.) Bongard vars. Map 388. Local, more common near North Shore and in eastern uplands. Brooks and springs, usually woods.

Var. ***isophylla*** Fern. The more common variety; more or less transitional to var. *calycantha*. (*S. borealis* var. *isophylla* Fern., Rhodora 16: 151. 1914).

Var. ***floribunda*** Fern. Brackley Point Road, Grand Tracadie; specimens approaching var. *calycantha*. from Wood Islands Station. (*S. borealis* var. *corollina* J.M. Macoun, Can. Record Sci. Jan. 1895, 4).

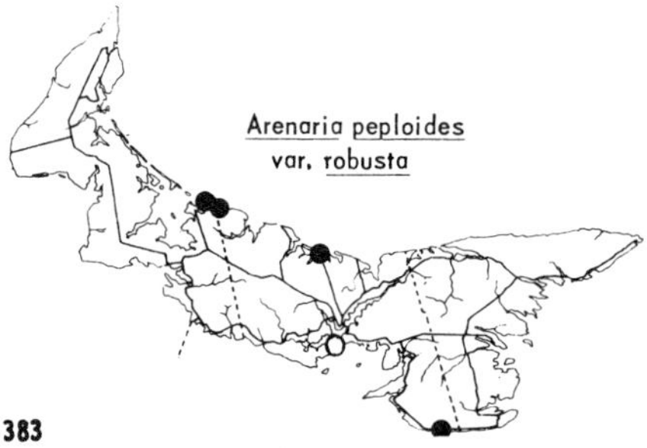

383

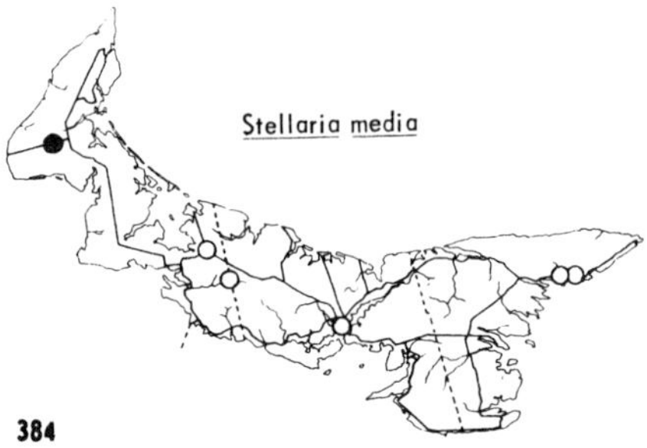

384

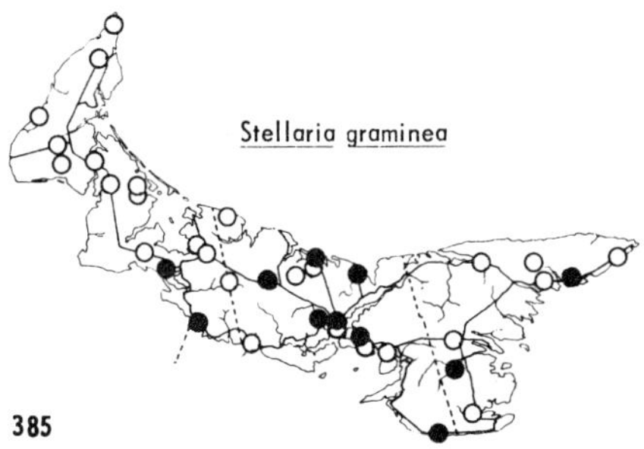

385

386

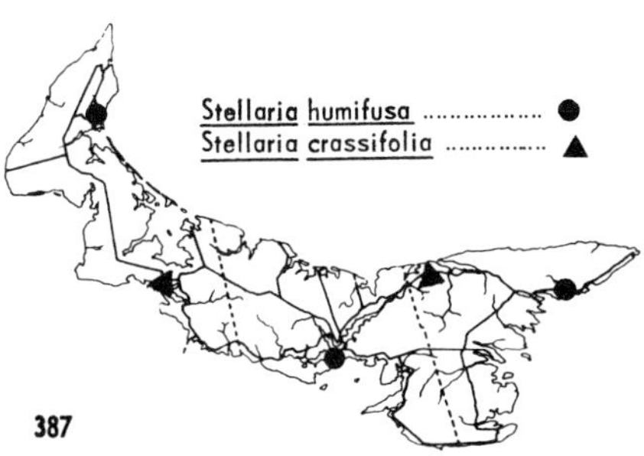

387

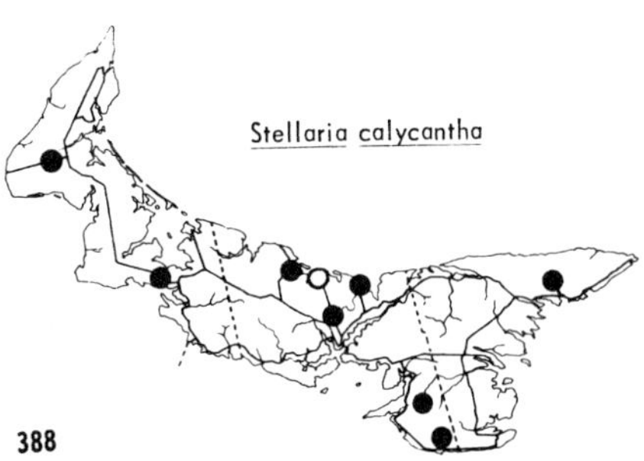

388

Cerastium vulgatum L. Map 389. Common, throughout. Weed of light or disturbed soil: fields, gardens, roadsides. 35 Lots (Groh), Lochhead. MacSwain and Bain, MacSwain, Hurst.

Cerastium arvense L. "WHITEWEED" Map 390. Scattered throughout; very common from Kensington to beyond Summerside. Naturalized. Locally abundant weed, pastures and edges of cultivation. 1 Lot (Groh who arrived late in July, too late for the conspicuous flowers); but "a bad weed in some parts of P.E.I." (Lochhead). MacSwain and Bain, MacSwain, Hurst.

[**Cerastium tomentosum** L. Souris, around dump on waste sand behind beach dunes. Introduced, but not naturalized.]

[*Cerastium viscosum* L. Not seen; report perhaps erroneous. MacSwain and Bain, MacSwain, Hurst.]

[**Agrostemma githago** L. Formerly a weed of grain fields, extinct since the use of pure seed. Poisonous. Prince County, 1 Lot (Groh). (*Lychnis Githago* (L.) Scop., MacSwain and Bain, MacSwain); Hurst.]

[**Lychnis chalcedonica** L. Rarely introduced garden plant; a single plant found escaped, at Tracadie Beach, 1901 (Churchill). Fernald 1950, Campbell.]

Lychnis alba Mill. Map 391. Rare: Summerside, Kensington, St. Peters. Naturalized, European. Weed of waste places and yards. 1 Lot, on the road to Georgetown (Groh). (*Melandrium* Garcke); Hurst.

Silene noctiflora L. Map 392. Rare weed: Summerside (1912), Brackley Point (1888), Hunter River (1952), Mount Herbert (1916). Naturalized, European. Waste or cultivated ground; 3 Lots (Groh). (*Melandrium* Fries); Hurst.

Silene cucubalus Wibel Map 393. Scattered persistent weed: Campbellton, Bideford, Borden, Hunter River, St. Peters. 6 Lots (Groh). MacSwain: (*S. latifolia* (Mill.) Britten & Rendle, Hurst); Campbell.

Saponaria officinalis L. Map 394. Around towns and villages, central Queens and Kings; introduced and naturalized, European. Spreading from old gardens and dumps. Double-flowered form usual; single in Charlottetown. 6 Lots (Groh). Hurst.

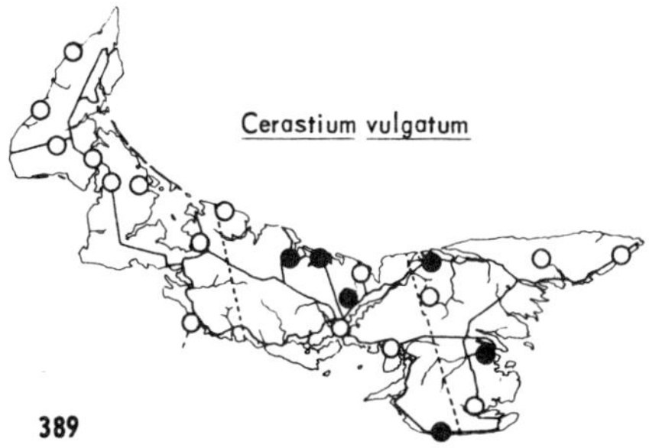

389

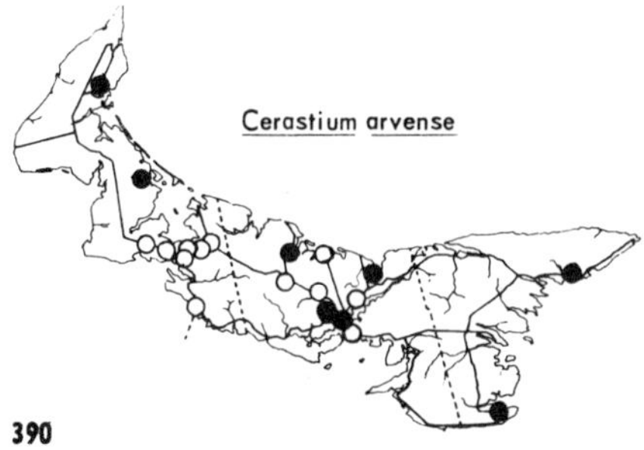

390

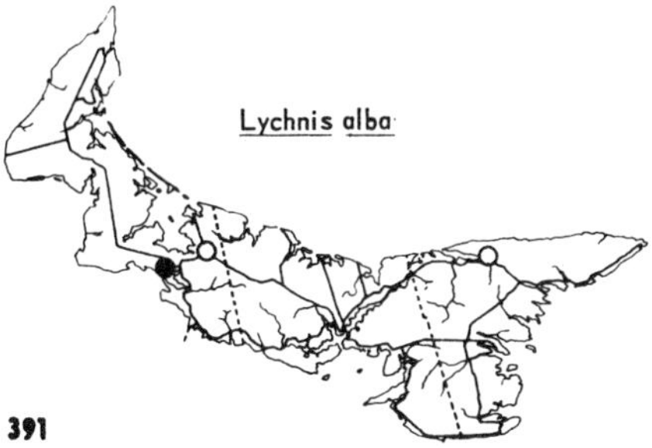

391

392

393

394

Dianthus armeria L. Map 395. Stanhope Beach, sandy soil, by hotel, 1944 (*Holman*); adventive, European. Groh and Frankton (Can. Weed Surv. 7: 68. 1949); Campbell.

Dianthus deltoides L. Panmure Island. European; naturalized for many years, spreading from an old garden. Reported as *D. barbatus* ? or *deltoides* in Can. Weed Surv. 7: 68. 1949. Hurst 1941.

CERATOPHYLLACEAE

Ceratophyllum demersum L. Map 396. Tracadie, barrier beach pond, in shallow water (*Fernald & St. John*); Pisquid Pond.

NYMPHAEACEAE

Nuphar variegatum Engelm. Map 397. Scattered from Alberton to Miscouche; bog ponds and lakes of east-central lowland. Shallow ponds and still streams. (*N. advena*, Bain 1892, MacSwain, Campbell; *Nymphaea advena* var. (Engelm.) Fern., Hurst).

* *Nymphaea odorata* Ait. Map 398. Rare: Lake Verde; pond on High Bank road, Murray River. Acid waters.

Nymphaea alba L. Charlottetown, Winsloe Road, Southport, New Zealand, East Baltic. European. Introduced at the Experimental Farm lily pond and distributed to owners of ponds; now established. (*Castalia odorata*, Hurst; *N. odorata*, Campbell).

RANUNCULACEAE

Ranunculus trichophyllus Chaix Map 399. Queens and Kings, on Alberry - Charlottetown soil series. Flowing streams with sandy bottom. (*R. aquatilis*, MacSwain and Bain; *R. aquatilis* var. *trichophyllus* (Chaix) Gray, MacSwain; var. *capillaceus* (Thuill.) DC., Hurst); Campbell.

Ranunculus cymbalaria Pursh Map 400. All around the coast. Salt marshes, sand flats, rarely fresh marsh at head of salt (as at Cape Wolfe, where forma *hebecaulis* Fern. prevails). MacSwain and Bain, MacSwain, Hurst.

Ranunculus gmelinii DC. var. *hookeri* (G. Don) Benson Map 401. Local; around Tignish, Grand Tracadie, Mount Stewart, Dundee, Albion Road and Wood Islands. Slow or still, acid (often peaty) waters. (*R. multifidus*, Macoun; *R. delphinifolius*, Hurst; *R. flabellaris*, Campbell; *R. Purshii*, Richards., Hurst; *R. Gmelini* var. *terrestrio* (Ledeb.) Benson, Roland); Campbell.

[*Ranunculus sceleratus* L. Not seen. Record not impossible. New Brunswick records. MacSwain, Hurst.]

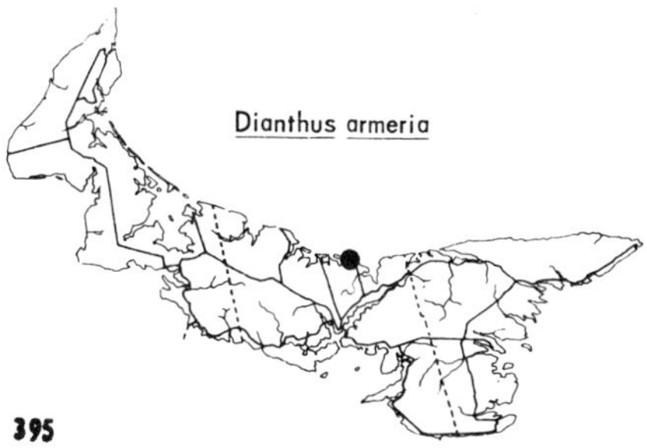

395

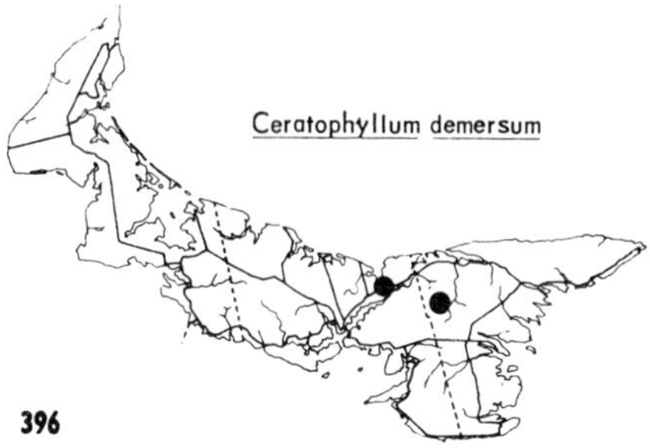

396

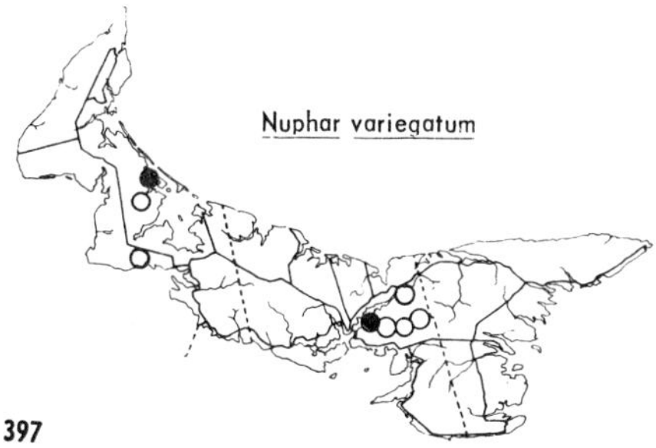

397

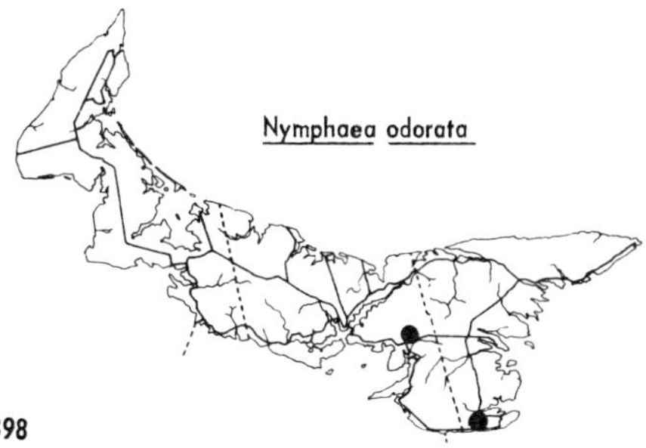

398

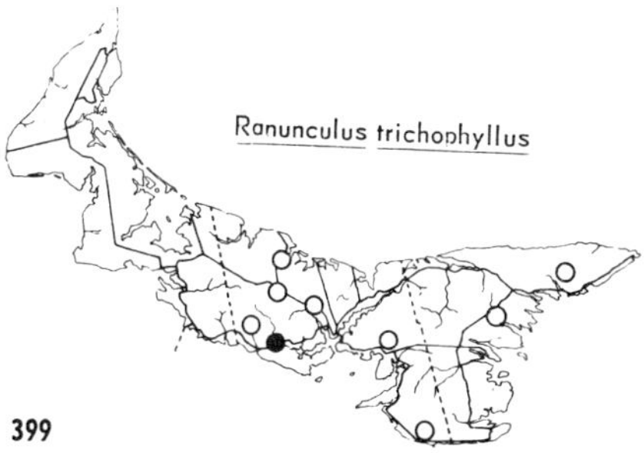

399

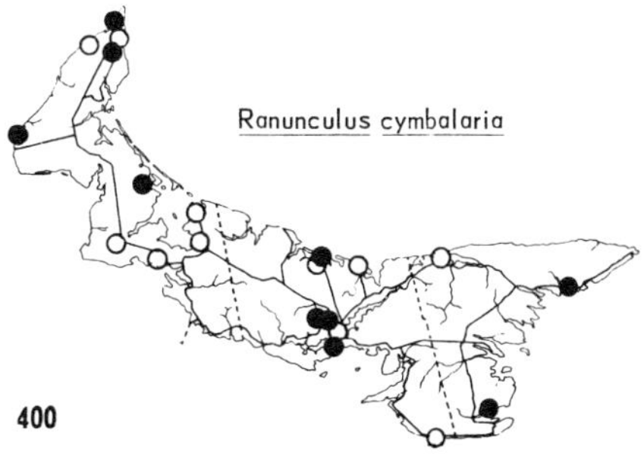

400

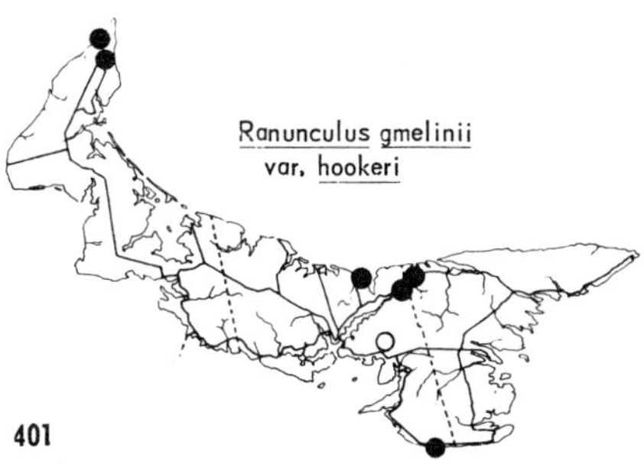

401

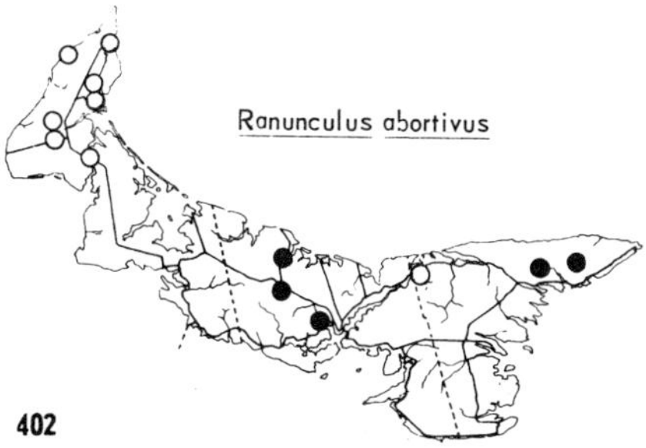

402

Ranunculus abortivus L. var. ***acrolasius*** Fern. Map 402. West Prince; central and northeastern uplands. Hardwood leaf litter. MacSwain, Hurst.

Ranunculus recurvatus Poir. Map 403. Harmony (K), rich deciduous woods (*Fernald & St. John*).

Ranunculus pensylvanicus L. f. Map 404. Around Tignish. Alder thickets, meadow.

Ranunculus septentrionalis Poir. Map 405. Rare in West Prince; Black River (Q). Wet woods. Macoun.

Ranunculus repens L. Map 406. Throughout, common except in thinly settled regions. Naturalized, European. Damp muddy ground, in open; in woods, only along streams, woods roads or in clearings. In shade or deep grass along streams, tall forms with few stolons (and these often from higher nodes). Common weed of damp fields; 23 Lots (Groh). MacSwain and Bain, MacSwain, Hurst.

Ranunculus acris L. Map 407. Common, throughout. Naturalized, European. Meadows and damper places; poisonous weed of pastures, spreading because ungrazed. 52 Lots (Groh).

Var. ***stevenii*** (Andrz.) Lange Summerside (Churchill 1902), according also to Boivin, Can. Field Nat. 65: 1-22. 1951. A late-in-the-season form. MacSwain and Bain, MacSwain, Hurst.

[***Ranunculus bulbosus*** L. Not seen; apparently less frequent in the Maritimes than formerly, even if record of MacSwain and Bain is correct.]

****Thalictrum confine*** Fern. Map 408. Rare: Campbellton, in alluvium of creek; possibly present elsewhere along estuaries. Specimen determined by B. Boivin.

Thalictrum polygamum Muhl., vars.

Var. ***polygamum*** Throughout, common. Thickets along streams. (***T. pubescens,*** Bain, MacSwain and Bain); MacSwain, Hurst.

Var. ***hebecarpum*** Fern. Mount Stewart (***Macoun***). Cited by Boivin (Rhodora 46: 478. 1944).

[***Hepatica americana*** (DC.) Ker, rare in N.S., seems a record without foundation. (***H. triloba,*** MacSwain and Bain).]

****Anemone canadensis*** L. Map 409. Damp clay soil of railway ditch and bank at edge of swampy woods halfway between O'Leary and Coleman. Probably not introduced.

Clematis virginiana L. Map 410. Local in Prince; rare eastward: Campbell's Mills near New Glasgow; Selkirk. Alder thickets; eastward along gravelly river-banks. Hurst 1940

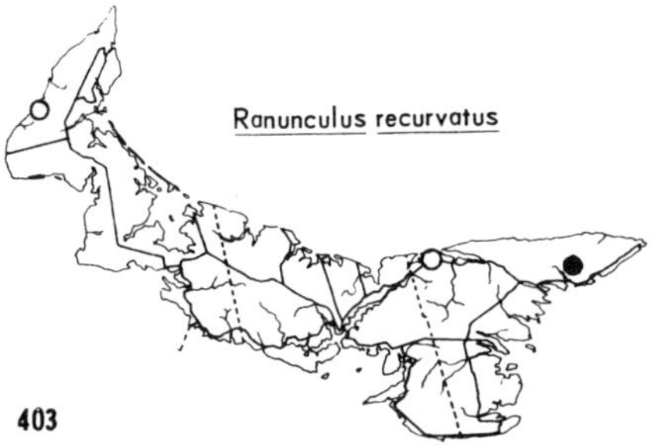

403

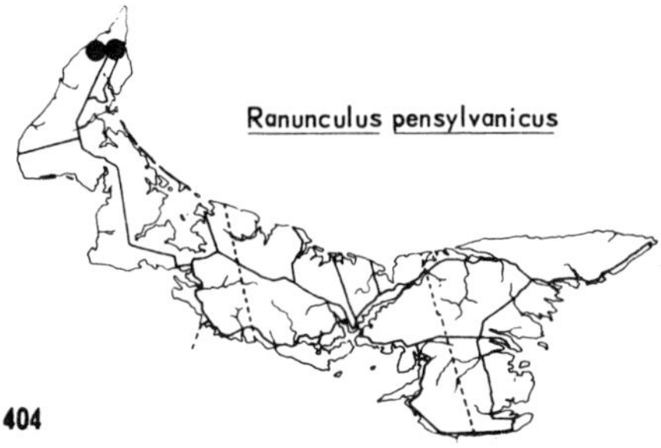

404

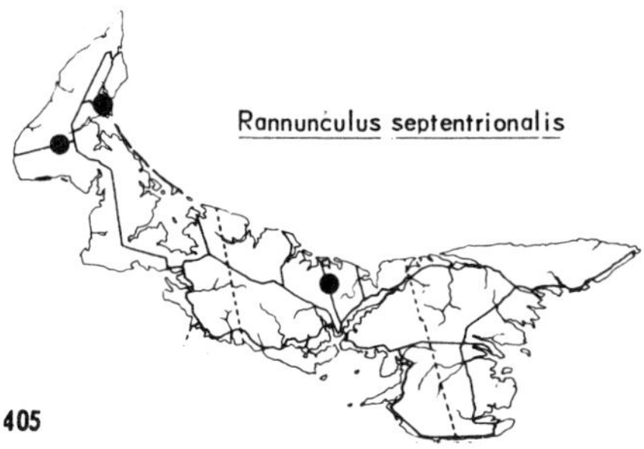

405

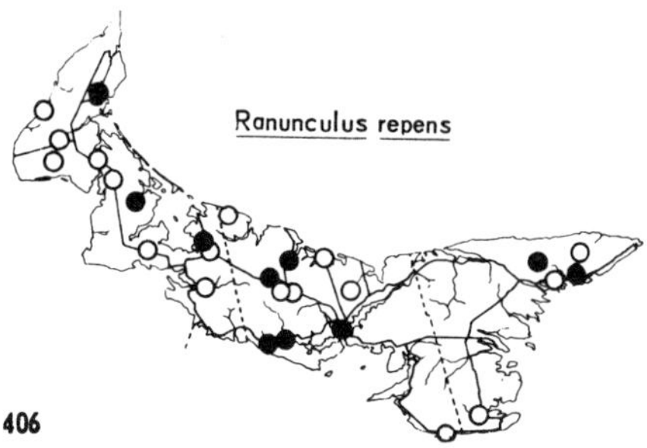

406

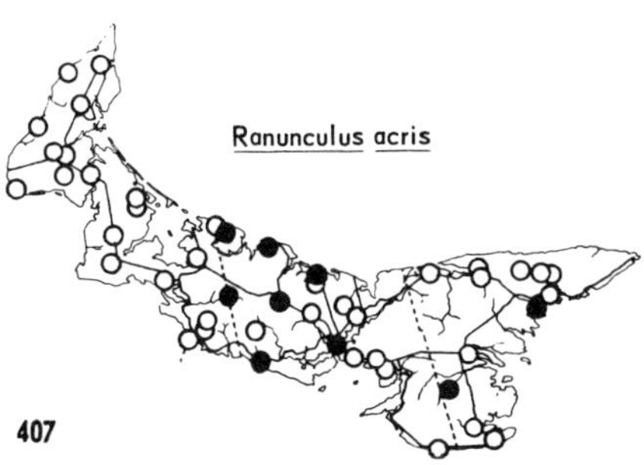

407

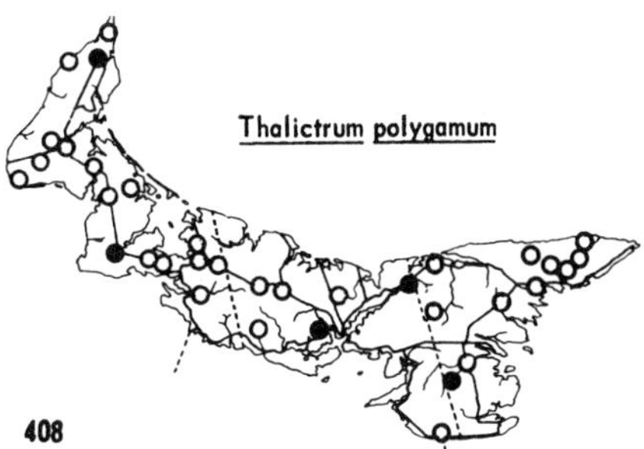

408

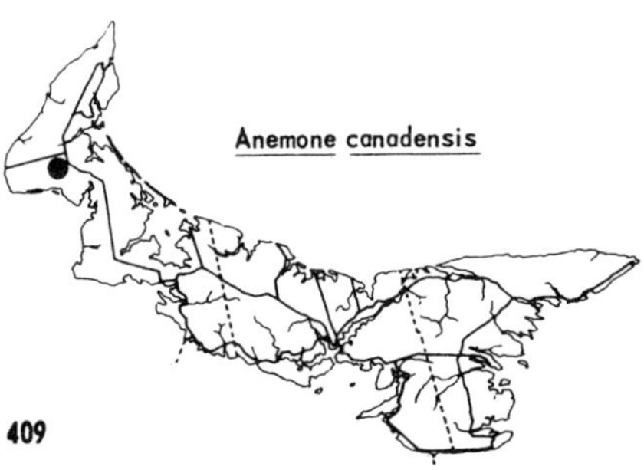

409

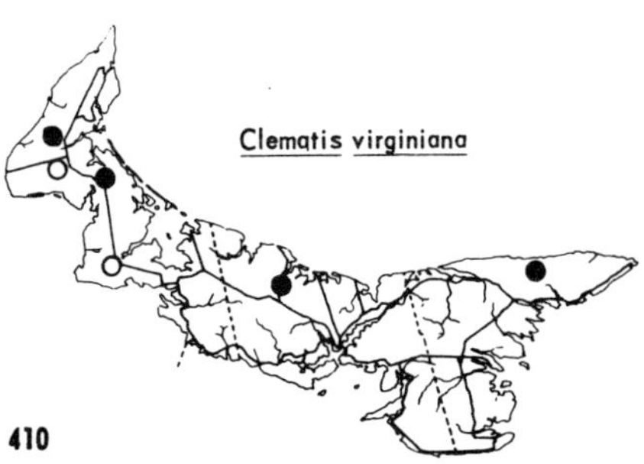

410

Caltha palustris L. Map 411. Throughout, common. Wet thickets in Prince County; stream banks eastward. Flowering in May and June, earliest in the east (Bain). MacSwain and Bain, MacSwain, Hurst.

Coptis groenlandica (Oeder) Fern. Map 412. Throughout; fairly frequent. Moss floor of coniferous woods; occasionally on bog hummocks. (*C. trifolia*, MacSwain and Bain, MacSwain, Hurst); Campbell.

Aquilegia vulgaris L. Map 413. Summerside, Hunter River, Brackley Beach, Suffolk, Souris. Introduced and locally naturalized garden plant, European. Grassy roadsides. Color variation rich, from purple to pink and white. Hurst 1940.

Actaea rubra (Ait.) Willd. Map 414. Scattered through Prince County and upland Queens; rare in the northeastern hills. Hardwoods, dry to damp sites. Poisonous; 4 Lots (Groh). (*A. Americana* Pursh, MacSwain and Bain; *A. spicata* var. *rubra* Ait., MacSwain); Hurst.

Forma ***neglecta*** (Gillman) Rob. Indian River, Dunk River, with the species.

Actaea pachypoda Ell. Map 415. Local in West Prince, central and northeastern uplands. Hardwood leaf mould. (*A. alba*, MacSwain, Hurst); Campbell.

BERBERIDACEAE

[***Berberis vulgaris*** L. BARBERRY Introduced as ornamental for hedges; formerly escaping but undesirable, being alternate host of wheat rust. About Charlottetown. 1 Lot (Groh). Hurst.]

[***Berberis thunbergii*** DC. Introduced as hedge plant, seldom escaping. Not harbouring wheat rust. Upton, Queens County.]

PAPAVERACEAE (including FUMARIACEAE, MacSwain and Bain, MacSwain, Hurst)

[*Sanguinaria canadensis* L. MacSwain and Bain, record abandoned.]

[*Chelidonium majus* L. Charlottetown, 1950. Adventive, European. Campbell.]

[**Eschscholzia californica* Cham. Introduced garden plant; scarcely escaping. Souris, not established.]

[*Dicentra cucullaria* (L.) Bernh. Apparently reliable reports of this species failed to produce any locality record. Requires substantiation. Hardwood forest floor. MacSwain, Hurst.]

Corydalis sempervirens (L.) Pers. Map 416. Scattered: not seen in Kings County. Weedy on rocks or thin soil, now in clearings, burns and roadsides in woods. (*C. glauca* Pursh, Bain, MacSwain and Bain, MacSwain); Hurst.

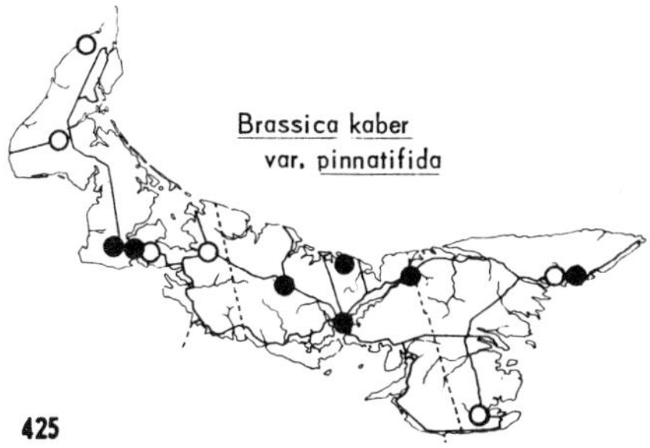

425

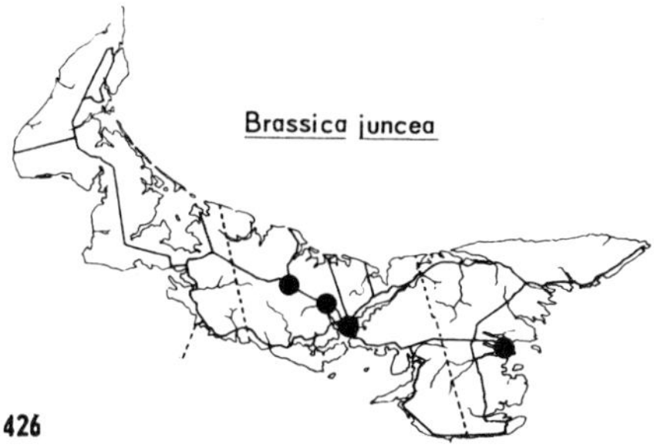

426

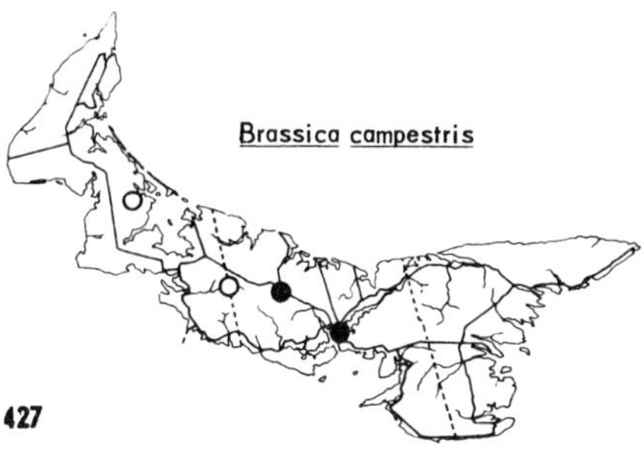

427

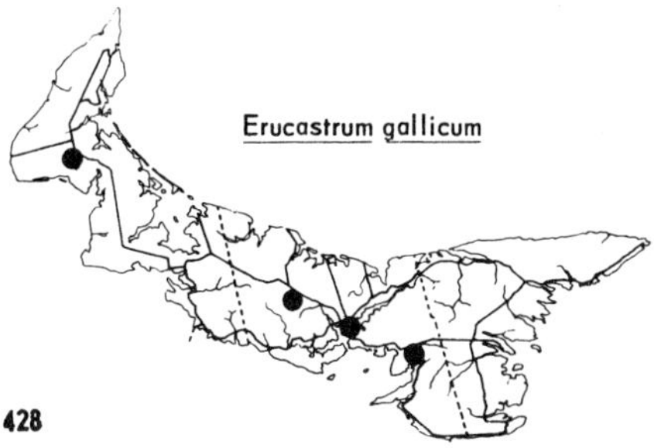

428

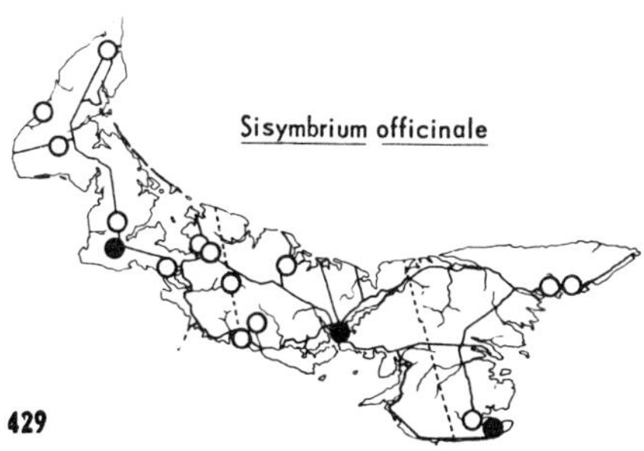

429

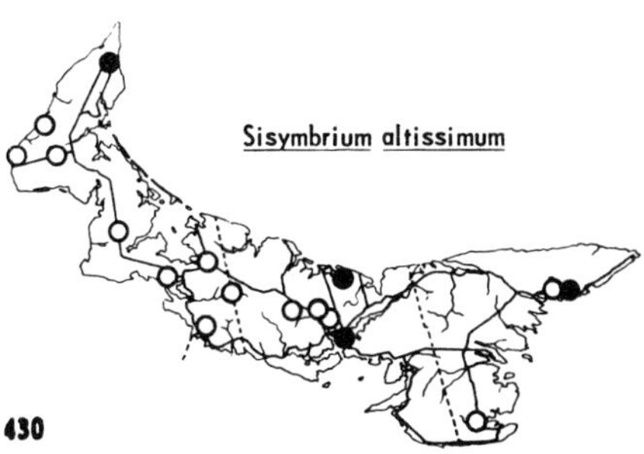

430

Sisymbrium altissimum L. Map 430. Scattered throughout. Naturalized, European. Waste ground, particularly along railway. Weed in 7 Lots (Groh). Hurst.

Descurainia sophia (L.) Webb Map 431. Scattered, Charlottetown (1926-present) to Ellerslie (1953). Naturalized, European. (*Sisymbrium Sophia* L., Groh, Hurst); Campbell.

Hesperis matronalis L. Map 432. Charlottetown Experimental Farm; Souris; O'Leary; Margate. Introduced European. Garden perennial, spreading to roadsides and open woods. Campbell.

Erysimum cheiranthoides L. Map 433. Scattered throughout; rather more common in Prince. First collected at Alberton (1912). Naturalized, European. Weed of cultivation, less often in waste places; 9 Lots (Groh). Hurst.

[****Rorippa sylvestris*** (L.) Bess. Southport (1951-2), introduced to garden in stable litter from Charlottetown, overrunning it and eradicated by use of 2,4-D. Adventive, European.]

Rorippa islandica (Oeder) Borbás, vars. Map 434. Local, Tignish to St. Peters. Creeks, pond margins, muddy fields. (*Radicula palustris* (L.) Moench, Groh, Hurst); Campbell.

Var. ***islandica*** Weed of wet fields, St. Peters; Charlottetown.

Var. ***fernaldiana*** Butters & Abbe Along creeks: Wellington, Tignish. (var. *microcarpa* (Regel) Fern., not (Beck) Thell, Roland).

Nasturtium officinale R. Br. Map 435. Local but throughout. Naturalized, European. Abundant in slow streams, where of trailing habit; tall erect plants in dry mud along road, Mount Herbert (Q). Introduced, 1 Lot (Groh). (*Redicula Nasturtium-aquaticum* (L.) Britten & Rendle, Groh, Hurst); Campbell.

Armoracia rusticana (Lam.) Gaertn., Mey. & Scherb. Map 436. Charlottetown. Naturalized, European. Introduced horse-radish, escaped and perennial. 3 Lots (Groh). (*Nasturtium Armoracia* (L.) Fries, MacSwain; *Radicula Armoracia* (L.) Rob., Groh, Hurst; *Armoracia lapathifolia* Gilib., Campbell).

Barbarea vulgaris R. Br. var. arcuata (J. & C. Presl) Opiz Map 437. Scattered, apparently absent from West Prince and rare in the east. Weed of waste places and fallow fields; not seen in August! (Groh). Hurst, Campbell. (*B. stricta*, Groh, Hurst; *B. orthoceras*, Campbell).

[*Barbarea verna* (Mill.) Asch. Reported from Iona (1937) by Hurst, 1940. Specimen not seen; presumably *B. vulgaris*.]

Cardamine pensylvanica Muhl. Map 438. Local, more common in upland regions. Damp, often shaded sites: stream margins, thickets, springy marshes. (*C. hirsuta*, MacSwain, Hurst); Groh, Hurst.

SARRACENIACEAE

Sarracenia purpurea L. Map 439. Western Prince; eastern Queens and Kings. Peat bogs. Bain, MacSwain and Bain, MacSwain, Hurst.

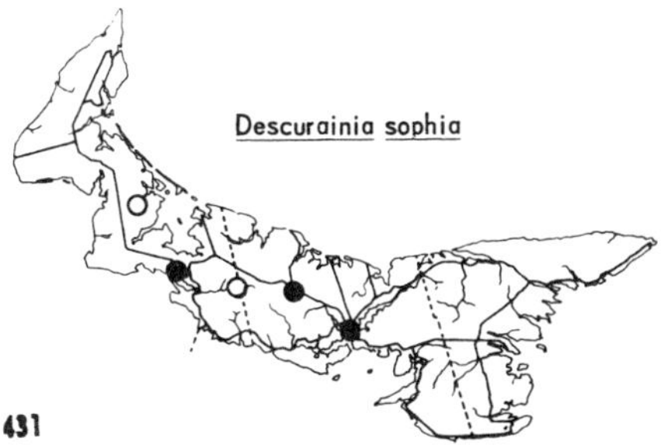

431

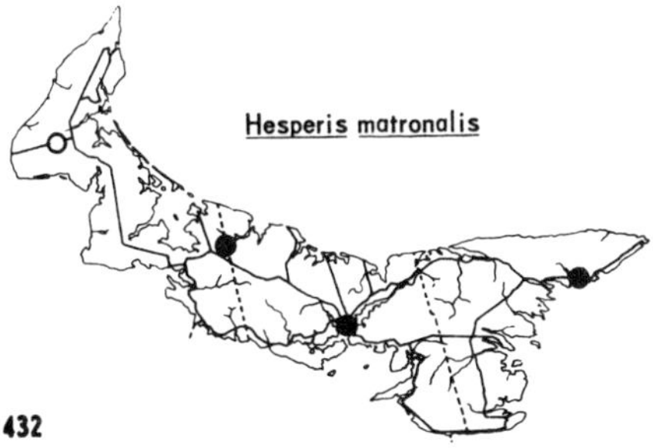

432

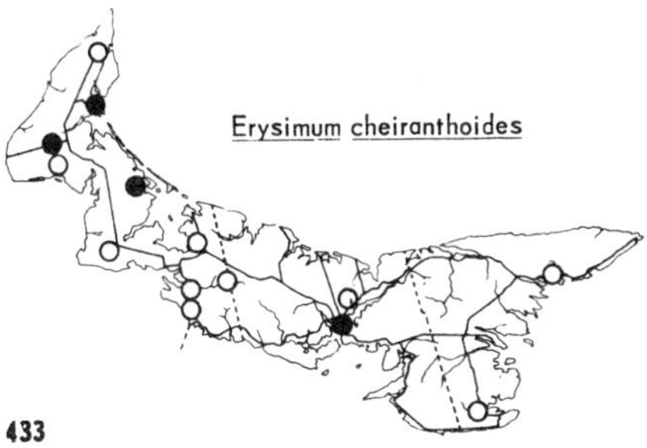

433

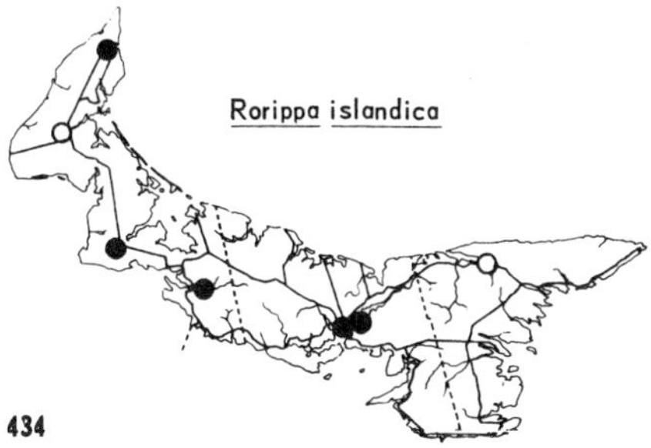

434

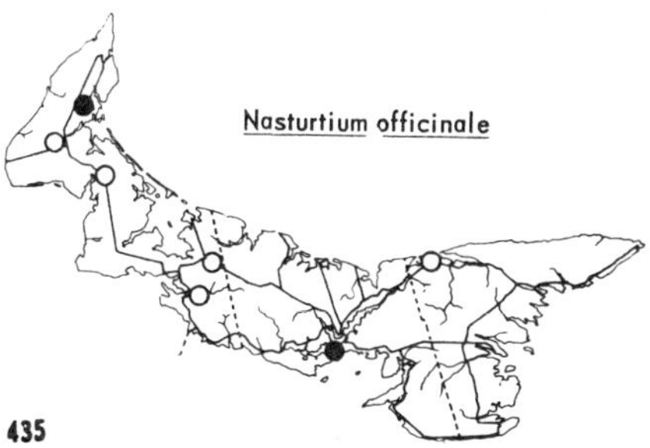

435

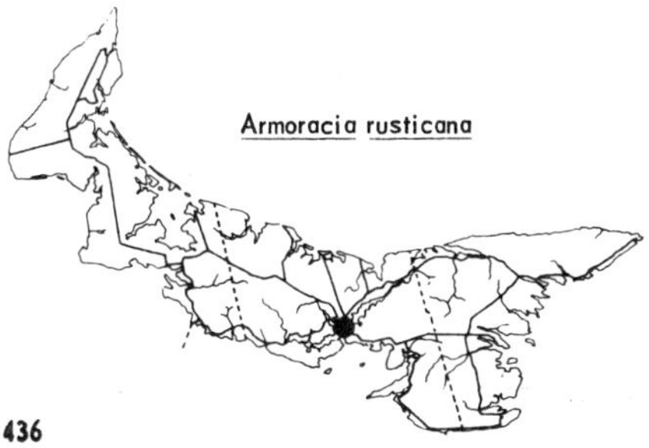

436

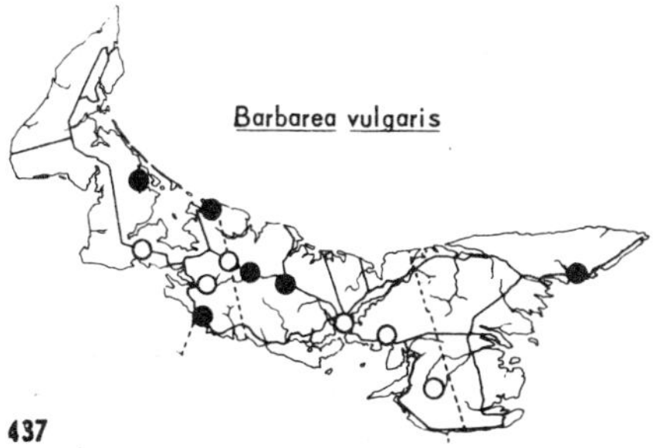

437

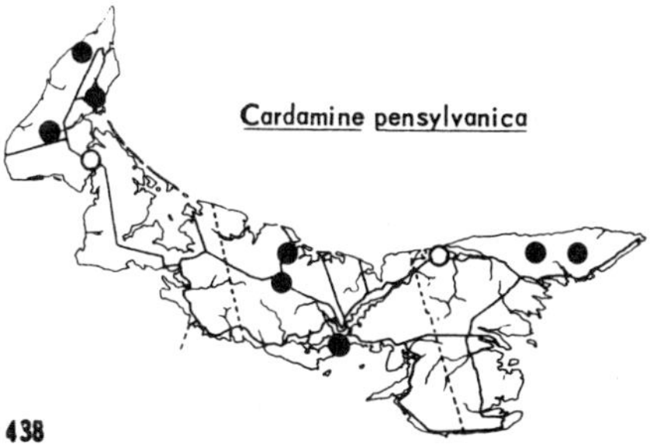

438

DROSERACEAE

Drosera rotundifolia L. Map 440. Lowlands and coasts. Peat: bogs, swamps, pond margins. MacSwain and Bain, MacSwain, Hurst.

Drosera intermedia Hayne Map 441. Bogs: Mount Stewart, Mermaid Lake.

CRASSULACEAE

Crassula aquatica (L.) Schönl. Map 442. Mount Stewart, Grand Tracadie. Muddy pond shores. (*Centunculus minimus*, Macoun; *Tillaea simplex* Nutt., J.M. Macoun, Can. Record Sci., Jan. 1895, MacSwain; *T. aquatica* L., Hurst, Campbell; including *T. Vaillantii*, Churchill 1902, J.M. Macoun, Ottawa Nat. 16:216. 1903, Adams, Hurst 1940).

Sedum purpurem (L.) Tausch "LIVE-FOREVER" Map 443. Scattered. Introduced and naturalized, European. Roadsides banks and ditches, weedy perennial. No number of lots (Groh). (*S. Telephium*, MacSwain); *S. triphyllum* (Haw.) S.F. Gray, Roland); Hurst.

[****Sedum acre*** L. Souris, around dump behind beach dunes. Introduced, adventive, European rock-garden flower.]

[****Sedum rupestre*** L. Souris, overgrown dump on dunes. Introduced, adventive, European rock-garden flower. Referable to the ssp. *rupestre*.]

SAXIFRAGACEAE

Mitella nuda L. Map 444. Common in Prince and upland Queens, rarer eastward. Damp shady sites: mossy woods along streams. MacSwain and Bain, MacSwain.

Chrysosplenium americanum Schwein. Map 445. Scattered throughout. Shady, wet sites: in and around shallow pools in brooks, springs, on thin mud. MacSwain, Hurst.

Parnassia parviflora DC. Map 446. West Prince coast: Cape Wolfe (collected here in 1893 by W.J. Wilson of the Canadian Geological Survey, then field assistant to Robert Chalmers; basis of record in Gray's Manual (1950), rediscovered 1953); Nail Pond. Damp, fresh sites exposed to the sea: springy clay bank; dune slacks. (*P. palustris*, MacSwain, Hurst, Campbell).

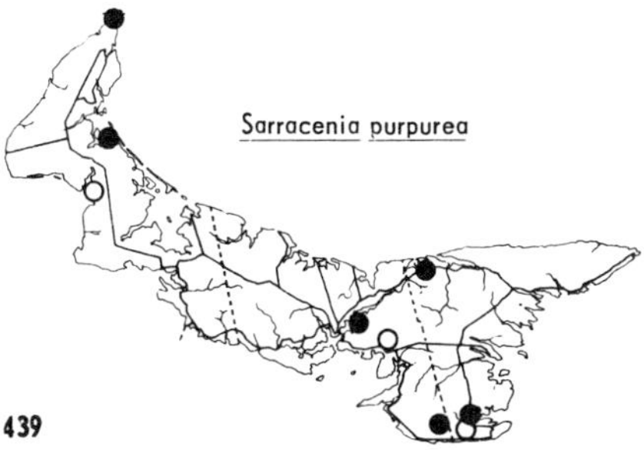

439

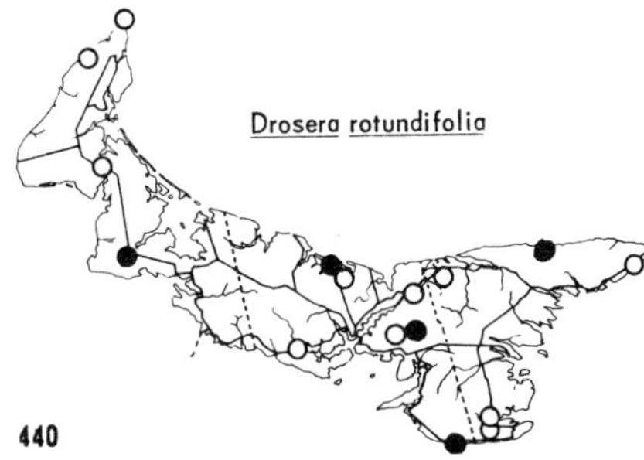

440

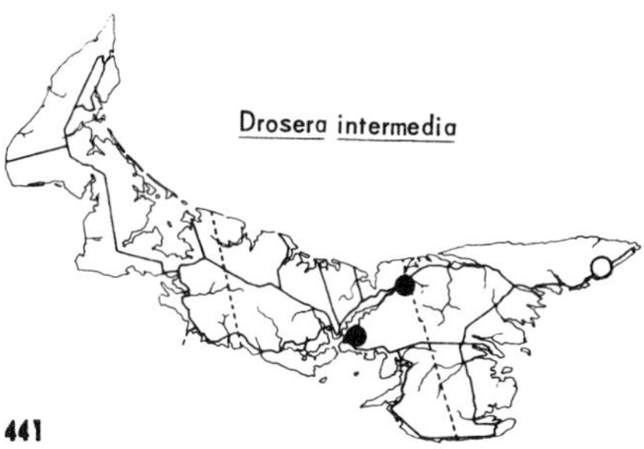

441

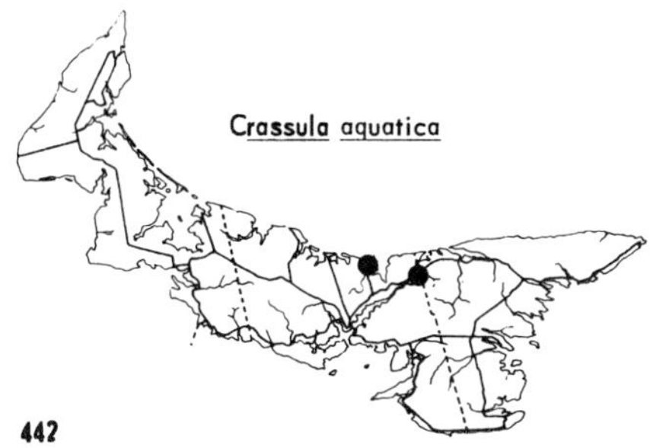

442

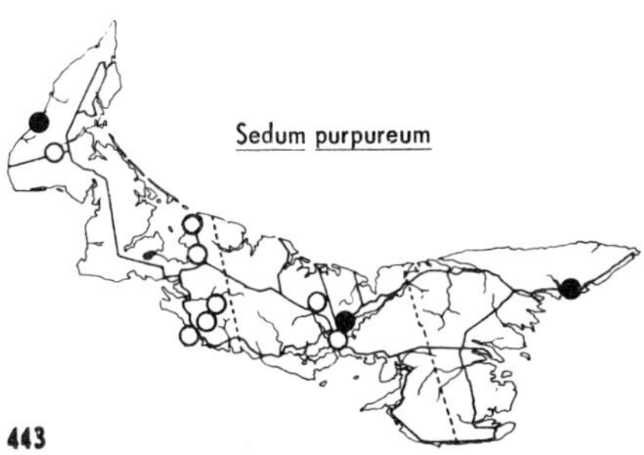

443

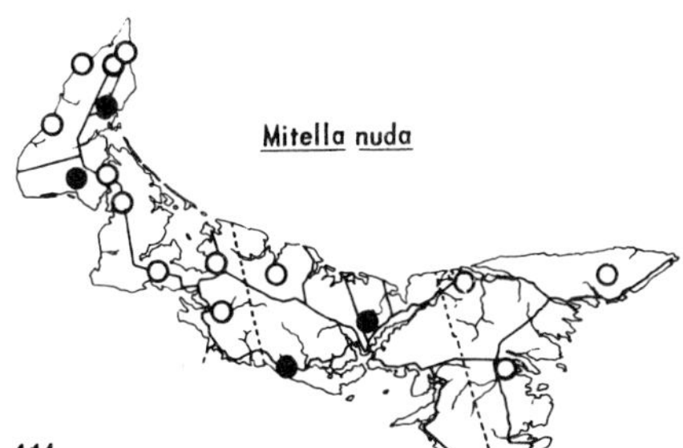

444

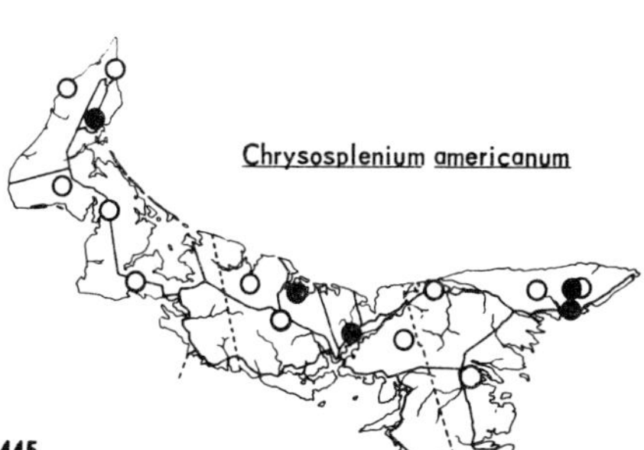

445

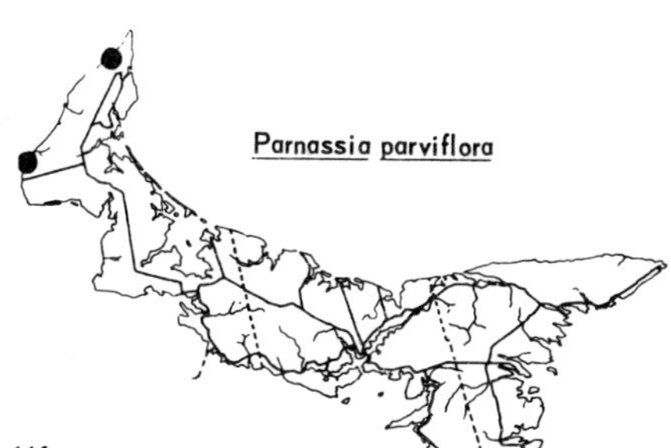

446

GROSSULARIACEAE

Ribes hirtellum Michx. var. ***hirtellum*** Map 447. Scattered: common in Prince, Kings; apparently absent from central uplands. Open sites, dry or damp: thickets, fencerows, stream banks. "The wild Gooseberry of French cleared lands, and earliest fruit for baking"-Stewart. (*R. Grossularia,* Stewart; *R. rotundifolium,* MacSwain and Bain; *R. oxyacanthoides,* MacSwain, Hurst, Campbell).

Ribes lacustre (Pers.) Poir. Map 448. Throughout, particularly common in Prince County. Damp, shady sites: mucky places, brook banks in hardwoods, never in peaty swamps. Fruit "harsh and disagreeable"-Stewart. (*R. nigrum,* Stewart; *R. floridum,* MacSwain and Bain, MacSwain, Hurst; *R. americanum,* Campbell); Hurst.

Ribes glandulosum Grauer Map 449. Common in West Prince, local eastward. Damp, open woods. (*R. rigens* Michx., MacSwain and Bain; *R. prostratum* L'Hér., Hurst); Campbell.

Ribes triste Pall. Map 450. Local in Prince County and north-eastern hills. Damp, shady sites: mossy mixed woods, alder thickets along streams. (*R. rubrum,* var. *subglandulosum* Maxim., MacSwain); Hurst.

Ribes sativum Syme RED CURRANT Charlottetown Experimental Farm. European, introduced fruiting bush. Occasionally escapes to hedge and woods; noted since 1912.

Ribes nigrum L. BLACK CURRANT Map 451. Rocky Point (*Fernald & St. John);* Bain's Creek near Tignish, thicket along stream. European, introduced for fruit and locally naturalized.

ROSACEAE

Spiraea latifolia (Ait.) Borkh. Map 452. Throughout, common except in the uplands. Damp open places; swamps, thickets, abundant and weedy in damp overgrazed pastures; 25 Lots (Groh). (*S. salicifolia,* MacSwain and Bain, MacSwain); Hurst.

Spiraea tomentosa L. Map 453. Common in southern Kings, very local in Queens, common in western Prince. Damp, acid places; in east in clearings, weed in 1 Lot (Groh); in west often in bogs. Hurst.

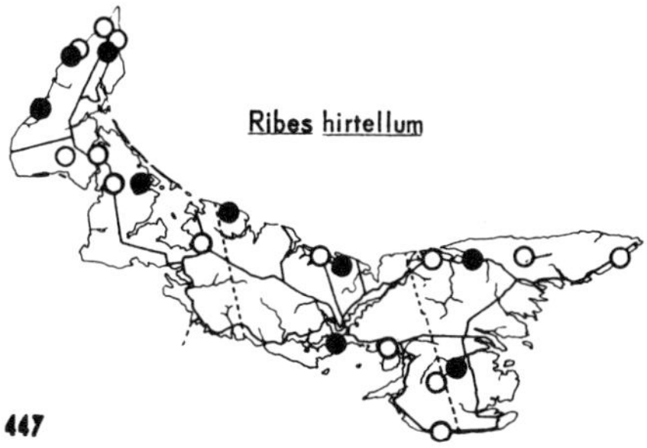

447

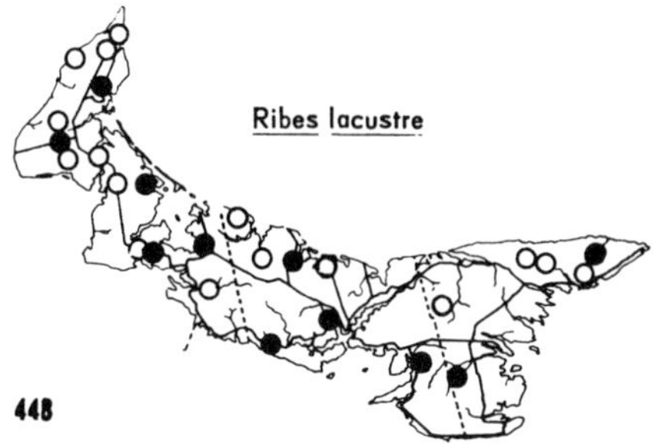

448

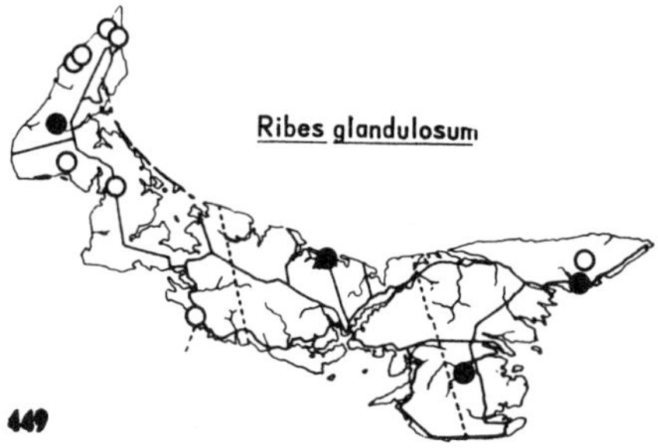

449

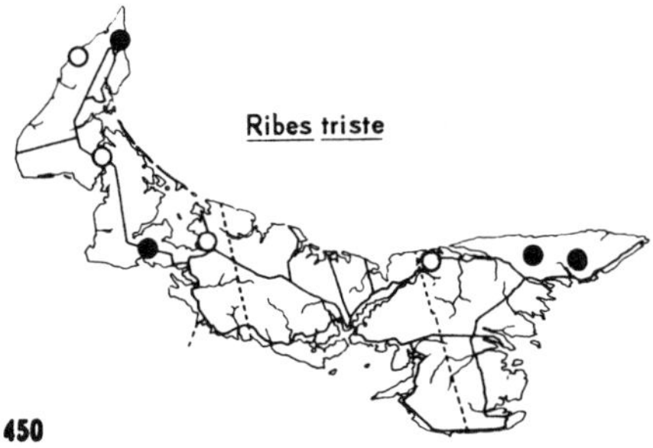

450

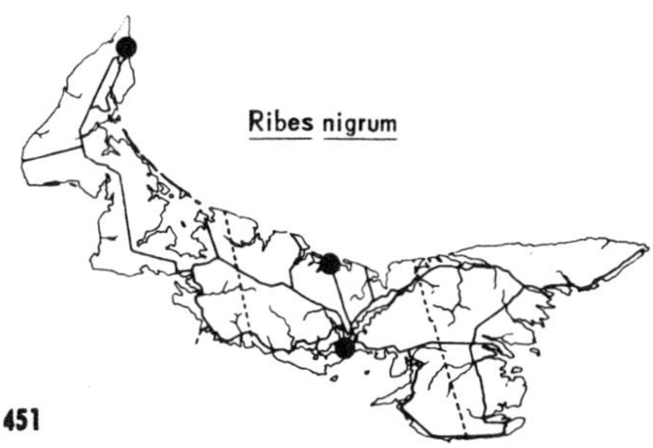

451

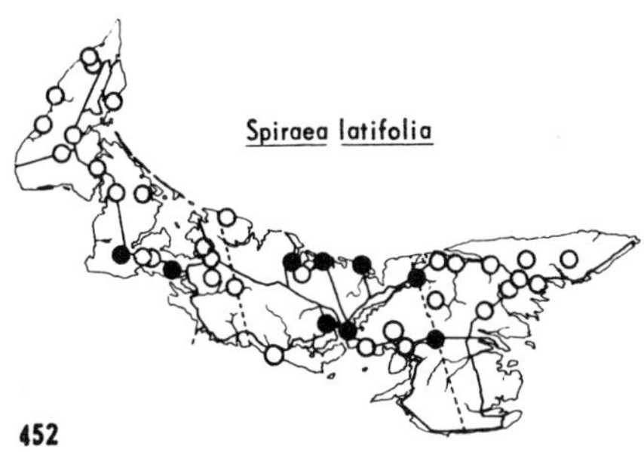

452

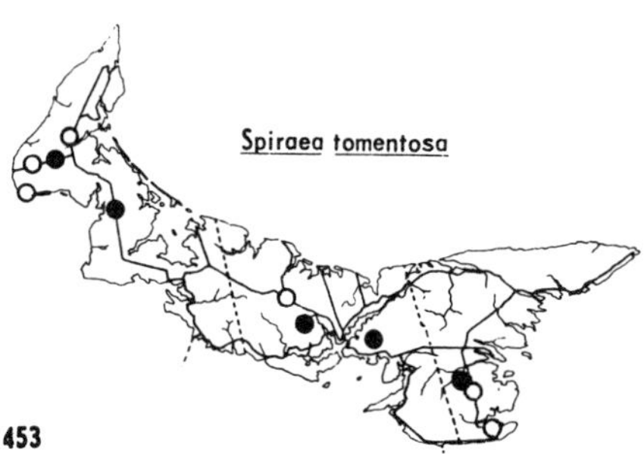

453

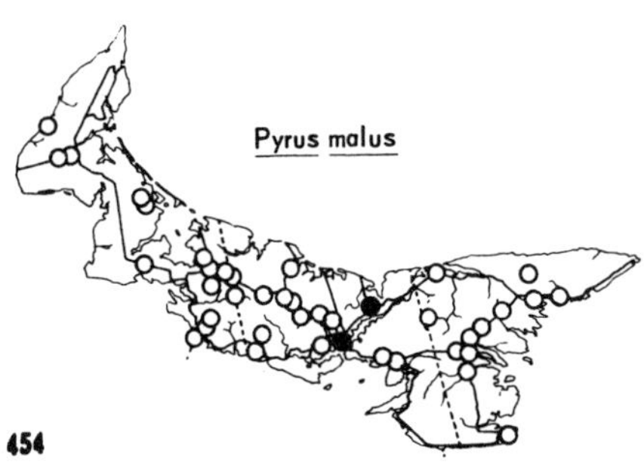

454

[**Sorbaria sorbifolia* (L.) A. Br. Spreading into lawns and persistent around old house sites: Charlottetown, Bonshaw. Ornamental shrub, till recently much planted. Asiatic.]

Pyrus malus L. Map 454. Seedlings established along roadsides throughout. European. 24 Lots (Groh). Formerly much planted around farmhouses; nowadays mostly neglected; the varieties of 1890 not replaced. Seedlings do not conform to the cultivated varieties. (***Malus pumila*** Mill.), Hurst.

Aronia prunifolia (Marsh.) Rehder Map 455. Scattered throughout, more frequent along North Shore. Thickets; swamps or acid sands. Fruit juicy but sour, considered edible by Bain (1890), known elsewhere as Chokeberry. (***Pyrus arbutifolia***, Bain, MacSwain and Bain, MacSwain, Hurst; *P. floribunda* Lindl.).

Hybrids with ***Sorbus*** spp. occur: O'Leary; also the "***Pyrus arbutifolia*** X ***sitchensis***" of Hurst.

Aronia melanocarpa (Michx.) Ell. Map 456. Rather local, in bogs, where the usual species; or swamps and acid sands as for *A.* ***prunifolia***. Forms transitional to *A.* ***prunifolia*** occur; e.g., at North Point, specimens had the glabrous foliage of *A.* ***melanocarpa***, the tomentose-hairy pedicels of *A.* ***prunifolia***. (***Pyrus melanocarpa*** (Michx.) Willd., Bain, MacSwain and Bain).

Sorbus aucuparia L. Map 457. Widely planted in older settlements: European introduced ornamental. Occasionally naturalized, seed spread by birds. (*P.* ***aucuparia*** (L.) Ehrh.), Bain 1892.

Sorbus americana Marsh. Map 458. Scattered throughout; frequent in Prince and along North Shore. Low sandy or swampy fir mixed woods. (***Pyrus americana***, MacSwain and Bain, MacSwain, Hurst; *P.* ***microcarpa*** (Pursh) DC.).

[***Sorbus decora*** (Sarg.) Schneid. Map 459. Not certainly present: material collected at Brackley Point and Charlottetown by Fernald and St. John determined as *S.* ***aucuparia*** by Jones. (***Pyrus sitchensis***, Hurst; *P.* ***decora*** (Sarg.) Hyland, Campbell).]

Amelanchier wiegandii Nielsen Map 460. Common throughout. Dry hilly or sandy sites: usually at wood-edges or in fencerows; at Bothwell, on the beach and cliff.

Amelanchier stolonifera Wieg. Rare: Dundee; Bristol (Lot 40 Station); Mermaid; Bothwell (?). At Bothwell, plants intermediate between *A.* ***stolonifera*** and both *A.* ***wiegandii*** and *A.* ***laevis*** occur, the latter determined by Wiegand as hybrids.

Amelanchier fernaldii Wieg. Recorded by Fernald (1950). Described originally as an apparent hybrid of *A.* ***stolonifera*** and *A.* ***laevis:*** this hybrid (as determined by Wiegand in 1914) occurring from Cavendish scattered to Bothwell. (*A.* ***ovalis***, Bain (?); *A.* ***alnifolia***, MacSwain (?). Identities doubtful).

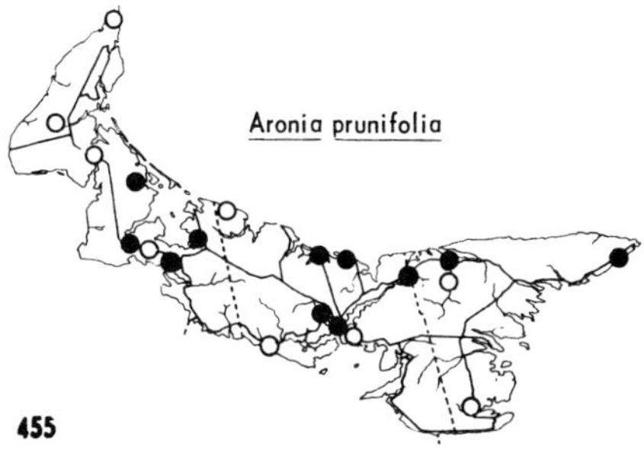

455

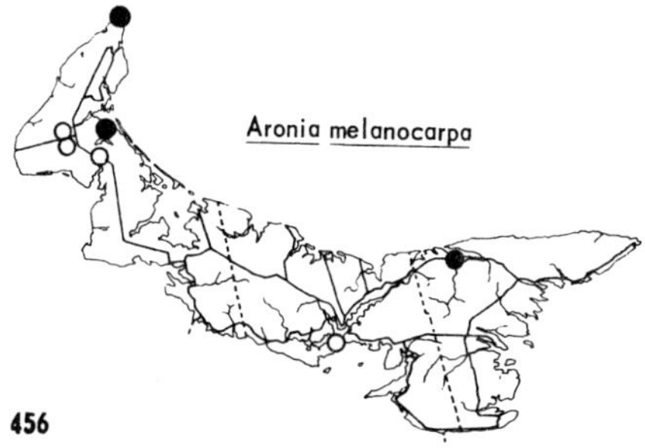

456

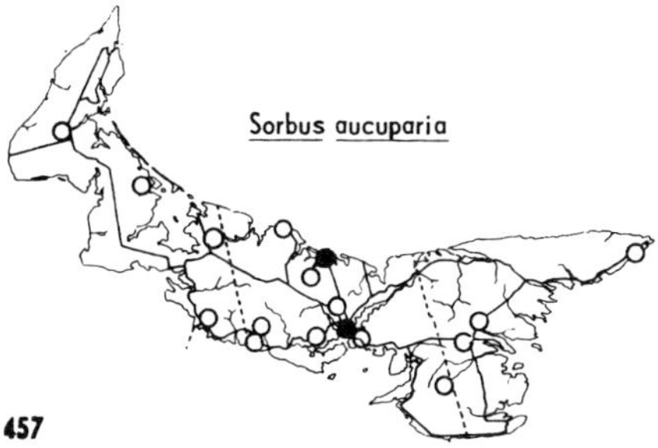

457

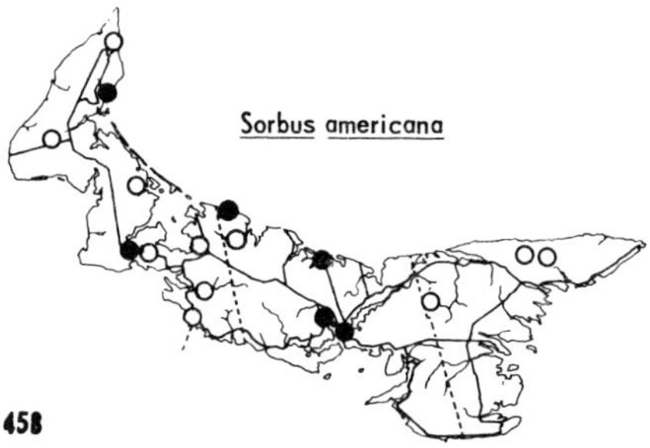

458

459

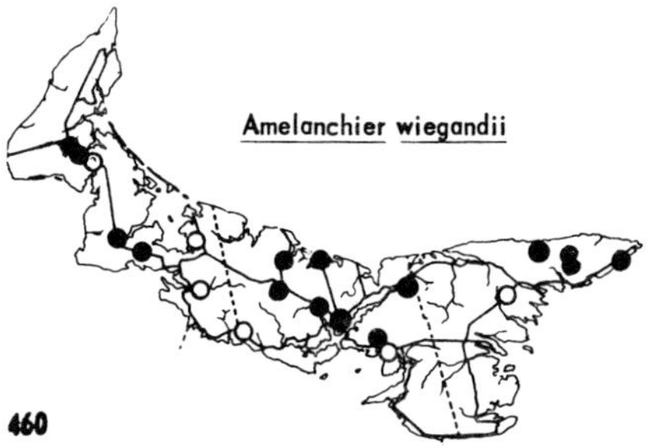

460

Amelanchier intermedia Spach Map 461. Local in Queens and Kings. Dry open hilly sites: in hedgerows or at wood edges. Similar small trees or tall shrubs in Prince County are more woolly.

Amelanchier laevis Wieg. WILD PEAR Map 462. Scattered from Prince to southern Kings, in upland regions. Dry sites: riverbanks in gravelly soils, or fencerows in hilly places. Always a tree or cluster of tall (12 foot) shrubs; the sepals usually woolly within but summit of ovary with scanty wool, only around the style base. Fruits edible, juicy, slightly sweet, often distorted by the rust *Gymnosporangium*. (*A. Botyrapium*, Bain; *A. canadensis*, MacSwain, Hurst; *A. oblongifolia*, MacSwain, Hurst).

Var. *nitida* (Wieg.) Fern. from Murray River.

X *Amelanchier neglecta* Eggleston (cited G.N. Jones, *Am. Sp. Amelanchier*, 1946) Charlottetown and Mount Stewart (*Fernald et al.*). Hybrid (sometimes allopolyploid?) of *A. bartramiana* X *laevis*.

Amelanchier bartramiana (Tausch) Roem. Map 463. Scattered in the Prince and east-central lowlands. Bogs and swampy thickets, only *Amelanchier* of damp or shady habitats.

Crataegus monogyna Jacq. Map 464. From Summerside to Charlottetown, more locally eastward. Introduced European ornamental (white "MAY" and rose-flowered), naturalized in fencerows and dry pastures, seeded by birds. 10 Lots (Groh). (*C. Oxyacantha*, MacSwain, Groh, Hurst, Campbell).

Crataegus chrysocarpa Ashe (?) Map 465. Mount Albion (*Fernald et al.*). Clumped shrubs. (*C. coccinea*, MacSwain, in part, queried by Hurst).

Crataegus succulenta Schrad. Map 466. Locally abundant in Queens and Kings. Fencerows and hedges, tall treelike shrubs. (*C. coccinea*, MacSwain, in part, questioned by Hurst). RED HAWS are found in West Prince also.

Fragaria virginiana Duchesne WILD STRAWBERRY Map 467. Throughout, common. Open dryish sites: pastures, banks, dry thickets. Old fields, open spots in forest, and indicator of land good for white clover (Stewart). Fruits edible, sweet and acid. Bain, MacSwain and Bain, MacSwain, Hurst. (Including *F. vesca*, Stewart; MacSwain and Bain, MacSwain, queried by Hurst).

The often recognizable var. *terrae-novae* (Rydb.) Fern. & Wieg. Rocky Point, Charlottetown, Brackley Beach, the last nearly glabrous.

[X *Fragaria ananassa* Duchesne, the cultivated strawberry, is considered to be a hybrid of *F. virginiana* and *F. chiloensis*. Plants from a railway embankment at O'Leary, approach *F. ananassa* closely; Experimental Farm, Charlottetown, wood edge.]

[*Potentilla fruticosa* L., MacSwain and Bain, MacSwain, Hurst, is to be excluded. Even in cultivation at Smelt Creek, it does very poorly.]

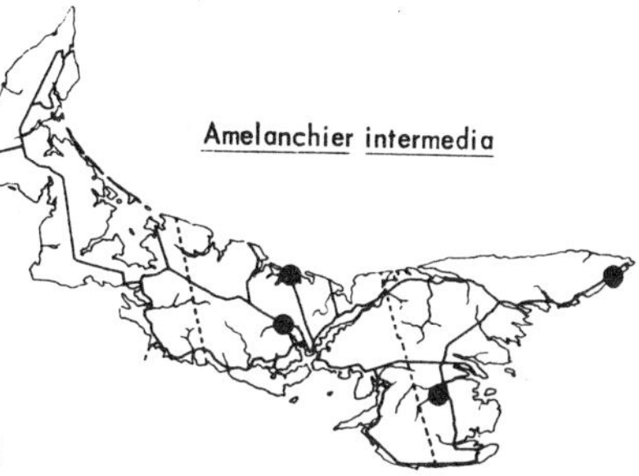

461

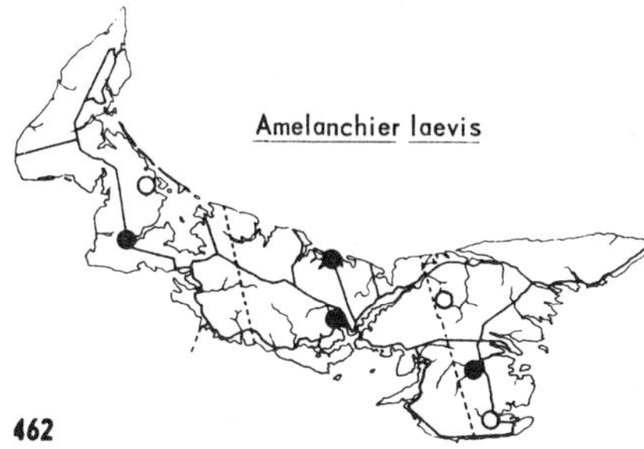

462

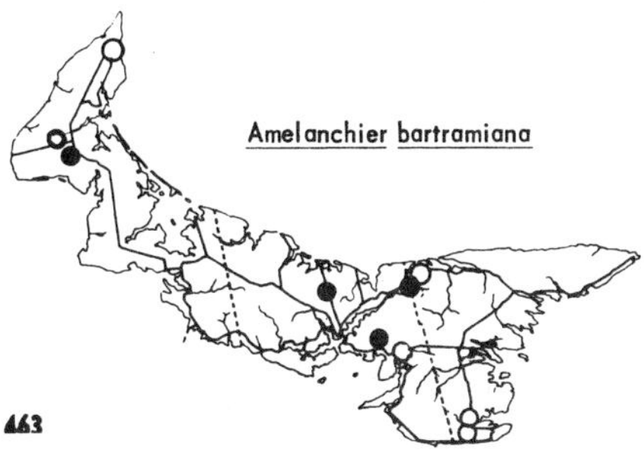

463

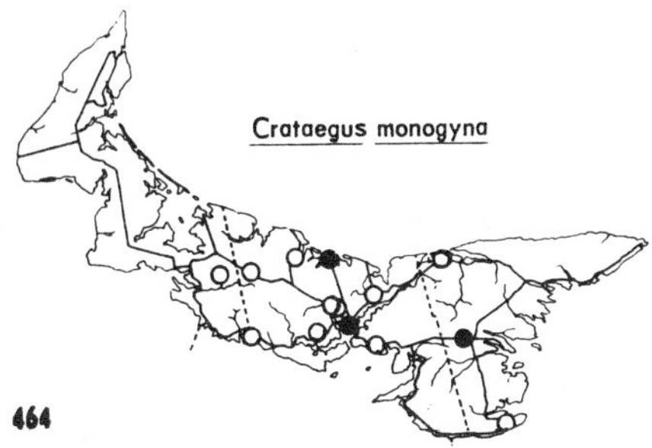

464

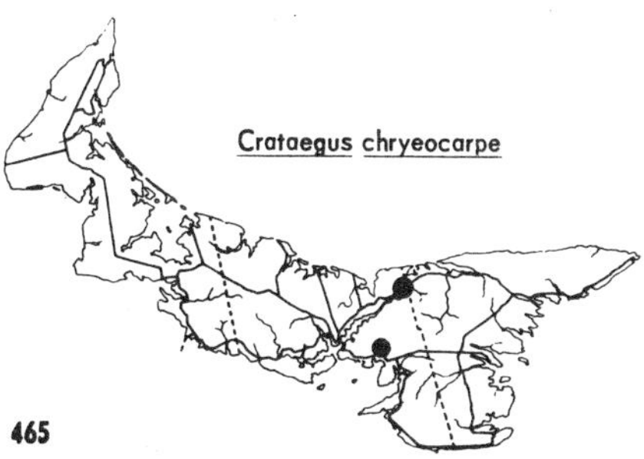

465

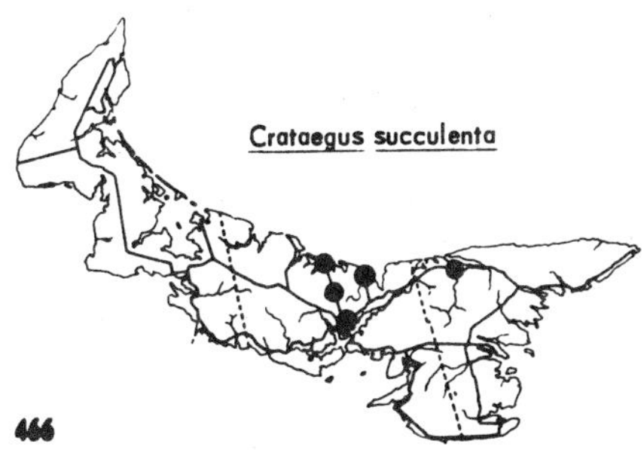

466

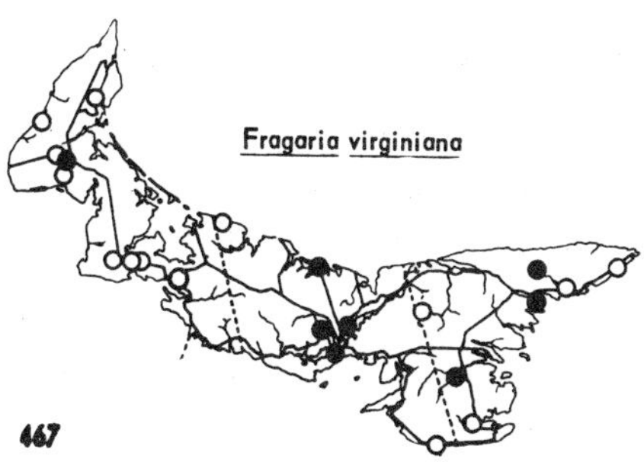

467

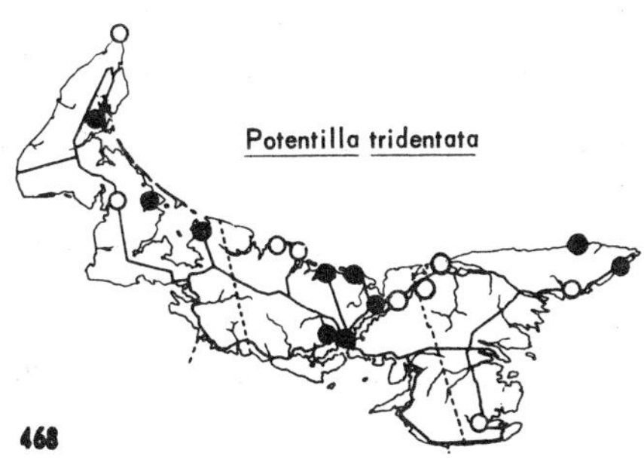

468

Potentilla tridentata Ait. Map 468. Common along the North Shore; locally throughout. Dry sandy or gravelly sites, among poor thin grass; exposure barrens on fixed dunes, old fields, banks in open woods. MacSwain and Bain, MacSwain, Hurst. (***Sibbaldiopsis*** Rydb.).

Potentilla palustris (L.) Scop. Map 469. Near North Shore, around to West Point, rarely inland, Watervale; Pisquid Pond; Dundee. Pond margins. MacSwain and Bain, Hurst. (***Comarum*** L.).

Potentilla argentea L. Map 470. Scattered throughout; more common near sandy shores. European. Waste sandy or gravelly soil: roadsides, banks, a weed in sandy fields and gardens. 5 Lots (Groh).

[**Potentilla gracilis* Douglas var. *gracilis* Brackley Point, a single perennial clump on grassy farm tracks (1953). Western North America, adventive.]

Potentilla recta L. var. *sulphurea* (Lam.) Lapeyr. Map 471. Cape Wolfe (1940), Ellerslie, Kinkora, New Glasgow (1938), Murray River, Souris. European, recently naturalized. Dry fields or pastures. (***P. sulphurea*** Lam., Hurst 1941; ***P. recta,*** Hurst 1940, Campbell). Var. ***sulphurea*** is probably correctly known as var. ***recta.***

Potentilla norvegica L. Map 472. Throughout. Dry unoccupied sandy soil, weed of roadsides, edges of fields. 21 Lots (Groh). MacSwain and Bain, MacSwain, Hurst (including ***P. monspeliensis*** L., Groh, Hurst).

Potentilla intermedia L. Map 473. Scattered throughout Prince and western Queens; seen at Bristol. European, recently naturalized (Plat River, near Summerside, 1914; New London, 1934). Roadsides, dry much-trodden pasture and fields. Groh, Hurst.

[**Potentilla inclinata* Vill. Bideford, in weedy farmyard (1953). Adventive, European.]

Potentilla simplex Michx. Map 474. Frequent in Kings County; scattered westward (Hunter River, O'Leary). Banks and thickets, open woods; weed in old fields. 2 Lots (Groh). (***P. canadensis,*** Groh, Hurst, Campbell).

Potentilla anserina L. Map 475. Throughout, particularly common in Prince and along the North Shore. Sands: seashores, inland along roadsides as weed. 12 Lots (Groh). Bain, MacSwain and Bain, MacSwain, Hurst.

Potentilla rolandii Boivin, in Can. Field–Nat. 65: 21. 1951. Map 476. Coastal only; in sheltered estuaries. Mud and salt marsh accompany its sites. (***P. pacifica,*** Roland, not Howell).

Filipendula ulmaria (L.) Maxim. Map 477. Introduced European ornamental; naturalized locally around old settlements in roadsides ditches or thickets. 4 Lots (Groh). Hurst.

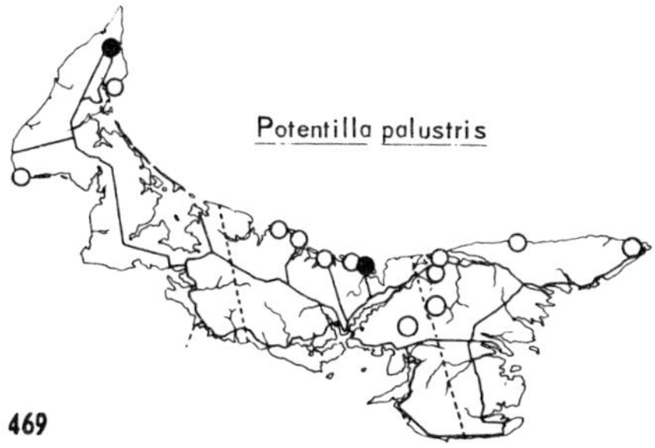

469

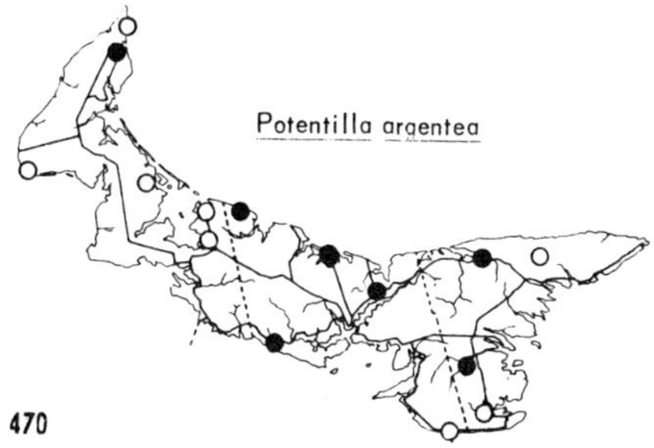

470

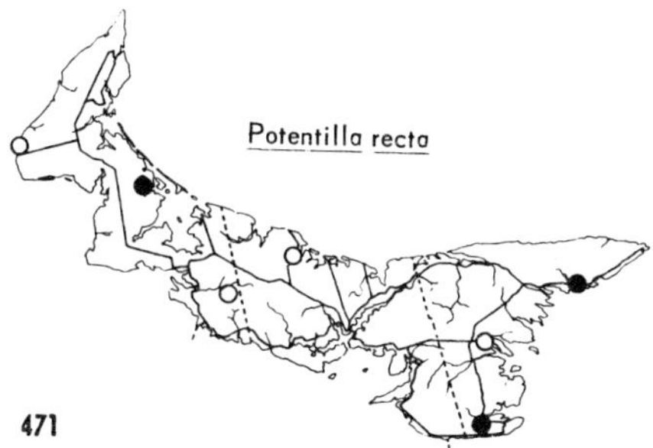

471

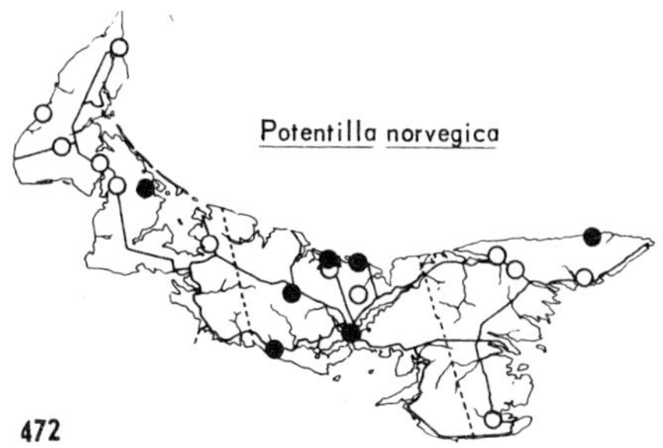

472

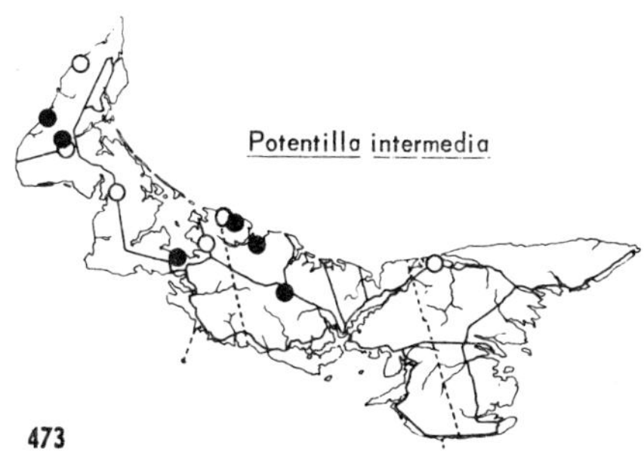

473

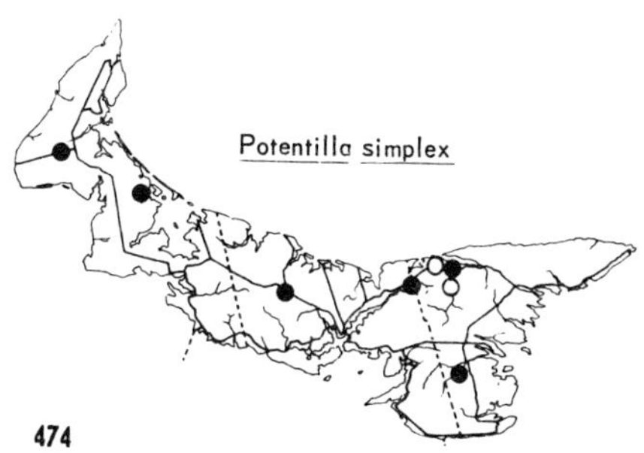

474

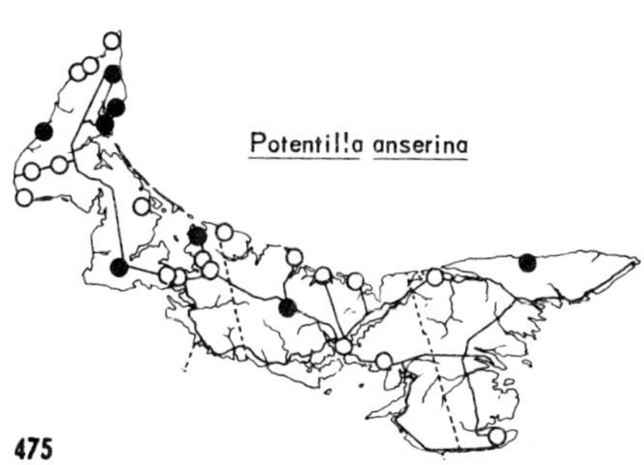

475

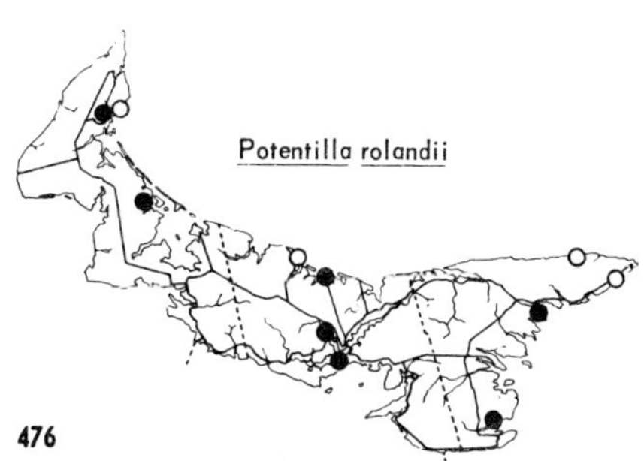

476

Geum laciniatum Murr. Map 478. West Prince; rare eastward: Winter River near York, creek-bottom alder thickets.

Geum aleppicum Jacq. var. ***strictum*** (Ait.) Fern. Map 479. Prince County. Alder thickets along creeks and banks; weedy where cut for roadsides or clearing. 5 Lots (Groh). (*G.* ***strictum***, MacSwain and Bain, MacSwain, Hurst; Campbell.

Geum macrophyllum Willd. Map 480. Local: West Prince; Hunter River valley; northeastern district. Damp, partially shaded sites; alder thickets, damp hardwoods. MacSwain and Bain, MacSwain, Hurst.

Geum rivale L. Map 481. Common in Prince; local, northern Queens, Kings. Marshy thickets or mossy stream banks. Hurst 1940.

Rubus chamaemorus L. YELLOWBERRY Map 482. Western Prince County bogs, in thick peat hummocks. Fruits eaten locally; few flowers set fruit. Hurst.

Rubus pubescens Raf. Map 483. Throughout; common particularly in West Prince. Characteristic of damp mixed woods, damp thickets and weedy along railroad banks. (*R.* ***triflorus*** Richards., MacSwain and Bain, MacSwain, Hurst), Campbell.

Rubus idaeus L., sspp. RASPBERRY Map 484.

Ssp. ***idaeus*** Bideford, colony along old road (now a cart track) through second growth spruce woods. Introduced from Europe for the large fruits.

Ssp. ***strigosus*** (Michx.) Focke Common throughout, openings in woods; very abundant in clearings, roadsides, thickets and on burns with fireweed. Noted on the "treeless lands" by Jacques Cartier. "After cutting, abundant". (*R.* ***strigosus*** Michx., Bain, MacSwain and Bain, MacSwain; *R.* ***idaeus*** var. ***aculeatissimus***, Hurst, Campbell; *R.* ***idaeus***, Stewart).

Var. ***strigosus*** (Michx.) Maxim. Common. The forma ***albus*** (Bailey) Fern. noted by Stewart from "a single station."

Var. ***canadensis*** Richards. Charlottetown Experimental Farm; Brackley Point.

[*Rubus* ***occidentalis*** L. reaches only to Quebec. Bain, MacSwain and Bain, Hurst.]

[*Rubus* ***trivialis*** Michx. reaches only to Maryland. Bain, Hurst.]

Rubus hispidus L. Map 485. East of Charlottetown. Damp, open thickets and clearings. Cold moist locations, not frequent nor abundant (Stewart). Fruit sour, edible. (*R.* ***moluccanus***, Stewart; *R.* ***canadensis***, MacSwain and Bain, Hurst); MacSwain and Bain, MacSwain, Hurst.

Rubus tardatus Blanchard South Lake near Bothwell, clay bank (***Fernald et al.***).

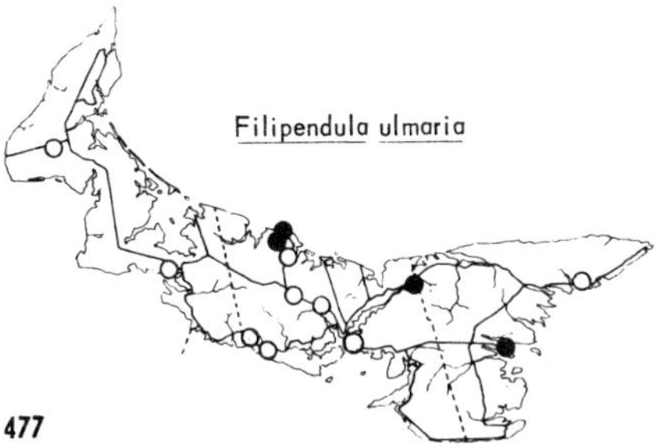

477

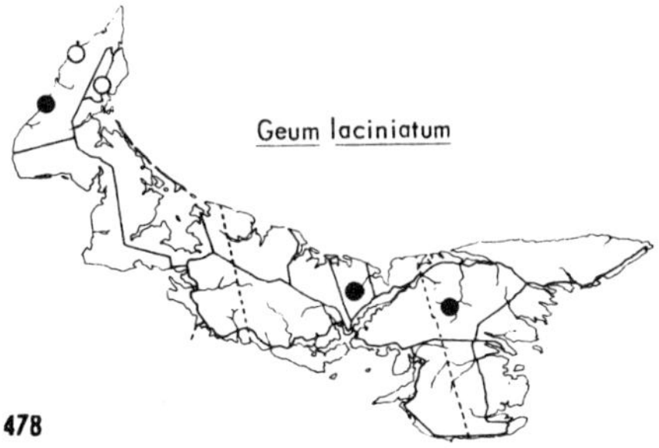

478

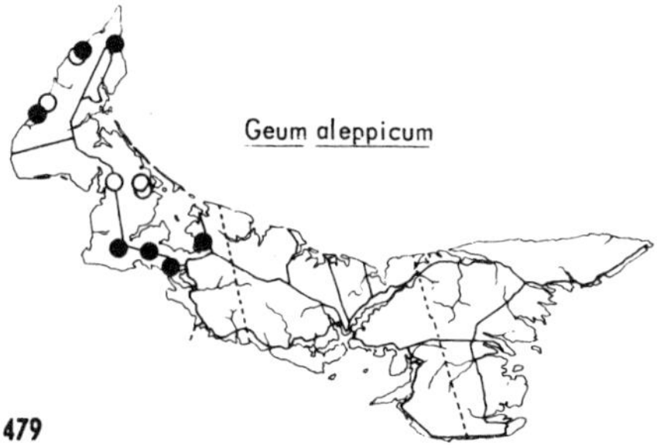

479

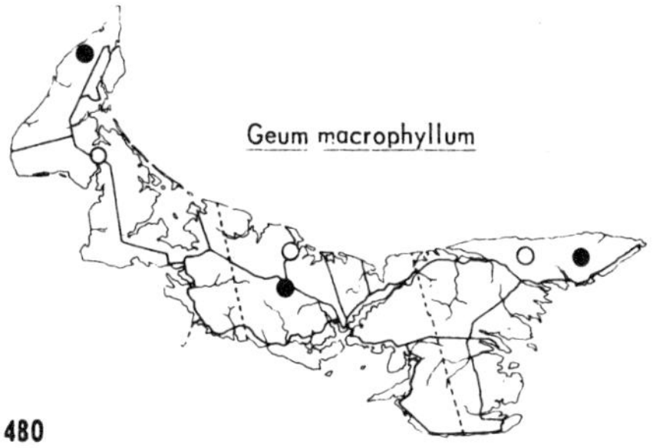

480

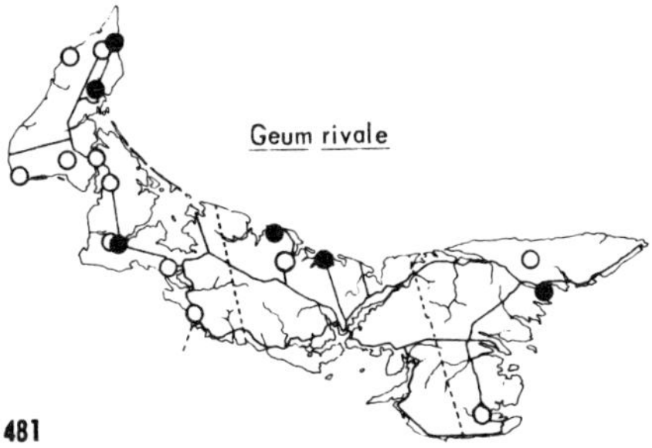

481

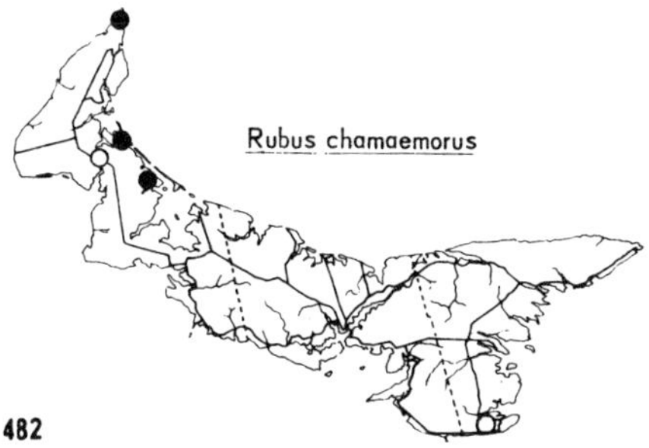

482

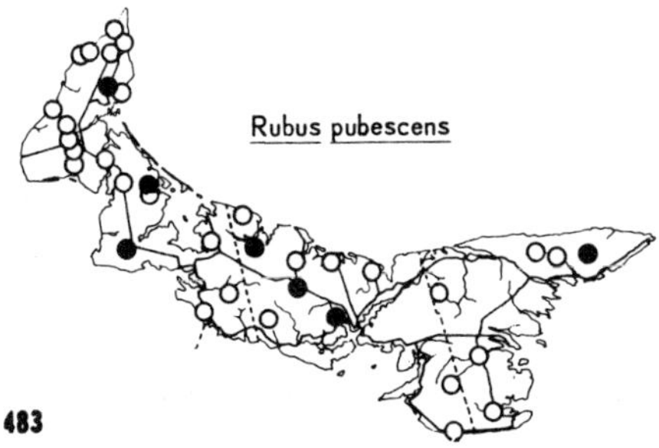

483

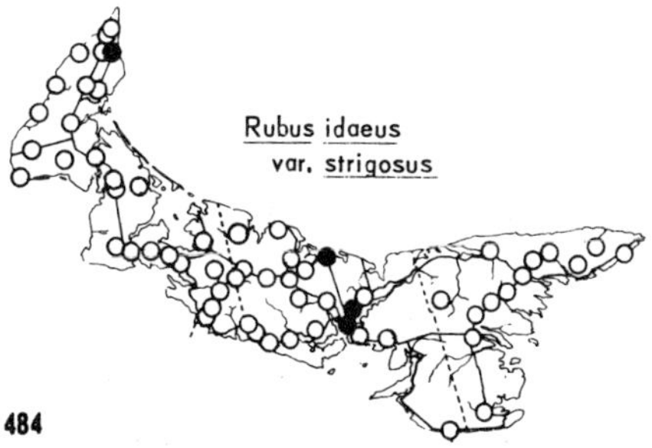

484

Rubus arcuans Fern. & St. John Map 486. Dundee (type locality), Mount Stewart, Tracadie Beach, Summerside, Miscouche. Dry open sites; in poor sandy soil.

Rubus vermontanus Blanchard Sandy lands of east-central lowland: "Tracadie, trailing amongst grass" (*Macoun*); Morell; Murray River. (*R. setosus*, Gray's Manual (1908), Hurst).

Rubus perinvisus Bailey Brackley Point Road (*Fernald et al.* 7650).

Rubus canadensis L. Map 487. Common. Abundant in openings in woods, thickets, and clearings, tolerant of some shade. Older records refer to a prostrate species.

Rubus elegantulus Blanchard Mount Stewart (*Fernald et al.*). Probably only a slender, pricklier version of the former.

Rubus amicalis Blanchard Cove Head (*Macoun*); North River Road, Charlottetown. Clearings, thickets.

Rubus kennedyanus Fern. Recorded from P.E.I. in Roland (1947). (*R. quaesitus* Bailey, Roland).

Rubus pensilvanicus Poir. Murray River (*Blanchard*); Huntley River near Alberton, steep bank. (*R. pergratus* Bailey).

[*Rubus frondosus* Bigel. as interpreted by Bailey, *R. canadensis* var. *pergratus* (Blanchard; Bailey, Roland), does not occur northeast of Massachusetts. Bain, Hurst.]

Rubus allegheniensis Porter Map 488. Our collections from east and central Prince. Relatively sandy soil, in thickets at border of woods. (*R. villosus*, MacSwain and Bain, MacSwain, Hurst).

[*Dalibarda repens* L. Warren's herbarium, accumulated at North River, includes one collection, possibly from the Island. No other record.]

Agrimonia gryposepala Wallr. Map 489. Local: vicinity of Tignish; Brackley Point; North River. Damp, open places, thickets. Flowers paler than in *A. striata*. (*A. Eupatoria*, in part, MacSwain and Bain, MacSwain); Groh, Hurst.

Agrimonia striata Michx. Map 490. West and central Prince, central Queens and northeastern Kings. Open sites: thickets, borders of woods, weedy on roadsides. 4 Lots (Groh). (*A. Eupatoria*, in part, MacSwain and Bain, MacSwain); Groh, Hurst.

[*Agrimonia odorata* Mill. Brackley Point, introduced from Wiltshire, England (1920) and abundantly persisting in Hurst's garden.]

[**Rosa centifolia* L. CABBAGE ROSE Rosebank in Southport, roadside. Persistent or escaped ornamental, introduced from Europe.]

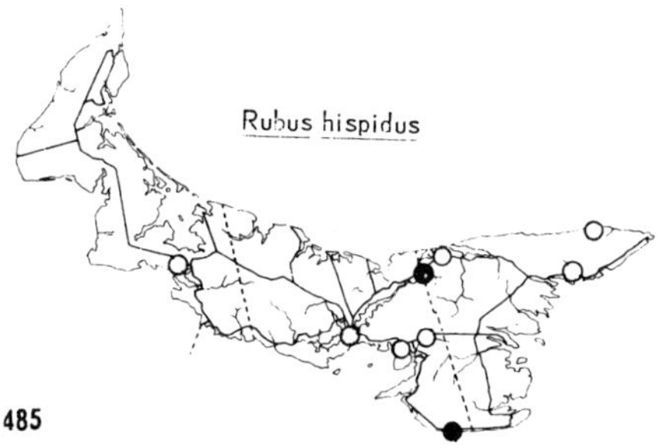

485

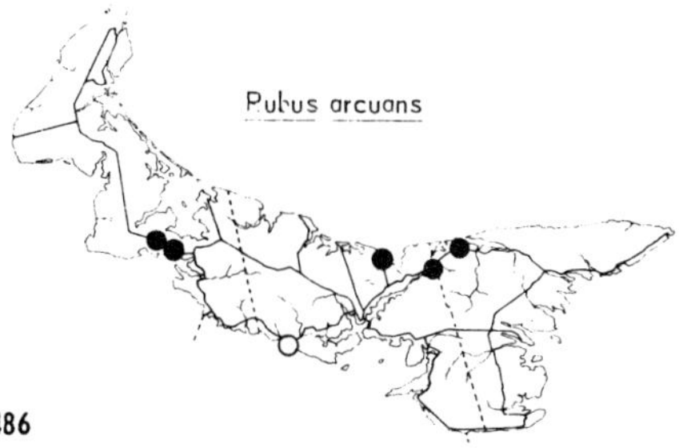

486

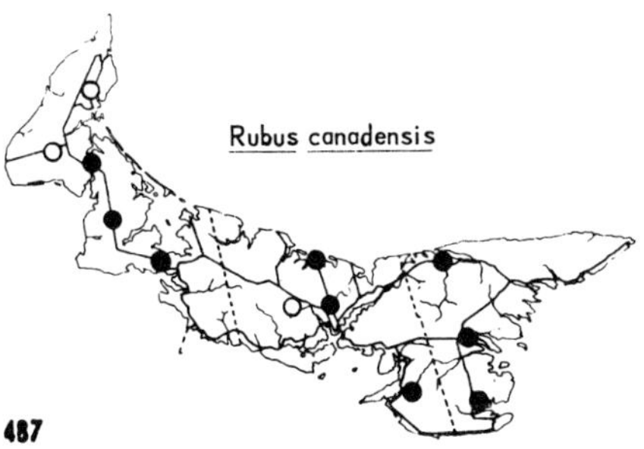

487

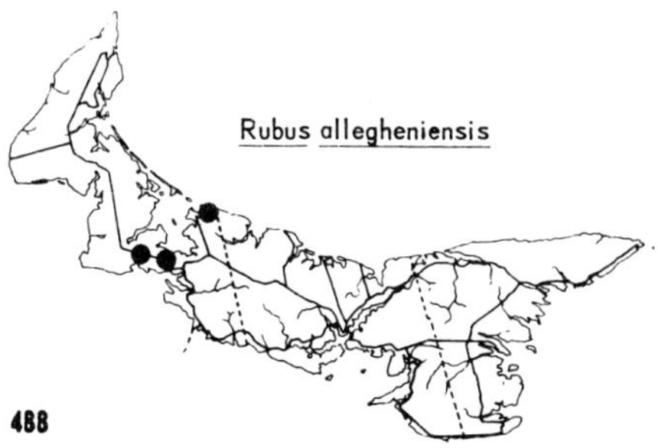

488

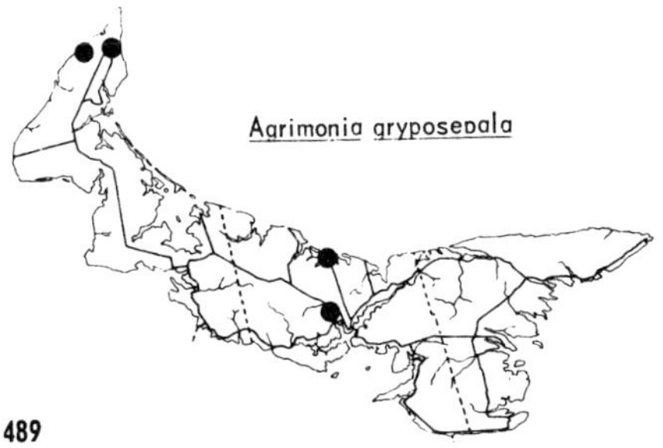

489

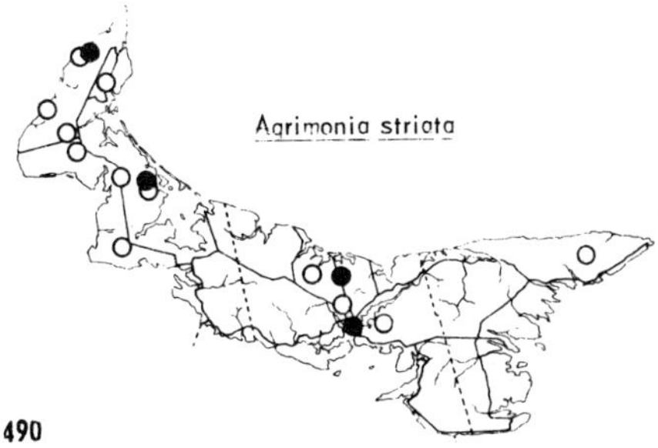

490

[*Rosa tomentosa* Sm. var. *globulosa* Rouy Southport to Keppoch, roadside thicket (*Fernald et al.*). European, introduced ornamental, escaped. Campbell.]

Rosa micrantha Sm. Map 491. SWEET BRIAR Alberton; Charlottetown (Groh); Brackley Point and Road; frequent in the southeast. Introduced from Europe; naturalized. No longer planted. Thickets and dry pastures. Characterized by the large brown galls like tufts of hair, called "Robin's pincushions". (*R. rubiginosa*, MacSwain and Bain, MacSwain, Hurst; *R. Eglanteria*, Campbell).

Rosa nitida Willd. Map 492. Mount Stewart region to Dundee; near De Sable, Murray River. Larch swamps and bogs.

Rosa virginiana Mill. Map 493. Common throughout. Damp, partly sunlit places: thickets, swamp margins, marshes. Very common component of *Calamagrostis* thicket. (*R. lucida* Ehrh., MacSwain and Bain, Churchill 1901, MacSwain); Hurst, including *R. palustris*.

Rosa carolina L. Map 494. North Shore: Rustico, Cavendish. Sandy headlands and fixed dunes. Apparently elsewhere, less common than in Nova Scotia, forming colonies in fencerows. MacSwain and Bain.

[*Rosa palustris* Marsh. Hurst's specimen seems referable to *R. virginiana.*]

[*Rosa spinosissima* L. Southport, introduced European ornamental. Roadside bank.]

[*Rosa rugosa* Thunb. Ornamental, introduced from Japan via Europe. Flowers white or pink. Escaping to roadside banks (e.g., Souris West).]

Rosa cinnamomea L. Map 495. Ornamental, introduced from Europe. Flowers double. Formerly much planted; extensive colonies established along the roadsides.

[*Rosa blanda* Ait. is not known to occur east of the Saint John Valley. Reported by MacSwain, Hurst.]

[*Prunus depressa* Pursh is absent, though known from New Brunswick river beaches. Reported by Bain, MacSwain, doubted (as *P. pumila*) by Hurst.]

Prunus pensylvanica L. f. WILD CHERRY Map 496. Common throughout, open dryish sites. "Follows clearings in abundance, and also fire, but lives only 15 to 20 years" (Stewart). "Fruits put into spirits for rum and brandy" (Stewart). (*P. virginiana*, Stewart); Bain, MacSwain and Bain, MacSwain, Hurst.

[*Prunus cerasus* L. SOUR CHERRY Garden fruit-tree, occasionally planted. Introduced from Asia Minor via Europe; escaped on the Brackley Point Road (*Fernald et al.*).]

[*Prunus serotina* Ehrh. appears to be absent, though found just across the Strait in N.S. and N.B. Reported by MacSwain and Bain, MacSwain, Hurst.]

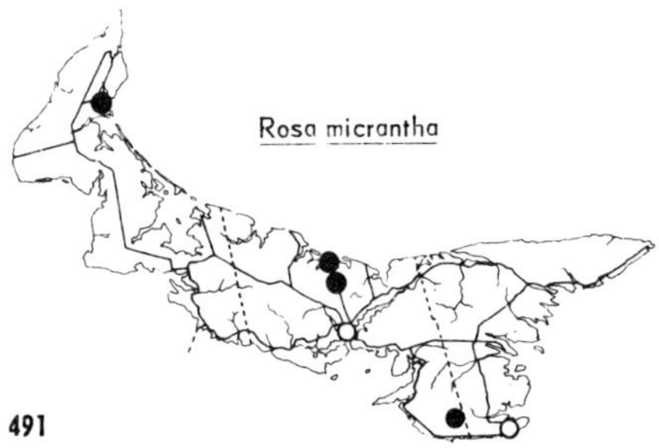

491

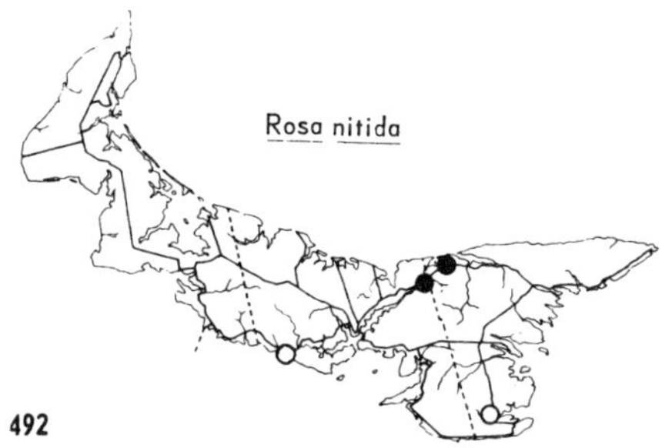

492

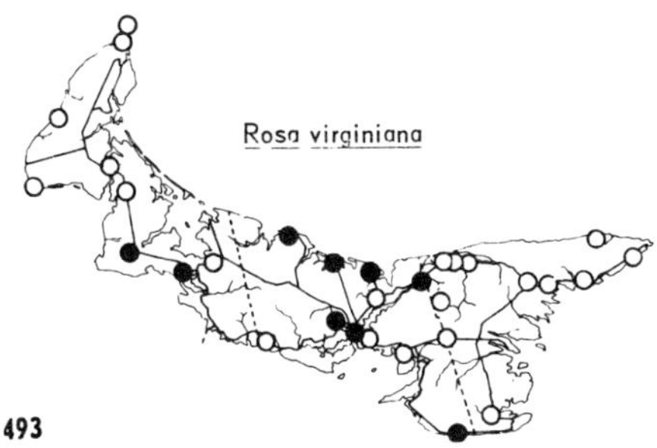

493

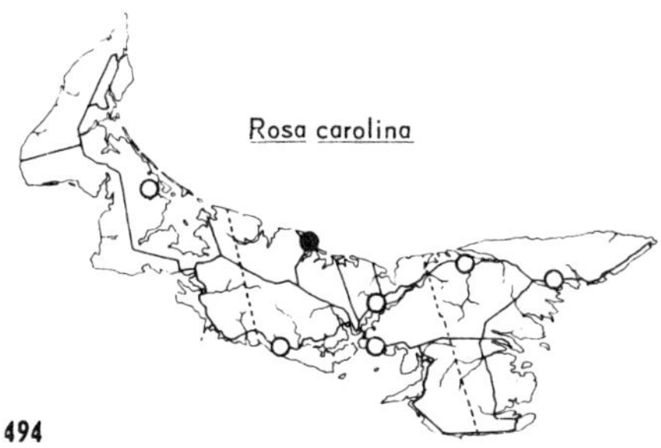

494

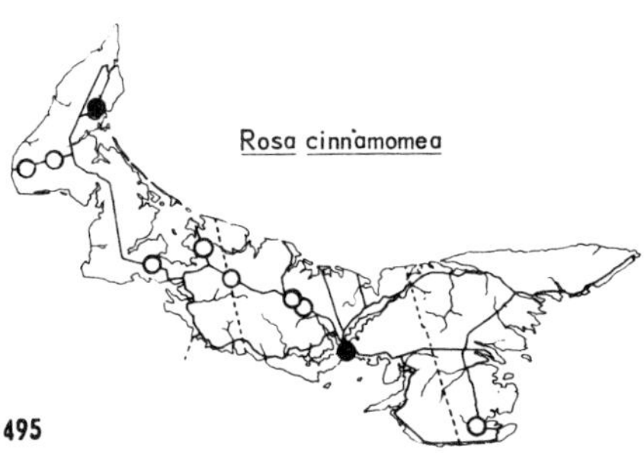

495

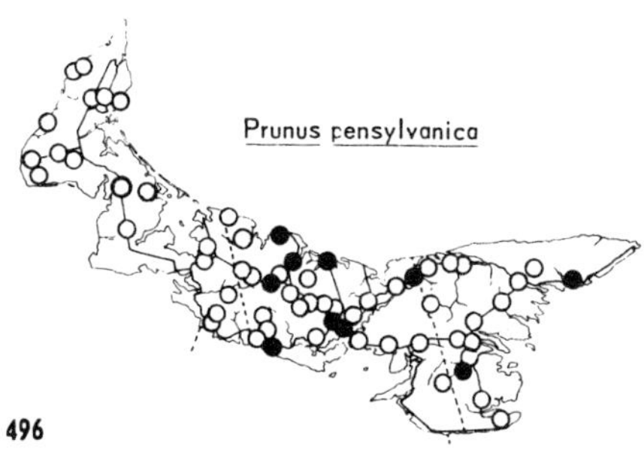

496

Prunus virginiana L. Map 497. CHOKECHERRY Common through Prince and Queens, mainly on the Alberry-Charlottetown sandy loams. Well drained, open sites; thickets and fencerows. Racemes occasionally very dense with short pedicels and rachis: Experimental Farm, Charlottetown. Fruits edible but drying and puckering the mouth. Bain, MacSwain and Bain, MacSwain, Hurst.

LEGUMINOSAE

[*Gleditsia triacanthos* L. Ornamental shade tree introduced from the eastern States. A single small seedling tree in Hunter River, 1941.]

[*Thermopsis mollis* (Michx.) M.A. Curtis Mount Herbert, naturalized along roadside near the Consolidated School, according to H.A. Messervy; if persistent, was in any case destroyed by road improvement. Introduced from southeastern United States.]

Cytisus scoparius (L.) Link Map 498. SCOTCH BROOM Pisquid Road at Watervale (discovered by E.S. Blanchard). Naturalized on Kayes farm, Georgetown; also spread to roadside banks where considerably cut back by roadside use of weed sprays; introduced presumably from Britain, limited to the milder eastern parts of the Island. Hurst 1941. (*Sarothamnus* Wimmer).

Lupinus polyphyllus Lindl. Map 499. LUPINE Garden ornamental, introduced from British Columbia. Naturalized since 1912 at least, in cemeteries and spread to roadsides in milder climates of the Island from Hunter River east. (*L. perennis*, Hurst 1940).

Trifolium arvense L. Map 500. Scattered: common around Tignish; rare, Ellerslie; from Charlottetown locally eastward through Kings especially along the railway. Naturalized from Europe. Open sites on light or disturbed soils; roadsides, railways, sandy fields. MacSwain and Bain, MacSwain, Hurst.

Trifolium pratense L. Map 501. RED CLOVER Throughout; common from East Prince through to Kings. Cultivated with *Phleum pratense* for clover hay, often in mixtures with *T. hybridum;* introduced from Europe; established on roadsides, rough pastures, waste places. Our plants are transitional between var. *americanum* Harz and var. *sativum* (Mill.) Schreb. or referable to the latter. MacSwain and Bain, MacSwain, Hurst.

[*Trifolium medium* L. Bay View: edge of field, Walter Simpson farm, perhaps as early as 1892 and as late as 1911, when examined by Malte; roots sent to Department of Agriculture, Ottawa, for trial introduction (1897). Introduced from Europe; earliest record from Maritimes. Bain 1892, MacSwain, Hurst.]

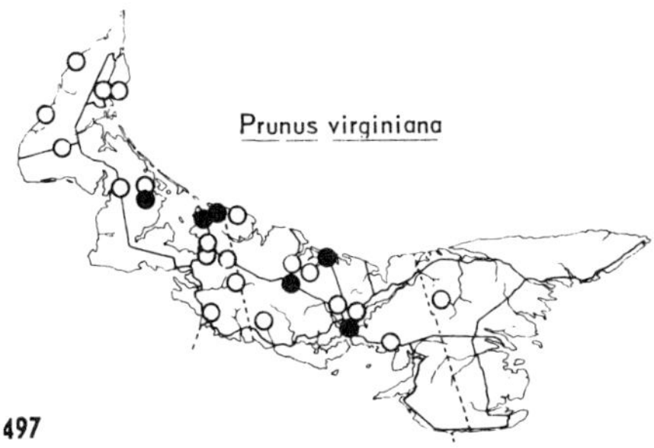

497

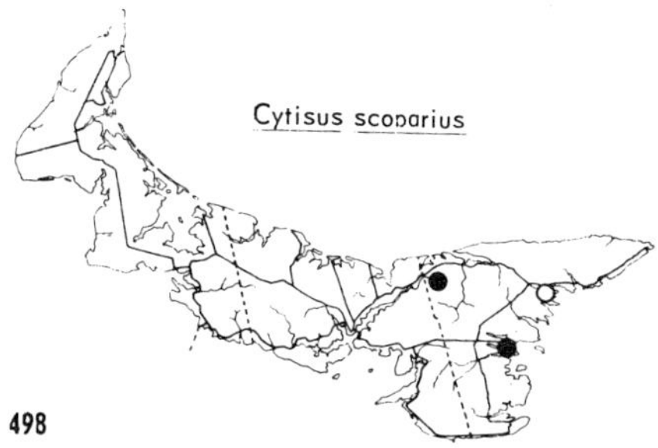

498

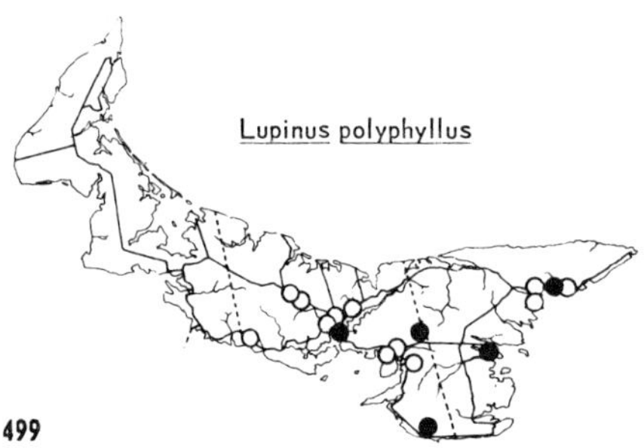

499

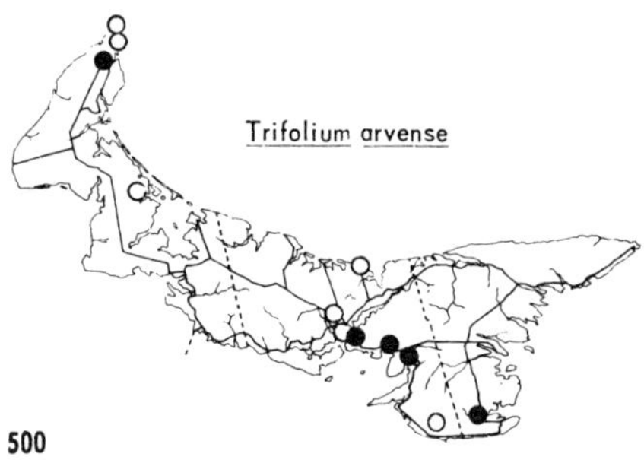

500

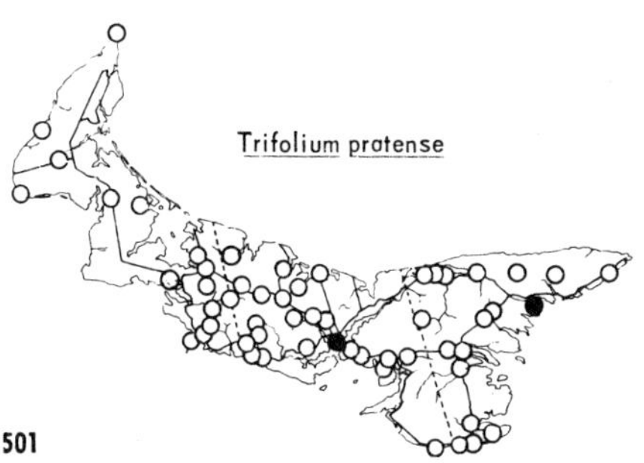

501

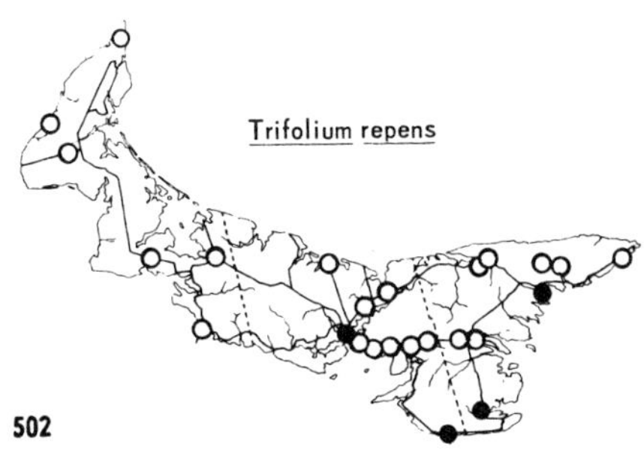

502

Trifolium repens L. Map 502. WHITE CLOVER Scattered throughout, common in lowlands. Introduced from Europe for pastures, naturalized. Partially sunlit sites with loamy soils, shaded by taller herbs: pastures, lawns, woods, roads. Bain distinguished White or Creeping Clover (which he supposed native) from White Dutch Clover (introduced), presumably influenced by the botany of England where the native English strains are considered superior to the introduced Dutch for pasturage. MacSwain and Bain, MacSwain, Hurst.

Trifolium hybridum L. Map 503. ALSIKE Throughout. Introduced from Europe; planted in clover-hay mixtures with ***Phleum pratense***, and naturalized: roadsides, fields. Our specimens belong to var. ***elegans*** (Savi) Boiss. MacSwain, Hurst.

Trifolium agrarium L. Map 504. YEO CLOVER Scattered throughout; common in central uplands. Introduced from Europe, naturalized. Roadsides, edges of fields, meadows. Local name from its introduction by Senator John Yeo. (***T. aureum*** Poll.); MacSwain and Bain, MacSwain, Hurst.

Trifolium procumbens L. Map 505. Common throughout. Naturalized from Europe. Weed of light or disturbed soils; very abundant on railways and fixed sand dunes; also along roadsides and in thin grass. 38 Lots (Groh). Ours belong to the variety called ***T. campestre*** Schreb. var. ***majus*** (Koch) Gremli in Europe. Bain 1892, MacSwain, Hurst.

Melilotus officinalis (L.) Lam. Map 506. Occasional throughout: common only in the central arable region, from Miscouche to Charlottetown. Introduced from Europe as fodder before 1850; naturalized as weed of fields and roadsides. Groh, Hurst 1940.

Melilotus alba Desr. Map 507. Scattered throughout: fairly common in arable Alberry-Charlottetown soils. Introduced from Europe as fodder, before 1888; naturalized as weed of fields and roadsides, often on cinders. It is only in recent years that intensified stock-raising has made these fodder plants common. 8 Lots (Groh). MacSwain, Hurst.

[***Medicago sativa*** L. Map 508. ALFALFA Scattered, mostly in low lands. Introduced from Europe as fodder plant, persisting in fields, rarely weedy on roadsides. 1 Lot (Groh). Groh, Campbell.]

Medicago lupulina L. Map 509. Scattered throughout. Naturalized European weed of roadsides and railway cinders. 11 Lots (Groh). Represented at some stations (e.g., Campbellton))by var. ***lupulina***, at others (e.g., Charlottetown) by var. ***glandulosa*** Neilr. MacSwain, Hurst.

Robinia pseudo-acacia L. Map 510. Ornamental tree, often planted around houses. Introduced from eastern United States. Forming sucker colonies along roadsides (Crapaud, DeSable, Charlottetown, Georgetown). 4 Lots (Groh). MacSwain and Bain, MacSwain, Hurst.

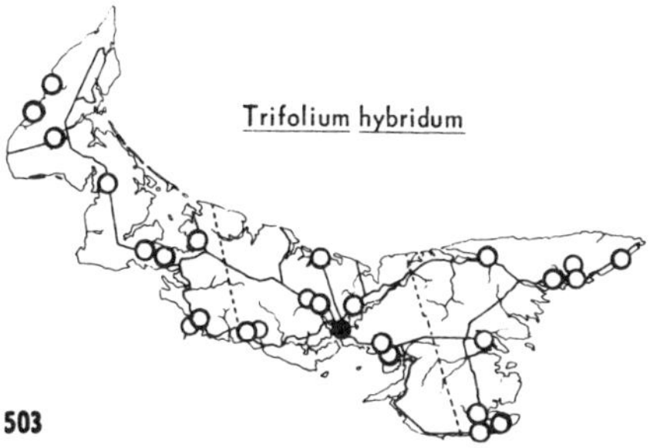

503

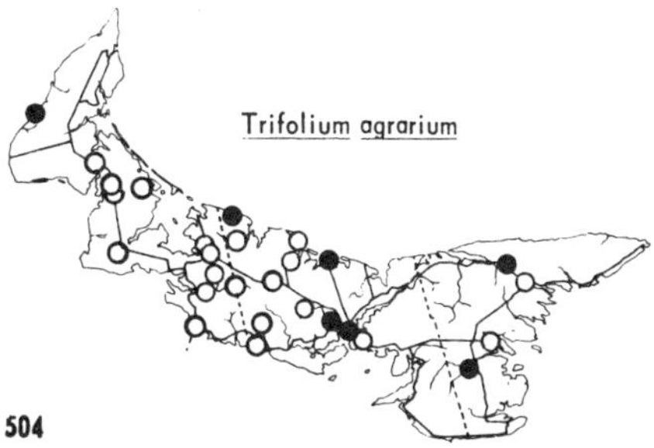

504

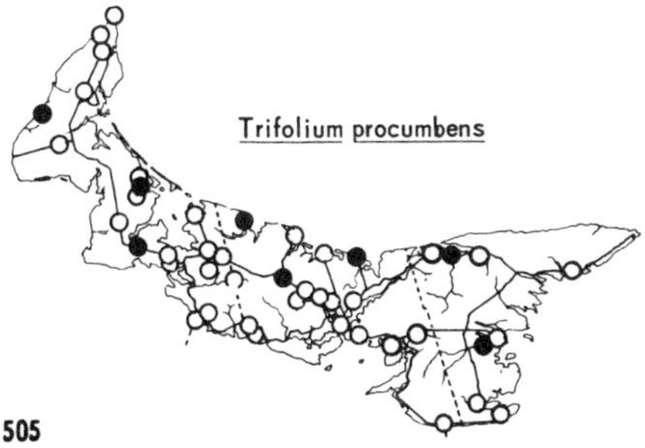

505

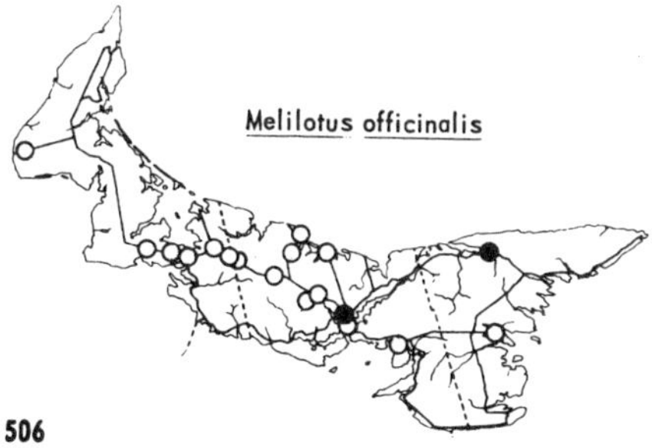

506

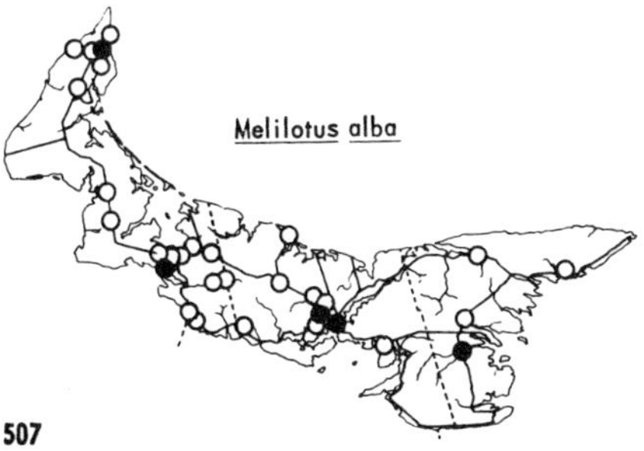

507

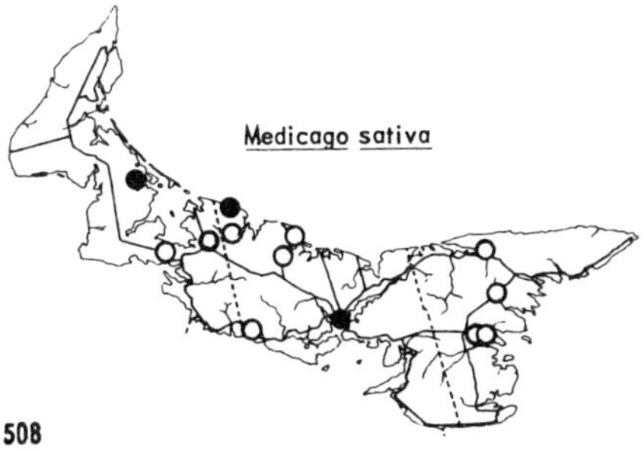

508

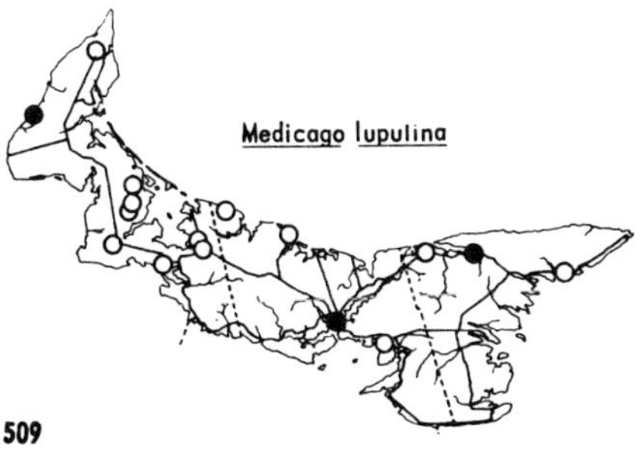

509

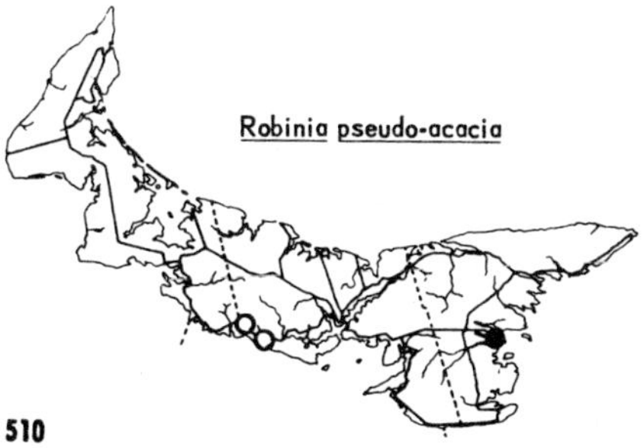

510

[*Robinia viscosa* Vent. Map 511. Ornamental shrub, occasionally planted. Introduced from eastern United States. Forming sucker colonies where planted: Hunter River, Appin Road (Messervy), and elsewhere. 2 Lots (Groh). MacSwain, Hurst.]

[*Robinia hispida* L., a smaller eastern American shrub, may be grown. Not seen. Reported by MacSwain and Bain.]

[*Vicia sativa* L. Map 512. Seldom established: western Prince. Sown with admixture of peas in oats for green fodder. Introduced from Europe, weed only in farmyards.]

Vicia angustifolia Reichard Map 513. Throughout, common on the North Shore. Naturalized from Europe; weed of fields, pastures and waste places, on sandy soil. Possibly the wild ancestor of *V. sativa.* 16 Lots (Groh). (*V. sativa,* MacSwain and Bain, MacSwain); Groh, Hurst.

Vicia sepium L. Map 514. Charlottetown, ditches and garden fencerows. Naturalized from Europe before 1926. 1 Lot (Groh). Hurst.

Vicia tetrasperma (L.) Moench. Map 515. Brackley Point (*Macoun*); Cherry Valley (1953). Naturalized from Europe, weed of grass verges. Groh, Hurst.

Vicia hirsuta (L.) S.F. Gray Map 516. Rare: Alberton (1912), Rusticoville, Bunbury, Vernon (1926), Kilmuir (Watson), Souris. Naturalized from Europe, weed of sandy waste land, roadsides, inner dunes, borders of fields. 10 Lots (Groh). MacSwain, Hurst.

Vicia cracca L., sspp. Map 517. Very common throughout. Naturalized from Europe. Weed of fields, waste places, roadsides. 46 Lots (Groh). MacSwain and Bain, MacSwain, Hurst.

Ssp. *cracca* Common everywhere.

Ssp. *tenuifolia* (Roth) Gaudin Charlottetown, waste ground (*Gorham*). Adventive from southern Europe.

[*Vicia americana* Muhl. Not known east of southern Quebec. Reported by MacSwain and Bain.]

Lathyrus japonicus Willd. var. *glaber* (Ser.) Fern. BEACH PEA Map 518. Around the coast; common on North Shore. Sand dunes and barrier beaches, observed by Jacques Cartier on the "treeless lands." (*L. maritimus* Bigel., Bain, MacSwain and Bain, MacSwain, Hurst, M.W. Smith); Campbell.

Lathyrus palustris L. var. *pilosus* (Cham.) Ledeb. Map 519. Scattered mainly coastal. Damp thickets, fresh borders of salt marshes and sand beaches, more characteristically *pilosus* in the latter habitats. (*L. palustris,* MacSwain and Bain, MacSwain; var. *linearifolius,* Hurst, Campbell).

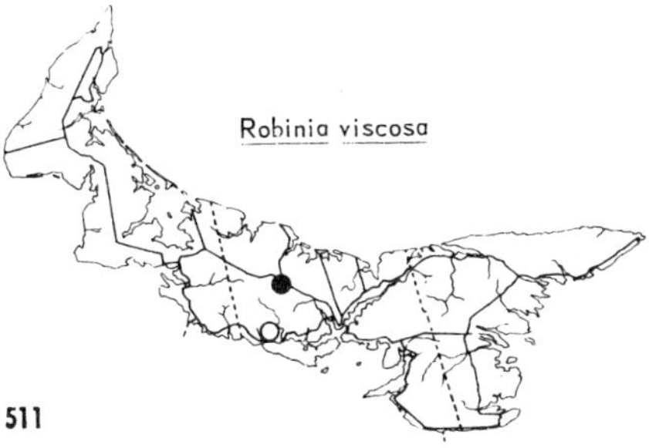

511

512

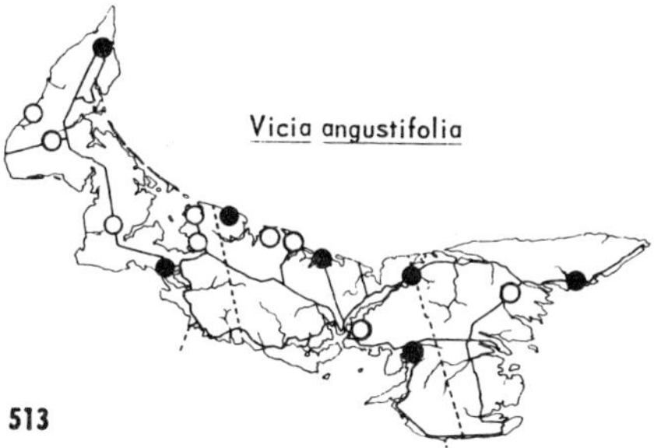

513

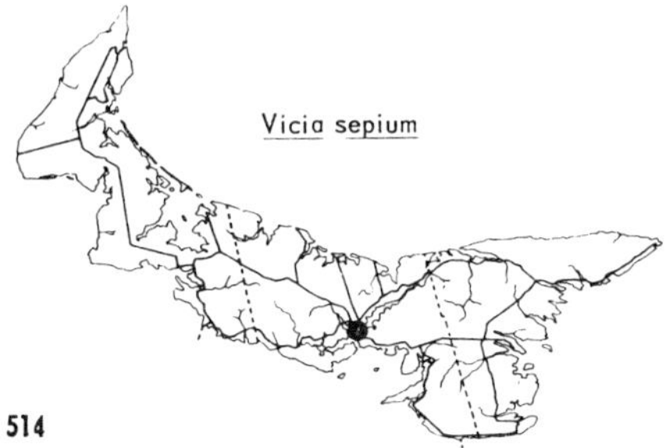

514

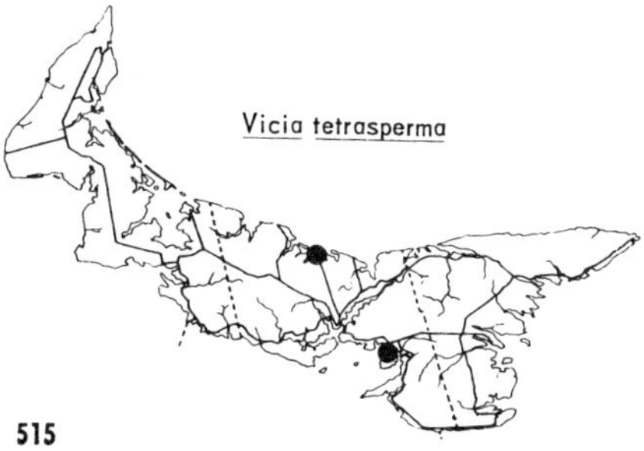

515

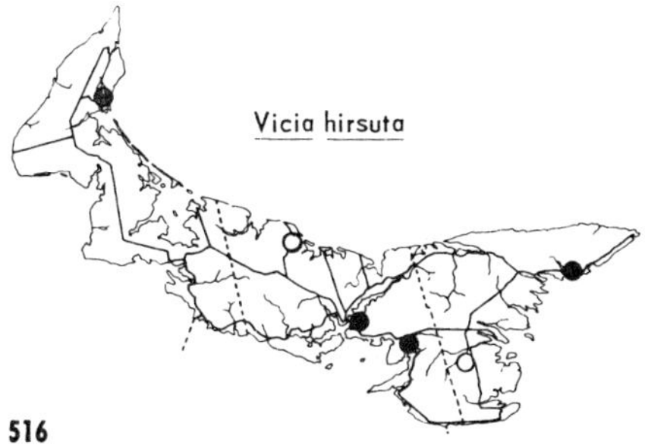

516

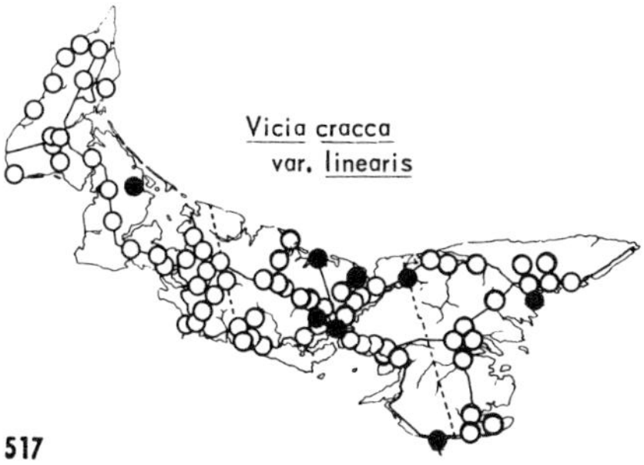

517

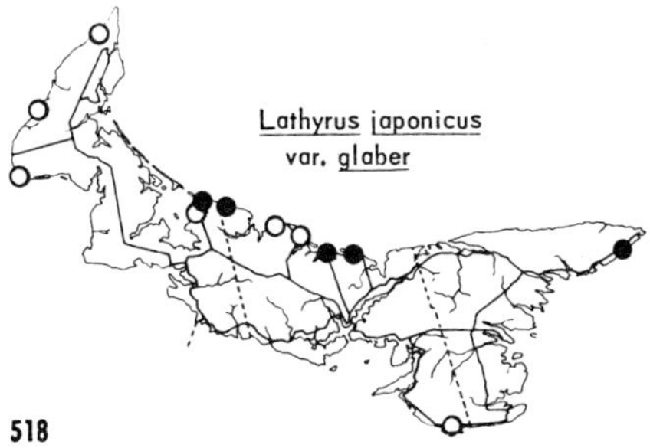

518

Lathyrus pratensis L. Map 520. Locally naturalized: Wellington, Borden (1 Lot Groh), Charlottetown and Rocky Point. European. Along railway bank, but usually in permanent grass: meadow, churchyard. MacSwain, Hurst.

[*Lathyrus sylvestris* L. Introduced from England; planted in his hedgerow by by Blythe Hurst, and still persistent. Campbell.]

LINACEAE

[*Linum usitatissimum* L. FLAX Introduced for linseed and fiber elsewhere in Canada, from Europe. Casual weed of railway yards. Charlottetown (1953); 1 Lot (Groh). Hurst.]

Linum catharticum L. Map 521. Souris (1952), on railway bank; naturalized as in northern Nova Scotia; European.

OXALIDACEAE
(included in GERANIACEAE, MacSwain and Bain, MacSwain)

Oxalis montana Raf. WOOD SORREL Map 522. Scattered through Prince and Queens; rare in Kings. Mossy floor of mixed or coniferous woods in lowlands or ravines. (*O. Acetosella*, Bain, MacSwain and Bain, MacSwain, Hurst), Campbell.

Oxalis stricta L. (Eiten, Taxon 4 (5): 99. 1955) Map 523. Scattered, in Prince and Queens. Common native weed: wood edges, stream banks, gardens. 9 Lots (Groh). MacSwain and Bain; (?) Groh; (*O. corniculata* var. *stricta* (L.) Gray, MacSwain, Hurst; *O. corniculata*, Campbell; *O. europaea* Jord., Roland).

The forms recognized by Wiegand under *O. europaea* occur: f. *cymosa* is common, f. *pilosella* collected at Hunter River, f. *villicaulis* at Bonshaw.

Oxalis dillenii Jacq. (Eiten, l.c.) Charlottetown, Montague. North American weed; gardens, railway cinders. (*O. stricta*, Gray's Manual (1950), including *O. stricta* var. *piletocarpa* Wieg.).

GERANIACEAE

[*Geranium pratense* L. Map 524. Established in the 1940's, not persisting, along roadsides at Andrew's Mills on Mayfield Road north of New Glasgow; Hunter River and Cymbria. Planted in gardens; introduced from Europe. 1 Lot (Groh). Hurst.]

Geranium robertianum L. Map 525. West Prince; northeastern uplands. Damp, partially shady sites: swampy woods, thickets.

POLYGALACEAE

**Polygala sanguinea* L. CANDYSTICK Map 526. Cavendish, in sandy old field at ranger's house, discovered by Sterling Campbell (1950-3).

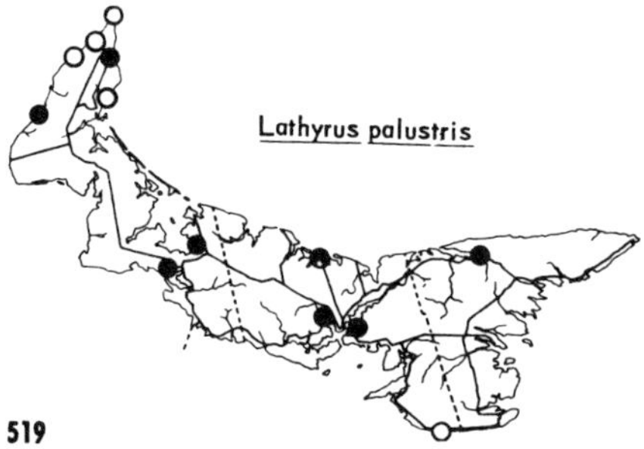

519

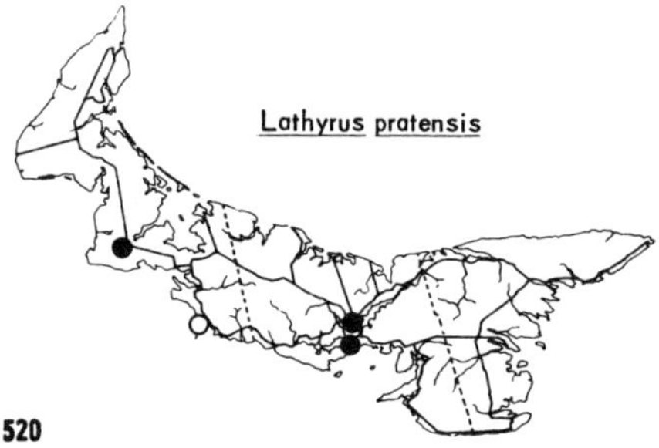

520

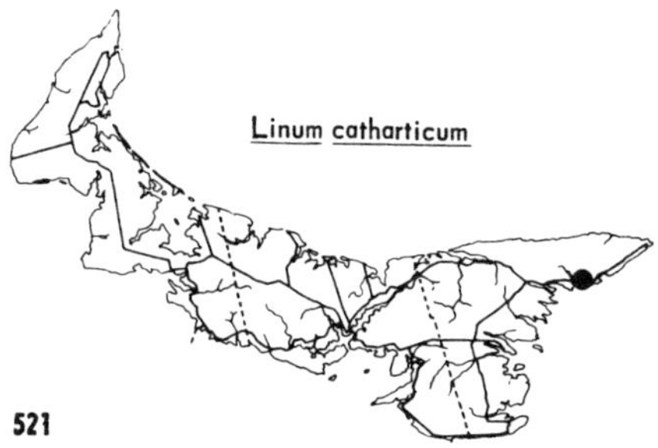

521

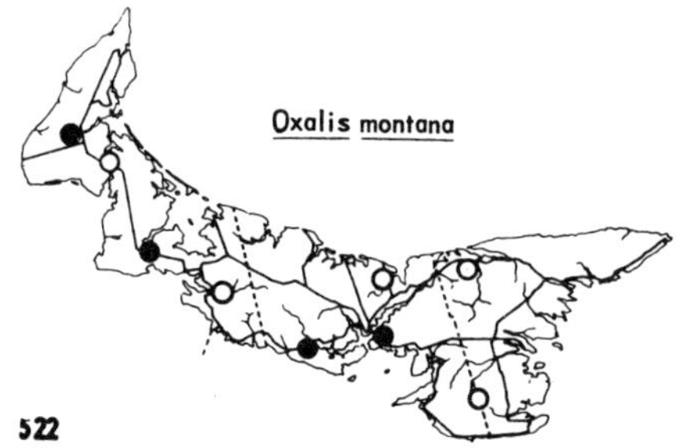

522

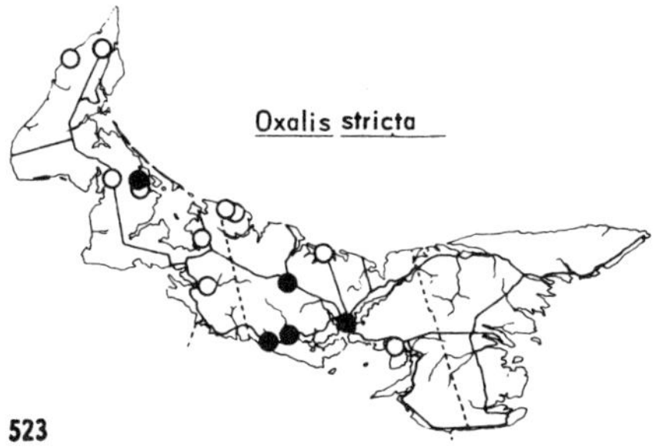

523

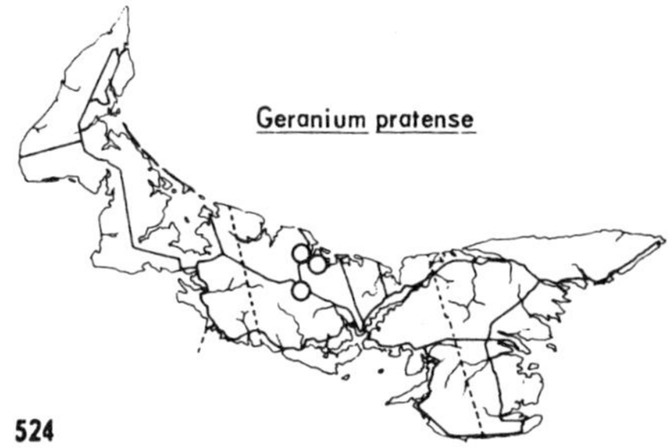

524

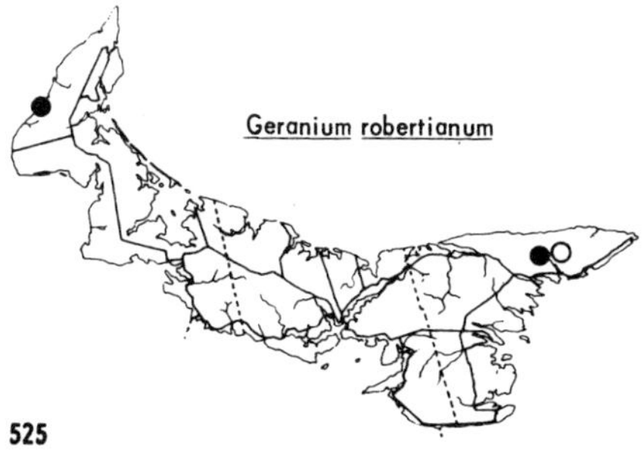

525

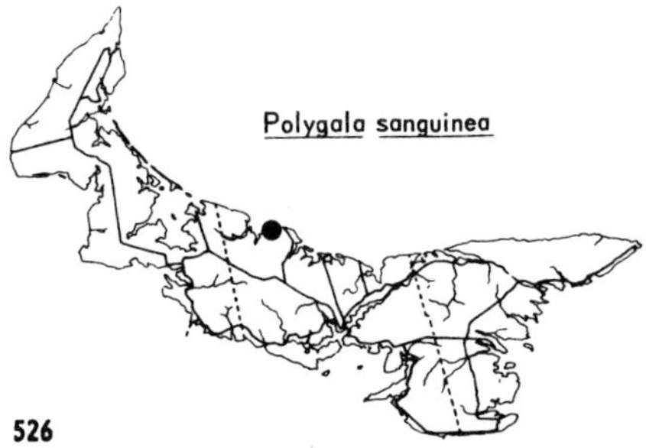

526

EUPHORBIACEAE

Euphorbia helioscopia L. Map 527. Occasional; around towns and on North Shore. Naturalized from Europe before 1888. Weed of railways, sandy gardens and roadsides. 6 Lots (Groh). MacSwain and Bain, MacSwain, Hurst.

Euphorbia cyparissias L. Map 528. Introduced from Europe to decorate gardens and graveyards; persistent and spreading to roadsides, especially in central region, towns. The tetraploid race, which sets seed, forms a large colony at Brackley Point. 1 Lot (Groh). MacSwain, Hurst.

*****Euphorbia esula*** L. York, established colony along bank on farm. Naturalized from Europe.

[***Euphorbia peplus*** L. Summerside, weed on dump. European; casual. 1 Lot (Groh). Hurst.]

*****Euphorbia supina*** Raf. Charlottetown, railway weed (1953). Naturalized from Canada. (***E. maculata,*** authors).

Euphorbia polygonifolia L. Map 529. North Shore; Wood Islands (Roland map). Sandy beaches. MacSwain and Bain, MacSwain, Hurst.

CALLITRICHACEAE
(included in HALORAGEAE by MacSwain)

Callitriche palustris L. Map 530. West Prince; east-central lowland. Aquatic or emersed; muddy border of ponds, streams, springs. (***C. verna*** L., MacSwain and Bain, MacSwain, "doubted" by Hurst); Hurst.

EMPETRACEAE

Empetrum nigrum L. Map 531. Along North Shore; inland: Mermaid bog; north of Souris, Portage, North Point. Cold sites: exposed dunes or headlands; inland in bogs. Fruit black, sour. MacSwain, Hurst.

Empetrum atropurpureum Fern. & Wieg. Map 532. North Shore; inland only around bog at North Point. Exposed sites: sand dunes. Fruit reddish purple, insipid. (***E. nigrum*** var. ***andinum,*** Hurst 1940); Hurst 1940, Campbell.

Corema conradii Torr. Map 533. Bothwell; Murray River; Bristol. Sand dunes; barren gravelly bank.

ANACARDIACEAE

Rhus typhina L. SUMAC Map 534. Very local throughout. Sandy or gravelly knolls or banks; often retained as a garden shrub around house sites. Used for tanning white leather (Bain). Bain, MacSwain and Bain, MacSwain, Hurst; (***R. hirta*** (L.) Sudworth, Hurst).

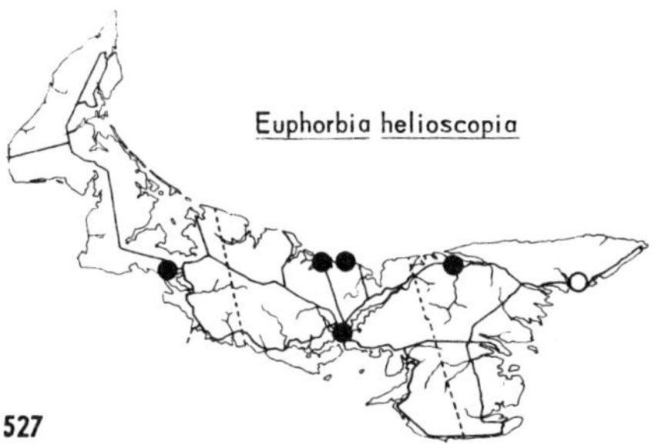

527

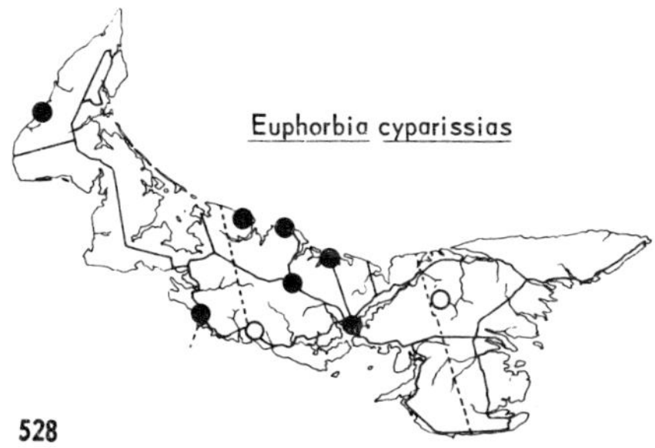

528

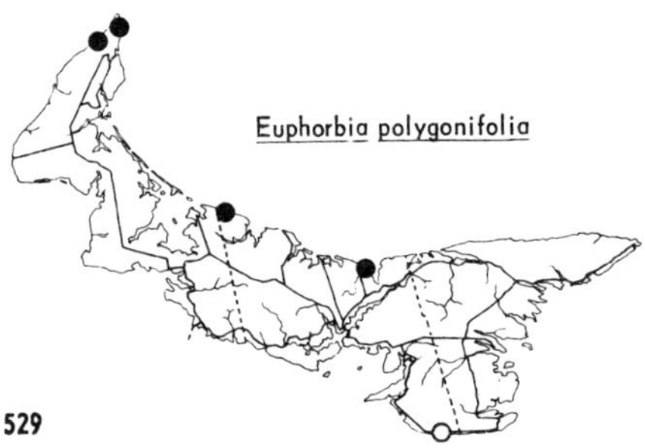

529

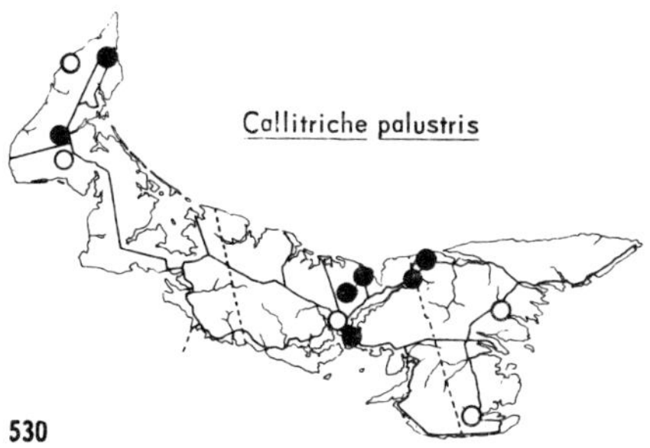

530

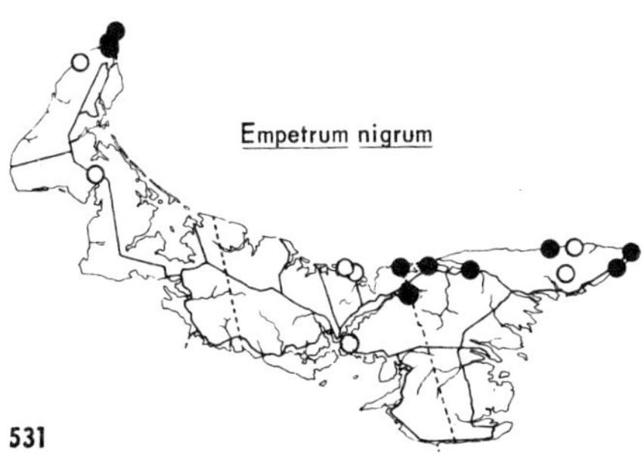

531

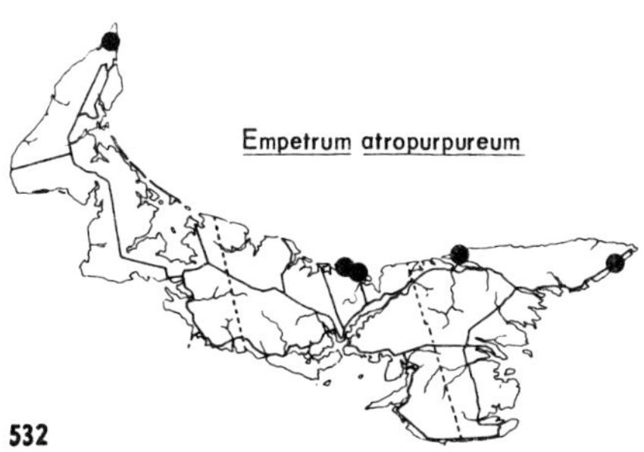

532

533

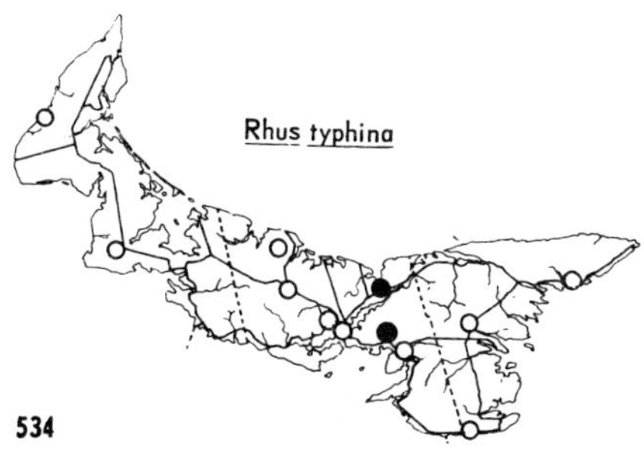

534

Black crowberry (*Empetrum nigrum*) mats the dune slacks along the North Shore; a clump of bayberry in the center foreground.

[*Rhus glabra* L. does not extend northeast of central Maine. Reported by MacSwain and Bain, MacSwain, Hurst.]

Toxicodendron radicans (L.) Ktze. Map 535. (Barkley, Ann. Mo. Bot. Gard. 24: 417. 1937) POISON IVY Local; usually near coast, most common in Prince. Damp, sunny sites: hollows among inner dunes, tension line around salt marsh, open swamp thickets: "dry rich woods" (Bain) inappropriate for Island occurrences. The irritating skin rash caused by contact with the oil of its stems and leaves seems like an allergic reaction, immunity often being lost only after several contacts with it. The oil may be removed by scrubbing with laundry soap if the rash has not set in, otherwise the only precaution to be taken is bandaging to prevent spreading the irritation; baking soda may be used as a palliative, perhaps because it dissolves the oil. Island specimens belong to the northern low variety, var. ***rydbergii*** (Small) n. comb., based on *Rhus rydbergii* Small, Mem. N.Y. Bot. Gard. 1: 268. 1900. (***Rhus Toxicodendron,*** Bain, MacSwain and Bain, Hurst; ***R. radicans*** L., Campbell).

AQUIFOLIACEAE (ILICINEAE, MacSwain)

Ilex verticillata (L.) Gray Map 536. Lowlands, throughout. Swampy streams, borders of damp woods, in shade the leaves thinner and with translucent spots, the "var. ***tenuifolia.***" (***Prinos verticillatus*** L., MacSwain and Bain); MacSwain, Hurst.

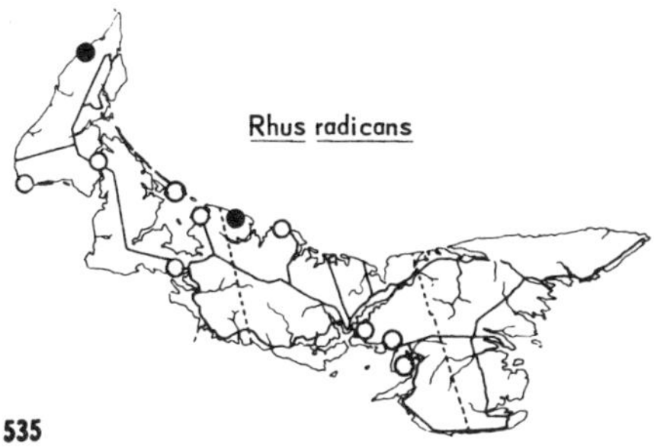

535

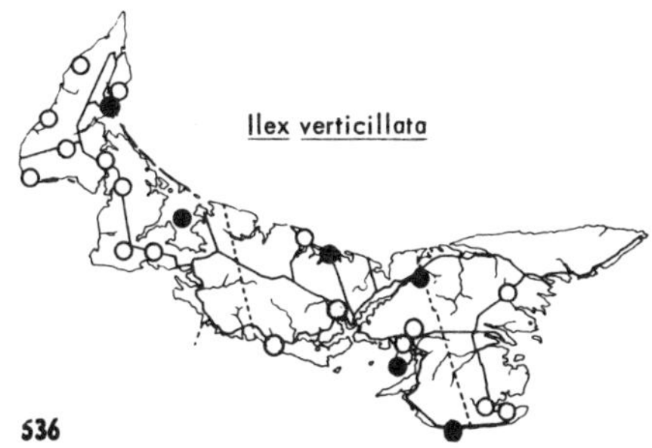

536

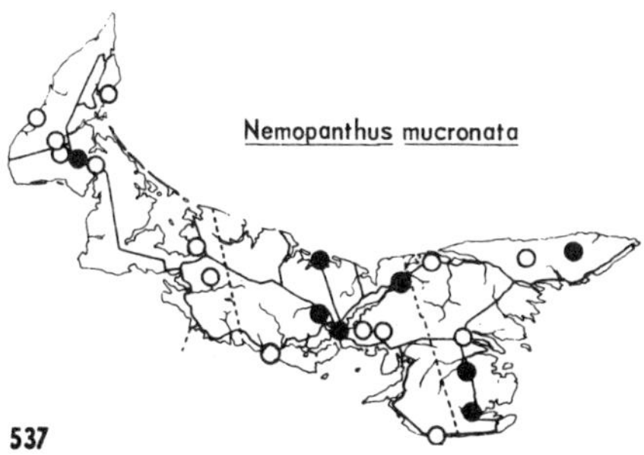

537

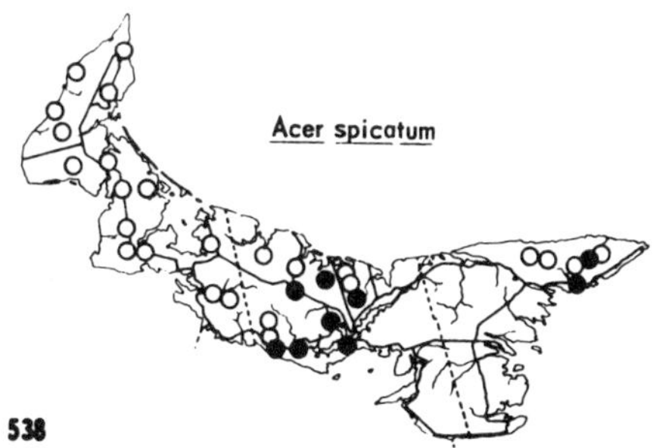

538

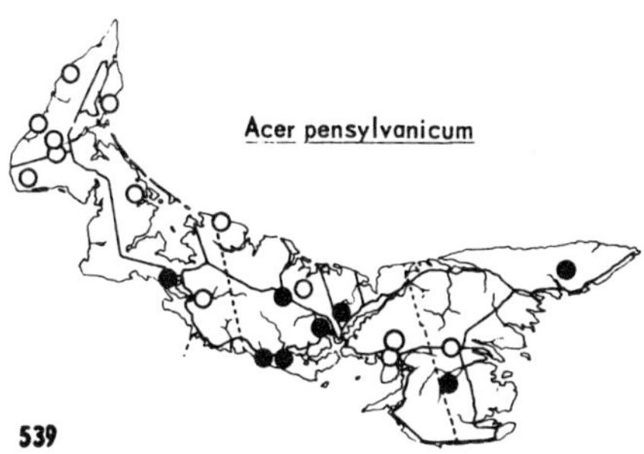

539

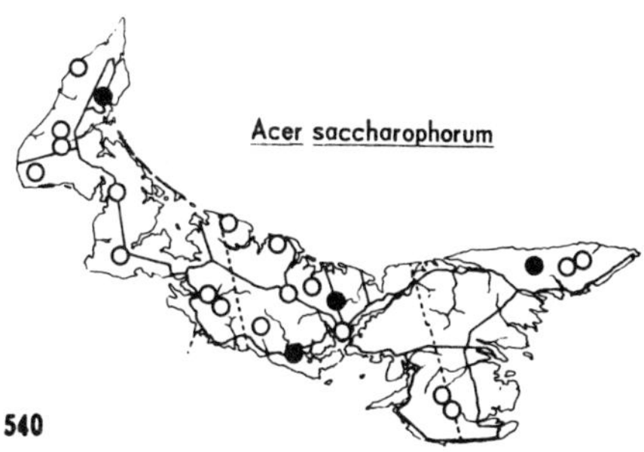

540

Nemopanthus mucronata (L.) Trel. Map 537. Scattered throughout. Swamps and thickets; in open bogs more compactly branching and gnarled. (*N. canadensis* DC., MacSwain and Bain; *N. fascicularis* Raf., MacSwain); Hurst.

ACERACEAE (SAPINDACEAE, MacSwain)

Acer spicatum Lam. Map 538. General in Prince and Queens; in the northeastern hills of Kings. Openings or understory of hardwood forest, on well-drained sites: knolls, upland or ravines. Bain, MacSwain and Bain, MacSwain, Hurst.

Acer pensylvanicum L. STRIPED MAPLE Map 539. Scattered throughout; common throughout and at borders of upland. Ravines, gullies and slopes; understory in mixed or hardwood forest. Possibly the "white maple (*A. Negundo*)" of Stewart? Bain, MacSwain and Bain, MacSwain, Hurst.

[*Acer saccharophorum* K. Koch (Rousseau, Contr. Inst. Bot. Univ. Montr. 35: 1-66. 1940) ROCK MAPLE, SUGAR MAPLE Map 540. Throughout. With Beech, forming the climax forest of well-drained sites, the uplands and knolls. The sweet sap, gathered between March 25 and April 10 (Stewart) is boiled for maple sugar, eaten as a candy and used as maple flavoring now that imported sugar is more economical. In 1890, Bain estimated the annual production as 25,000 pounds. No commercial production now. The curled grain of the wood was used in ornamental woodwork as "bird's-eye maple" (Stewart). Extensively planted as a shade tree. (*A. saccharinum*, Stewart, Bain, MacSwain and Bain, MacSwain; *A. saccharum*, Hurst, Campbell).]

[*Acer saccharinum* L., is occasionally planted but is not native east of the Saint John. (*A. dasycarpum* Ehrh., MacSwain); Hurst.]

Acer rubrum L. RED MAPLE Map 541. Common throughout. Hardwood forest (well-drained sites), often pure on uplands after clearing, due to stump-sprouting; in damp sites, very abundant, with spruce forming mixed woods. Often planted as a shade tree, though more brittle than *A. saccharum*. Its red fall color is our best. Stewart, Bain, MacSwain and Bain, MacSwain, Hurst.

The forma *tomentosum* (Desf.) Dansereau present (Bonshaw).

Acer negundo L. MANITOBA MAPLE, BOX ELDER Map 542. Introduced from the Prairies as a shade tree, though very brittle; naturalized along Indian River, frequent before 2,4-D spray (1952) along roadsides near Alberton. (*Negundo aceroides* Moench, MacSwain); Hurst.

Forma *sanguineum* L. Martin Montague.

BALSAMINACEAE (GERANIACEAE, MacSwain and Bain, MacSwain, both in part)

[*Impatiens pallida* Nutt. Not seen, but to be sought. MacSwain, Hurst.]

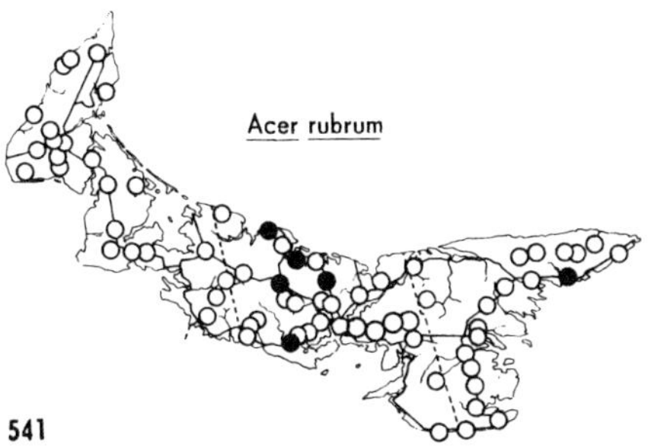

541

542

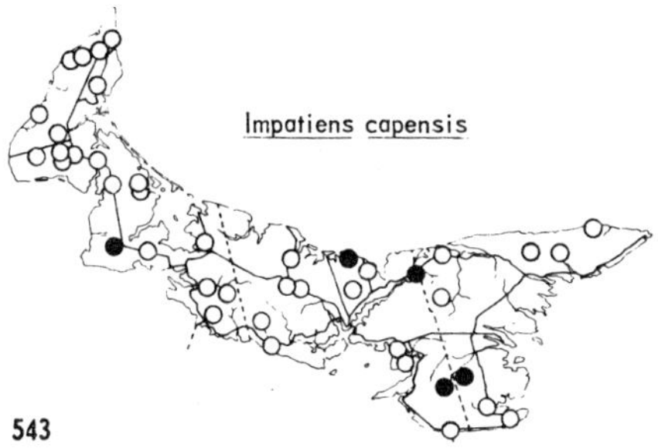

543

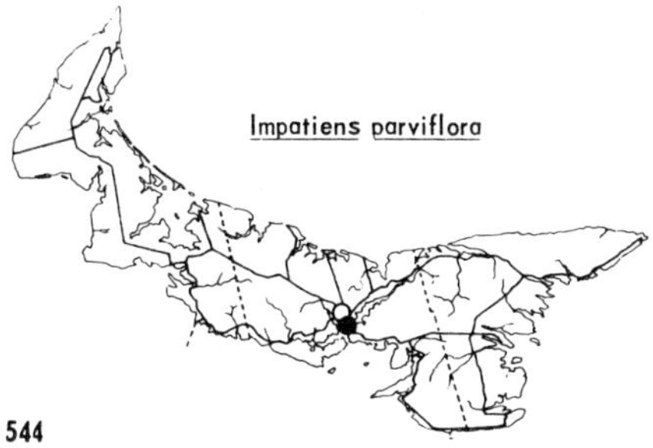

544

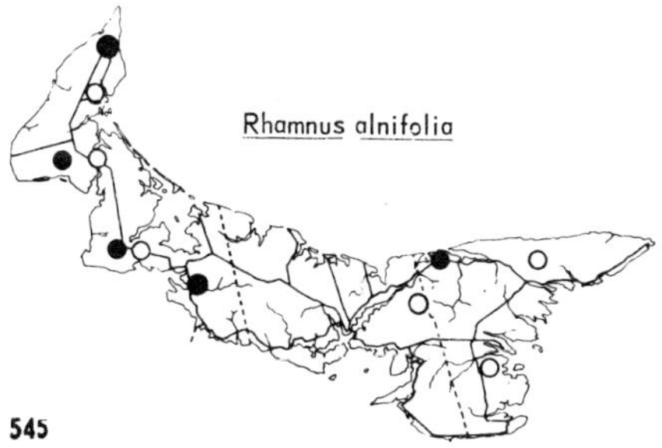

545

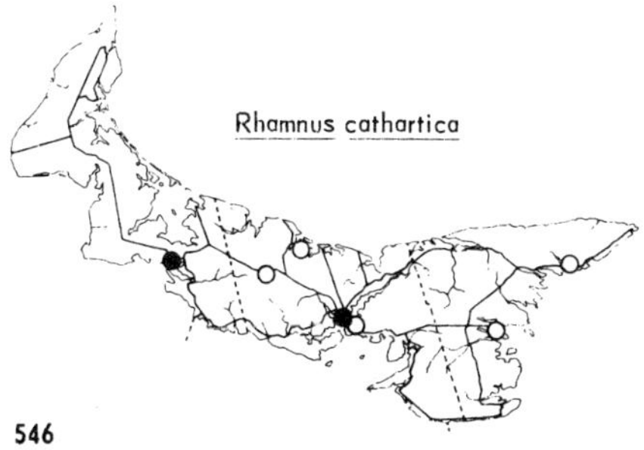

546

Impatiens capensis Meerb. Map 543. Common throughout. Damp thickets along streams. (*I. fulva* Nutt., Bain, MacSwain and Bain, MacSwain; *I. biflora* Walt., Hurst); Campbell.

Impatiens parviflora DC. Map 544. Charlottetown and environs: naturalized from Europe (originally central Asiatic) before 1912. Covering dampish ground in shrubbery; ditches, thickets. 1 Lot (Groh). Hurst.

RHAMNACEAE

Rhamnus alnifolia L'Hér. Map 545. Local, western and eastern lowlands. Swamps and wet woods, in shade rarely fruiting. Harbors the crown rust of oats, *Puccinia coronata*, in aecial stage. Gorham 1944, Campbell.

[*Rhamnus cathartica* L. BUCKTHORN Map 546. Introduced from Europe for hedges; occasionally escaped (e.g., Summerside). As alternate host of crown rust of oats (*Puccinia coronata*), its other stations plotted by Gorham (1944): Fredericton, Cymbria, Charlottetown, Georgetown; also in Souris, at all these stations planted. 3 Lots (Groh). Hurst.]

[*Rhamnus davurica* Pall. "Introduced in Prince Edward Island about twenty-five years ago [at the Experimental Farm] and spreading in a limited fashion on farm [the same] near Charlottetown," as *R. cathartica* var. *davuria*, Gorham.]

**Rhamnus frangula* L. European; introduced for hedges on the Experimental Farm, Charlottetown, and naturalized there in the hedgerows and wood edges. (*Frangula alnus* Mill.).

VITACEAE

Parthenocissus quinquefolia (L.) Planch. Map 547. VIRGINIA CREEPER Eastern North American, introduced as a creeper for houses; persistent, and at Souris escaped to roadsides.

[*Parthenocissus inserta* (Knerr) K. Fritsch, as to Island records probably refers to the former species. Fernald (1950).]

TILIACEAE

**Tilia europaea* L. Dunstaffnage, edge of woods behind school. European, introduced for avenues, and escaping.

MALVACEAE

[*Malva neglecta* Wallr. Map 548. Around Savage Harbour since 1930's (Messervy); Albany (1950). Naturalized from Europe. Waste places (*M. rotundifolia*, Hurst 1940; Campbell).]

[*Malva verticillata* L. var. *crispa* L. Indian River mill, planted under lilacs; two plants escaped to roadside. European, introduced.]

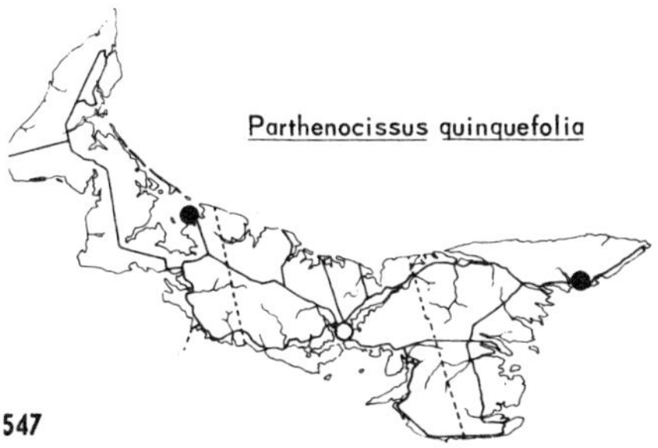

547

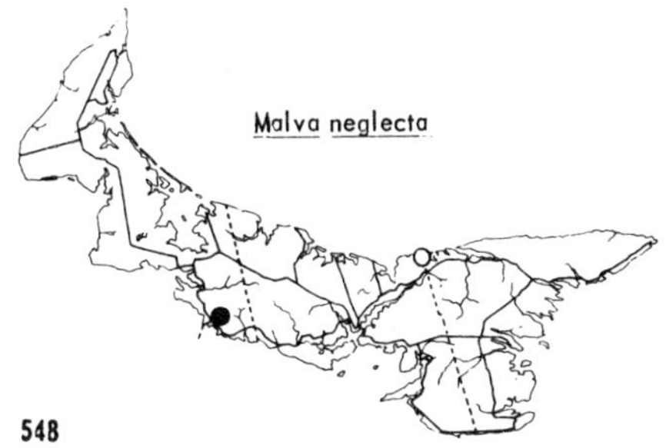

548

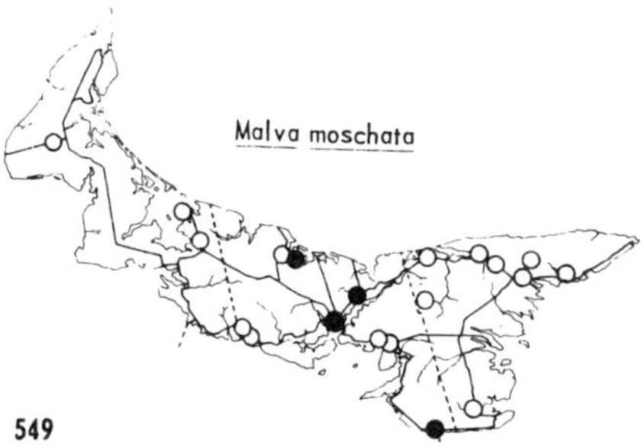

549

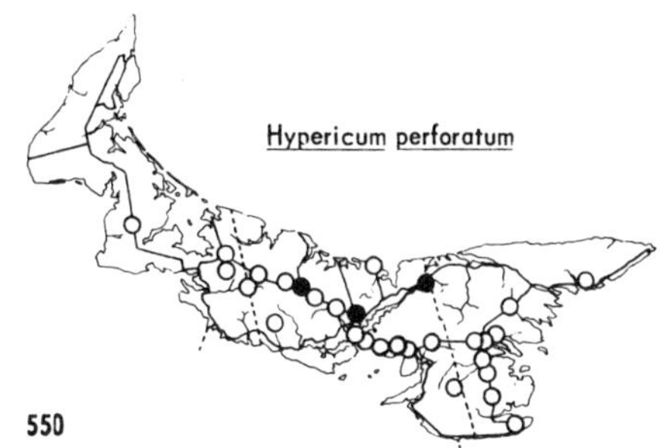

550

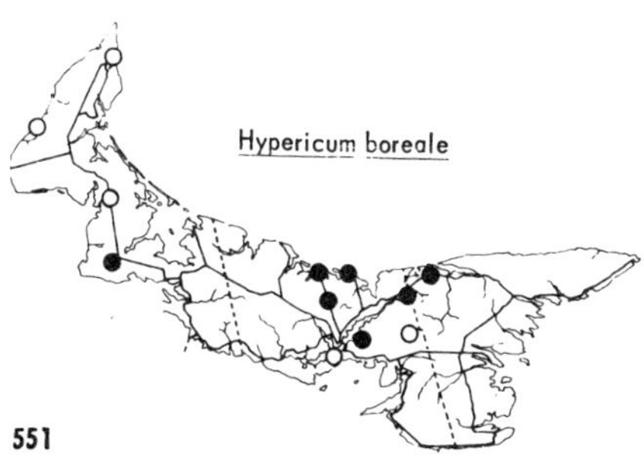

551

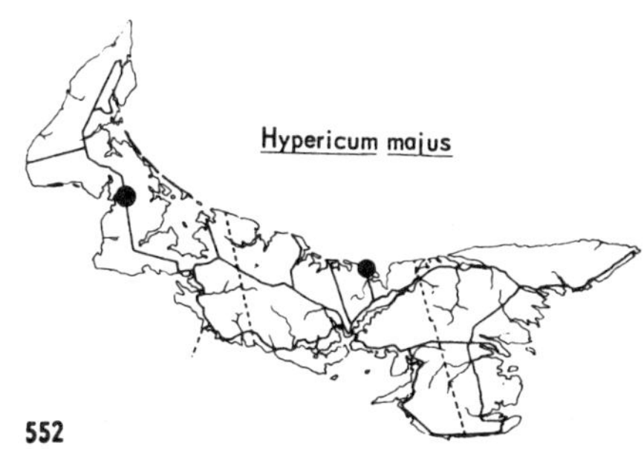

552

Malva moschata L. Map 549. Scattered throughout; planted as ornamental around houses, cemeteries; naturalized commonly along roadsides. Pink or white petals. MacSwain and Bain, MacSwain, Hurst.

[****Abutilon theophrasti*** Medic. Charlottetown, railway yards, adventive from central Canada; originally from India.]

HYPERICACEAE
(GUTTIFERAE, Fernald 1950)

Hypericum perforatum L. Map 550. Becoming common along roadsides and in sandy pastures, central and eastern regions; Richmond its western limit (1953). European, naturalized. Poisonous to stock. MacSwain and Bain, MacSwain, Hurst.

Hypericum boreale (Britton) Bickn. Map 551. Prince County and east-central lowlands. Muddy shores and ditches, often in boggy places. (*H. mutilum*, MacSwain and Bain, MacSwain, Hurst.

****Hypericum majus*** (Gray) Britton Map 552. Local: McNeills Mills corner (P); Dalvay House. Damp sand: roadside pit and dune slacks. At both stations, as in Cape Breton, mixed with *H. canadense*.

Hypericum canadense L. Map 553. East-central and southeastern region and western Prince County. Damp open sites; muddy fields, shores, bogs. MacSwain, Hurst.

Hypericum virginicum L. var. ***fraseri*** (Spach) Fern. Map 554. Scattered throughout lowlands. Damp sites: pond shores, dune slacks, creek banks. (*Elodes virginica* (L.) Nutt., MacSwain and Bain; *E. campanulata* Pursh, MacSwain); Hurst.

ELATINACEAE

Elatine minima (Nutt.) Fisch. & Mey. Map 555. Glenfinnan Lake, Lake Verde; submerged (in six to eighteen inches of water) on sandy shores, though gathering fine clay.

CISTACEAE

Hudsonia ericoides L. Map 556. Bothwell, sand dunes. Forma *leucantha* Fern. originally described from here (Fernald, Rhodora 19: 76. 1917).

X ***Hudsonia intermedia*** (Peck) n. comb. Bothwell, sand dunes. Always associated with *H. ericoides* clumps amid the more abundant *H. tomentosa*. (*H. tomentosa* var. *intermedia* Peck, Rep. N.Y. State Mus. 45: 86. 1893). Fernald (1950).

Hudsonia tomentosa Nutt. Map 557. North Shore, locally from Bothwell to Malpeque. Sand dunes. MacSwain and Bain, MacSwain, Hurst.

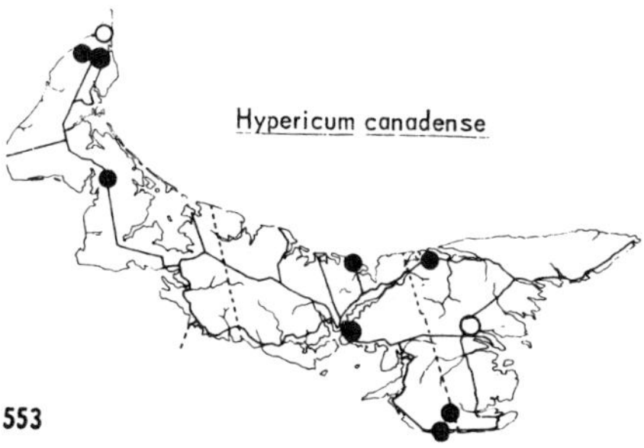

553

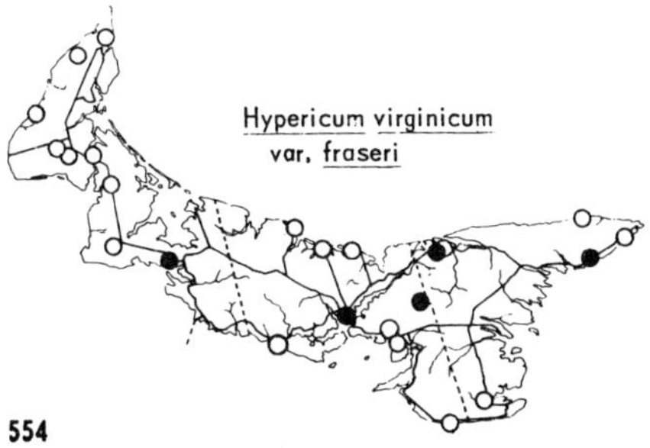

554

555

556

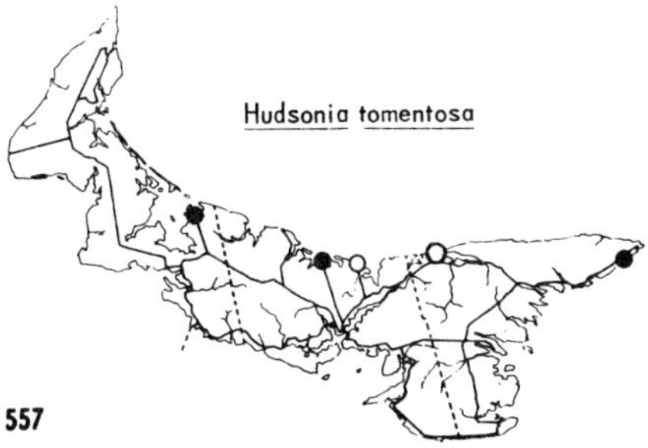

557

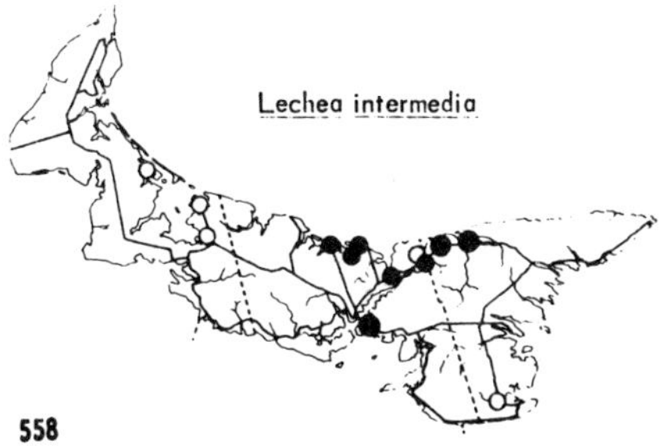

558

Lechea intermedia Leggett Map 558. Mermaid (Q); Murray River; near North Shore, Morell to East Bideford. Sandy open sites: weedy on sandy roadsides, spruce or pine woods and exposure barren. 2 Lots (Groh). (*L. minor*, MacSwain, doubted by Hurst, Campbell); Groh, Hurst.

VIOLACEAE

Viola cucullata Ait. BLUE VIOLET Map 559. Common throughout. Damp sites, shady or open, often weedy in damp fields, stream banks. Pale-flowered forms were referred to *V. Watsoni* by Greene. Bain, MacSwain and Bain; (*V. palmata* var. *cucullata* (Ait.) Gray, Watson in MacSwain; doubted by Hurst).

X ***Viola melissaefolia*** Greene. Brackley Point Road (*Fernald et al.*), is *V. cucullata* X *septentrionalis.*

Viola nephrophylla Greene Map 560. West Prince, very characteristic of cedar swamps; also in ravine woods.

Viola fimbriatula Sm. Map 561. Vicinity of Charlottetown, open sandy woods. Rocky Point, specimens referred by Fletcher to "*V. fimbriatula* X *nesiotica*". (*V. dentata*, J.M. Macoun, Ott. Nat. 13: 159. 1899).

Viola septentrionalis Greene Map 562. Wellington Centre, Summerside, Hunter River. Well-drained sites: dry fields and banks. Hybrid with *V. cucullata* from Brackley Point Road. Robinson & Fernald 1908, Hurst.

Viola pallens (Banks) Brainerd WHITE VIOLET Common throughout. Damp sites in woods, along brooks, in marshes. (*V. blanda*, Bain, MacSwain and Bain, Watson in MacSwain, Hurst; *V. Macloskeyi* ssp. *pallens* (Banks) M. Baker).

Viola incognita Brainerd Map 564. Scattered throughout. Well-drained (rarely damp) sites in coniferous woods; characteristic under second-growth spruce. (*V. alsophila*, Watson in MacSwain; not referable by Hurst).

Viola renifolia Gray Map 565. Common in West Prince, scattered eastward. Woods: ravine slopes, stream banks or cedar swamp.

Viola pensylvanica Michx. var. ***leiocarpa*** (Fern. & Wieg.) Fern. Map 566. YELLOW VIOLET Rare: near Alberton (Watson in MacSwain); scattered in the central hilly region. Open hardwoods. (*V. pubescens*, Watson in MacSwain, Hurst).

Viola adunca Sm. var. ***minor*** (Hook.) Fern. Original station of plants cited by Brainerd in Rhodora 15: 109. 1913 as received from Watson, not recorded. (*V. Labradorica* Schrank, Watson in MacSwain, Hurst; *V. adunca* var. *glabra* Brainerd); Campbell.

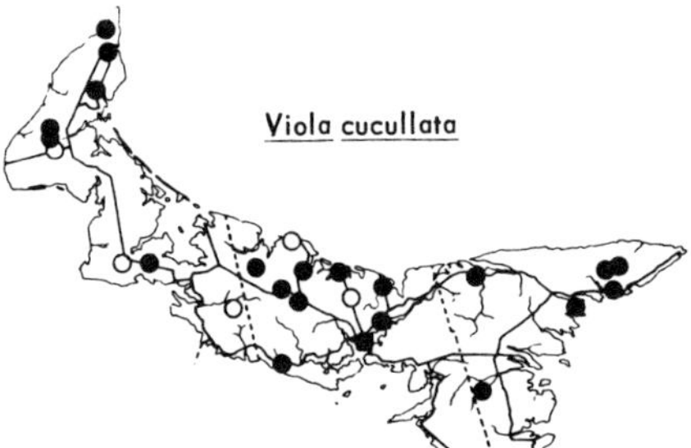

559

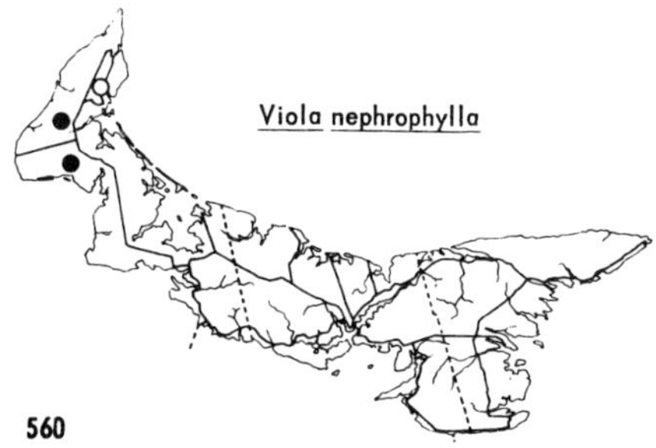

560

561

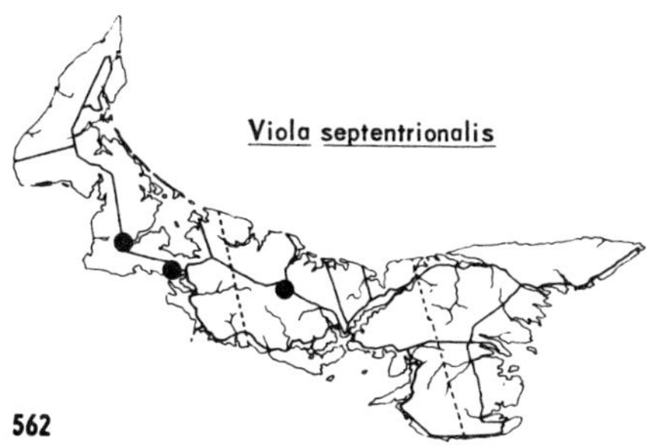

562

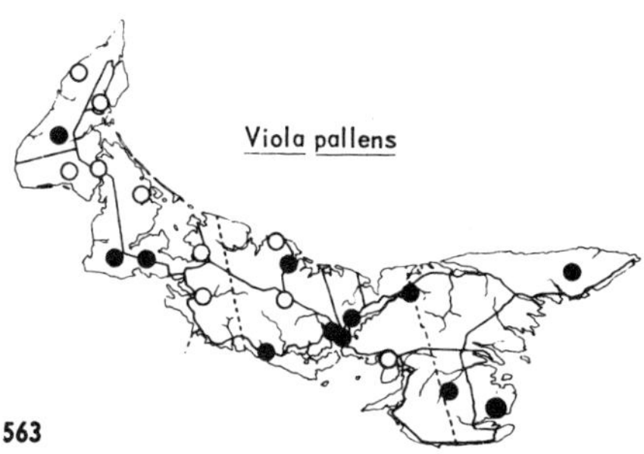

563

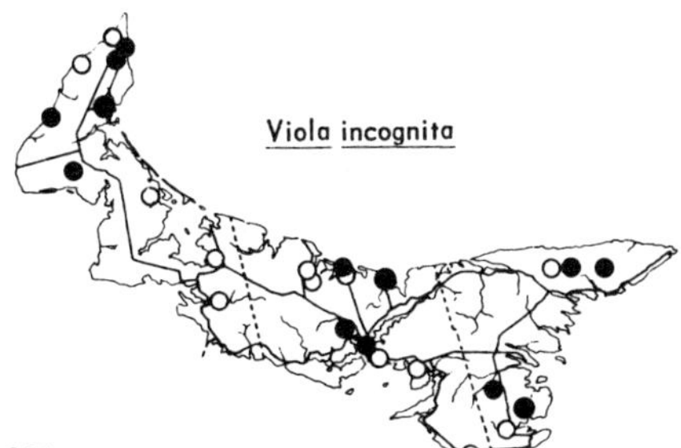

564

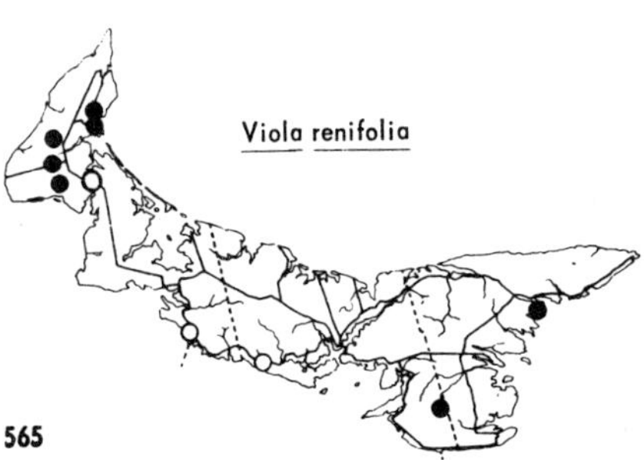

565

566

Viola arvensis Murr. Map 567. Most common in Queens County, records since 1919 (at Kinkora). Weed of fallowed grain fields, rarely roadsides; flowering early. Naturalized from Europe. Hurst 1940; (*V. tricolor* "not well naturalized," from 1 Lot, Groh, Hurst).

THYMELAEACEAE

Daphne mezereum L. DAPHNE Map 568. New Annan, 1953 (*Poole*); Hunter River; Dunstaffnage (Messervy). Flourishing roadside colonies were apparently wiped out in the winter of 1944-5 (Hurst in *Guardian*, 21 June 1945), but it has escaped to woods at Hunter River. European, introduced ornamental. Hurst.

LYTHRACEAE

Lythrum salicaria L. Map 569. Scattered roadside stations, in damp ditches. Formerly (till 1955) most abundant in the marsh (now drained) at the Experimental Farm, Charlottetown, the parent colony. Introduced- from the St. Lawrence?; European. 1 Lot (Groh). Hurst.

ONAGRACEAE (OENOTHERACEAE, Bain)

Epilobium angustifolium L. FIREWEED Map 570. Common throughout. Open, usually dryish land; characteristic of the clearings and burns of coniferous woods. 49 Lots (Groh). (*E. spicatum* Lam., MacSwain and Bain); MacSwain, Hurst.

Epilobium strictum Muhl. Map 571. Local: East Point, Bristol, Mount Stewart, Bunbury, Grand Tracadie, Summerside, Bloomfield. Peat bogs, marshes. Bain 1892, J.M. Macoun, Can. Record Sci., Apr. 1894, 82.

Epilobium leptophyllum Raf. Map 572. Common in western Prince County; scattered in lowland Queens; Murray River. Marshy spots, including dune slacks, thickets, pond borders, bogs. (*E. rosmarinifolium* Pursh (not Haenke), MacSwain and Bain; *E. densum*, Hurst; *E. strictum*, Campbell).

Epilobium palustre L. Map 573. Western Prince, east-central Queens. Swamp or boggy sites; dune slacks. MacSwain and Bain, MacSwain, Hurst.

Var. ***palustre*** Tignish (*Macoun*).

Var. ***oliganthum*** (Michx.) Fern. Usual form. (var. *monticola*, Roland, not Haussk.).

Epilobium adenocaulon Haussk., vars. Map 574. Rather common, throughout. Damp, open spots; thickets; roadside ditches. 8 Lots (Groh). (*E. coloratum*, MacSwain and Bain, MacSwain, Hurst); Groh, Hurst.

Var. ***adenocaulon***. Common. White-petalled forms occur: York. (*E. glandulosum* var. *adenocaulon* (Haussk.) Fern., Campbell).

Var. ***occidentale*** Trel. Tracadie. (*E. glandulosum* var. *occidentale* (Trel.) Fern., Roland: map).

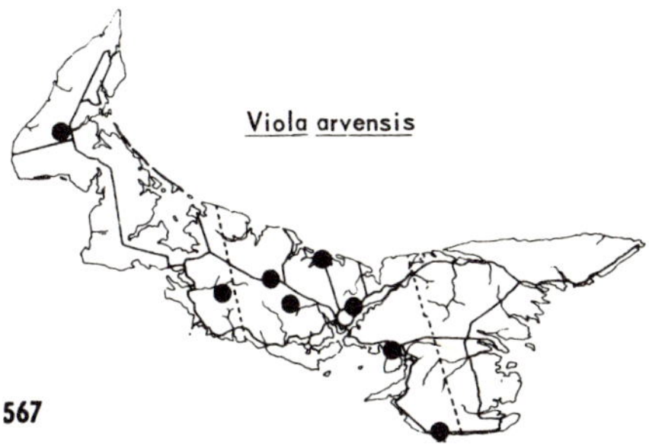

567

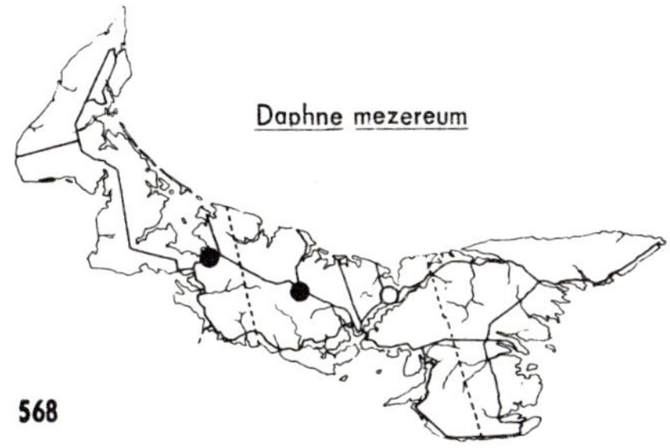

568

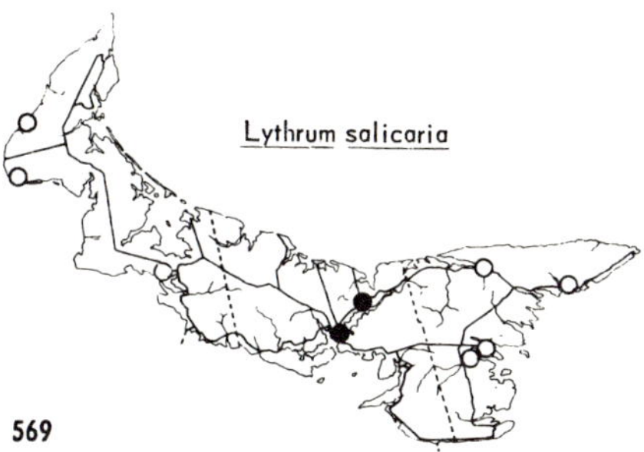

569

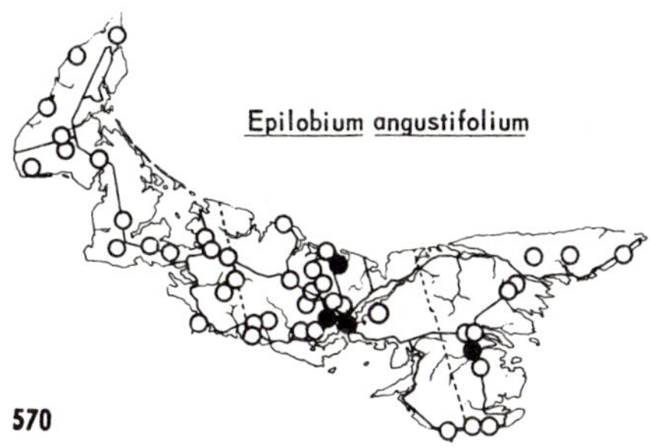

570

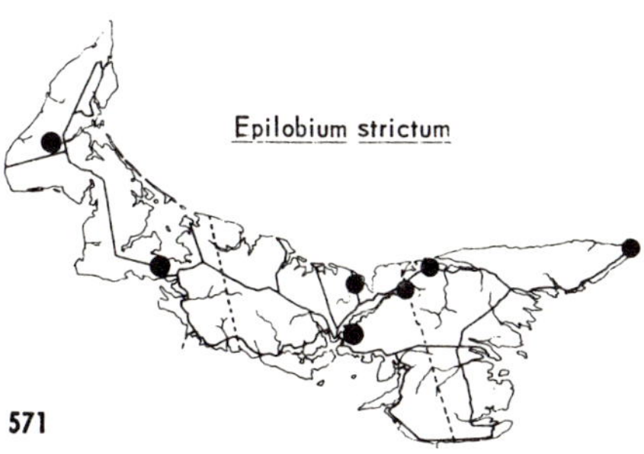

571

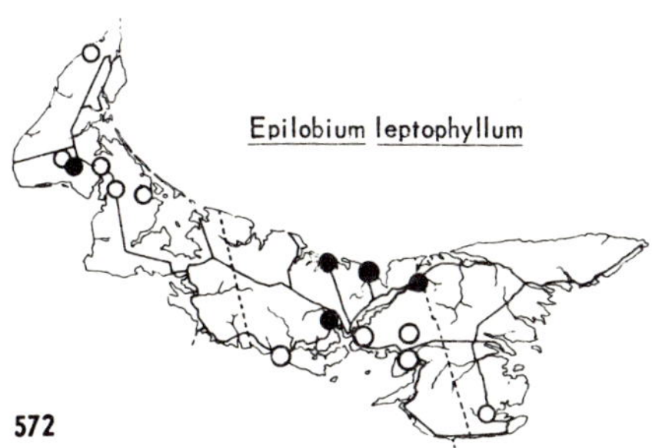

572

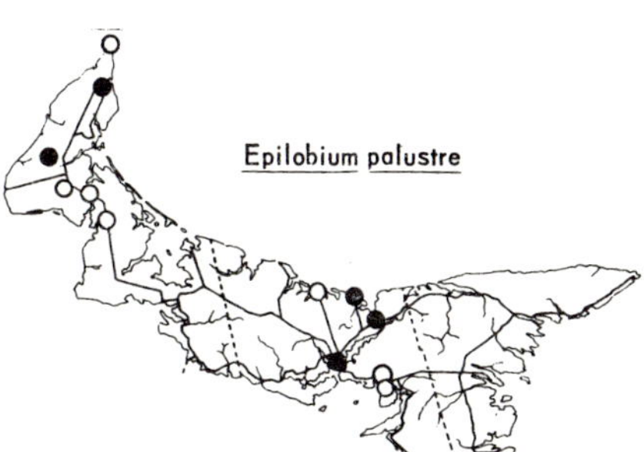

573

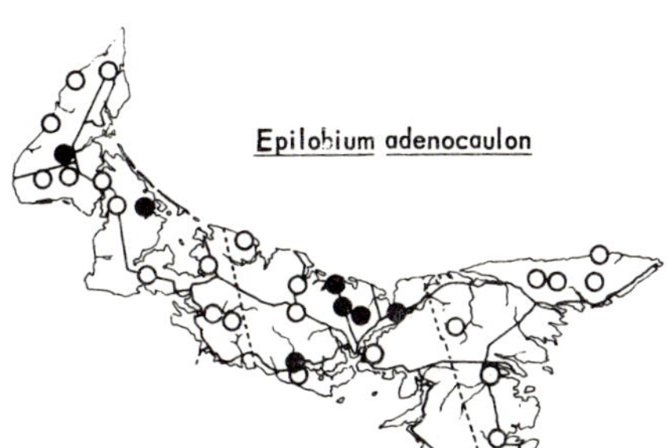

574

Epilobium ciliatum Raf. Brackley Point, "salt marsh" (probably dune slacks), (*Macoun*). Annotated by Trelease as "perhaps *E. ciliatum* Raf., treated by me as a dwarf crisp-pubescent form of *E. adenocaulon*." J.M. Macoun, Can. Record Sci., Apr. 1894, 83.

Oenothera biennis L. (*Oe. novae-scotiae* Gates) Map 575. Common weedy species, particularly in the central uplands. Dry sandy places: roadsides, barren pastures, and waste places. (Very variable in stem and leaf color, in width and shape of leaves and length of internode, in presence or absence of red papillae as bases of hairs, in abundance of strigose hairs on capsule, contrasting forms often occurring together). 34 Lots (Groh). (*Oe. muricata*, Groh, Hurst in part; *Oe. parviflora*, Campbell in part); Bain, MacSwain and Bain, MacSwain, Hurst.

Var. *strigosa* (Rydb.) Piper, Fl. State Washington, 407, 1906. Tignish (var. *hirsutissima*, Fernald, 1950 not Gray).

Oenothera cruciata Nutt. Map 576. Charlottetown, Brackley Beach, Fortune Bridge. Dry sandy roadsides. On the railway siding below the Charlottetown Experimental Farm, its dark green, glossy, lanceolate leaves, in size only gradually reduced to bracts of the inflorescence, and the red cortex of its roots, contrasted with the light green, oval-oblong, more denticulate leaves abruptly decreasing into the bracts of the inflorescence, and the brown root-cortex of *Oe. biennis*. But there were also plants with linear petals of various widths, whose vegetative characters were those of *Oe. biennis*, though no intermediates with *Oe. biennis* floral characteristics were found. Gleason (1952), on the other hand, reduces *Oe. cruciata* to *Oe. parviflora* L. Hurst.

Oenothera parviflora L. (*Oe. ammophiloides* Gates & Catcheside) Map 577. Common coastwise, rare inland. Beach bars, dune sand. (*Oe. muricata*, Groh, Hurst in part); Campbell in part.

[*Oenothera fruticosa* L. Report in 1891 may have intended *Oe. tetragona*, but neither occurs east of southern Nova Scotia. MacSwain and Bain.]

Oenothera perennis L., vars. Map 578. Common in the uplands, scattered throughout. Weed of old fields, sandy stream margins, rarely of gardens. 4 Lots (Groh). (*Oe. pumila* L., MacSwain and Bain, MacSwain, Groh, Hurst); Campbell.

Var. ***perennis*** Throughout.

Var. ***rectipilis*** Blake Concentrated between Brackley Beach and Mount Stewart.

Circaea alpina L. Map 579. Scattered throughout. Damp shady sites: streambank woods, swamps. MacSwain and Bain, MacSwain, Hurst.

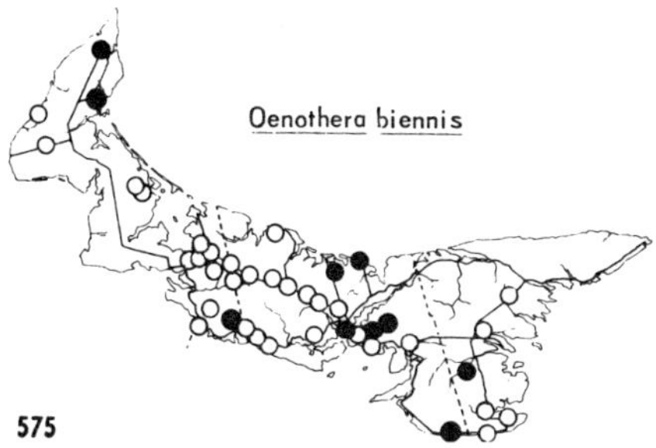

575

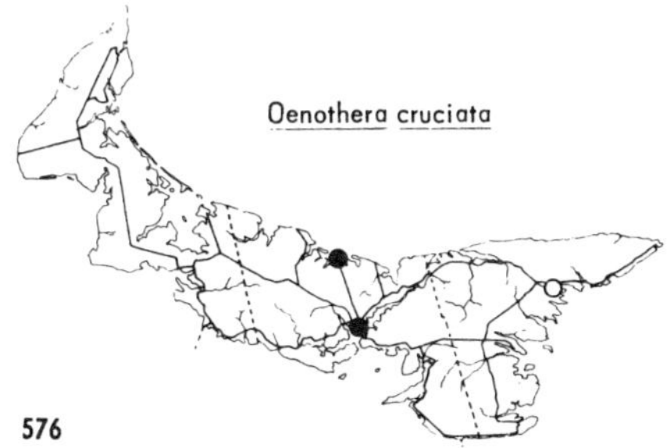

576

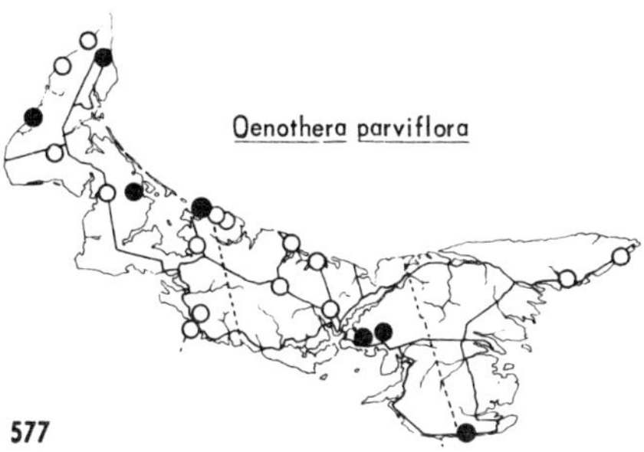

577

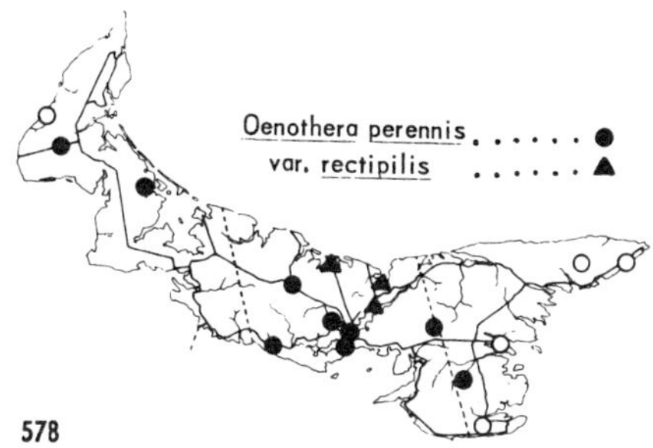

578

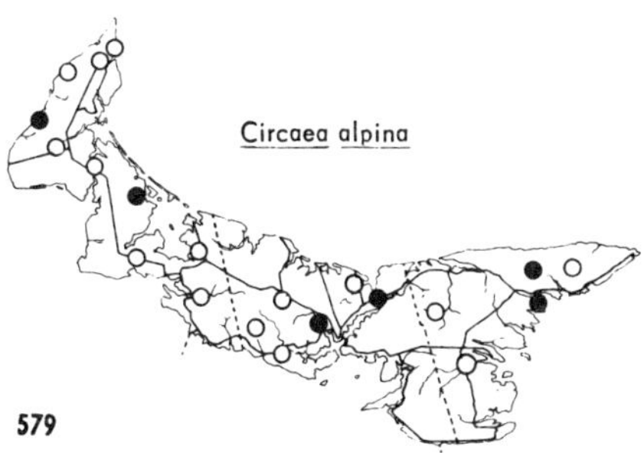

579

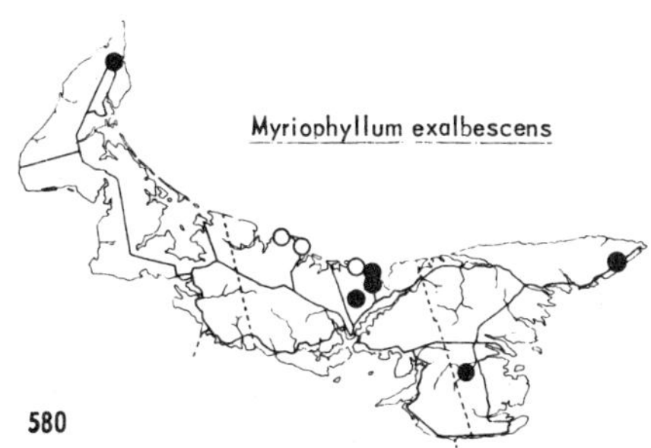

580

HALORAGIDACEAE

Myriophyllum exalbescens Fern. Map 580. North Shore ponds and streams; also near Montague. Fernald cited specimens from Tignish and Black Pond (Rhodora 21: 121. 1919). Cool waters; well-aerated (usually on sandy bottom); on the North Shore, common in the cooler ponds, elsewhere only at influx of streams (M.W. Smith 1946). (*M. spicatum*, MacSwain, Hurst), Campbell.

Myriophyllum verticillatum L. var. *intermedium* Koch Brackish millpond, Tignish (*Fernald et al.*). (Fernald 1950).

HIPPURIDACEAE

Hippuris vulgaris L. Map 581. Local: frequent near Tignish; Wellington; frequent in ponds in National Park; Cambridge (K). Muddy shores in shallow water". . .only in cold spring water"; agreeing with Fassett's observation that *Hippuris* is usually a species of cold waters (M.W. Smith 1946). Bain 1892, MacSwain, Hurst.

ARALIACEAE

Aralia racemosa L. Map 582. Rare, central upland: Dunk River, Hunter River. Deciduous or mixed woods, sandstone cliffs or knolls. Bain, MacSwain and Bain, MacSwain, Hurst.

Aralia hispida Vent. SARSAPARILLA Map 583. Scattered throughout, very common in Kings. Open, sandy woods, dry or damp; weedy in clearings and slash or burn. "Sarsaparilla" noted by Stewart as more abundant here than anywhere on the mainland north of Boston. The drupes collected for a tonic "wine." (*Aralia* sp., Stewart); Bain, MacSwain and Bain, MacSwain, Hurst.

Aralia nudicaulis L. Map 584. Common throughout. Well-drained shady sites, in hardwoods or coniferous. "Ginseng in great plenty in forest, in large timber and good soil"—Stewart. (Not ginseng, however). (*Panax trifolium*, Stewart); Bain, MacSwain and Bain, MacSwain, Hurst.

Panax trifolius L. Map 585. Northeastern uplands only: Mount Stewart (Roland map); Fortune Bridge to Glencorradale. Well-drained hardwood sites with rich leaf mold, and spring sunshine; plants wilting and dying back about July 7th. (A dwarf ginseng). Hurst 1940.

UMBELLIFERAE

Hydrocotyle americana L. Map 586. Local: Prince and Queens, east to Watervale, Tracadie. Damp ground, partial shade: thickets, stream banks. Hurst 1940.

Sanicula marilandica L. Map 587. Very local: West Prince; Hunter River; Dundee. Open damp woods and stream-bank thickets.

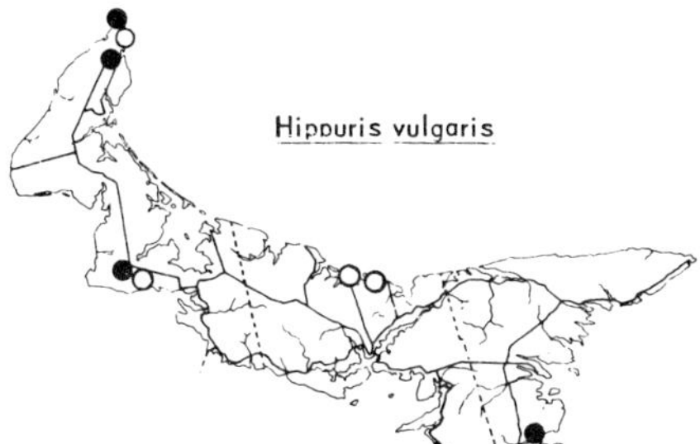

581

582

583

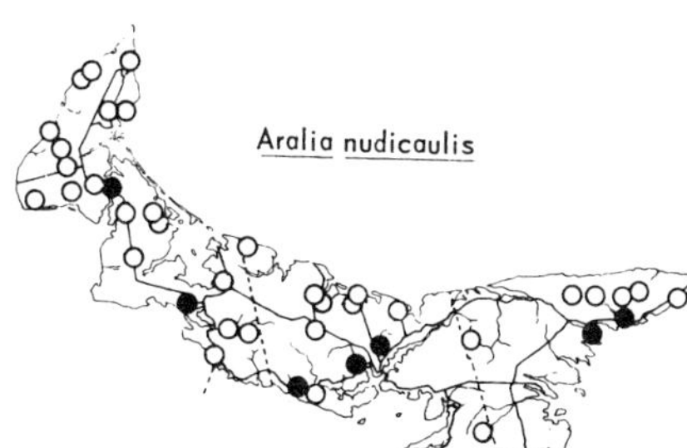

584

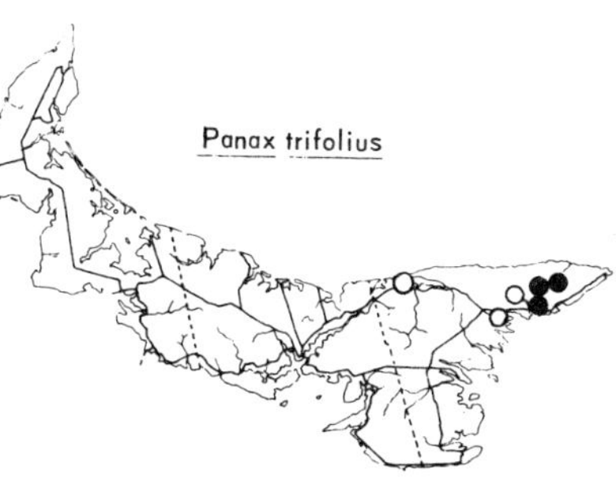

585

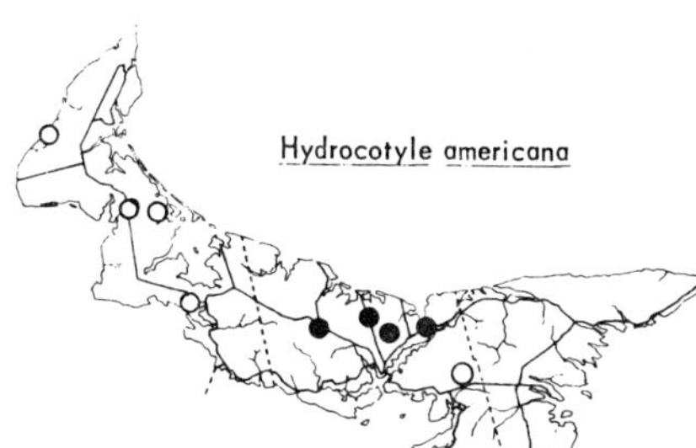

586

587

588

Osmorhiza claytonii (Michx.) C.B. Clarke Map 588. Common in northeastern upland; very local westward: Bonshaw, Portage, Pleasant View. Slopes, hollows or runs in hardwood forest. (*O. brevistylis* DC., J.M. Macoun Can. Record Sci., Oct. 1895, MacSwain); Hurst.

Cicuta bulbifera L. Map 589. Scattered throughout, usually near the coast in lowland: swamps or marshes. Poisonous. 1 Lot (Groh). MacSwain, Hurst.

Cicuta maculata L. Map 590. Common in Prince and east-central lowlands; Murray Harbour. Marshes: often in roadside or railway ditches. Poisonous. 9 Lots (Groh). Bain, MacSwain and Bain, MacSwain, Hurst.

Carum carvi L. CARAWAY Map 591. Common in Prince County and around old settlements eastward. European; introduced for the fruits ("seeds") used as a flavoring (in bread, cake, etc.), now naturalized. Waste places, fields. Bain, MacSwain and Bain, MacSwain, Hurst.

[*Aegopodium podagraria* L., collected at Brackley Beach (*Hurst*), is the var. *variegatum* Bailey, introduced as garden plant, persistent but not naturalized. Groh & Frankton. Can. Weed Surv. 5: 44. 1948.]

Sium suave Walt. Map 592. Common throughout, except in uplands. Wet places: river banks, pond margins, marshy ditches. Poisonous. 14 Lots (Groh). (*S. latifolium*, MacSwain and Bain; *S. lineare* Michx., MacSwain and Bain; *S. cicutaefolium* (Schrank) J.F. Gmel., MacSwain, Hurst); Campbell.

Ligusticum scothicum L. Map 593. Coastal: frequent in Kings, rarer westward at least to Ellerslie. Tension line at head of beach or salt marsh. 2 Lots (Groh). (*Coelopleurum Gmelini*, Macoun); MacSwain, Hurst.

*__Conioselinum chinense__ (L.) BSP. Map 594. Pleasant View (P), along shady banks of creek gullied into clay beds.

*__Coelopleurum lucidum__ (L.) Fern. Wood Islands, turf along top of low cliff. Macoun's specimens called *C. Gmelini* from Brackley Point belong to *Ligusticum scothicum*.

Angelica atropurpurea L. Map 595. West Prince, marshy thickets or open swamps in woods. MacSwain, Hurst.

Pastinaca sativa L. PARSNIP Map 596. Scattered, in central Queens, around Summerside and West Prince. European, introduced and cultivated for the edible roots; naturalized and weedy in waste places. 7 Lots (Groh). MacSwain and Bain, MacSwain, Hurst.

Heracleum lanatum Michx. (Rickett, Rhodora 46: 389. 1944) Map 597. Local: Western Prince County; Hunter River to Rustico Island; northeastern region; and Belle River (K). Damp, open places: grassy banks along streams or above shores. (*H. maximum*, Campbell); MacSwain and Bain, MacSwain, Hurst.

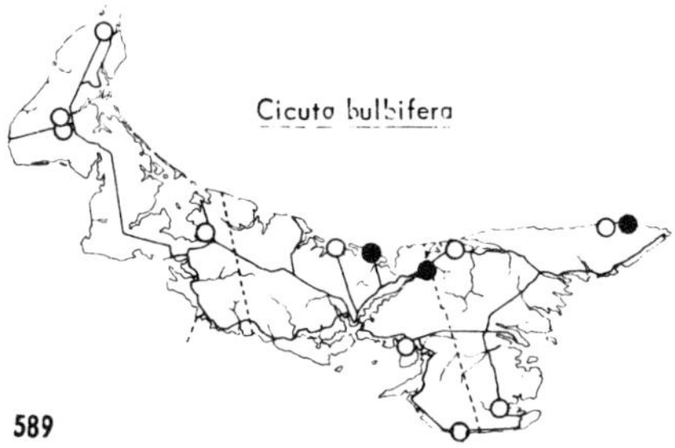

589

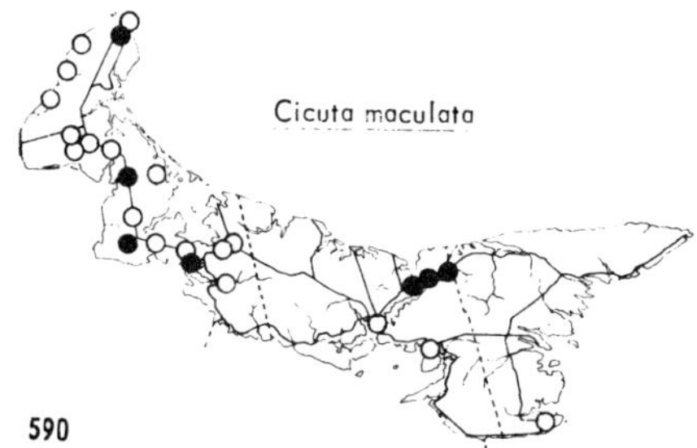

590

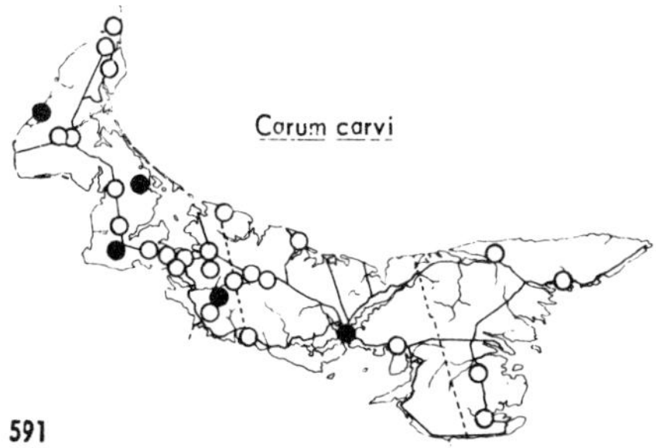

591

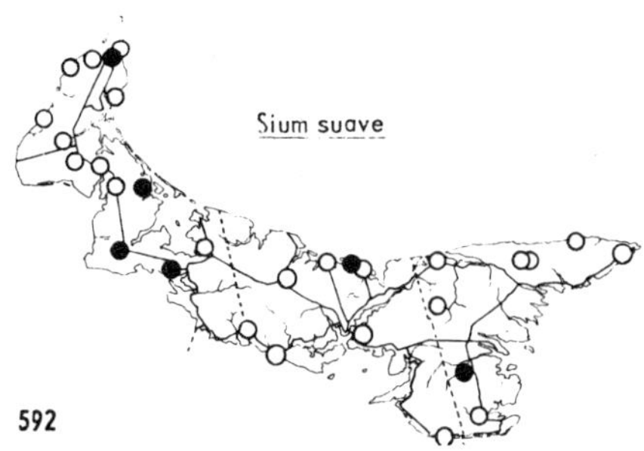

592

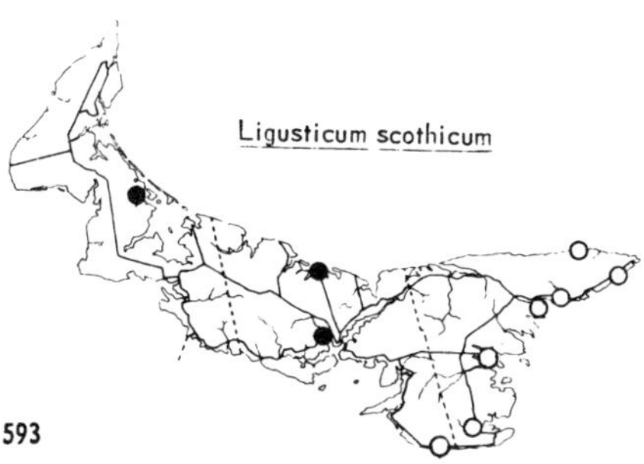

593

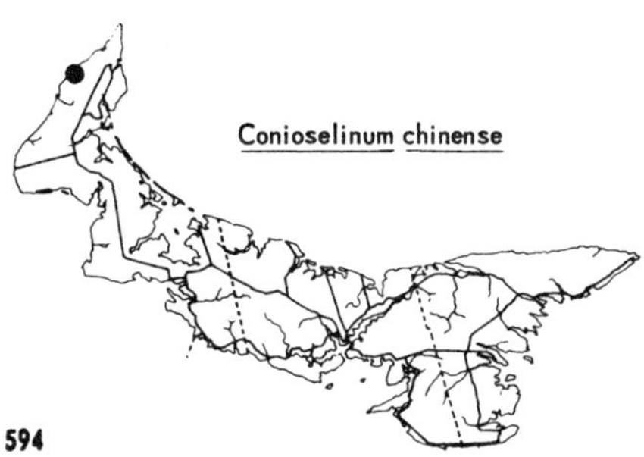

594

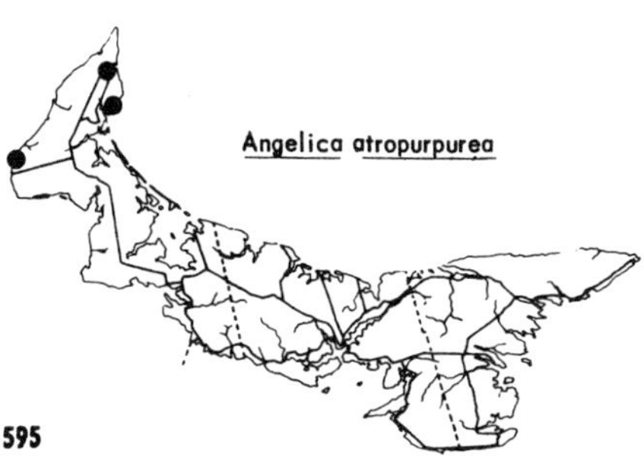

595

596

Daucus carota L. WILD CARROT Map 598. Sparsely scattered throughout. European; naturalized. Open, dryish sites: fields, roadsides and waste places. MacSwain and Bain, MacSwain, Hurst. Var. *sativa* is the cultivated carrot.

CORNACEAE

Cornus alternifolia L. f. Map 599. Local in northeastern and central uplands and on Prince County knolls. Well-drained, hardwood sites, flourishing at the open edges, and in hardwood forest often failing to flower. MacSwain and Bain, MacSwain, Hurst.

[*Cornus rugosa* Lam., recorded by Bain (1890), is not present, perhaps due tc absence of rocky sites (*C. circinata* L'Hér., Bain, doubted by Hurst, Campbell.]

Cornus stolonifera Michx. Map 600. Common in Prince County, scattered eastward along the rivers, and on the northeastern uplands. Damp, open groudn; often forming thickets. (*C. sericea,* MacSwain and Bain); MacSwain and Bain, MacSwain, Hurst.

[*Cornus florida* L. Northeastern limit in Maine; included by wishful thinking. MacSwain and Bain.]

Cornus canadensis L. PIGEONBERRY Map 601. Common throughout. Woodlands, particularly characteristic of the floor of coniferous woods but also weedy in brush and along roadsides. The pair of leaves below the "whorl" may or may not develop. The bright scarlet berries," mawkish sweet," fatten fowl fast --Stewart. Bain, MacSwain and Bain, MacSwain, Hurst.

ERICACEAE
(including PYROLACEAE, Fernald (1950) and VACCINIACEAE, Bain)

Chimaphila umbellata (L.) Barton var. ***cisatlantica*** Blake Map 602. Local: uplands of West Prince, center and southeast. Dry mixed or coniferous woods, rarely in dry old fields. MacSwain and Bain, Hurst.

Orthilia secunda (L.) House, vars. Map 603. Western Prince; central and northeastern regions. (*Pyrola secunda* L., MacSwain and Bain, MacSwain, Hurst, Campbell).

Var. ***secunda*** Dry mixed or coniferous woods, frequent.

Var. ***obtusata*** (Turcz.) House Bloomfield, Dundee, in larch swamps; Harmony, mossy spruce forest.

Moneses uniflora (L.) Gray Map 604. Scattered, in upland regions, West Prince, central, southeastern and northeastern. Mossy coniferous woods, in ravines or around swamps. (*M. grandiflora* Salisb., MacSwain and Bain, MacSwain); Hurst.

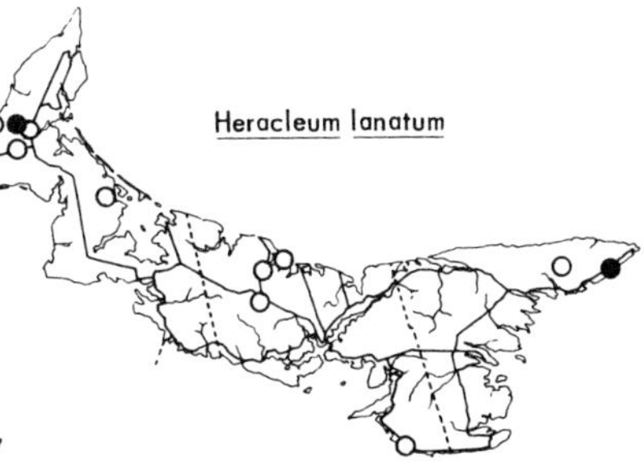

597

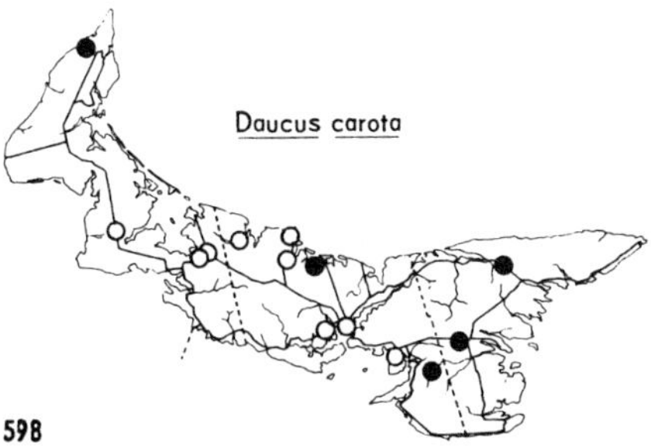

598

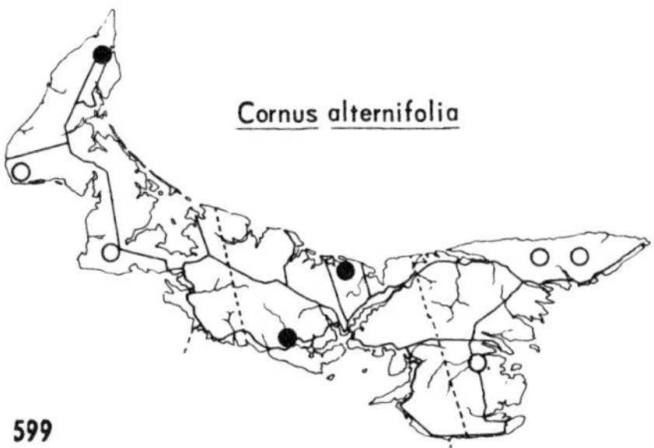

599

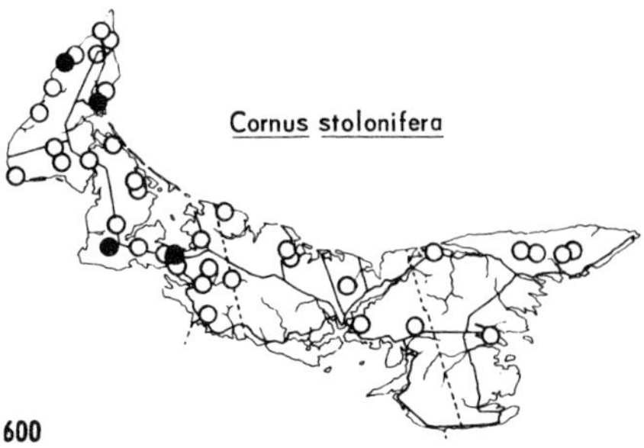

600

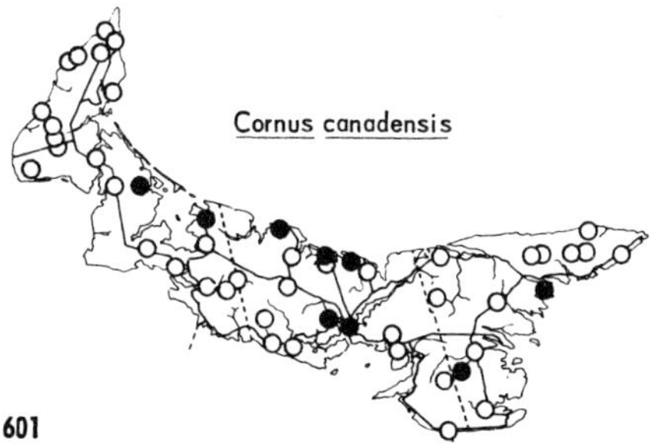

601

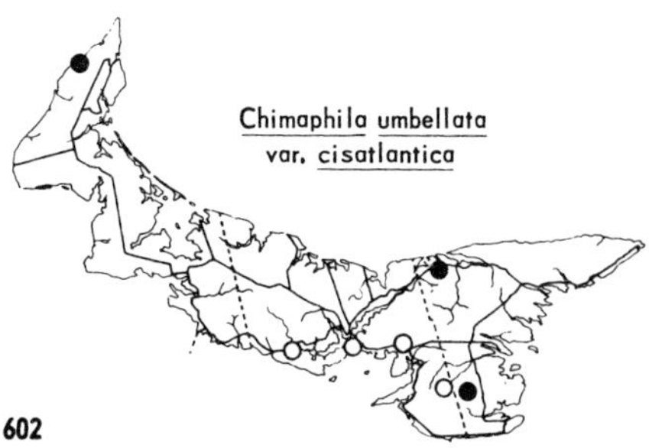

602

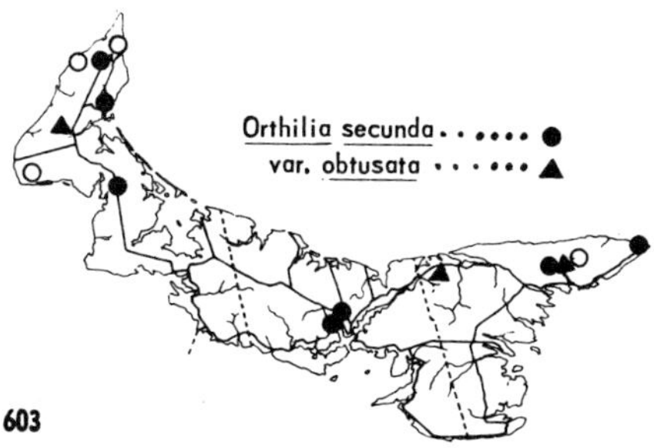

603

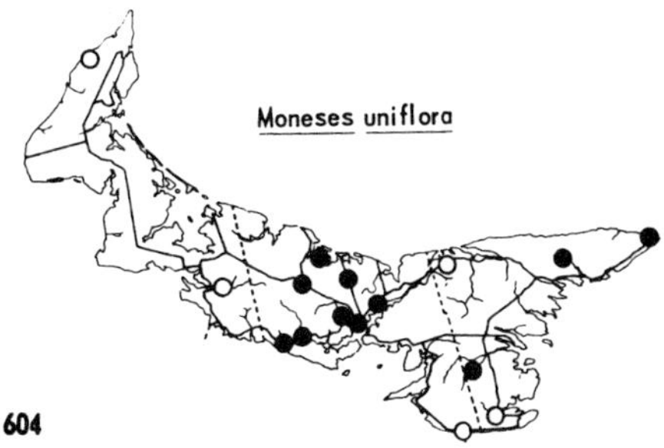

604

Pyrola minor L. Map 605. Rare: Hunter River (*Macoun*); Harmony (K) (*Fernald & St. John*). Damp fir woods.

Pyrola chlorantha Sw. Map 606. West Prince. Coniferous woods. MacSwain, Hurst; (*P. virens* Schweigg., Campbell).

Forma *paucifolia* (Fern.) Camp One of the original collections is from Alberton.

Pyrola elliptica Nutt. Map 607. Throughout, the most common *Pyrola*. Dry woods, on uplands, knolls or sandy soil. MacSwain and Bain, MacSwain, Hurst.

Pyrola rotundifolia L. var. *americana* (Sweet) Fern. Map 608. Scattered; Prince County, Charlottetown area and the southeast. Open dry or slightly damp sites: barrens due to burning or clearing; among the *Betula populifolia, Vaccinium angustifolium*, etc. (*P. rotundifolia*, MacSwain and Bain, MacSwain; *P. americana* Sweet, Hurst); Campbell.

Pyrola asarifolia Michx., vars. Map 609. Rare: west and east, woods and swamps.

Var. *asarifolia* Pleasant View (P), rich thicket along stream in clay beds.

Var. *purpurea* (Bunge) Fern. Dundee, larch swamp (*Fernald et al.*); report from Mermaid Lake (Hurst), needs confirmation, as acid peat is an unfavourable habitat for it. (*P. uliginosa* Torr., Hurst); Campbell. Fernald (Rhodora 50: 212-3. 1948) remarked that Roland had erroneously mapped var. *asarifolia* as present in the Island and var. *incarnata* (i.e., *purpurea*) from Nova Scotia, while the text indicated the contrary; Fernald (1950) made the same mistake, recording var. *asarifolia* from P.E.I. and var. *purpurea* from Nova Scotia.

Monotropa uniflora L. INDIAN PIPE Map 610. Scattered throughout. Dry woods, more commonly under conifers (as a saprophyte tolerating continual shade). MacSwain and Bain, MacSwain, Hurst.

Monotropa hypopithys L. var. *americana* (DC.), Domin Farwell, Am. Mid. Nat. 10: 39. 1926. Map 611. Bedeque (*Fernald & St. John*); Brackley Point (*Macoun*); Kilkenny Road (Hurst); Harmony (*Fernald & St. John*). Woods, often under conifers., var. *rubra* (Torr.) Farwell, Roland. (*Hypopithys monotropa* Crantz, Copeland); MacSwain, Hurst.

Pterospora andromedea Nutt. Prospect Creek, pine woods. July 1888 (*Macoun*) J.M. Macoun, Can. Record Sci., 1897, 280; MacSwain and Bain, Robinson & Fernald, Hurst.

Ledum groenlandicum Oeder Map 612. Common in the lowlands. Acid swamps, bogs and barrens due to fire or cutting. (*L. latifolium* Ait., Bain, MacSwain and Bain, MacSwain); Hurst.

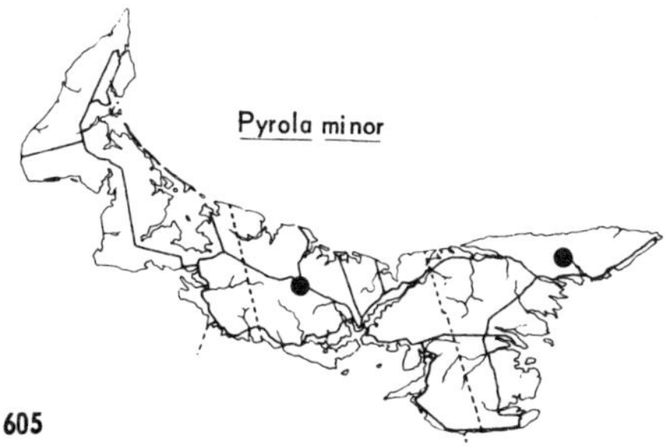

605

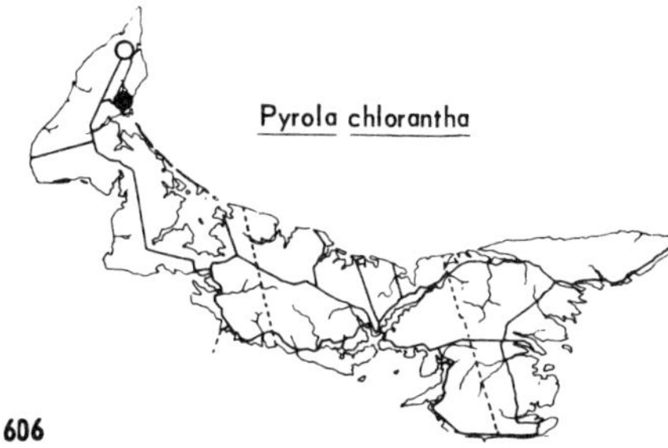

606

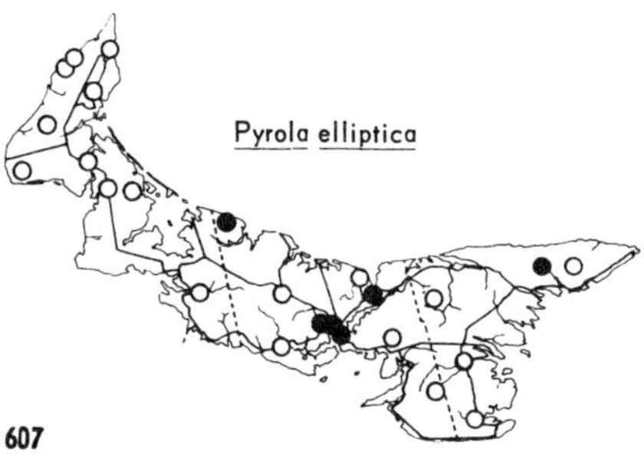

607

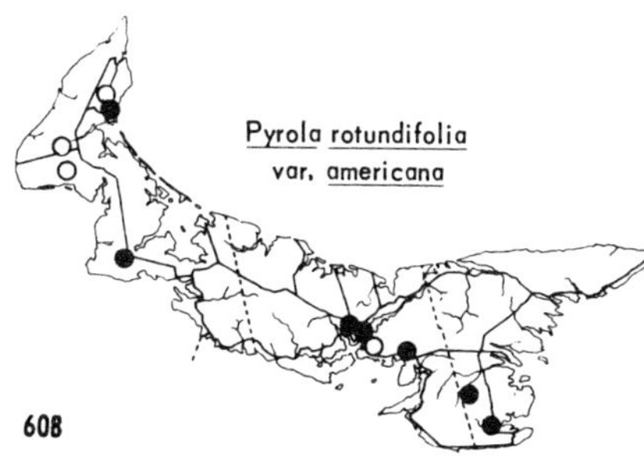

608

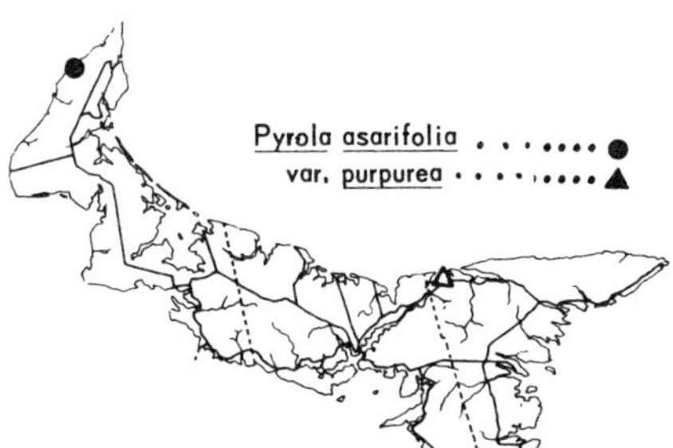

609

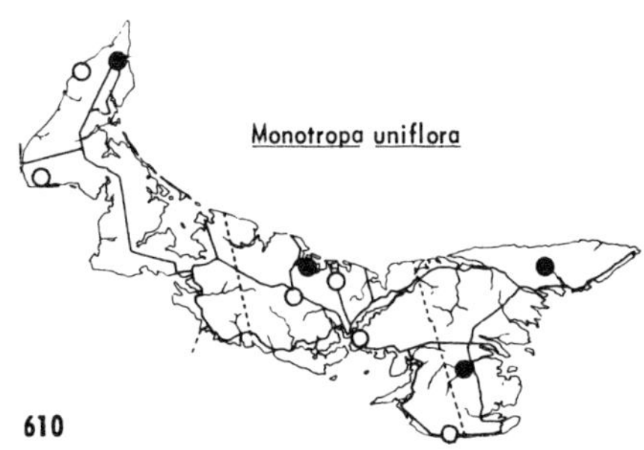

610

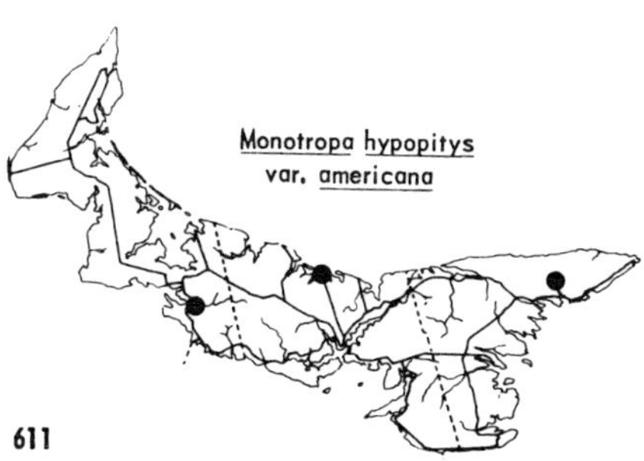

611

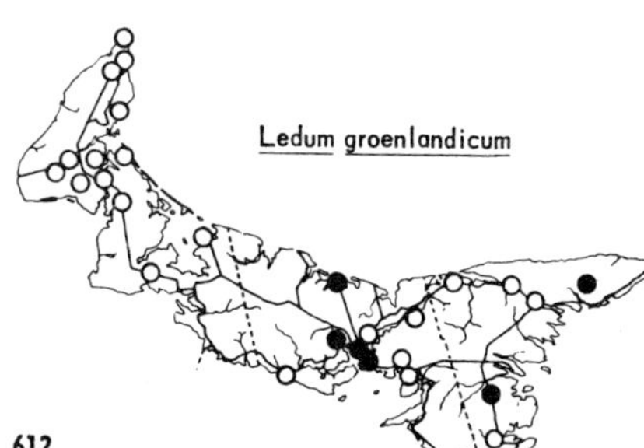

612

Azalea canadensis (L.) Ktze. RHODORA Map 613. Common except in the uplands. Acid barrens, bogs, swamps. Flowering May to early June in Queens; till the end of June at North Point. (*Rhodora canadensis* L., MacSwain and Bain; *Rhododendron Rhodora* G. Don, MacSwain; *Rh. canadense* (L.) Torr., Hurst, Campbell).

[*Azalea viscosa* L., not known east of Maine. MacSwain and Bain.]

Kalmia angustifolia L. DWARF LAUREL Map 614. Common throughout. Bogs, open or cleared swampy and acid woodland (barrens), on fire barrens in pine stumpland (Stewart). Poisonous. 44 Lots (Groh). MacSwain and Bain, MacSwain, Hurst.

Kalmia polifolia Wang. Map 615. Scattered: West Prince, central Queens and eastern regions. Peat bogs. (*K. glauca* Ait., MacSwain and Bain, MacSwain); Hurst.

**Andromeda glaucophylla* Link Map 616. Western Prince: Waterford, East Bideford peat bogs. Abundant at edge of floating mat on bog-ponds, later persisting amid peat. (*A. polifolia*, MacSwain and Bain ?).

Chamaedaphne calyculata (L.) Moench var. *angustifolia* (Ait.) Rehder Map 617. Common in lowlands. Open, acid sites: bogs, or "barrens" due to burning or clearing of woods. (*Cassandra calyculata* (L.) D. Don, MacSwain and Bain, MacSwain); Hurst.

[*Lyonia ligustrina* (L.) DC. extends northeast only to southern New England. (*Andromeda ligustrina* L., Bain), 1892.]

Epigaea repens L. var. *glabrifolia* Fern. MAYFLOWER Map 618. Scattered: western Prince, central Queens and southeastern uplands. Acid, bushy barrens or mossy floor of the succeeding second-growth mixed woods. MacSwain and Bain, MacSwain, Hurst.

Gaultheria procumbens L. CHECKERBERRY, BOXBERRY Map 619. Common east of Charlottetown, local in West Prince. Open, dry acid sites: dry peat or bare patches in open woods, barrens and old fields. Fruting calyx edible, wintergreen flavored. MacSwain and Bain, MacSwain, Hurst.

Chiogenes hispidula (L.) T. & G. Map 620. Scattered: West Prince; central and eastern regions. Covering mossy hummocks in coniferous woods at edge of swamp or bog. Fruit edible, wintergreen-flavored. (*Ch. serpyllifolia* Salisb., MacSwain and Bain, MacSwain); Hurst; (*Gaultheria hispidula* (L.) Bigel., Campbell).

Arctostaphylos uva-ursi (L.) Spreng. var. *coactilis* Fern. & MacSwain. Map 621. Very local, coastal: West Point, Dalvay, Bothwell. Sandy open sites: dunes, cliff slopes, barrens, persisting in shade of succeeding spruce. (*Arbutus Uva-ursi* L., Bain, MacSwain and Bain); MacSwain, Hurst.

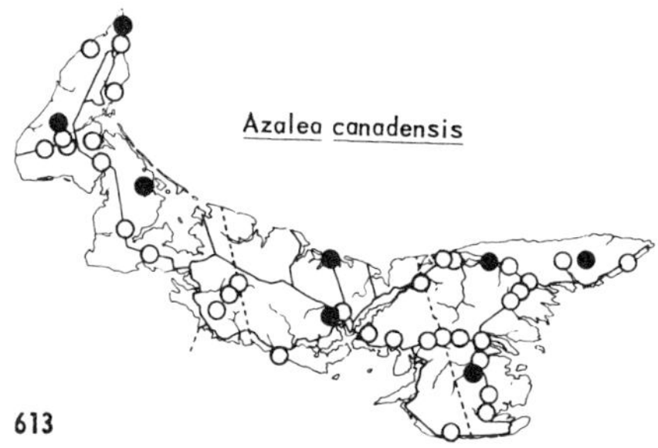

613

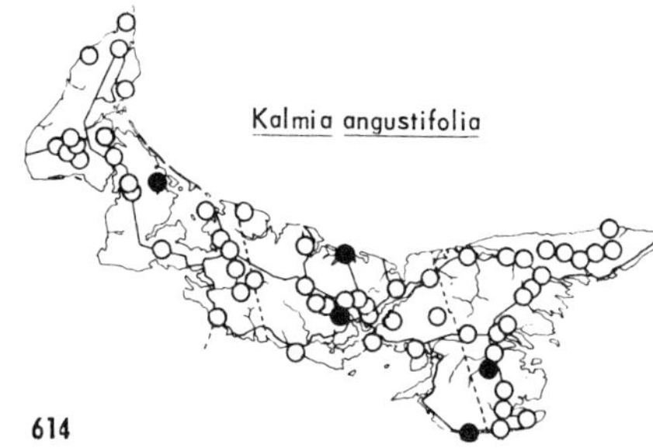

614

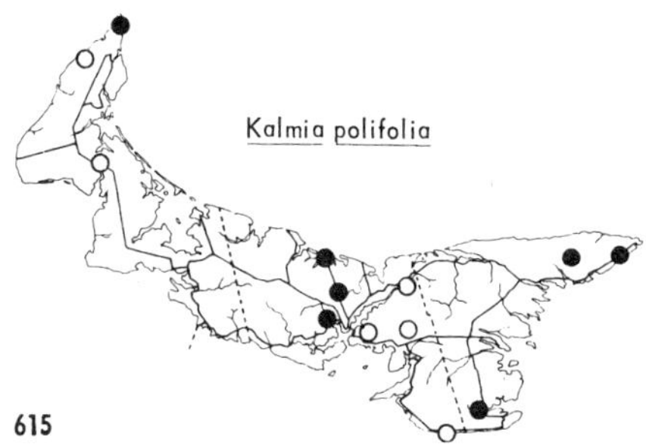

615

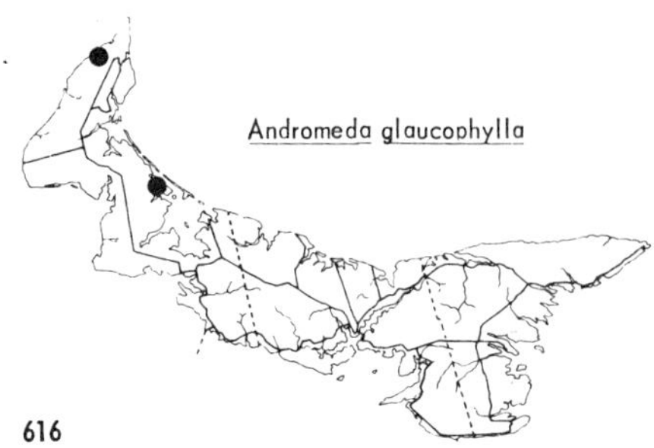

616

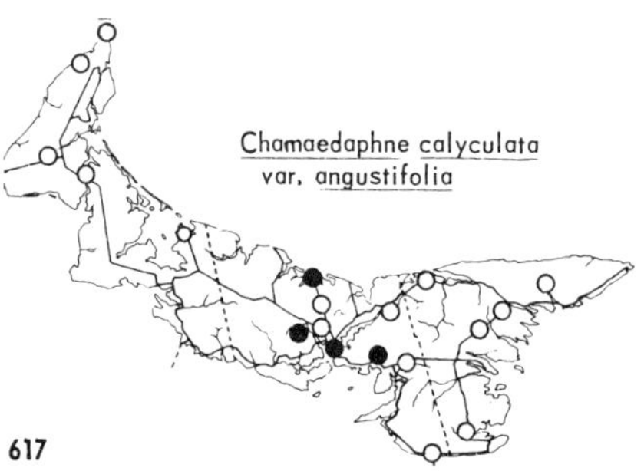

617

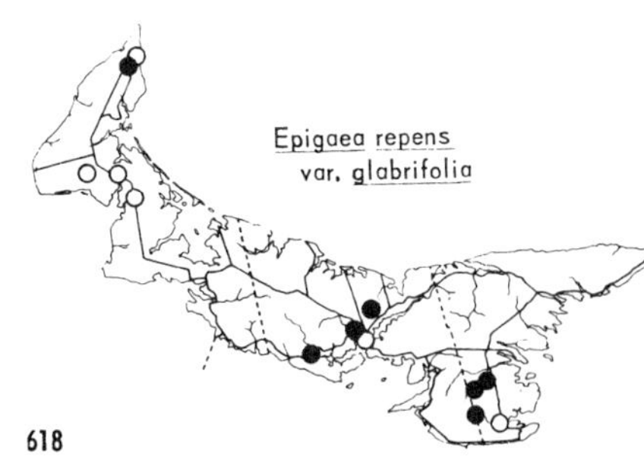

618

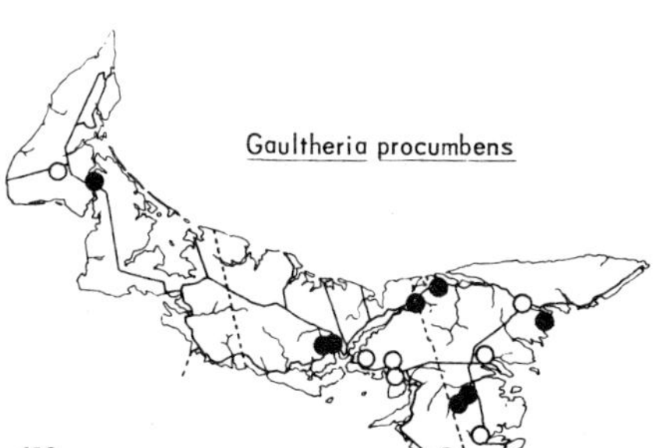

619

620

[*Calluna vulgaris* (L.) Hull, the Scotch Heather, possibly introduced but not seen. See also ***Limonium***. MacSwain, Hurst.]

Gaylussacia dumosa (Andr.) T. & G. var. ***bigeloviana*** Fern. Map 622. Rare: Black Banks; Village Green; near Murray River. Peat bogs. Fruit edible but insipid compared to the next. Hurst 1940.

Gaylussacia baccata (Wang.) K. Koch Map 623. Scattered, locally abundant, in Prince County, central and southeastern regions. Sandy barrens, dry old fields, bogs. The f. ***glaucocarpa*** (Rob.) Mackenz. reported by Messervy (Hurst). (***Vaccinium resinosum*** Ait., MacSwain and Bain; ***G. resinosa*** (Ait.) T. & G., MacSwain); Hurst.

****Vaccinium uliginosum*** L. var. ***alpinum*** Bigel. Map 624. Waterford (West Prince), peat bog hummocks. Sea-level records are rare south of the Gulf.

Vaccinium myrtilloides Michx. BLUEBERRY Map 625. Scattered throughout. Common in West Prince and the eastern half of the Island. Dry acid sites: "barrens" from clearing or fire; more shade-tolerant than ***V. angustifolium,*** persistent in open second-growth woods. Fruit less sweet than those of ***V. angustifolium,*** but abundant and edible. (***V. canadense*** Kalm ex Torr., MacSwain and Bain, MacSwain, Hurst); Campbell.

[***Vaccinium tenellum*** Ait., of the southern States, is an erroneous record based on ***V. tenellum*** sensu Pursh, presumably ***V. angustifolium*** var. ***nigrum.*** MacSwain and Bain.]

Vaccinium angustifolium Ait., vars. BLUEBERRY Map 626. Common throughout. Dry acid sites: clearings, barrens, old fields. Following fire, extensive even-aged stands reach maximum fruit production the second year (as in Lot 12 after the fire of 1840, the bushes still persisting in 1870, or on the famous fire barrens of Tracadie, extending toward St. Peters). The most valuable native small fruit. "So abundant locally as to afford swine feed for weeks. A gallon of gin can be made from a bushel" (Stewart).

Var. ***laevifolium*** House Common throughout, variable in leaf form, twig color and fruit size. (***V. corymbosum,*** Stewart; ***V. pensilvanicum*** Lam., not Mill., MacSwain and Bain, MacSwain, Hurst); Campbell.

Var. ***nigrum*** (Wood) Dole Map 627. Scattered in West Prince, Kings; colonies with leaf form of var. ***laevifolium.*** (***V. tenellum,*** MacSwain and Bain; ***V. Brittonii*** Porter; ***V. pensilvanicum*** var. ***nigrum*** Wood, Hurst); Campbell.

Vaccinium vitis-idaea L. var. ***minus*** Lodd. Map 628. North and east coasts; rarely inland: Keefe's Lake. Exposed sites: inner sand dunes and sandblown barrens; open sandy woods, old fields and banks near sea (***V. Oxycoccos,*** MacSwain and Bain?); Churchill 1902, Adams, Hurst 1940.

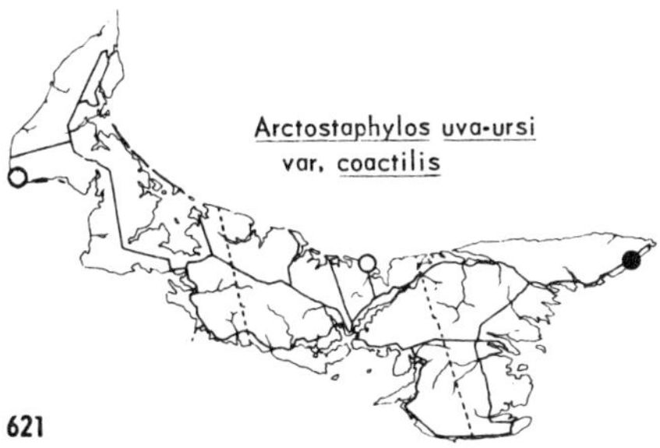

621

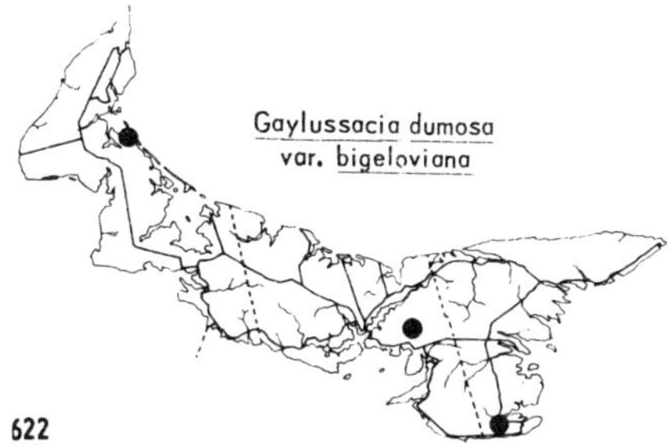

622

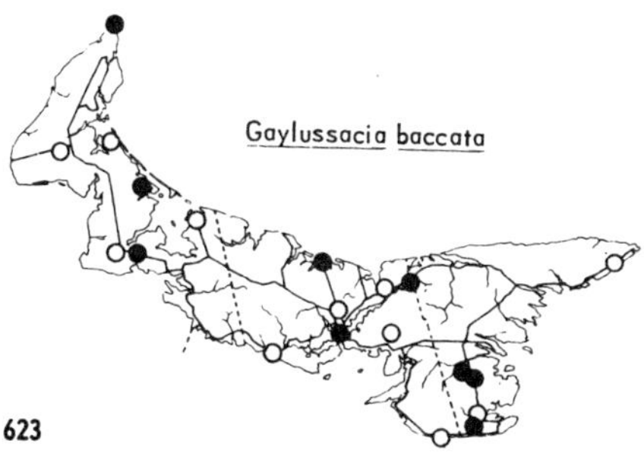

623

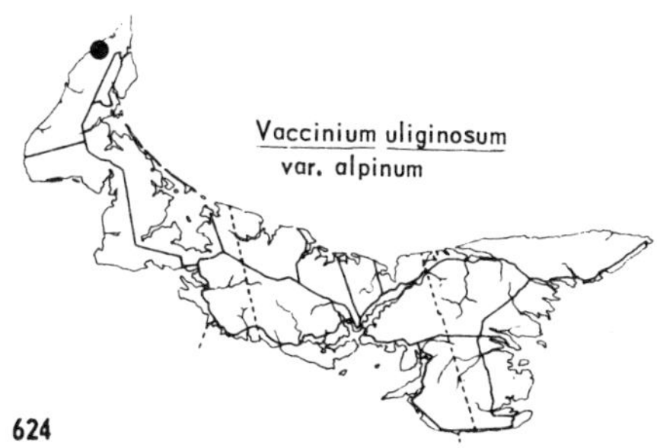

624

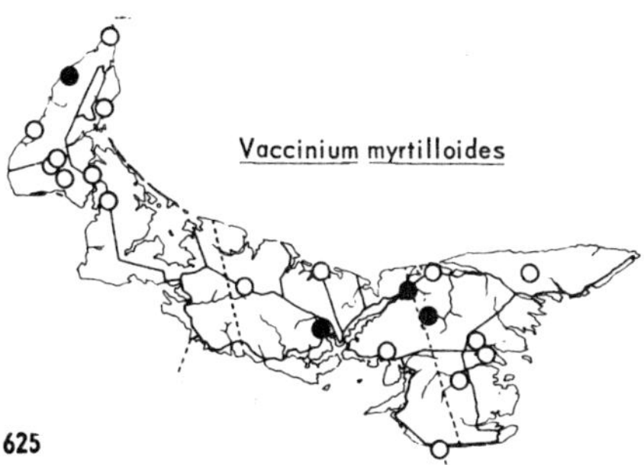

625

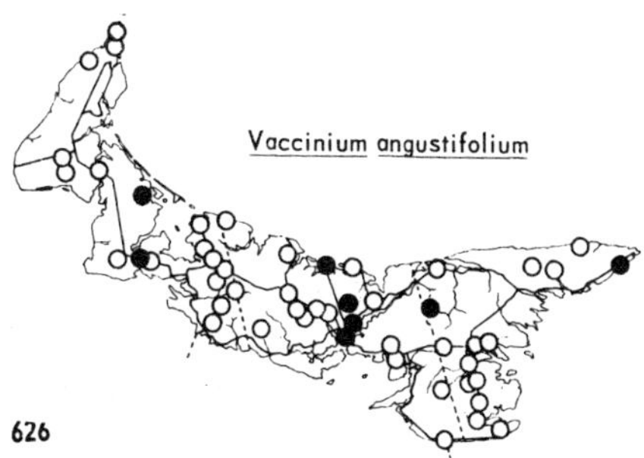

626

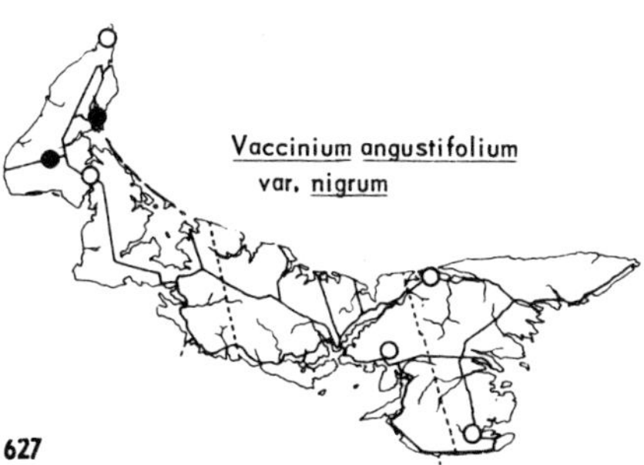

627

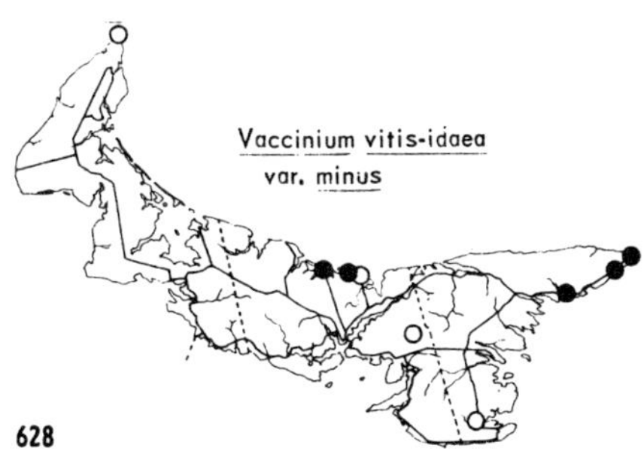

628

Vaccinium* oxycoccos** L. CRANBERRY Map 629. Locally, throughout, in peat bogs. Smaller fruit then the following. Ours are referable to, or transitional to var. ***microphyllus (Lange) Rousseau & Raymond (in Nat. Can. 79:(2): 82 1952). (***Oxycoccos vulgaris*** Bong., MacSwain and Bain; ***O. palustris*** Pers.); MacSwain, Hurst.

Vaccinium macrocarpon Ait. CRANBERRY Map 630. Common, central and eastern regions, largely coastal. "Marshes adjoining upland; low wet poor sandy land" (Stewart), dune slacks and barrier-beach pond bogs, occasionally inland peat bogs, then particularly in floating mats. Fruit large, high in benzoic acid, used for sauce and flavoring milder fruits. "Exported, and used for sauce, the overwintered berries less acid" (Stewart). (*V. Oxycoccos*, Stewart; *Oxycoccos macrocarpus* (Ait.) Pers., MacSwain and Bain); MacSwain, Hurst.

On the peaty sand borders of Mermaid Lake, cranberries (*Vaccinium macrocarpon*) form mats among the *Scirpus cyperinus* tussocks, scattered *Eriophorum virginicum* and *Rhynchospora alba*. Where the peat has been trampled (see cart track in background) *Juncus* spp. form small clumps.

PRIMULACEAE

[***Lysimachia vulgaris*** L. Charlottetown, rubbish heap (*Fernald & St. John*). European, garden plant, not established.]

[***Lysimachia punctata*** L. var. ***verticillata*** (Bieb.) Boiss. Map 631. Murray River, dump by river, cultivated in village. European, garden plant.]

Lysimachia terrestris (L.) BSP. Map 632. Scattered, chiefly coastal but inland around ponds and lakes. Muddy or marshy pond shores, stream banks and ditches. Hybridizing with *L. thyrsiflora*. (*L. stricta* Ait., Bain 1892); Hurst.

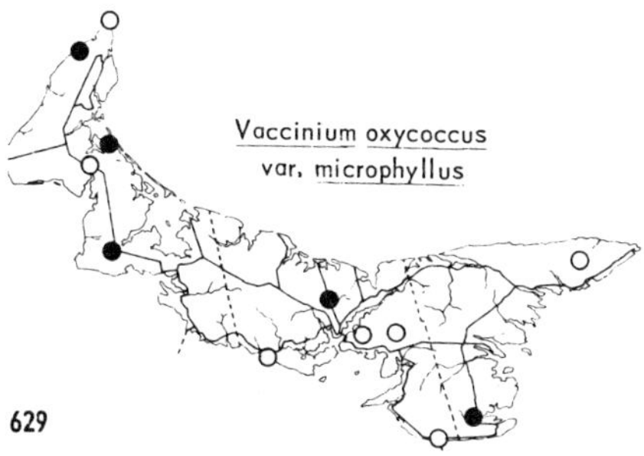

629

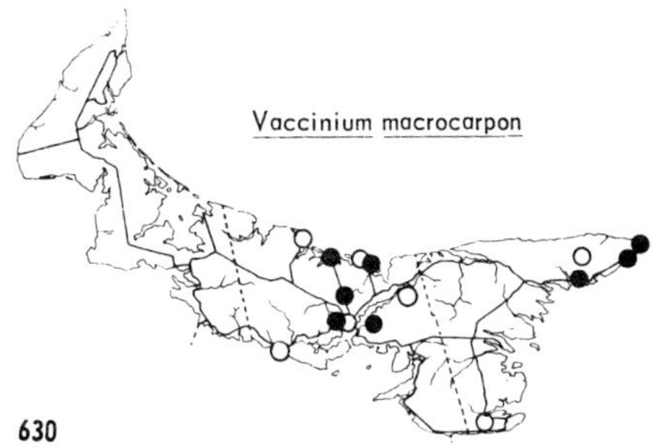

630

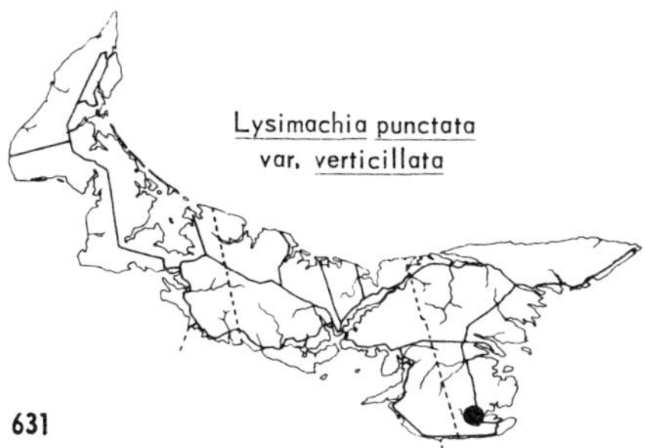

631

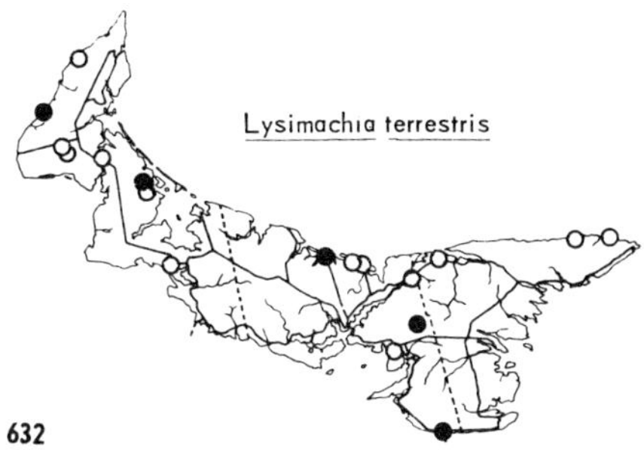

632

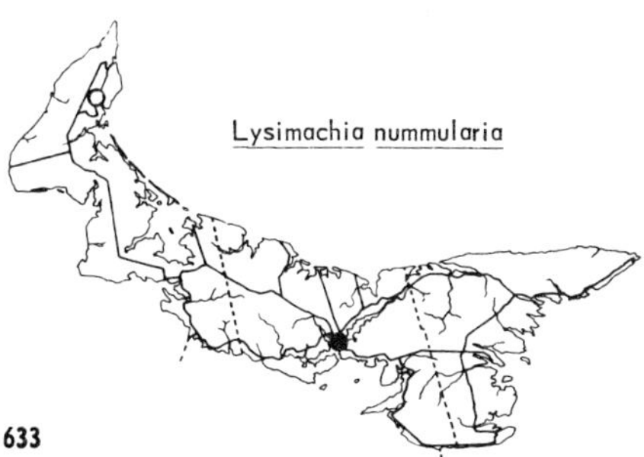

633

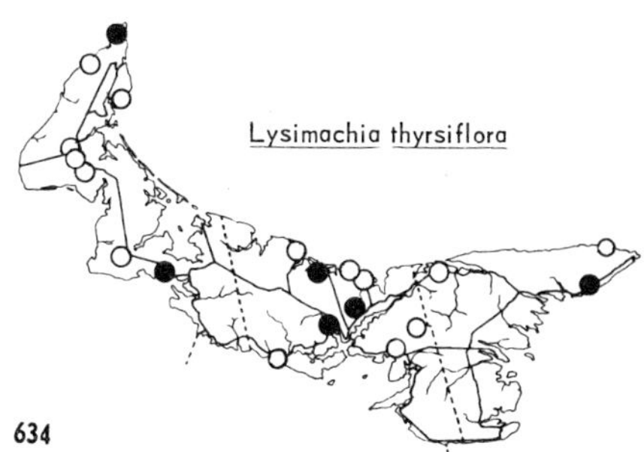

634

Lysimachia nummularia L. Map 633. Rare: Montrose (P); Charlottetown; North River (?), 1905 (*Warren*). European, naturalized. Ditches and stream banks. Campbell.

Lysimachia thyrsiflora L. Map 634. Scattered, coastal and lowland. Damp, open ground: shores, marshes, pond margins. Hybridizing with *L. terrestris*. MacSwain, Hurst.

X ***Lysimachia commixta*** Fern. in Rhodora 52: 199. 1950 (*L. terrestris* X *thyrsiflora*). North Lake (K) (*Fernald et al.*).

[****Lysimachia ciliata*** L.Wellington, a few plants among grass in churchyard (1952). Adventive (or introduced ?) from Nova Scotia.]

Trientalis borealis Raf. Map 635. Well-drained, woodland or mixed wood sites, where competition reduced by shade. Forma ***ramosa*** Vict. Upton, near Charlottetown. (*T. americana* (Pers.) Pursh, Bain, MacSwain and Bain, MacSwain, Hurst); Campbell.

[***Anagallis arvensis*** L. Map 636. North River, marine sand beaches, 1891 (*Bain*). European, adventive. MacSwain, Hurst.]

[*Centunculus minimus* L. Island records, including Fernald (1950), based on Macoun's misidentification of *Crassula aquatica*.]

Glaux maritima L. var. ***obtusifolia*** Fern. Map 637. Coastal, frequent. Sandy beach heads and salt marshes, and salt-sprayed meadow above cliffs, where plants depressed (by trampling ?) and much branched. (*G. maritima*, MacSwain and Bain, MacSwain, Hurst, Campbell).

Samolus parviflorus Raf. Map 638. Selkirk, shore of brackish pond (*Fernald & St. John*). (*S. Valerandi*, MacSwain and Bain; *S. Valerandi* var. *Americanus* Gray, MacSwain; *S. floribundus* HBK., Fassett 1928, Hurst); Campbell.

[*Hottonia inflata* L., to be excluded. MacSwain and Bain.]

PLUMBAGINACEAE
(included in PRIMULACEAE, Bain)

Limonium nashii Small "SEA HEATHER" Map 639. Coastal, common. Salt marsh mud., (var. *trichogonum* Blake, Roland; *Statice Limonium*, Bain, MacSwain and Bain, MacSwain; *L. carolinianum*, Hurst, Campbell).

OLEACEAE (ORNACEAE, Bain)

Fraxinus americana L. WHITE ASH Map 640. West Prince; occasionally planted elsewhere as a shade tree. "In good land but in no great quantity" (Stewart). Thickets on clay stream banks. (*F. excelsior*, Stewart); MacSwain, Hurst.

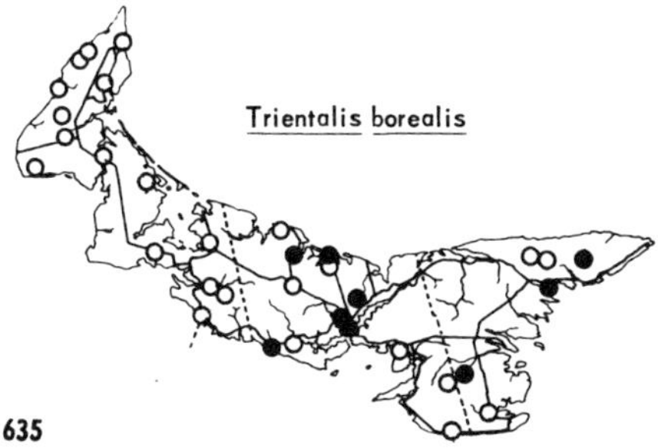

635

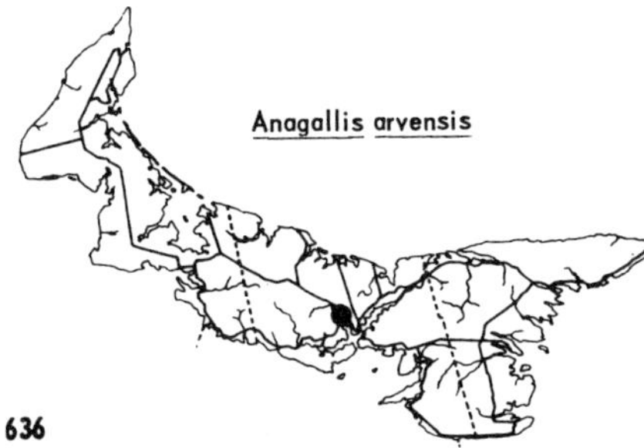

636

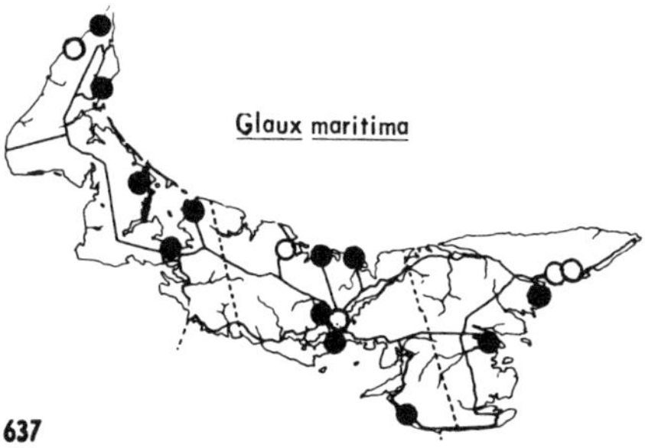

637

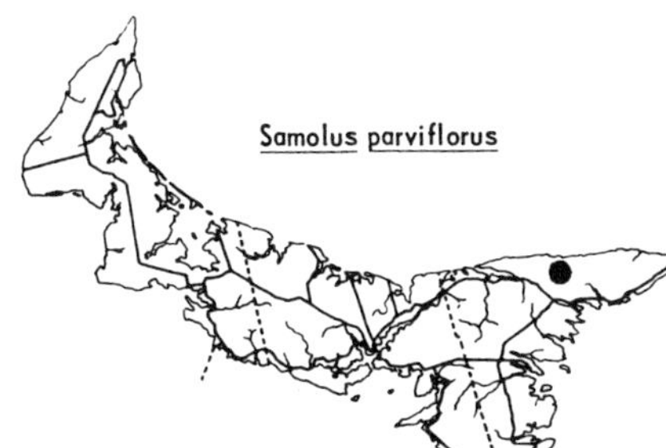

638

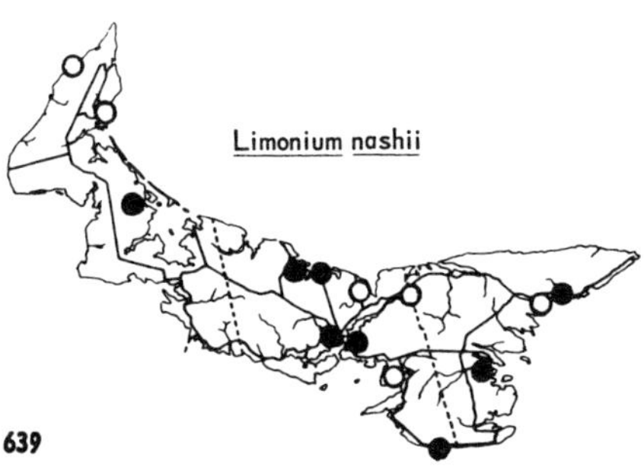

639

640

Fraxinus nigra Marsh. BLACK ASH Map 641. Scattered in western Prince; Brackley Point area. Swamps and clay thickets. Wood of no value; twigs used for baskets and brooms (Stewart). (*F. americana*, Stewart; *F. sambucifolia* Lam., MacSwain); Hurst.

Syringa vulgaris L. LILAC European, ornamental, persisting around old house sites. 12 Lots (Groh). Hurst.

MENYANTHACEAE

Menyanthes trifoliata L. Map 642. Very local in Prince: Nail Pond; Stanhope; locally abundant in east-central lowlands. Mucky wet thickets and pond-borders. (Including var. *minor* Michx.), Hurst.

APOCYNACEAE

Apocynum androsaemifolium L. Map 643. Common in West Prince, around central uplands and at Murray River. Weedy in open soil: clearings, railway banks, and roadsides. 12 Lots (Groh). MacSwain, Hurst.

ASCLEPIADACEAE

****Asclepias incarnata*** L. Avondale (Q), alder swamps. Sets seed.

Asclepias syriaca L. MILKWEED Map 644. Georgetown, abundant, fruiting, and spreading along railway (located by Messervy); Brackley Point, introduced from New Brunswick (1931) and spreading in Hurst's lawn, but not setting seed. Campbell.

CONVOLVULACEAE

[***Convolvulus sepium*** L. Map 645. Throughout, common in Prince and Queens. Open places, twining and climbing on tall herbs: tension line of seashore, weedy on damp roadsides and railway banks. (Represented by the pink-flowered forma ***coloratus*** Lange), (var. *americanus* Sims, Roland). 18 Lots (Groh). MacSwain, Hurst.]

Convolvulus arvensis L. "Near New Glasgow, July". Adventive, European, (Messervy, ca. 1930). Hurst.

Cuscuta gronovii Willd. Map 646. Plat River (P), Pleasant View (P), parasitic on herbs in clay-bottom alder thicket.

POLEMONIACEAE

[***Collomia linearis*** Nutt. Map 647. Milton, 1950 (*Bassett*); adventive by C.N.R. from region of concentration in Bay of Chaleur region. Campbell.]

BORAGINACEAE

Symphytum asperum Lepechin Map 648. Charlottetown (1912); Springvale, near Milton (1952). Naturalized along streams by roadside; introduced garden plant from the Caucasus. Macbride, Rhodora 18: 24. 1916.

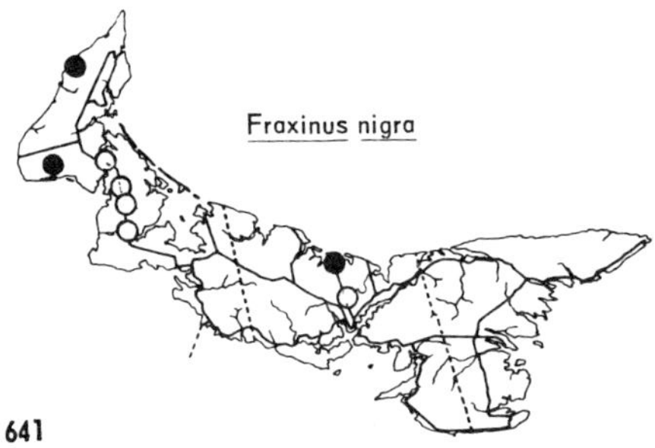

641

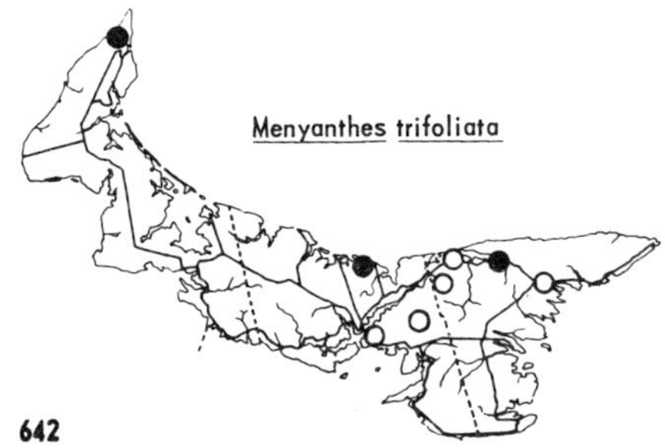

642

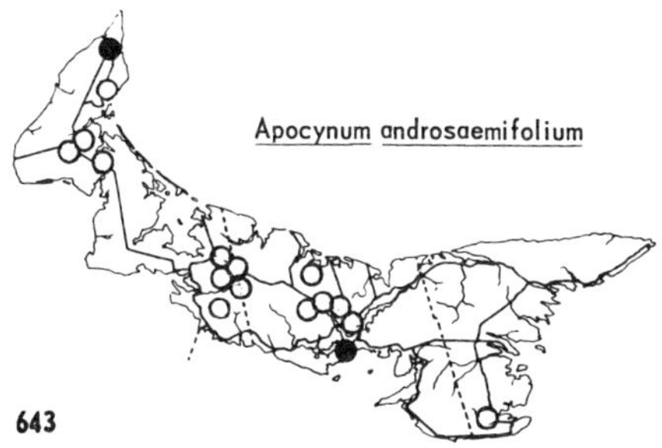

643

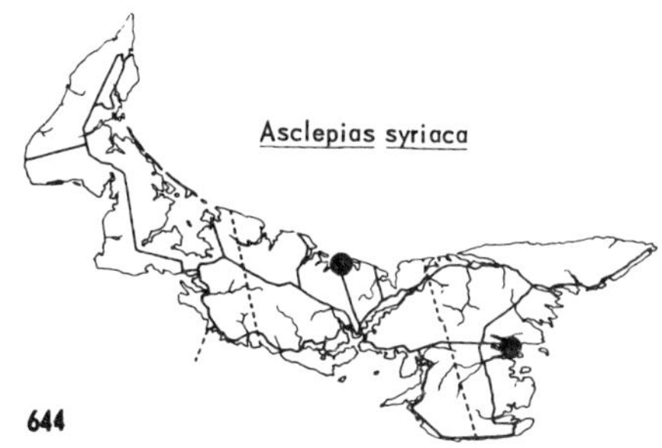

644

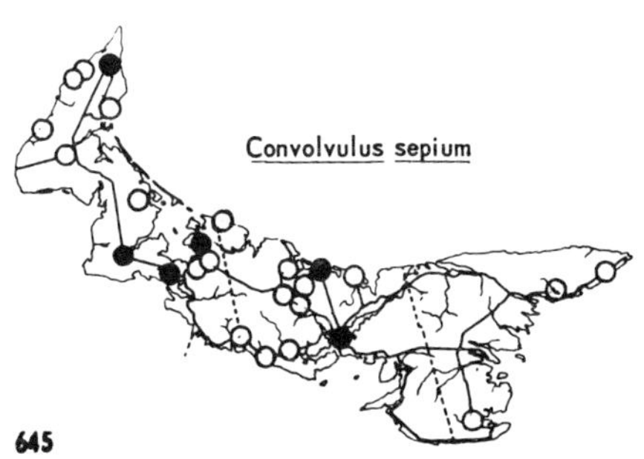

645

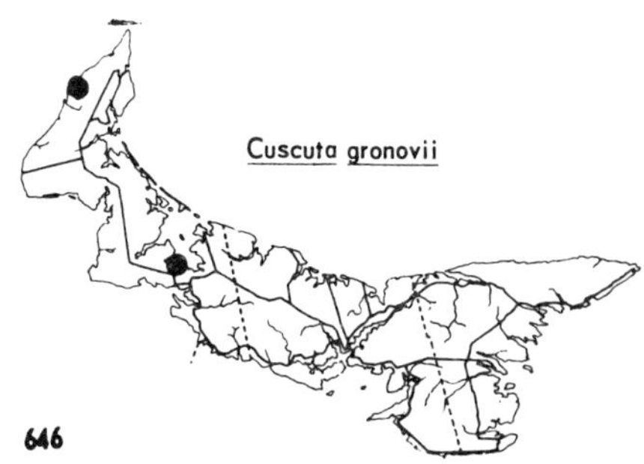

646

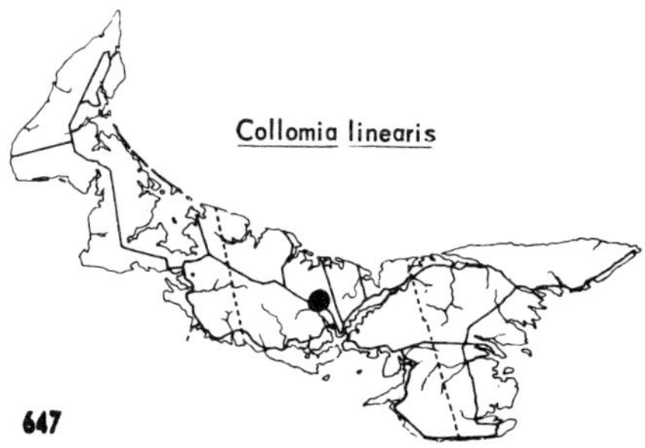

647

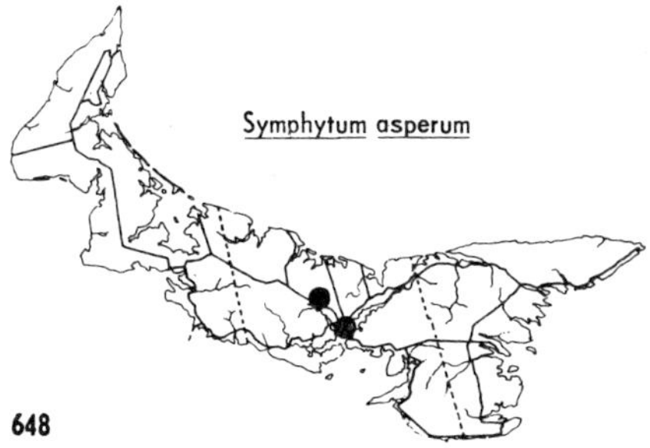

648

[*Symphytum officinale* L., introduced to gardens from Europe, weedy in 2 Lots (Groh). Without specimens, may refer to the preceding.]

Lycopsis arvensis L. Map 649. Sporadic weed, naturalized from Europe: Ellerslie (1948), Rustico (before 1884, 1952), Commercial Cross (1954), Lower Montague (1909); roadsides, hayfields, gardens on sandy soil. Macoun 1884, Adams, Hurst, Groh in Can. Weed Surv. 2: 32. 1944.

[***Echium vulgare*** L. Map 650. Iona (Q), 1915. European, weedy. Groh, Can. Weed Surv. 3: 42. 1946.]

Myosotis scorpioides L. FORGET-ME-NOT Map 651. Southern rivers of the central upland from Dunk River to DeSable and Kelly's Cross. European, naturalized along stream beds. Campbell.

Myosotis laxa Lehm. FORGET-ME-NOT Map 652. Very local, western Prince and eastern regions. Marshy, perennial streams; often in thickets. MacSwain, Hurst.

Myosotis arvensis (L.) Hill Map 653. Charlottetown, 1914 to present; Souris. European, naturalized. Weedy in lawns and waste sandy places. Groh, Hurst.

Lappula myosotis Moench Map 654. Becoming frequent, West Prince (1912), Borden (1926), central Queens (1950). European, naturalized. Weed of railways, grain fields and chicken runs, probably introduced in grain. 2 Lots (Groh). Johnston Contr. Gray Herb. 70: 47. 1924. (***L. echinata*** Gilib., Groh, Hurst, Campbell).

LABIATAE

[***Ajuga reptans*** L. Garden plant introduced from Europe. Not naturalized! Hurst.]

Scutellaria lateriflora L. Map 655. Common in Prince, scattered in east-central lowlands. Damp shady sites: thickets, swampy woods. MacSwain and Bain, MacSwain, Hurst.

Scutellaria galericulata L. var. ***pubescens*** Benth. Map 656. Western Prince, lowland Queens and much of Kings, probably throughout. Damp, more open sites than the former: thickets, swamps, abundant at pond margins. (***S. galericulata,*** MacSwain and Bain, MacSwain, Hurst; ***S. epilobiifolia*** A. Hamilton, Campbell).

Nepeta cataria L. CATNIP Map 657. Infrequent: Bedeque, Brackley Beach, Wood Islands. European, formerly planted around houses and usually remaining near them. Very attractive to cats. MacSwain, Hurst.

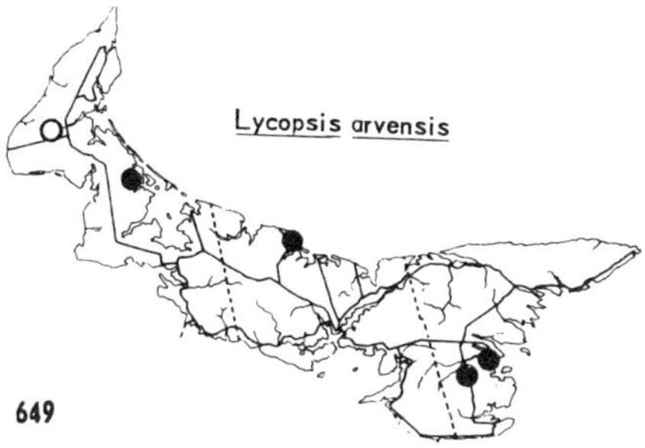

649

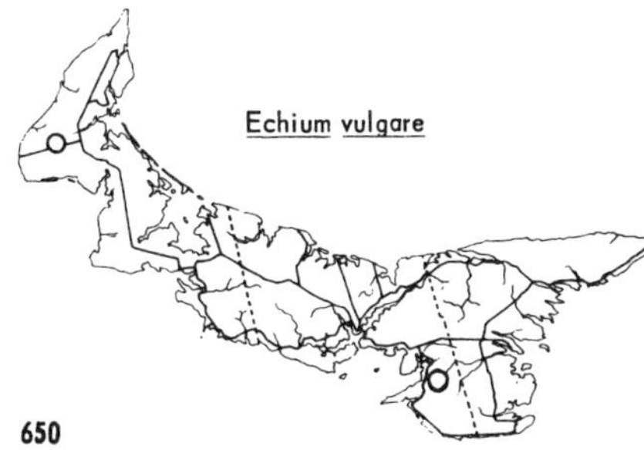

650

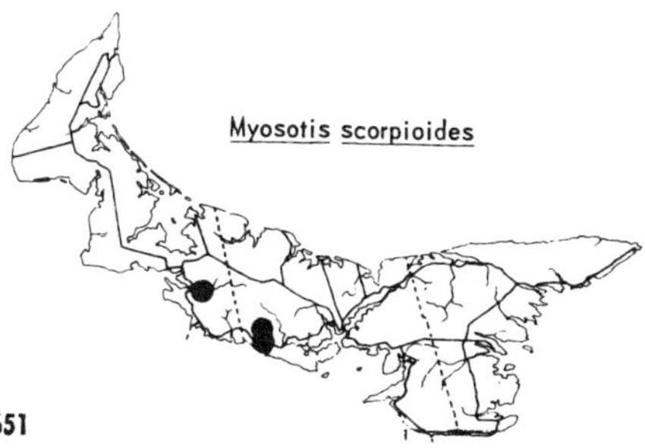

651

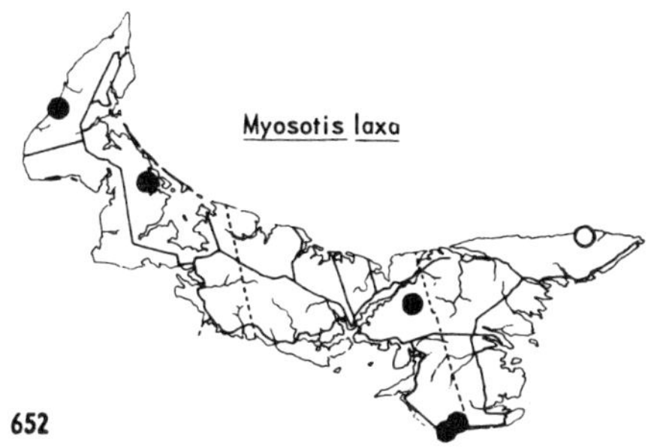

652

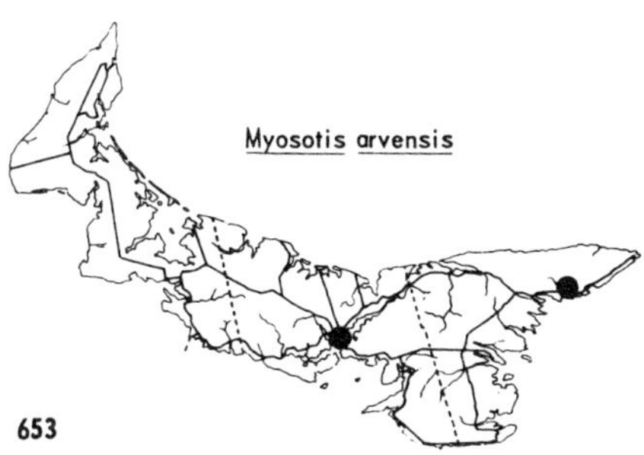

653

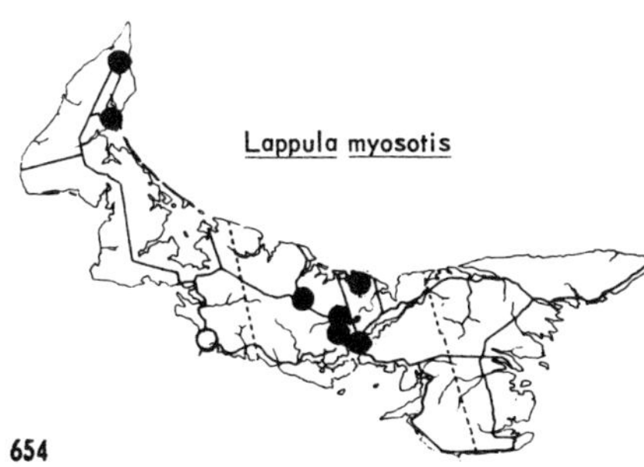

654

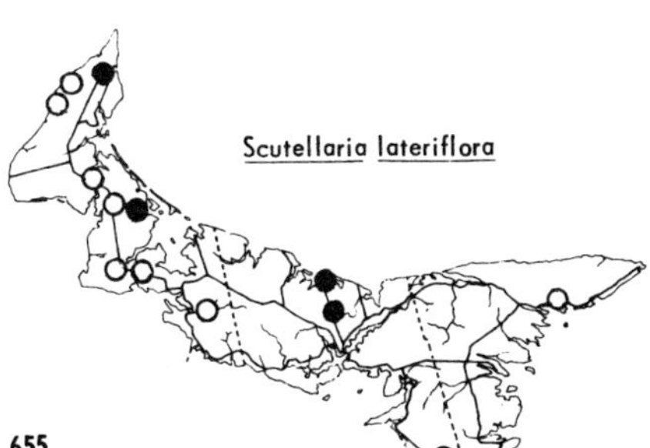

655

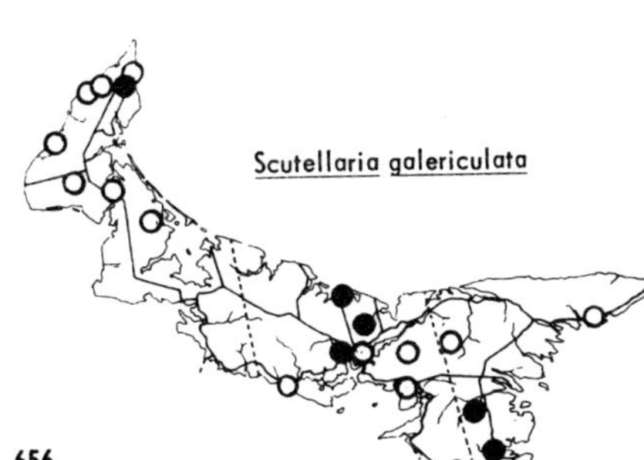

656

Glechoma hederacea L., vars. Map 658. Rather common, at least in Prince and western Queens. European. Weed of human occupation: dooryards, gardens, hedges and roadside banks. 6 Lots (Groh). MacSwain and Bain; (*Nepeta Glechoma* Benth., MacSwain; *N. hederacea* (L.) Trev., Hurst); Campbell.

Var. ***hederacea*** Charlottetown (1914), (*Fernald & St. John*).

Var. ***micrantha*** Moricand The usual form. (*Nepita hederacea* var. *parviflora* (Beuth.) Druce, Roland).

Prunella vulgaris L. Map 659. Common, throughout. Open places: weedy in pastures, fields, roadsides, thickets. 30 Lots (Groh). Apparently the native var. ***lanceolata*** (Barton) Fern. Forma ***rhodantha*** Fern., Bear River and doubtless elsewhere. MacSwain and Bain, MacSwain, Hurst.

Leonurus cardiaca L. Map 660. Formerly planted around houses*, persistent or escaping to hedgerows, as at Brackley Beach. European. 2 Lots (Groh). MacSwain, Hurst.

Galeopsis tetrahit L., vars. Map 661. Common, throughout. European. Weed of waste places, roadsides, and cattle-trodden thickets. 27 Lots (Groh). MacSwain and Bain, MacSwain, Hurst.

Var. ***tetrahit***. Rare: Kilmuir (***Bruce***).

Var. ***bifida*** (Boenn.) Lej. & Court. Common.

[***Lamium hybridum*** Vill. Map 662. Charlottetown 1912 (*Fernald et al.*); Summerside, 1935 (*Hurst*). Adventive, European. (*L. purpureum*, Hurst 1940, Groh in Can. Weed Surv. 2: 33. 1944).]

[***Lamium purpureum*** L. Map 663. Summerside, gardens and dumps (1951), probably not established. European. Hurst's records based on *L. hybridum* specimen.]

[*Stachys arvensis* L. At or near Campbell's Mills, New Glasgow, 1938 (S. Campbell). No specimen to vouch for this record. European, adventive. Hurst 1940.]

Stachys palustris L. Map 664. Wellington, Summerside, near DeSable, North River to Charlottetown. Naturalized, European. Forming colonies in ditches by roads. 3 Lots (Groh). MacSwain and Bain, MacSwain, Hurst.

[*Stachys aspera* Michx., record to be excluded. MacSwain and Bain.]

[*Stachys officinalis* (L.) Trev. Map 665. European, introduced in Blythe Hurst's garden and escaped to fence line. Campbell.]

Satureja acinos (L.) Scheele Map 666. Georgetown 1941 (Hurst); Portage, 1953; St. Eleanors, 1954. On the railway, adventive from central Canada. European. Hurst 1941.

***Like *Nepeta Cataria*, never in gardens but always in the bare shady ground around houses, these species regarded by Victorin (1935) as "commensals" of man, not as introductions.**

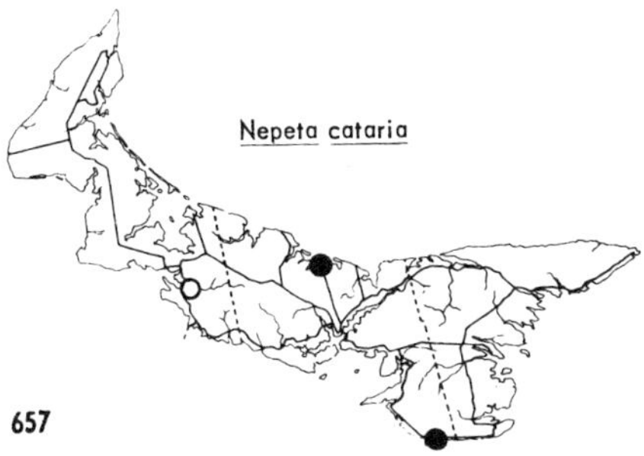

657

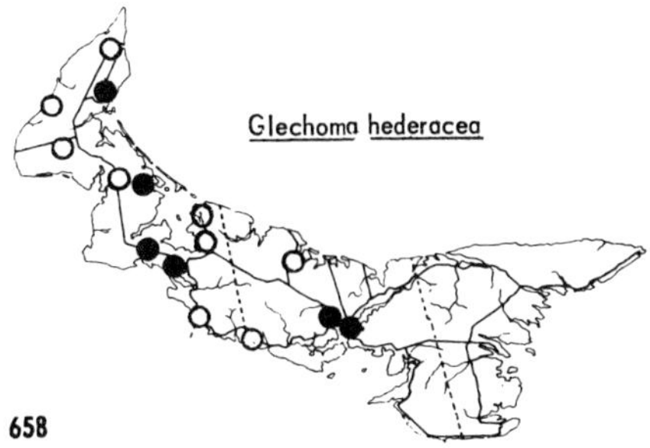

658

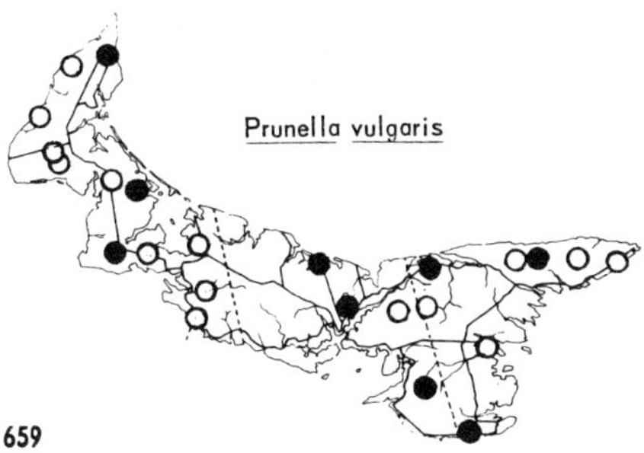

659

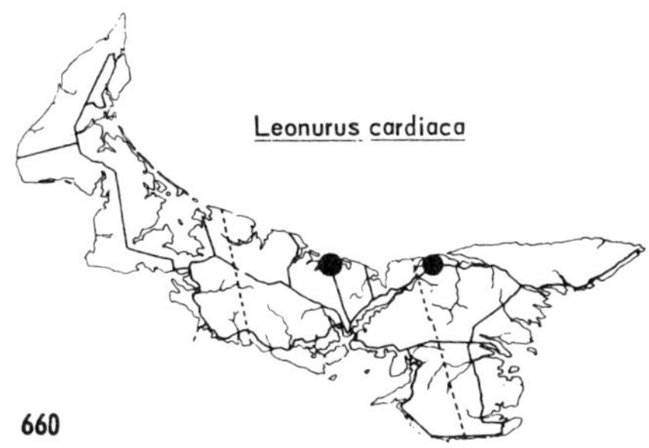

660

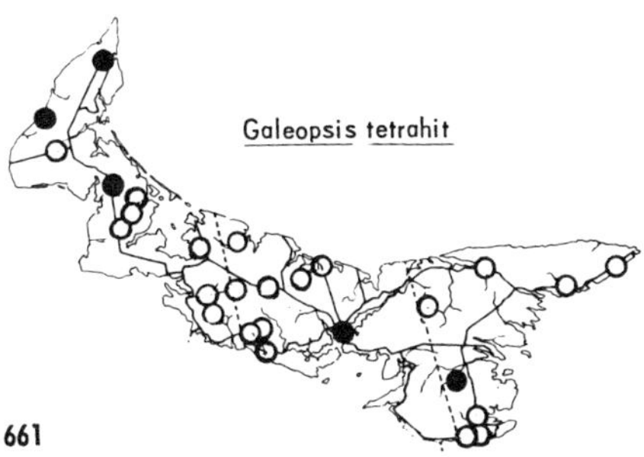

661

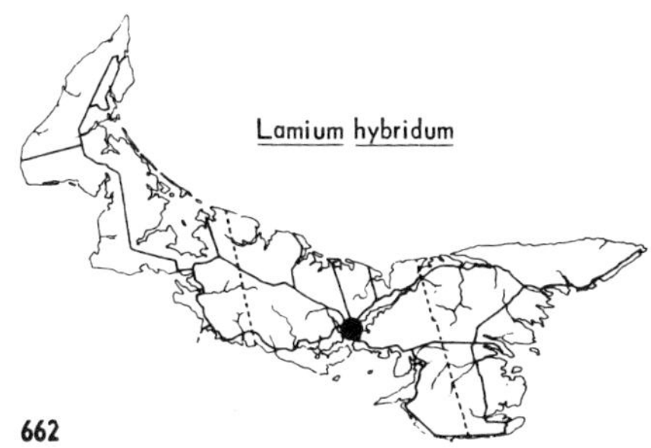

662

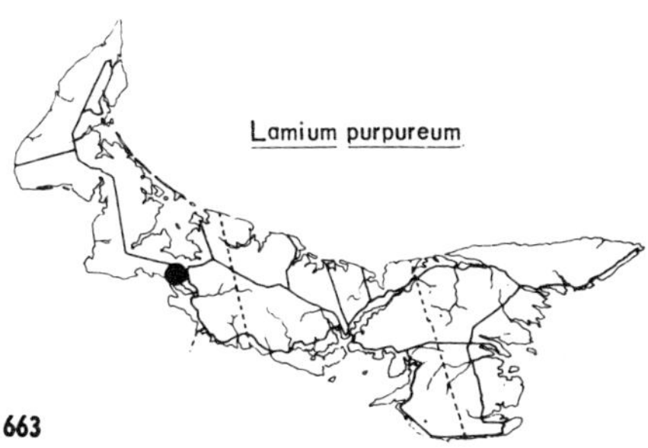

663

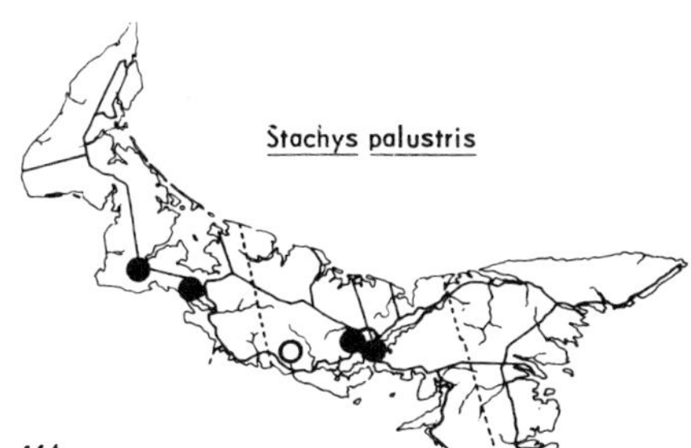

664

Origanum vulgare L. MARJORAM Map 667. Abundant, along road between New Glasgow and Rusticoville, colonies spreading vegetatively only, discovered 1941 by W.L. Holman. Naturalized, European. Hurst 1941.

Thymus serpyllum L. ssp. ***chamaedrys*** (Fries) Vollman (Hegi, Ill. Fl. Mit. Eur. 5: 4323. 1928) THYME Map 668. Established in West Prince, near DeSable, along St. Peters road northeast of Charlottetown, and in the Park. 1 Lot (Groh). Naturalized European. Light sandy soil, forming clumps. White-flowered form, Dalvay (***Bassett***). (***T. pulegioides*** L.; ***T. Serpyllum***, Groh, Hurst).

Lycopus uniflorus Michx. Map 669. Scattered throughout. Damp open or partially shady sites: fields, pond margins, thickets. (***L. virginicus***, MacSwain and Bain, MacSwain, doubted by Hurst); Hurst.

Lycopus americanus Muhl. Map 670. Scattered throughout in lowlands. Damp, open fields, swamps and stream banks. (***L. sinuatus*** Ell., MacSwain); Hurst.

[***Mentha rotundifolia*** (L.) Huds., not seen. Bain 1892.]

Mentha spicata L. SPEARMINT Map 671. Roadsides; Richmond, Charlottetown, Stanhope Road (***Macoun***). European, introduced and naturalized.

Mentha piperita L. PEPPERMINT Map 672. West Prince; Charlottetown, Peakes Station, Bothwell. European. Naturalized along roadside ditches.

Mentha gentilis L. Map 673. Royalty Junction, 1901; near wharves, Charlottetown, 1912 (***Fernald & St. John***); "frequent about Tracadie" (Churchill). (***M. rubra*** Huds., Churchill, Hurst, doubted by Campbell); Groh, Hurst, Campbell.

Mentha arvensis L. var. ***villosa*** (Benth.) S.R. Stewart. Map 674. Rather common, throughout. Native. Damp open sites: stream banks, ditches, pond margins. (***M. canadensis*** L., MacSwain and Bain, MacSwain; ***M. arvensis*** var. ***canadensis*** (L.) Briq., Hurst); Campbell.

Var. ***arvensis***, often cultivated in gardens for flavoring; introduced from Europe.

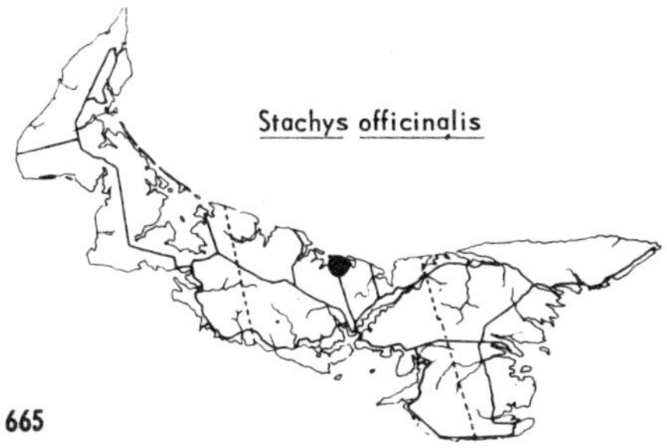

665

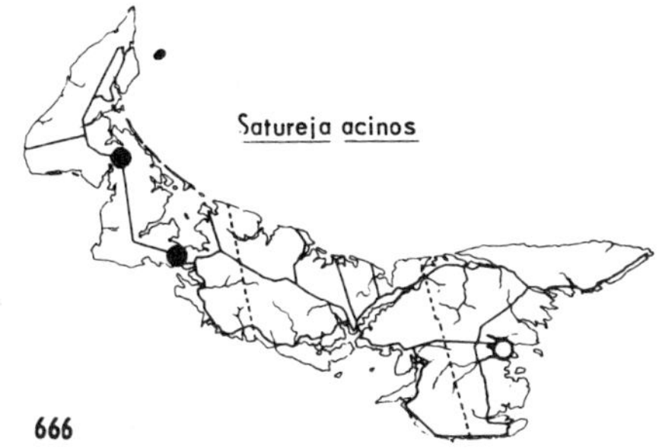

666

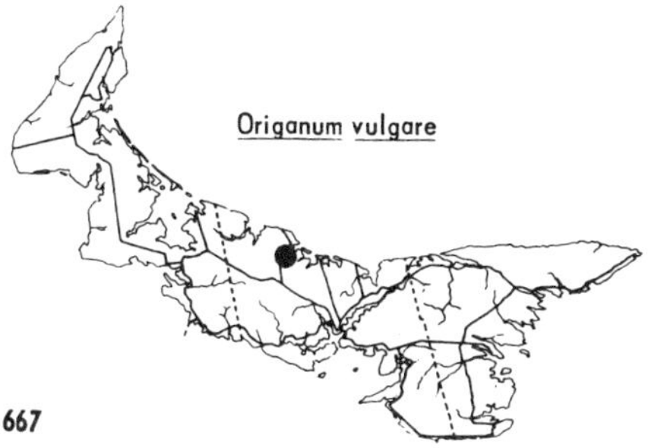

667

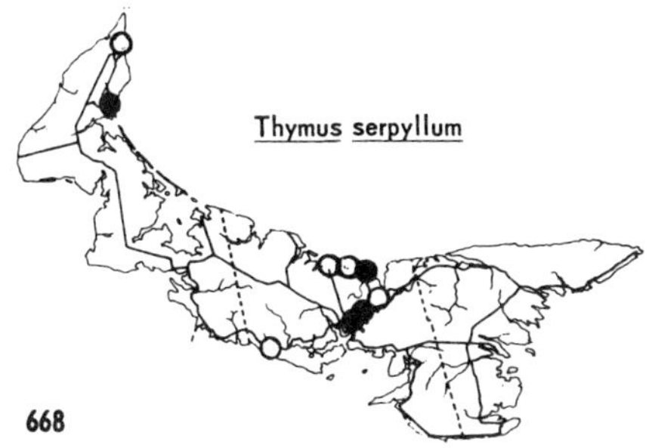

668

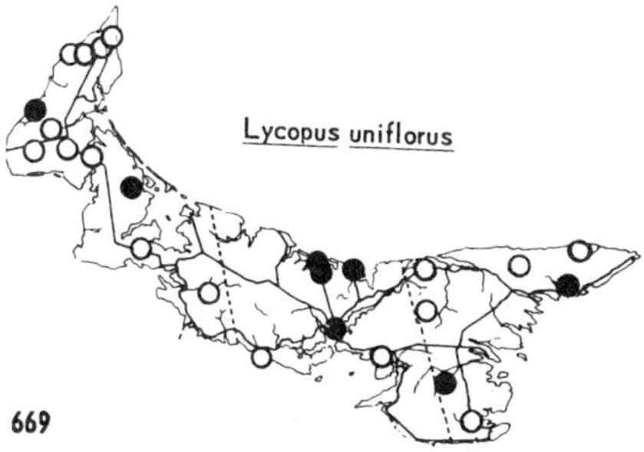

669

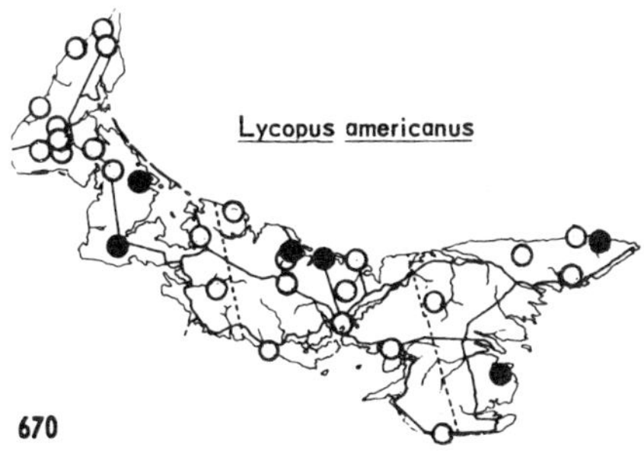

670

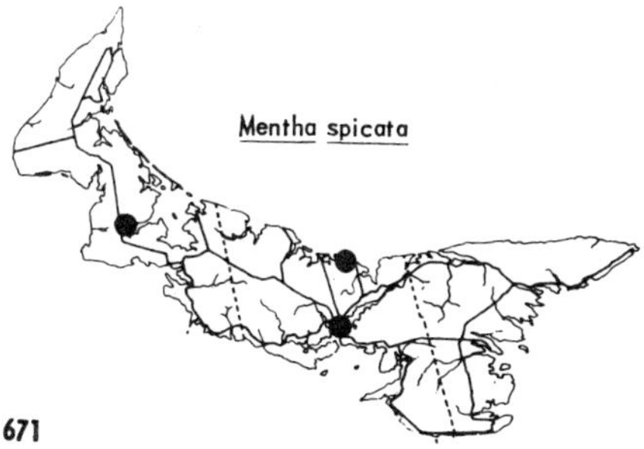

671

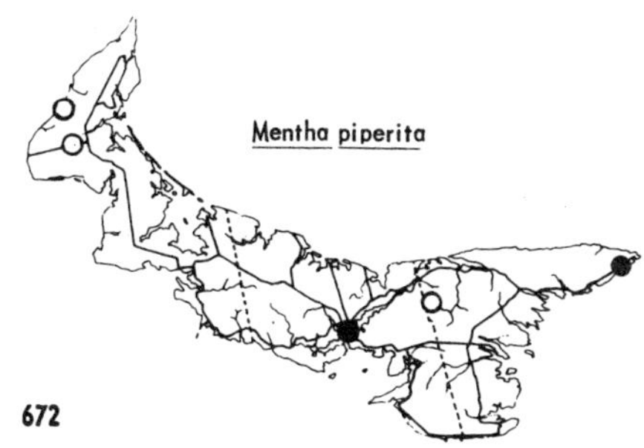

672

SOLANACEAE

Solanum dulcamara L. Map 675. Very local: Ellerslie, Charlottetown near Hillsborough Bridge since 1912, Murray River. Naturalized, European. Damp waste ground, often along streams. Poisonous. Campbell.

[***Solanum tuberosum*** L. POTATO Most important crop of the Island. South American. Springing up on dumps, never established.]

[***Solanum nigrum*** L., frequently recorded: ''Too common'' (Stewart), ''common nightshade'' (Bain); Charlottetown, 1943 (R.R. Hurst), cultivated or casual in garden. Not seen. European. MacSwain, Hurst.]

[***Solanum rostratum*** Dunal ''Casual'', in Charlottetown. Hurst 1940.]

[***Nicandra physalodes*** (L.) Pers. Map 676. Charlottetown, 1935. South American, introduced garden plant, casual escape. MacSwain, Hurst.]

[****Hyoscyamus niger*** L. Map 677. Miminegash, roadside, 1919 (*Perry*). Adventive; naturalized from Europe around Bay of Chaleur.]

[***Datura stramonium*** L. Map 678. Occasional in Charlottetown; New Haven (1946). Asiatic, adventive from southward. Wharves, railways, roadsides.]

SCROPHULARIACEAE

Verbascum thapsus L. MULLEIN Map 679. Scattered throughout, more common eastward. European, naturalized. Dry or stony open sites: abundant in old fields, especially overgrazed pastures. MacSwain and Bain, MacSwain, Hurst.

Linaria vulgaris Hill Map 680. Throughout, very common in central region. European, naturalized. Weed of dry waste land along roads and railways, rarely in pastures. 17 Lots (Groh). Forma ***leucantha*** Fern. Emerald Junction (*Bassett*). MacSwain, Hurst.

[***Linaria maroccana*** Hook. f. Brackley Beach (1927), casual introduction in Hurst's garden. African. (*L. reticulata*, Groh & Frankton, Can. Weed Surv. 7: 100. 1949).]

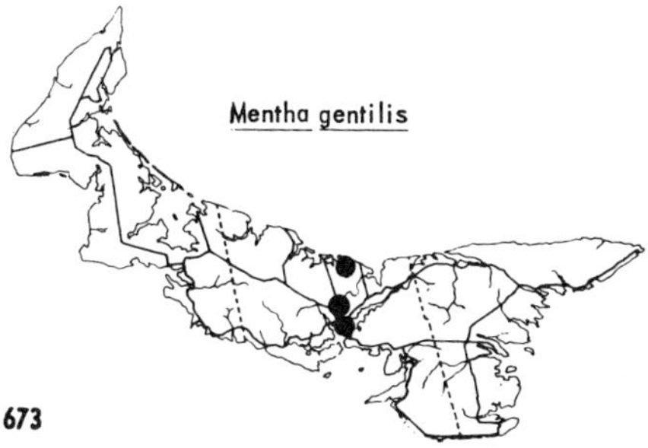

673

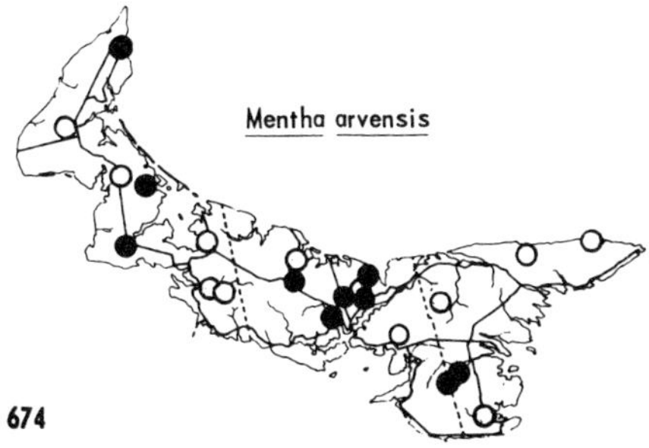

674

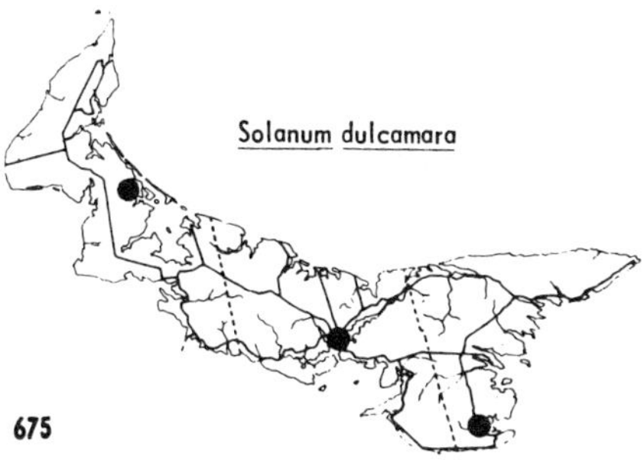

675

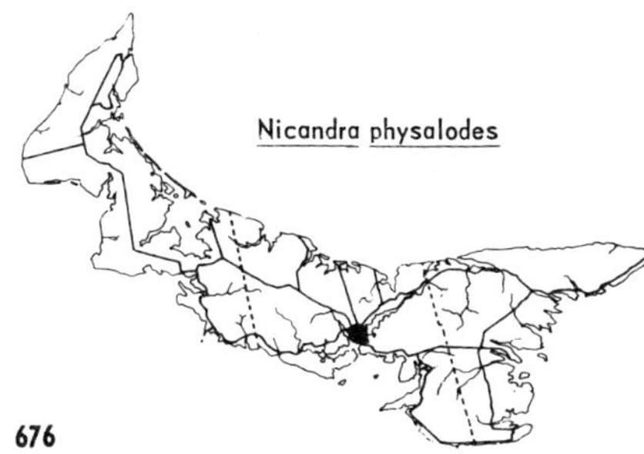

676

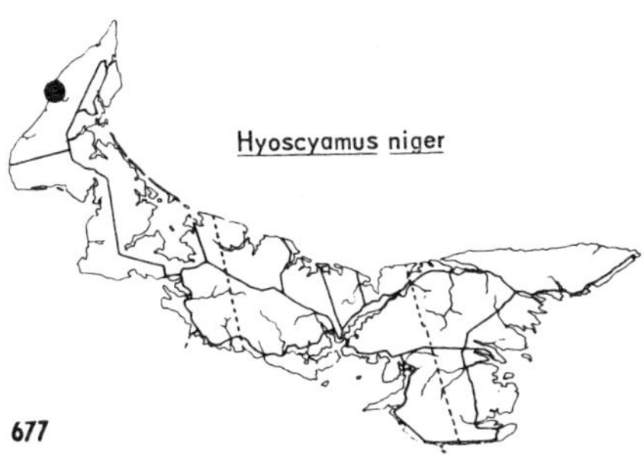

677

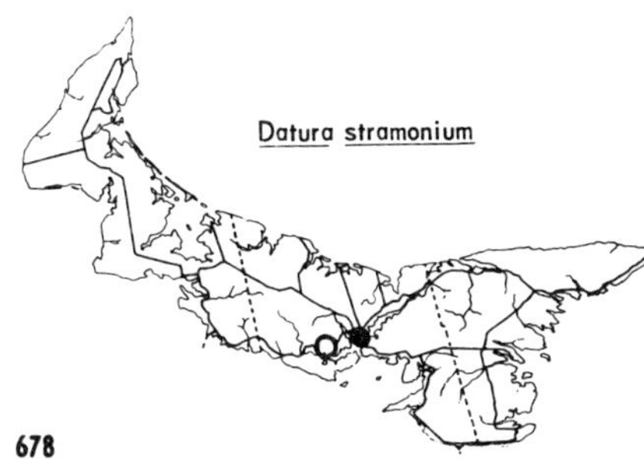

678

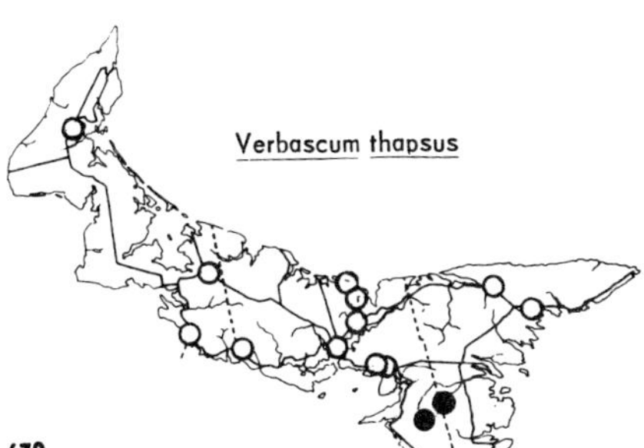

679

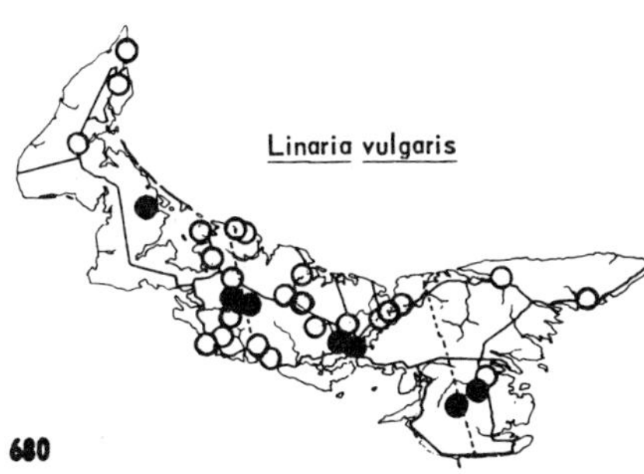

680

Chaenorrhinum minus (L.) Lange Map 681. C.N.R., now throughout. European, naturalized. Weed of railway cinders; 2 Lots, 1926 (Groh). (***Linaria minor*** (L.) Desf., Groh, Hurst), Campbell.

Chelone glabra L., vars. Map 682. Fairly common throughout in lowlands and along streams. Damp ground: marshy fields, thickets. MacSwain and Bain, MacSwain, Hurst.

Var. ***glabra*** Throughout; the only variety from southeast of Charlottetown.

Var. ***dilatata*** Fern. & Wieg. Prince County and North Shore: Bloomfield, Mount Stewart (*Fernald et al.*), Stanhope (*Bassett*).

[***Pentstemon hirsutus*** (L.) Willd. Not known from the Maritimes. (*P.* ***pubescens*** Soland., Bain 1892, MacSwain); Hurst.]

Mimulus ringens L. Map 683. Local in Prince and lowlands of Queens; rare eastward. Damp muddy stream banks, pond margins. MacSwain, Hurst.

Mimulus moschatus Douglas Map 684. Locally in West Prince; abundant around Bonshaw; Dundas. Pacific North American; naturalized. Springs, stream borders. Hurst.

Limosella subulata Ives Map 685. North and eastern coasts. Estuaries and barrier ponds, brackish muddy sand. (*L.* ***aquatica*** var. ***tenuifolia***, Macoun, Hurst); Campbell.

[***Veronica longifolia*** L. Map 686. Summerside, 1912 (*Fernald et al.*). European, introduced to gardens, casual escape. Referred by Pennell, Rhodora 23: 11. 1921, to his *V.* ***maritima*** forma *E.*, by the collectors to var. ***media*** (Schrad.) Hartm.; not the common or garden *V.* ***maritima*** forma *A.* of Pennell.]

Veronica serpyllifolia L. Map 687. Scattered. Prince and Queens; throughout? European, naturalized. Damp grassy places along upland streams, wood roads, in old fields. Bain, MacSwain and Bain, MacSwain, Groh, Hurst. (var. ***nummularioides*** Lec. & Lem.).

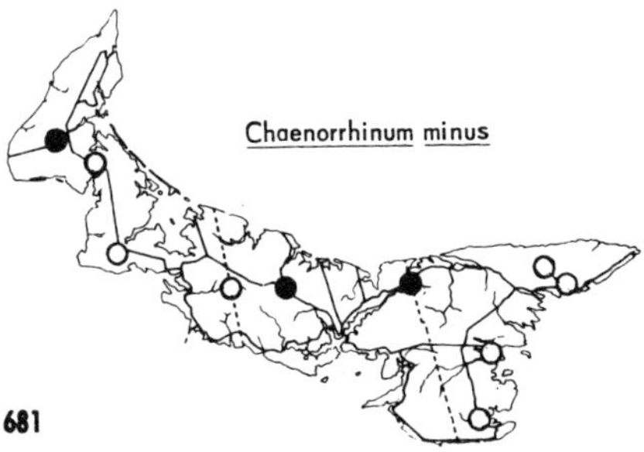

681

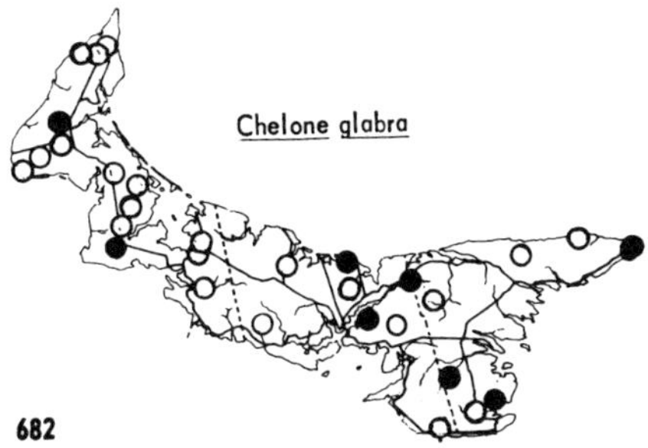

682

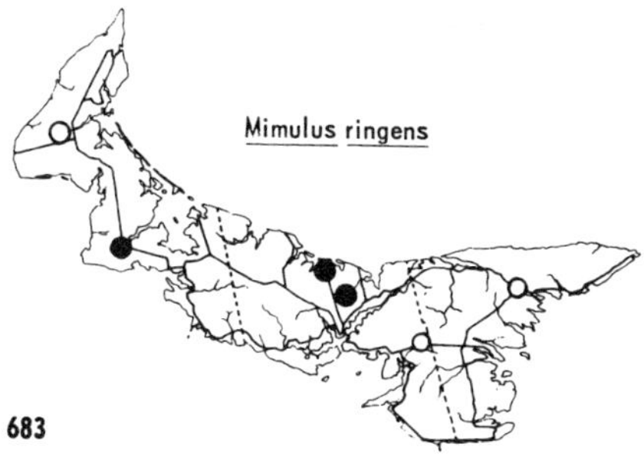

683

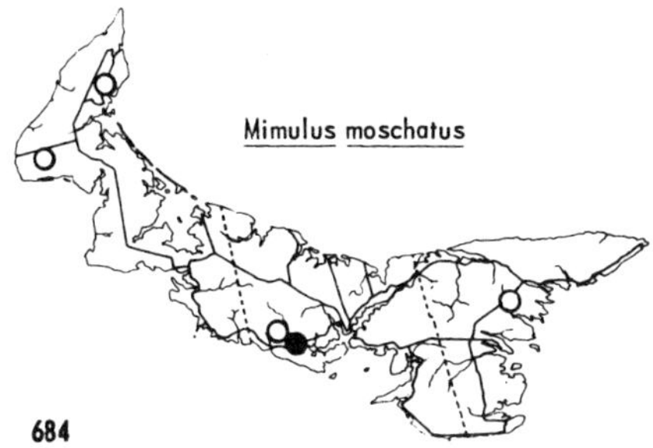

684

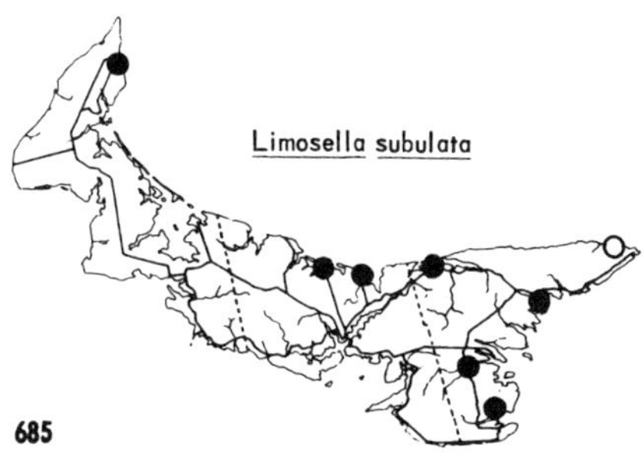

685

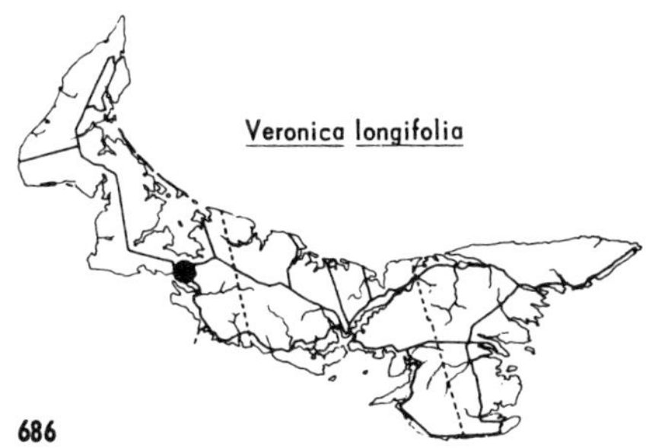

686

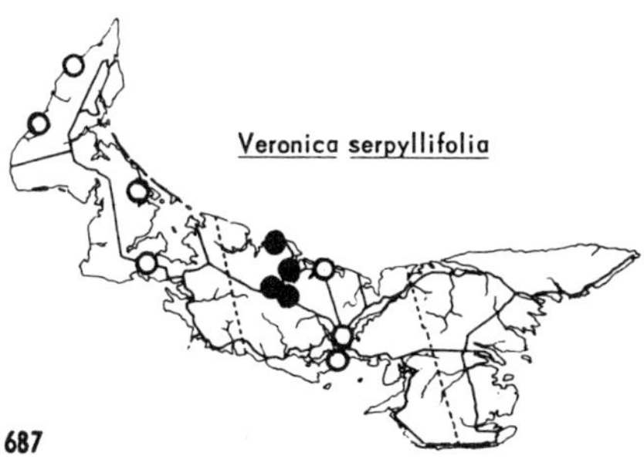

687

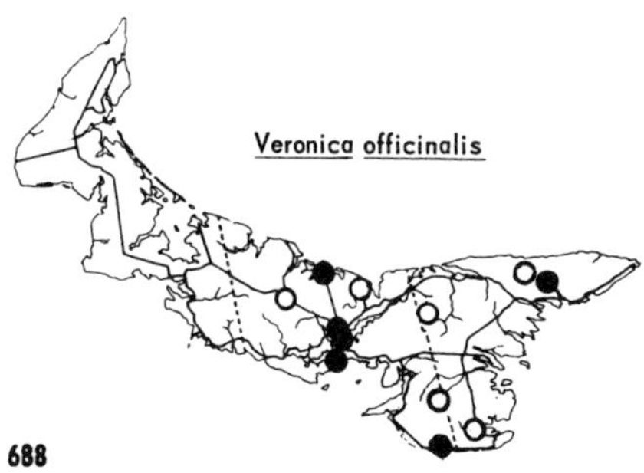

688

Veronica officinalis L. Map 688. Frequent in Queens and Kings, mainly in uplands. European, naturalized (not native). All material seems referable to var. ***tournefortii*** (Vill.) Reichenb. MacSwain and Bain, MacSwain, Hurst.

Veronica chamaedrys L. Map 689. Sherwood cemetery; Brackley Point Road, mainly in uplands. European, naturalized before 1912. Hurst 1940.

Veronica scutellata L. Map 690. Local in Prince County west of Summerside. Mucky pond shores and pools in swamps. (*V.* ***Anagallis,*** MacSwain and Bain, MacSwain; *V.* ***Anagallis-aquatica,*** Hurst).

Veronica americana (Raf.) Schwein. Map 691. Common in West Prince and the northeast; scattered in center and southeast. Around shallow, still water: streams, ditches, pools. MacSwain and Bain, MacSwain, Hurst.

[*Veronica* ***Anagallis-aquatica*** L. records probably refer to *V.* ***scutellata.***]

[*Veronica* ***arvensis*** L. was probably listed on no authority as Hurst only mistakenly attributed it to MacSwain.]

[****Veronica peregrina*** L. Bunbury, weed in the Cotton Memorial Nursery (1956), adventive from central Canada.]

Veronica persica Poir. Map 692. Charlottetown, 1912; 1 Lot, 1926 (Groh). Garden weed, from the Middle East. (*V.* ***Buxbaumii*** Ten., MacSwain; *V.* ***Tournefortii,*** Groh, Hurst); Campbell.

[*Veronica* ***agrestis*** L. Dundas, 1952 (G.C. Warren). Garden weed, European. Record possibly referable to the preceding. MacSwain, Groh, Hurst.]

Melampyrum lineare Desr., vars. Map 693. Common, central and eastern regions. Dry, open sites (hemiparasitic on grass roots): barrens, old fields and succeeding wire-birch woodland. (*M.* ***pratense,*** MacSwain and Bain; *M.* ***Americanum*** Michx., MacSwain); Hurst.

Var. ***lineare*** Common. (*M.* ***lineare*** Desr., MacSwain and Bain).

Var. ***americanum*** (Michx.) Beauverd Mount Stewart, Tracadie, Charlottetown, Wood Islands (growing among var. ***lineare***). (*M.* ***Americanum*** Michx., MacSwain and Bain).

Euphrasia randii Rob., vars. Map 694. North Shore: in thin turf, exposed sands or headlands along coast. Specimens cited by Fernald and Wiegand, Rhodora 17: 187-9. 1915.

Var. ***randii*** Swampy shore of Campbell's Pond, Sea View (*Fernald & St.John*).

Var. ***reeksii*** Fern. Dry brackish sand, Cape Aylesbury (*Fernald et al.*).

Var. ***farlowii*** Rob. Grass, East Point (Macoun).

[*Euphrasia* ***rigidula*** Jordan Cozen's Pond near Malpeque (*Fernald et al.* 8027), cited by Fernald & Wiegand (Rhodora 17: 197. 1915) as *E.* ***stricta*** HBK., non Host (this species). E.O. Callen considers this material incorrectly determined, and referable to the following.]

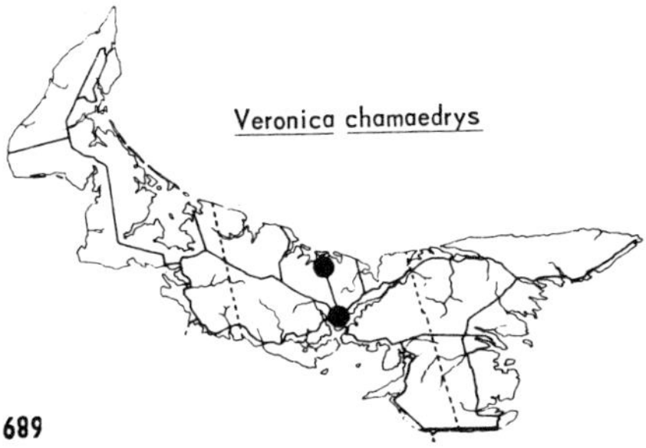

689

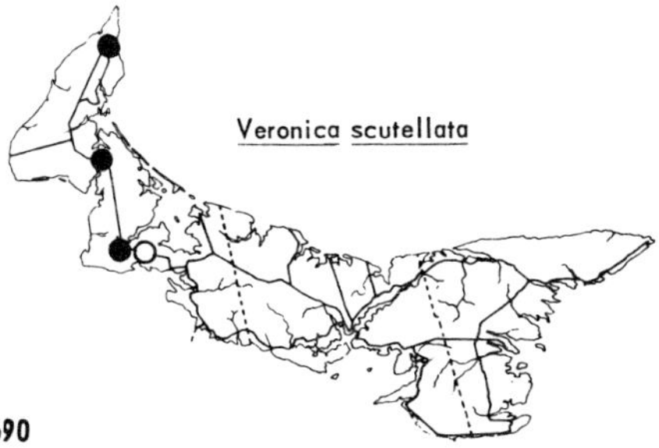

690

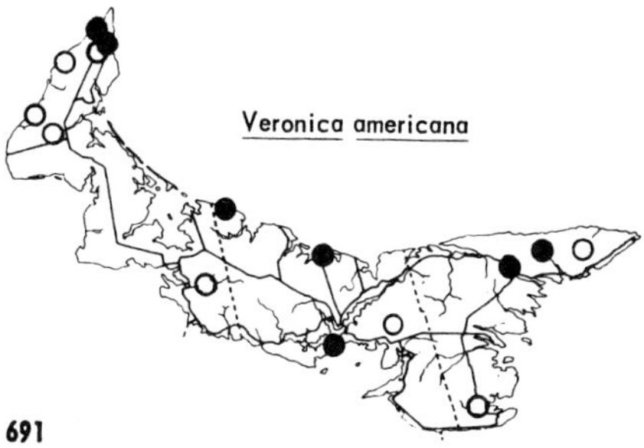

691

692

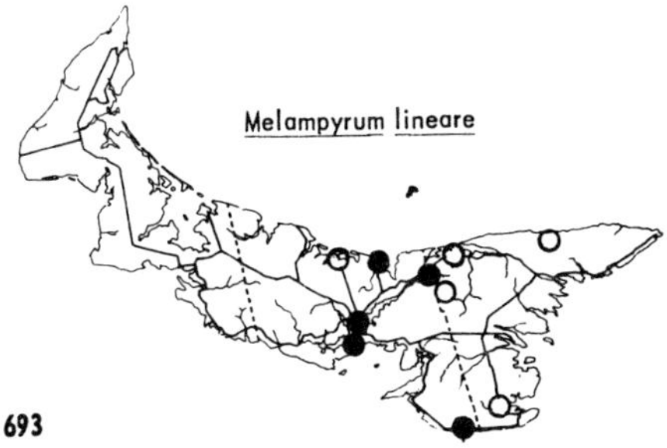

693

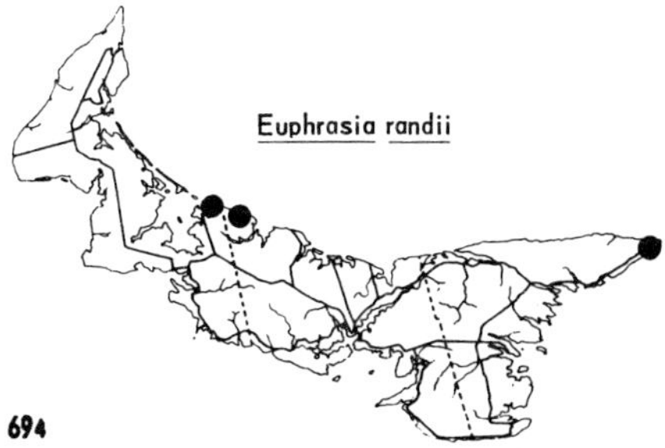

694

Euphrasia canadensis Townsend Map 695. Malpeque (*Fletcher*), cites as of Fowler by Fernald & Wiegand (Rhodora 17: 195. 1915), Cousins Pond (det. E.O. Callen); certainly at other stations along the North Shore and perhaps elsewhere.

X ***E. aequalis*** Callen (*E. "strictiformis"?)*, a hybrid of *E. americana* and *E. canadensis*, occurs in disturbed habitats where both species are present. Mount Royal near O'Leary, roadsides (det. Callen).

E. americana Wettst. Map 696. Common, widespread and variable; dry open fields and exposed turfy situations. (*E. officinalis*, MacSwain and Bain, MacSwain, Hurst).

Odontites serotina (Lam.) Dum. Map 697. Central and southwestern regions, west to Summerside. European, naturalized at least by 1900. Weedy, in turf of fields, on dry roadsides or sandy places; hemiparasite on grass roots. (*Bartsia Odontites* Huds., MacSwain; *O. rubra* Gilib., Groh, Hurst, Groh in Can. Weed Surv. 4: 38. 1947); Campbell.

Rhinanthus crista-galli L., vars. Map 698. Scattered near North Shore from Covehead eastward and elsewhere in Kings (?); near Tignish. Dry fields; headlands, tension line of salt marsh. 5 Lots (Groh). MacSwain and Bain, MacSwain, Hurst.

Var. ***crista-galli*** The form lacking anthocyanins, occasional near Tignish.

Var. ***fallax*** (W. & G.) Druce Only from eastward.

OROBANCHACEAE

Epifagus virginiana (L.) Barton Map 699. Recorded from West Prince and vicinity of Charlottetown; perhaps co-extensive with *Fagus*. Parasite on roots of beech.

Orobanche uniflora L. Map 700. Langley Beach near Southport, solitary specimen, June 1928 (Messervy). Root-parasitic on herbs, in damp meadows.

LENTIBULARIACEAE

Utricularia minor L. Map 701. Watervale (Q), marshy creek; Mount Stewart, pools in marsh. Macoun, Hurst.

Utricularia cornuta Michx. Map 702. Rare: East Bideford; Mermaid Lake. Mucky spots in acid peat bogs.

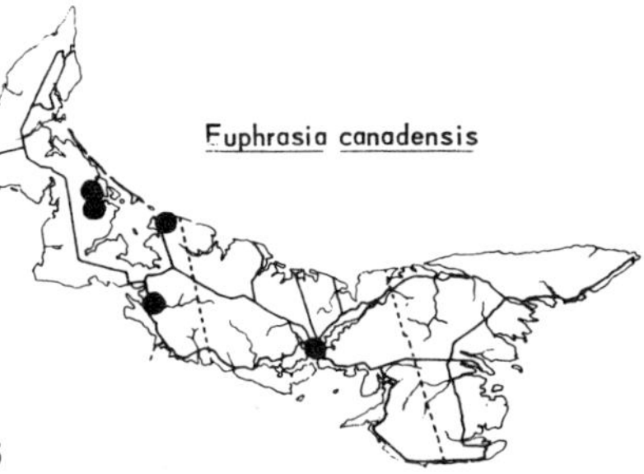

695

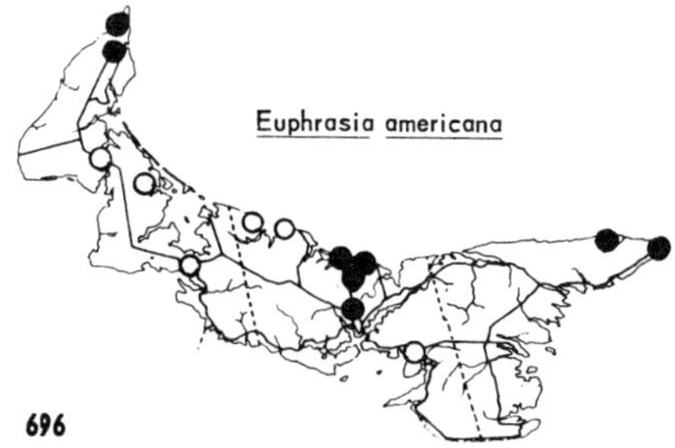

696

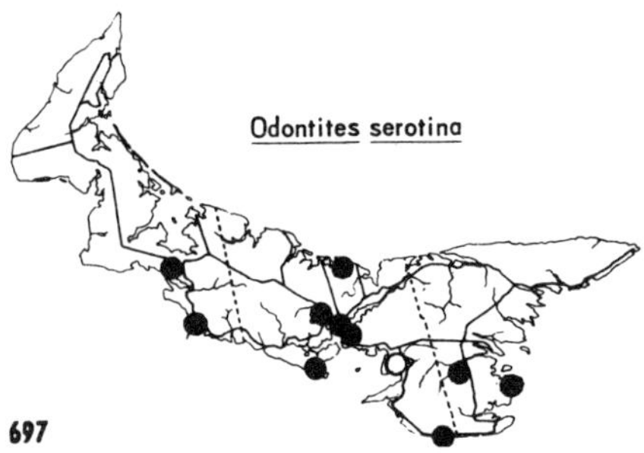

697

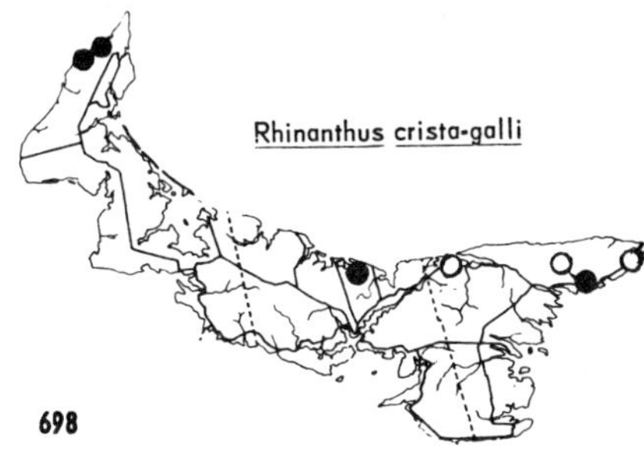

698

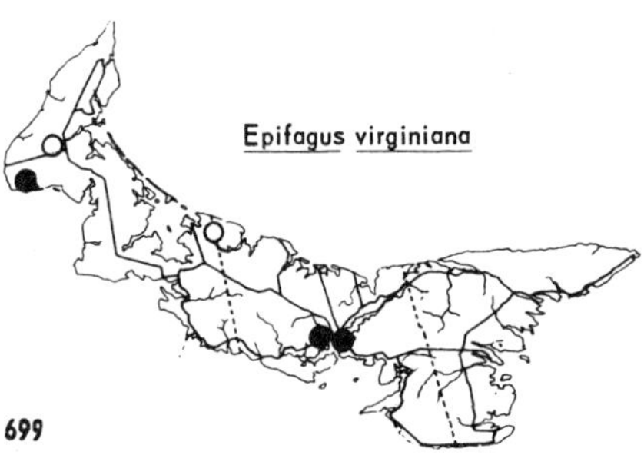

699

700

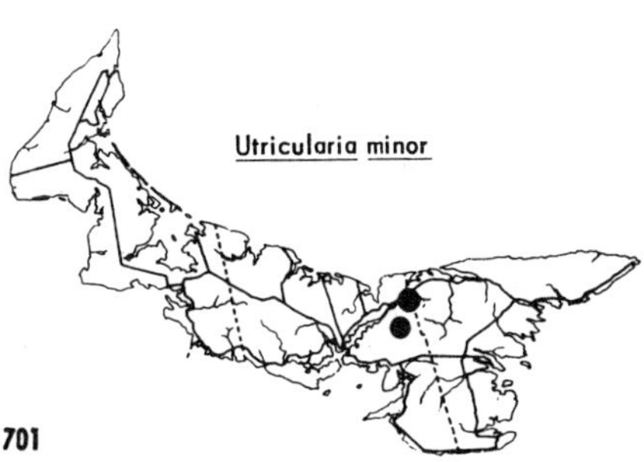

701

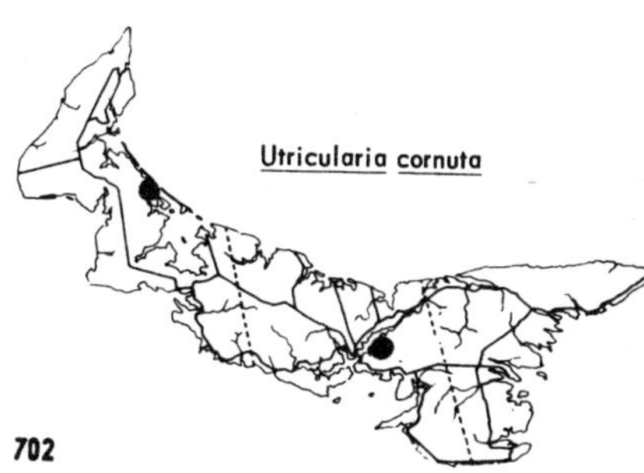

702

PLANTAGINACEAE

Plantago major L., vars. Map 703. Common throughout. Bain, MacSwain and Bain, MacSwain, Hurst. 37 Lots (Groh).

Var. ***major*** Naturalized from Europe. Weed of waste ground (yards, roadsides) and lawns. Forma ***intermedia*** (Gilib.) Pilger, the glabrous plants, Charlottetown; forma ***paniculata*** Domin, the plants with lateral spikes from lower axils of main spike, Pownal (*Malte*).

Var. ***scopulorum*** Fries & Broberg Brackley Point, Grand Tracadie, Wood Islands; maritime sands. Similar but more elongated plants from the muddy bottom of the former millpond at Wellington had the purple petiole bases of *P. rugelii*.

[*Plantago rugelii* Dcne. Campbellton (Watson specimen determined by Macoun, not preserved). Should be sought.]

Plantago juncoides Lam. var. ***decipiens*** (Barnéoud) n. comb., based *Plantago decipiens* Barnéoud, Mon. Plantag. 16 (1845), (not seen). Map 704. Scattered, maritime. Sandy headlands, beach heads and salt marshes. (*P. maritima*, Bain; *P. maritima* var. *juncoides*, MacSwain and Bain; *P. marina*, MacSwain and Bain; *P. decipiens* Barnéoud, MacSwain, Hurst; *P. juncoides* var. *decipiens* (Barnéoud) Fern., Campbell, all in part only).

Var. ***decipiens*** Common form. F. *pygmaea* (Lange) Rousseau (Nat. Canad. 59: 234. 1943), East Point, exposed headland, in sand and sandy cart track, while f. ***decipiens*** grew among the grasses.

Var. ***laurentiana*** Fern. Wood Islands (Fernald, Rhodora 27: 102. 1925), Souris; North Point; transitional form at Campbellton. Headlands.

Plantago oliganthos R. & S. Map 705. Coastal: Murray River, Souris, Charlottetown, Brackley Beach, Ellerslie. Salt marshes. (Wisely not distinguished from *P. juncoides* by local botanists).

Plantago lanceolata L. Map 706. Queens and Kings Counties, scattered. European, naturalized. Open dry or waste ground, weed in pastures, fallow fields and roadsides. 3 Lots (Groh). MacSwain and Bain, MacSwain, Hurst.

RUBIACEAE

Galium triflorum Michx. Map 707. Throughout, in Alberry-Charlottetown soils. Well-drained sites on floor of mixed or hardwood forest. MacSwain and Bain, MacSwain, Hurst.

****Galium mollugo*** L. var. ***erectum*** (Huds.) Briq. Southport; Charlottetown Experimental Farm, weed in hay and grain fields. European, naturalized. (*G. erectum* Huds.).

Galium palustre L. Map 708. Common throughout. Open marshy places; along streams and ditches. 10 Lots (Groh). Hurst.

703

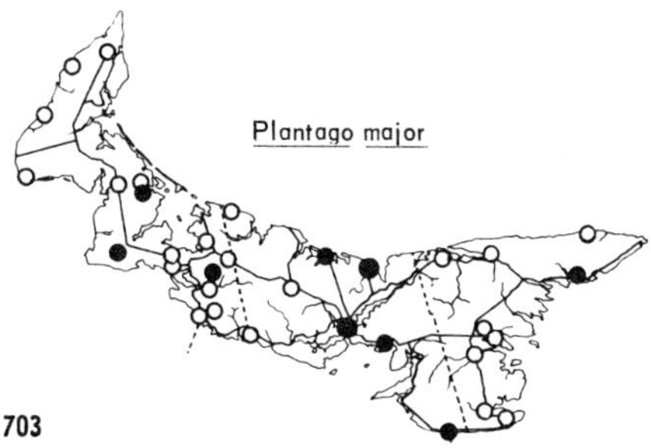

704

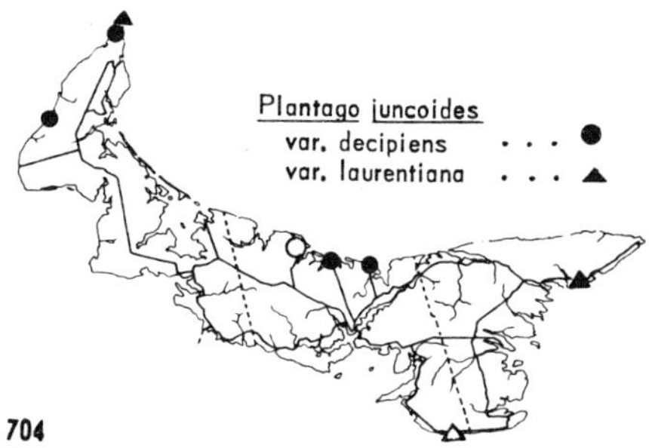

705

706

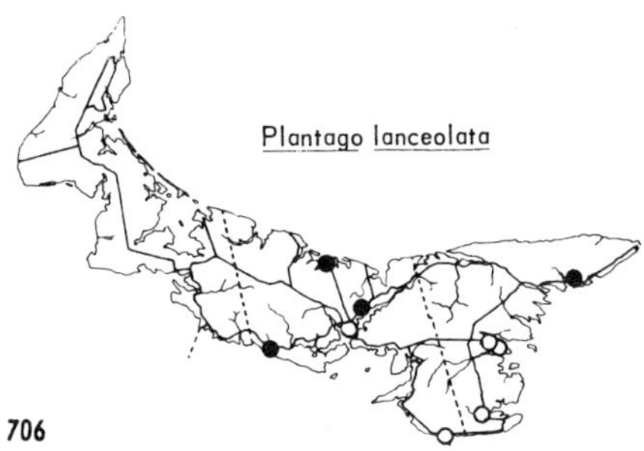

707

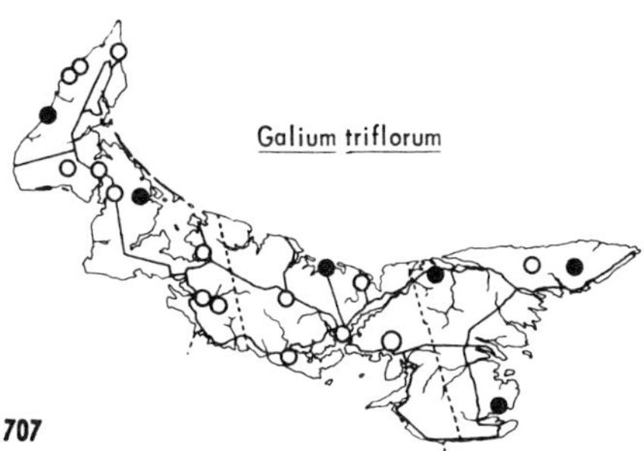

708

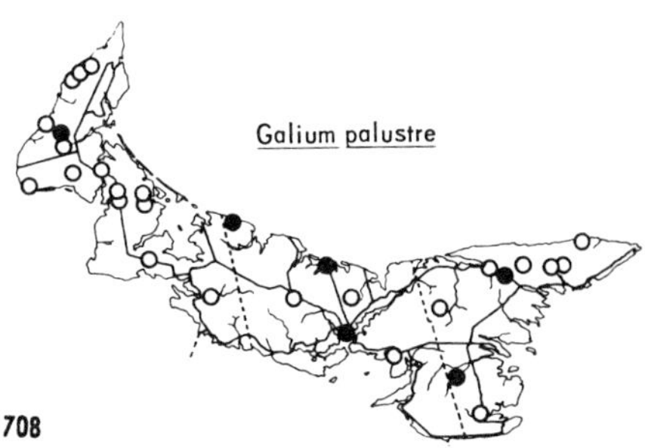

Galium trifidum L. Map 709 Scattered around the coast. Wet open places: marshy pond, dune slacks, springs. MacSwain and Bain, MacSwain, Hurst.

Var. *halophilum* Fern. & Wieg. Tracadie Beach (*Churchill*). Intermediate material from Bothwell.

Galium tinctorium L. Map 710. Scattered, Prince County and east-central lowland. Damp muddy spots: pond margins and thickets. (*G. Claytoni* Michx.); Hurst 1940 (?).

Galium labradoricum Wieg. Map 711. In dune slacks at Nail Pond (West Prince); bogs, Mount Stewart to St. Peters. Fernald 1950, Campbell.

Galium asprellum Michx. Map 712. Fairly common in Prince County and east-central lowland. Damp alder thickets; marshy ground along streams. MacSwain and Bain, MacSwain, Hurst.

**Mitchella repens* L. Portage, mossy hummock in damp mixed woods.

[*Cephalanthus occidentalis* L. Not known east of Queens Co., N.S. MacSwain and Bain.]

CAPRIFOLIACEAE

Diervilla lonicera Mill. Map 713. Most frequent in central regions; open, dry or upland soils at border of thickets. (*D. trifida* Moench, MacSwain); Hurst.

Lonicera villosa (Michx.) R. & S., vars. Map 714. Local in Prince County and east-central lowlands. Swamps and bogs. (*L. caerulea*, Hurst); Campbell.

Var. *solonis* (Eaton) Fern. Usual type in the east-central region.

Var. *calvescens* (Fern. & Wieg.) Fern. Miscouche bog, associated with the last.

Lonicera canadensis Bartr. Map 715. Scattered in Prince (especially in West Prince) and northeastern upland. Stream banks and ravine slopes in mixed woods. (*L. ciliata* Muhl., MacSwain); Hurst.

Symphoricarpos albus (L.) Blake Introduced for facing house-walls and for hedges, persisting but seldom escaping to roadsides. 4 Lots (Groh). Ours all var. *laevigatus* (Fern.) Blake. (*S. rivularis* Suksd., Roland). (*S. racemosus* Michx., Groh, Hurst); Campbell.

Linnaea borealis L. var. *americana* (Forbes) Rehder Map 716. Scattered throughout, lowlands and ravines. "Mossy-floored fir woods", often covering hummocks. (*L. borealis*, MacSwain and Bain, MacSwain); Hurst.

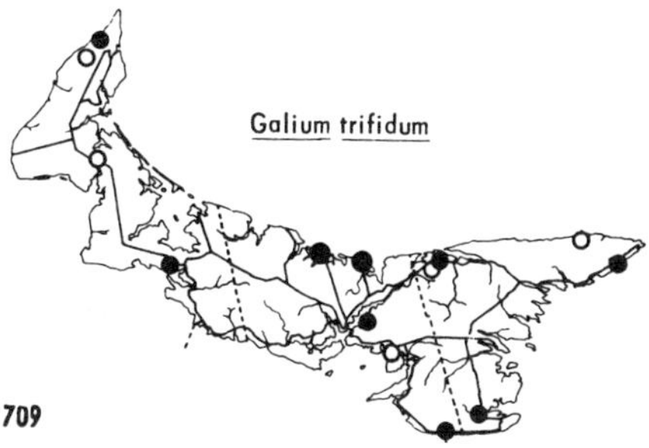

709

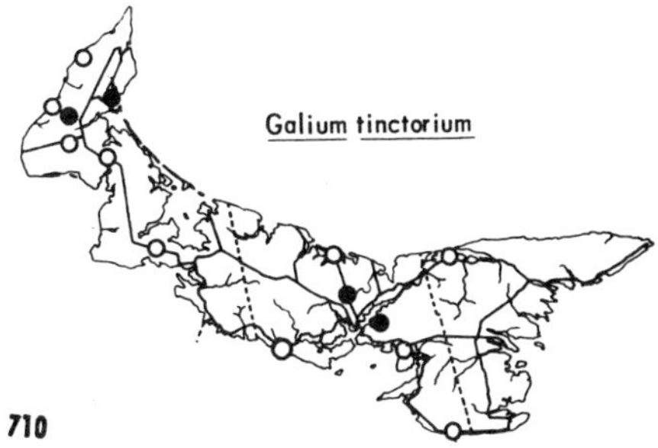

710

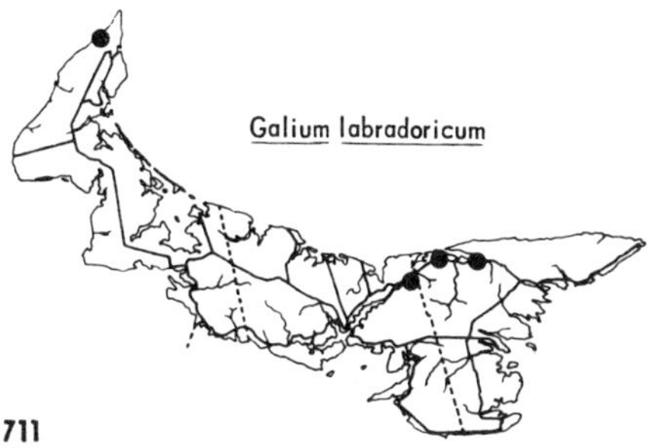

711

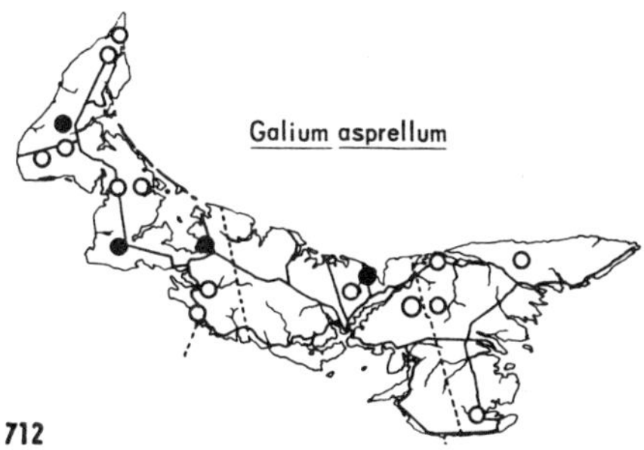

712

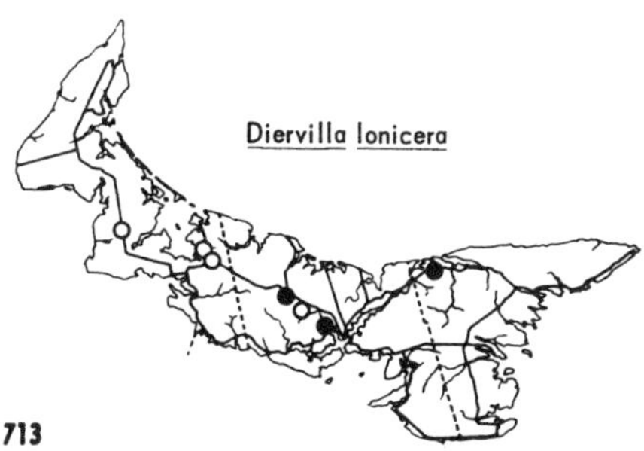

713

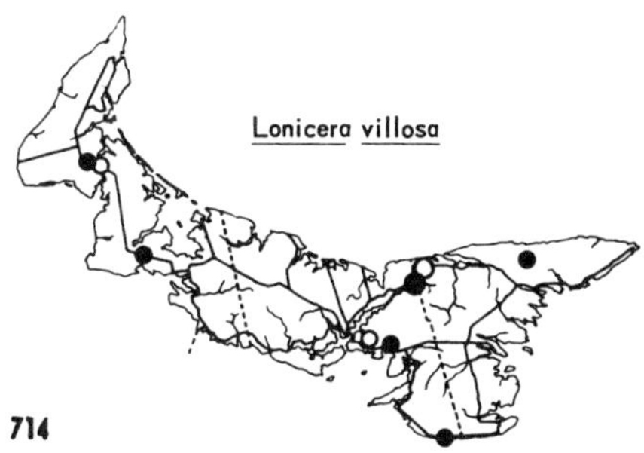

714

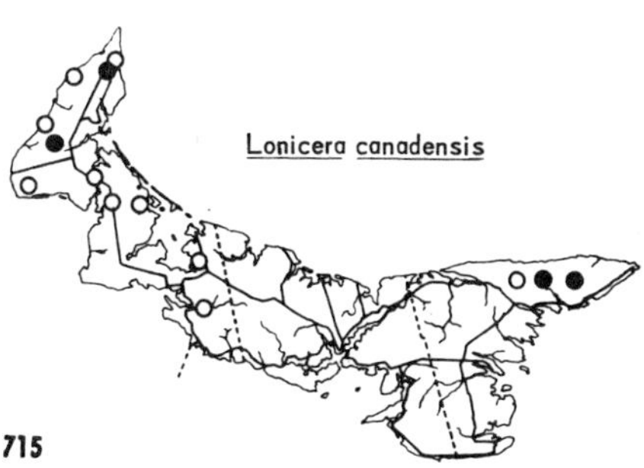

715

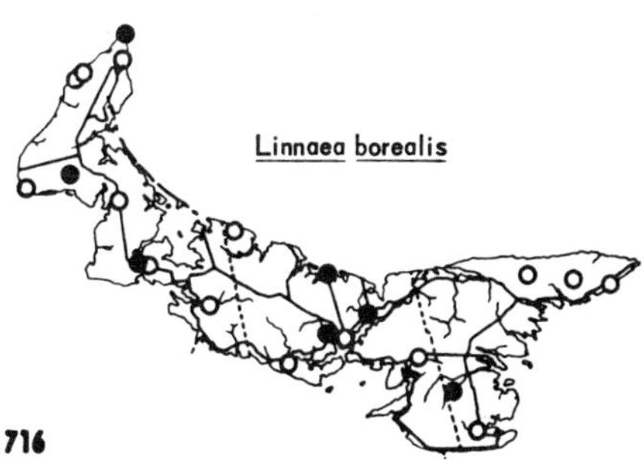

716

Viburnum alnifolium Marsh. Map 717. Very rare and confined to southeastern upland: Surrey, mixed woods in ravine (*Fernald & St. John*); Lewes (Profitt).

Viburnum cassinoides L. Map 718. Common throughout, except central upland. Swamps, damp acid barrens and clearings. (*V. nudum*, Bain, MacSwain and Bain); MacSwain, Hurst.

Viburnum trilobum Marsh. HIGH-BUSH CRANBERRY Map 719. Frequent in Prince County; very local eastward: Rustico Island (Hurst), Montague (Hurst), Bangor (MacDougall). Sometimes planted as ornamental, and escaped: road to Charlottetown airport (*R.R. Hurst*). Damp thickets and open swampy woods. Cranberries "smaller less pleasant on very ornamental bushes" (Stewart). (*V. acerifolium*, Bain, MacSwain and Bain, Hurst, Campbell; *V. Opulus*, Bain 1892; *V. Opulus* var. *americanum* Ait., Hurst); Campbell.

Sambucus canadensis L. ELDER Map 720. Common in Prince and Kings; local in Queens. Moist, open places along streams, swamps. 12 Lots (Groh). Fruit the basis of elderberry wine. Foliage poisonous. MacSwain and Bain (with doubt), Groh, Hurst.

Sambucus pubens Michx. Map 721. Common throughout. Well-drained sites, clearings or open hardwoods. Red berries unpleasant. Bain, MacSwain and Bain; (*S. racemosa*, MacSwain, Hurst); (*R.R. Hurst*).

CUCURBITACEAE

Echinocystis lobata (Michx.) T. & G. WILD CUCUMBER Map 722. Introduced from central Canada as a trellis plant for screens. Often abundant in waste land around dumps in towns: Charlottetown, Kensington. 3 Lots (Groh). Hurst.

CAMPANULACEAE (including LOBELIACEAE, Bain, MacSwain and Bain, Hurst)

Campanula rapunculoides L. Map 723. European, commonly introduced as a self-perpetuating garden plant. Waste lots in towns and escaped around settlements. 7 Lots (Groh). Hurst.

Campanula rotundifolia L. BLUEBELL (of Scotland) Colville Station, a few plants formerly on the railway banks (Messervy), adventive from mainland? Brackley Point, introduced by Hurst, still vigorous in lawn, 1953.

Lobelia spicata Lam. Charlottetown, 1910. Field weed, adventive from eastern Canada? MacSwain and Bain, MacSwain, Hurst, Groh & Frankton in Can. Weed Surv. 7: 106. 1949.

Lobelia inflata L. Map 724. Scattered, central and southeastern districts. Dry open land: weed of old fields, clearings. 3 Lots (Groh). Bain, MacSwain and Bain, MacSwain, Hurst.

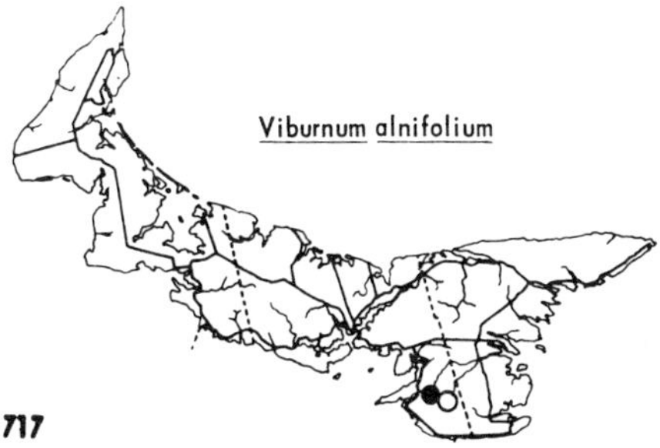

717

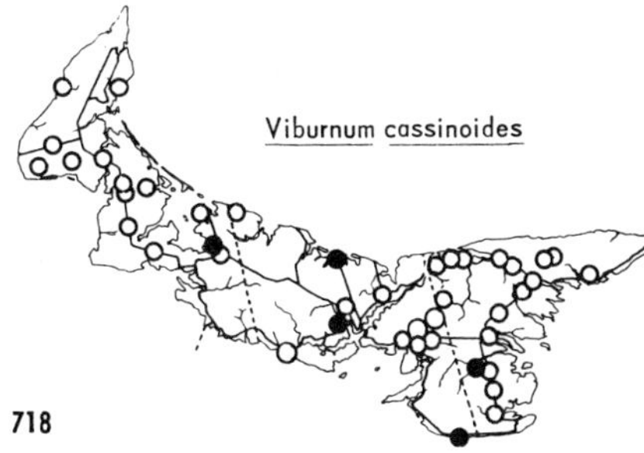

718

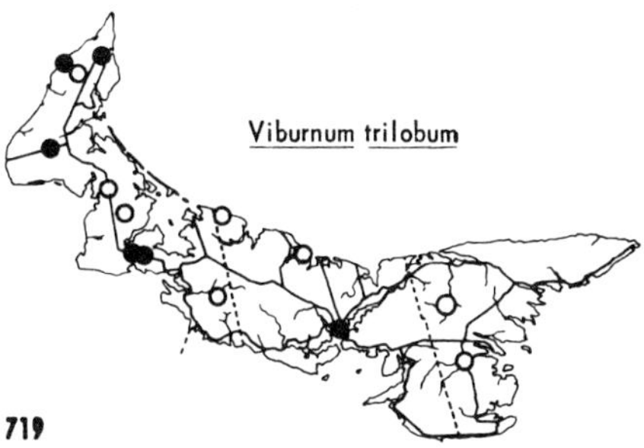

719

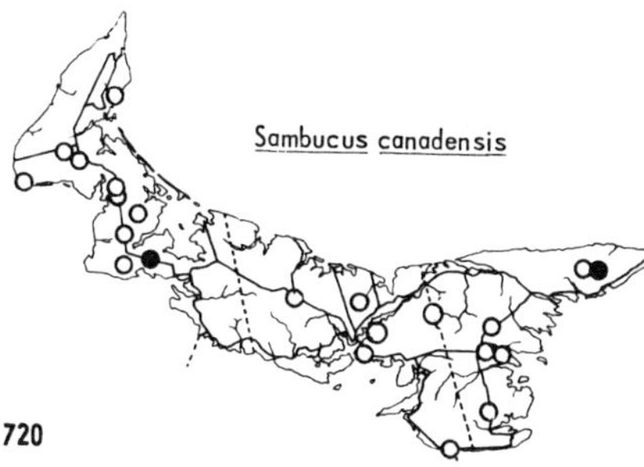

720

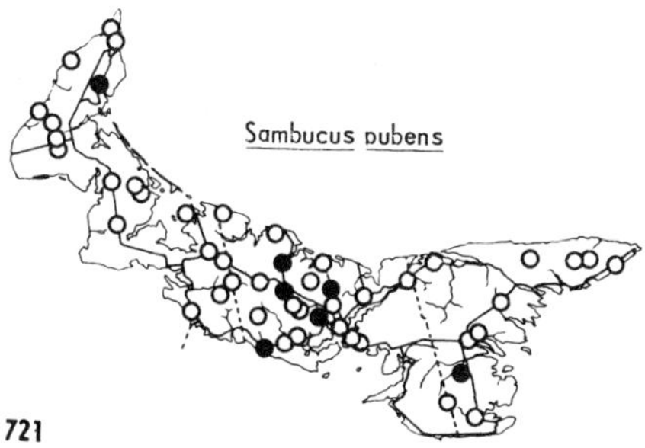

721

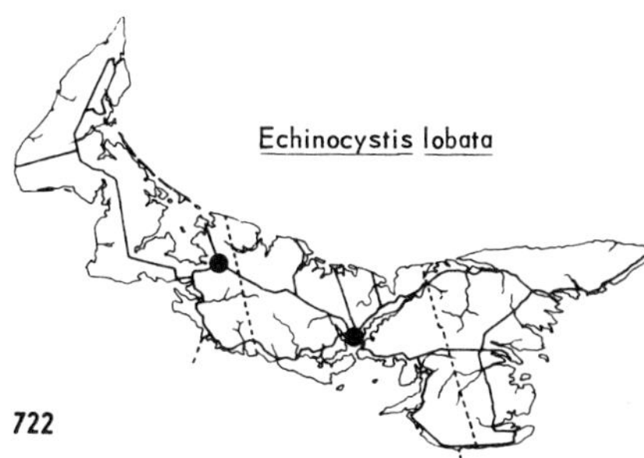

722

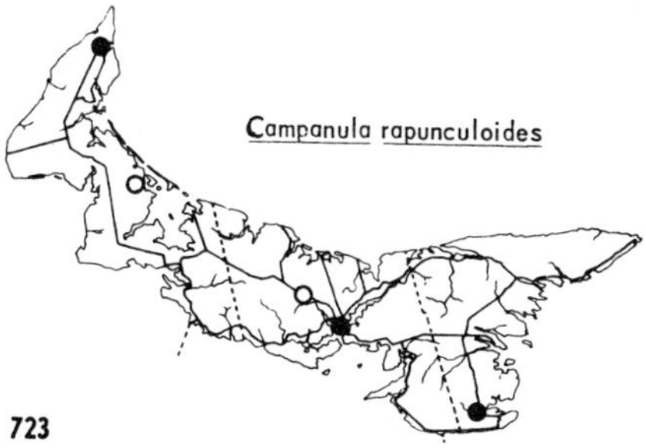

723

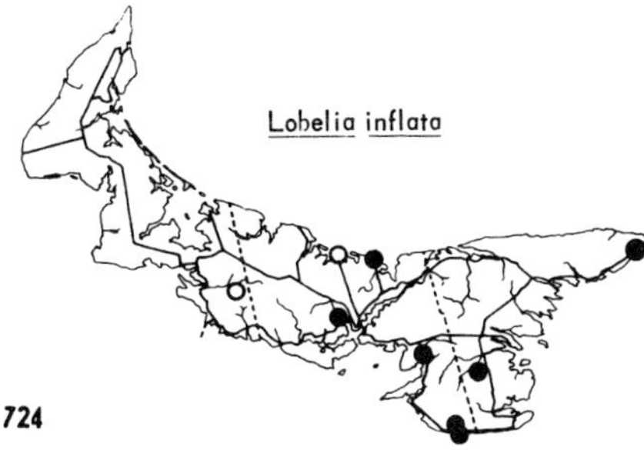

724

[*Lobelia kalmii* L. Report probably based on *L. spicata* or *L. inflata*, for Sterling Campbell found it as a field weed.]

Lobelia dortmanna L. Map 725. Keefe's Lake, Glenfinnan Lake, Lake Verde; shallow water, leaves submerged, flowers aerial.

COMPOSITAE

Eupatorium maculatum L. Map 726. Common throughout Prince County; by North Shore rivers and eastern swamps. Moist, open places: stream banks, thickets, marshy meadows (Bain); ditches. 31 Lots (Groh). Formerly in medicine, perhaps unwarrantably. (*E. purpureum*, MacSwain and Bain, MacSwain, Hurst, Campbell).

Var. **foliosum** (Fern.) Wieg. Bloomfield (*Fernald & St. John*), possibly the common form in West Prince.

Eupatorium perfoliatum L. Map 727. Scattered in Prince County; North River (*Warren*). Damp grassland along streams, damp thickets, roadside ditches. MacSwain, Hurst.

***Solidago flexicaulis** L. Northeastern upland: Glencorradale corner hardwood forest.

Solidago macrophylla Pursh Tignish (*Fernald et al.*); Harmony (*Fernald & St. John*). Mossy spruce woods.

Solidago bicolor L. Map 728. Restricted to region east of Charlottetown. Dry, open sandy land; old fields, barrens due to clearing or fire, roadsides. MacSwain, Hurst.

[*Solidago hispida* Muhl. "Coast north of St. Peters, together with *S. bicolor*," (MacSwain) probably refers to *S. nemoralis* which in that area grows commonly with it. (*S. bicolor* var. *concolor* T. & G., MacSwain and Bain, MacSwain); Hurst.]

Solidago puberula Nutt. Map 729. Scattered throughout. Dry, open sites: sandy roadsides, old fields, clearings; in burnt land often twice its usual height. MacSwain, Hurst.

Solidago sempervirens L. Map 730. Maritime, scattered. Salt marsh, dunes and seaward dune slacks. Hybridizes with *S. canadensis* and *S. rugosa*. (*S. laevigata* Ait., Bain, MacSwain and Bain); MacSwain, Hurst.

***X Solidago asperula** Desf. (*S. sempervirens* x *rugosa*) North River Bridge, sandy roadside gutter just above salt marsh; one clump (1952).

***Solidago juncea** Ait. Map 731. Kilmuir (K), sandy knoll. Surprisingly rare considering its abundance across the Strait in northern Nova Scotia.

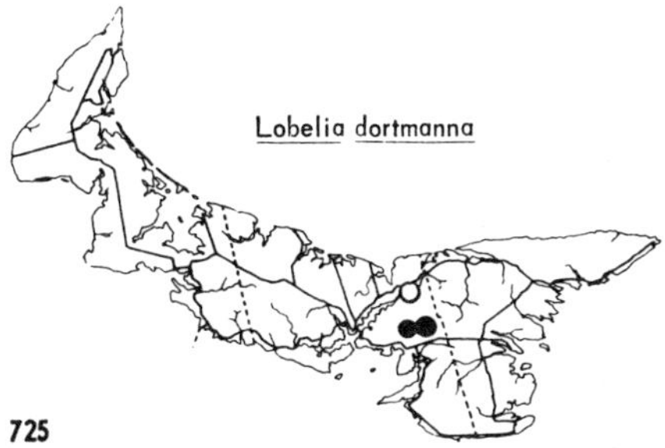

725

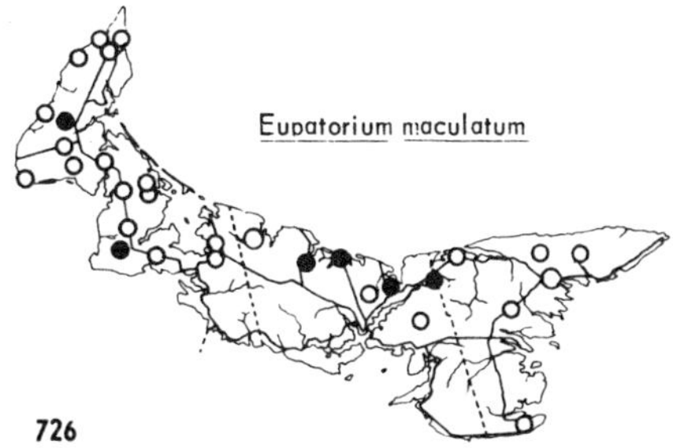

726

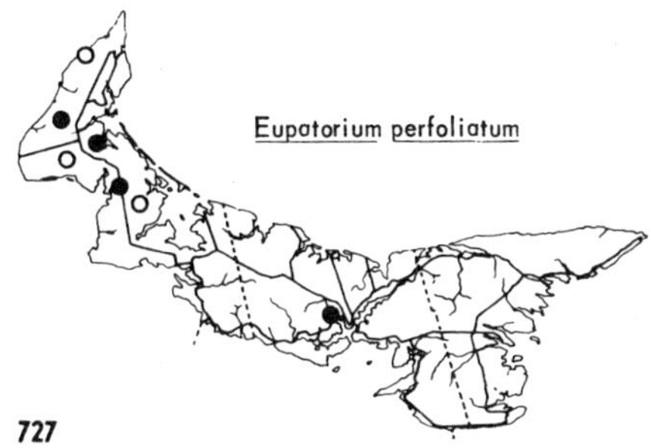

727

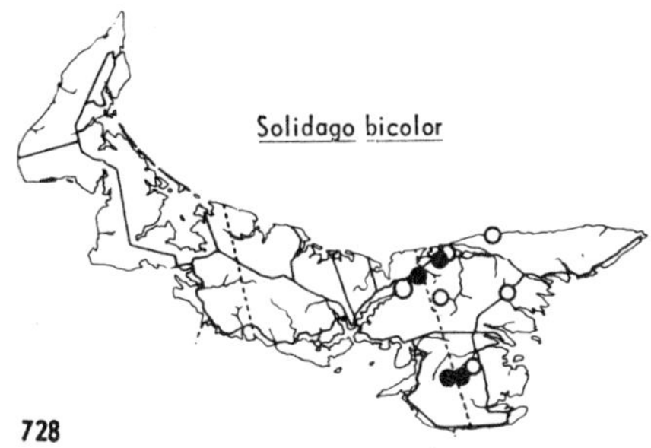

728

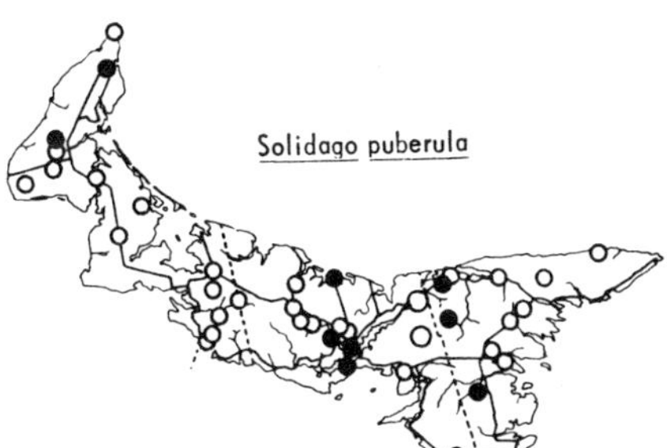

729

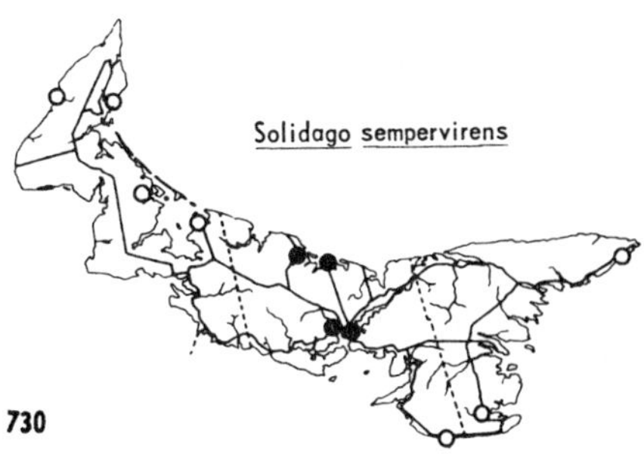

730

731

732

Solidago uliginosa Nutt. var. ***linoides*** (T. & G.) Fern. Map 732. Scattered: Frequent in Prince County and east-central lowland. "Heath lands" (Bain); bogs and barrens. (*S. stricta*, Bain, MacSwain and Bain, MacSwain, Hurst with doubt; *S. uliginosa*, MacSwain, Hurst; *S. uniligulata* (DC.) Porter, Roland).

Solidago nemoralis Ait. Map 733. Coastal; North Shore from Malpeque eastward, round to Wood Islands, on sand dunes. Inland in central Queens and southern Kings in the driest stony open sites, barrens, old fields and roadsides. More slender and with rounder, shorter panicle branches than the type common in Nova Scotia. Adams, Hurst 1940.

Solidago rugosa Mill. Map 734. Common throughout. Open, damp to dryish sites: borders of woods, thickets. One of the four important "yellowweeds" of fields. Hybridizes with *S. sempervirens* (X *S. asperula*) and with *S. canadensis* (near Charlottetown, *Malte*). (*S. altissima*, Bain, MacSwain and Bain, MacSwain, Hurst with doubt); MacSwain, Hurst.

Solidago canadensis L. Map 735. Common throughout. Open dryish land: thickets, roadsides. A "yellowweed" of fields, densely infesting near Summerside. Dark autumn honey is goldenrod honey. Pollen locally reputed to cause hay fever; however, *Solidago* is not wind-pollinated, unlike *Ambrosia* which causes most serious pollen allergy. Hybridizes with *S. rugosa* (Malte) and *S. sempervirens*. MacSwain and Bain, MacSwain, Hurst; (including *S. serotina*, MacSwain, Groh).

Solidago canadensis* X *sempervirens Shoulder of road across salt flat, Wood Islands; a single clump, there more seriously rusted than the parental species.

[*Solidago altissima* L. Records probably refer to *S. rugosa* (q.v.) as Bain calls it the broadest-leaved species.]

Solidago gigantea Ait. Map 736. West Prince only. Damp thickets; banks and ditches. Leaves of P.E.I. specimens are as broad as the extreme in any other area. (*S. serotina* Ait., not Retz., Bain (?)); MacSwain and Bain. MacSwain, Groh records based, according to Fernald, quoted by Adams (1937), on *S. canadensis*).

Solidago graminifolia (L.) Salisb. "POVERTYWEED" Map 737. Common throughout. Damp to dry open sites: thickets, dune slacks. The most pernicious "yellowweed" of old fields, pastures and fallow in Kings County; said to be reduced by fertilizing of the soil. (*S. lanceolata* L., Bain, MacSwain and Bain, MacSwain); Hurst. Var. *graminifolia* and var. *nuttallii* (Greene) Fern. not distinguished in the field. Glabrous plants, the latter variety, collected at Kilmuir (*Bruce*).

Aster macrophyllus L. Map 738. Very local, around Malpeque Bay: St. Eleanors, Indian River, French River; and at Dalvay (Hurst). Dry, open mixed woods or wood edges, on light soil. MacSwain and Bain, MacSwain, Hurst.

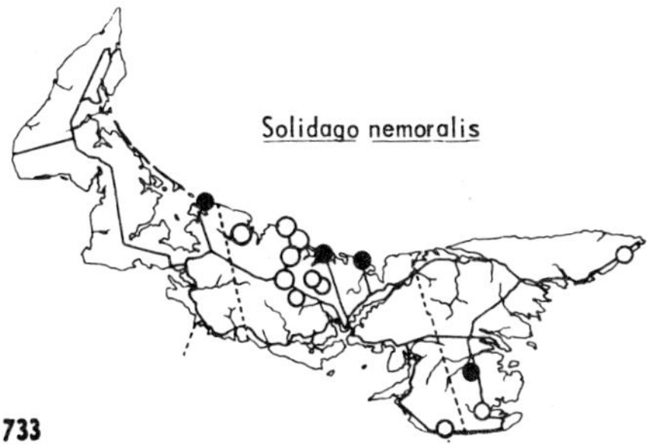

733

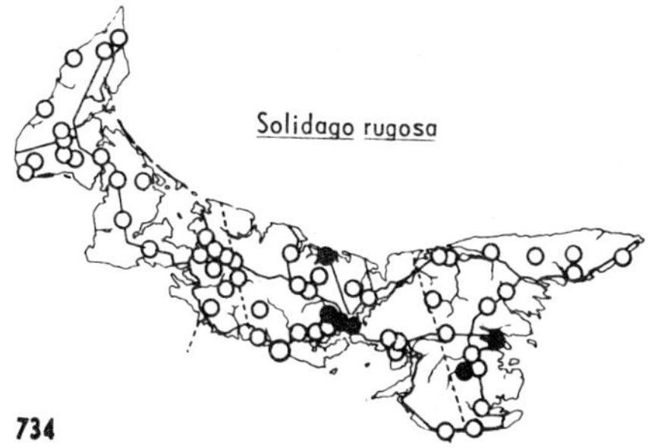

734

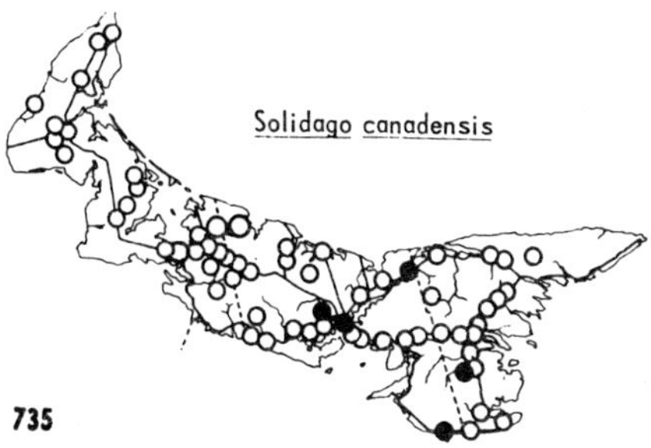

735

736

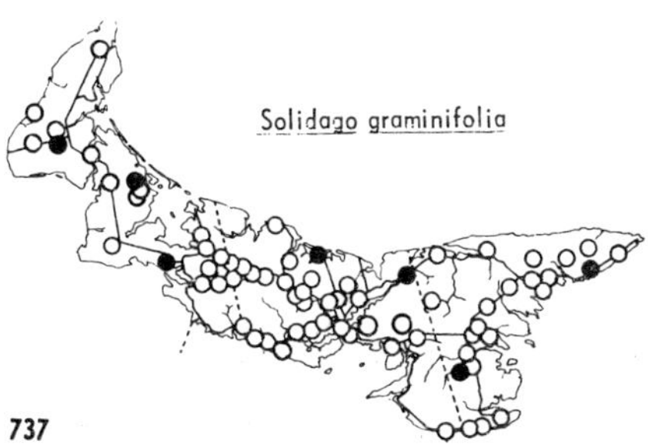

737

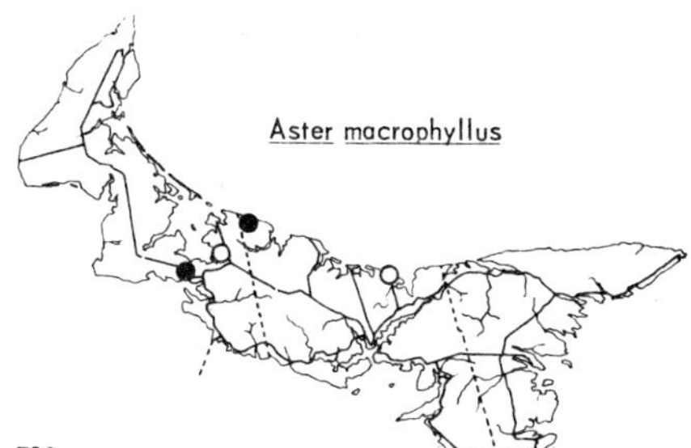

738

Aster cordifolius L. Map 739. Scattered: West Prince; Malpeque Bay to central uplands. "Common near North Shore" (Bain); thickets and edge of woods. MacSwain and Bain, MacSwain, Hurst.

Var. **racemiflorus** Fern. Malpeque, one of the original collections cited in Rhodora 19: 155. 1917.

Aster puniceus L. Map 740. Frequent throughout. Damp thickets along rivers and slow streams. "Six feet" (Bain); seldom much over four feet. Hybridizes with *A. cordifolius*. Bain, MacSwain and Bain, MacSwain, Hurst; (*A. amplexicaulis*, MacSwain and Bain).

X **Aster tardiflorus** L. (*A. cordifolius* X *puniceus*) Brackley Beach (Hurst).

[*Aster laevis* L., records to be excluded. MacSwain and Bain, MacSwain, Hurst].

Aster radula Ait. Map 741. Local in West Prince; Dingwells Mills (K). Boggy or mucky swampland.

Aster lateriflorus (L.) Britton Map 742. Scattered throughout. Dry, open land; acid soil in thickets and old fields. Plants from York represent a hybrid with *A. novi-belgii* (Malte). (*Aster miser* Ait., MacSwain and Bain; including *A. divergens* Ait., MacSwain and Bain; *A. multiflorus*, Bain, MacSwain and Bain, MacSwain, Hurst with doubt; *A. foliolosus*, MacSwain and Bain); Groh, Hurst.

Var. **tenuipes** Wieg. Type from Dundee, in larch swamp (*Fernald et al.*), described in Rhodora 30: 174. 1928. (*A. acadiensis* Shinners) Map 743.

[*Aster praealtus* Poir. Records to be excluded. (*A. salicifolius* Ait., not Lam., Bain, MacSwain and Bain, MacSwain, Hurst); Campbell.]

[*Aster simplex* Willd. To be expected, as present in Cape Breton and northern Nova Scotia; but was not seen. (*A. dracunculoides* Willd., not Lam., Bain, MacSwain and Bain); Bain 1892; (*A. paniculatus* Lam., not Mill., Roland).]

Aster junciformis Rydb. Map 743. Vicinity of Mount Stewart and Dundee, bogs and larch swamps. Fernald 1950.

Aster johannensis Fern. Western Prince: Tignish, Ellerslie, McNeills Mills; also in the northeast: St. Andrews, Grand Tracadie, Black Pond. Damp open thickets. Fernald 1950; (*A. foliaceus*, Roland, largely, not Lindl.)

Aster novi-belgii L., vars. Map 744. Scattered throughout.

Var. *novi-belgii* Stream banks and thickets, edge of woods. (*A. mutabilis*, MacSwain and Bain; *A. Rolandii* Shinners).

Var. **litoreus** Gray Coastal: Brackley Beach, Souris, Borders of salt marsh. Macoun, Hurst, Campbell.

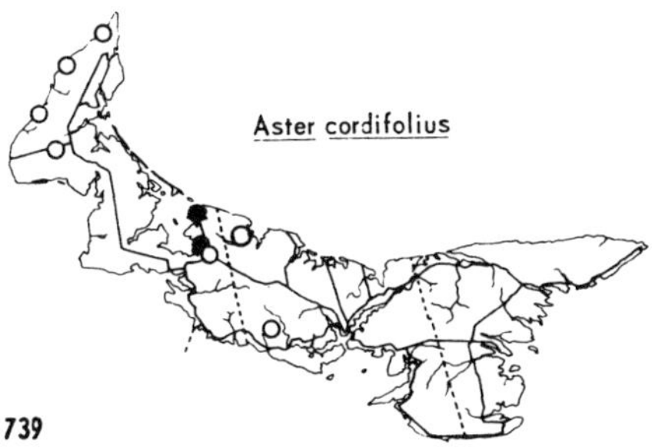

739

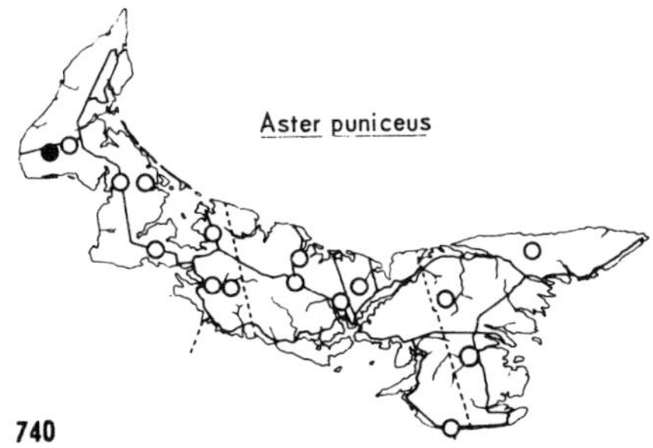

740

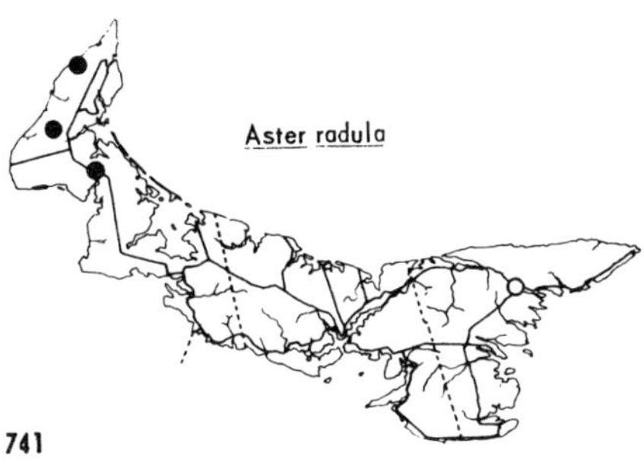

741

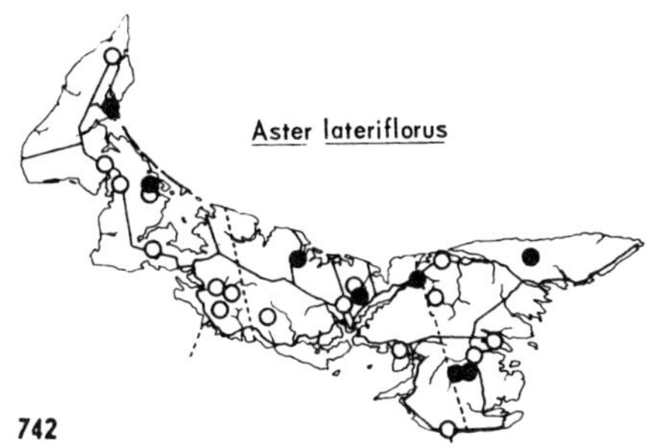

742

743

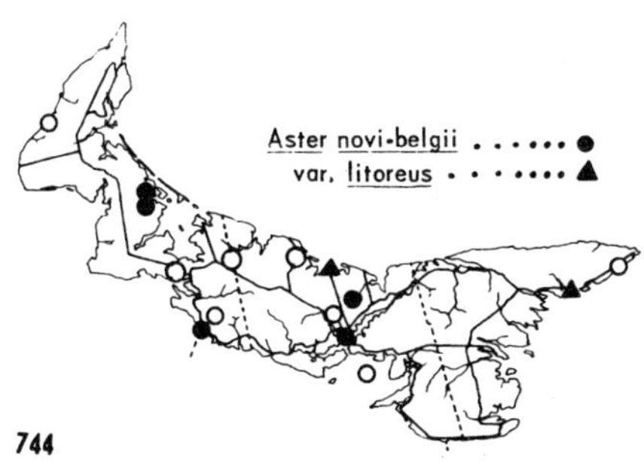

744

Aster nemoralis Ait. Map 745. Rare in West Prince: Bloomfield and near Peterville. Open or cleared boggy swampland.

Aster acuminatus Michx. Map 746. Fairly common throughout. Well-drained or, along streams, mossy, woodland sites, under hardwood or spruce, largest in clearings or borders of woods, where early advantage due to stolons. MacSwain and Bain, MacSwain, Hurst.

Aster umbellatus Mill. Map 747. Common throughout; tall and white. Damp thickets, ditches and borders of woods. MacSwain, Hurst.

Aster laurentianus Fern. var. ***laurentianus*** Map 748. North Shore: Tignish, Brackley Beach (type locality, see Rhodora 16: 59. 1914), Dalvay. Damp brackish sand around temporary ponds in dune slacks. Endemic to the Gulf of St. Lawrence; related to the western *A. frondosus*. (*A. subulatus*, Macoun, Hurst, Campbell; *A. frondosus*, Adams 1937, Hurst 1940, Campbell).

Erigeron annuus (L.) Pers. Map 749. White Sands, sandy field (*Fernald & St. John*). Weedy, perhaps naturalized from southward. Bain 1892 (?), Adams 1937, Hurst 1940.

Erigeron strigosus Muhl. vars. Map 750.

Var. ***strigosus*** Common throughout, native or invading from mainland. Weed of old fields, roadsides. 46 Lots (Groh). MacSwain and Bain, MacSwain (*E. ramosus*, Hurst); Campbell.

Var. ***septentrionalis*** (Fern. & Wieg.) Fern. Local: cited in original description: Rhodora 15: 60. 1913, from Tignish and Mount Stewart.

****Erigeron philadelphicus*** L. Local in western Prince: Pleasant View, McNeill's Mills corner. Alder thicket alluvium and banks. Ligules pale mauve, drying purple. Slender extremes of the species.

Erigeron canadensis L. Map 751. Scattered throughout. North American, naturalized. Weed of open land; railways, cultivation, hen yards and dunes. 22 Lots (Groh). Lochhead 1904, MacSwain, Hurst. (*Leptilon canadense* (L.) Britton).

Antennaria neodioica Greene, vars. Map 752. Common throughout. Dry open sites: Old fields, clearings, banks in open woods. 23 Lots (Groh, in part).

var. ***neodioica*** Common.

var. ***attenuata*** Fern. Alberton, O'Leary (and elsewhere ?)

var. ***chlorophylla*** Fern. Morell, Brackley Point, Southport, O'Leary.

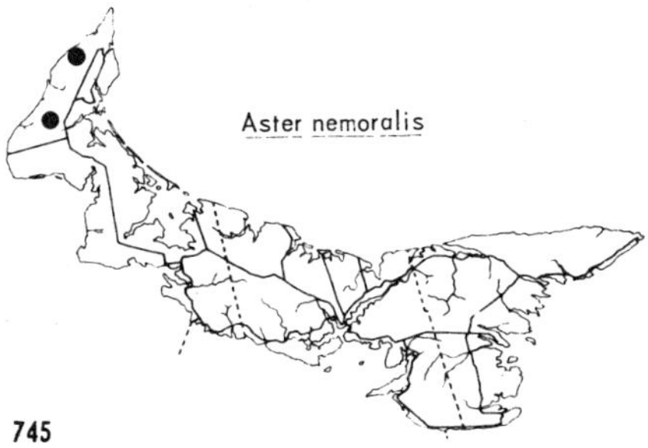

745

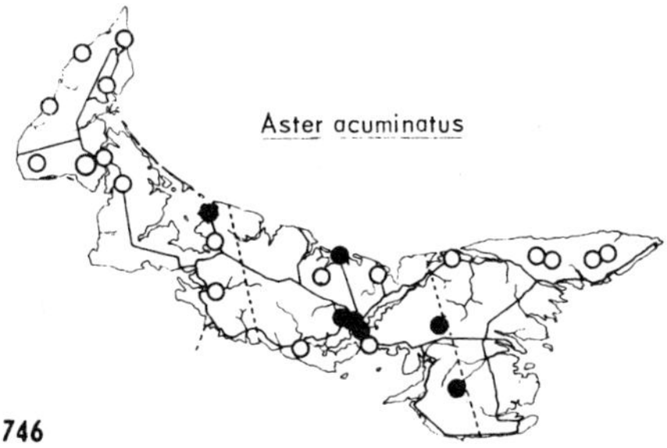

746

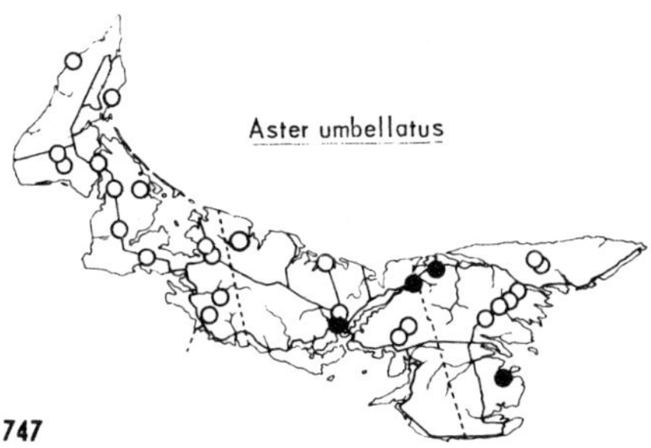

747

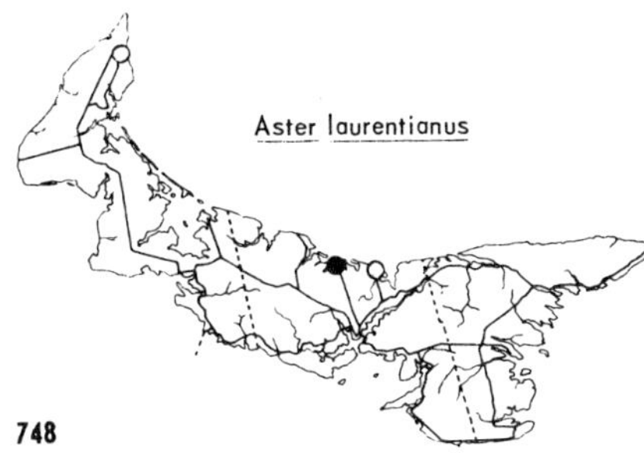

748

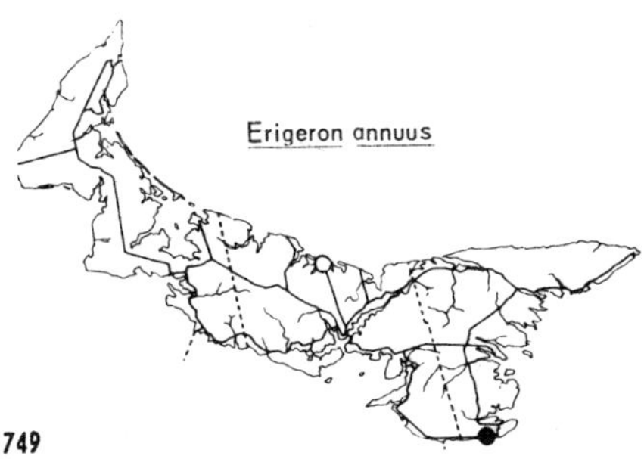

749

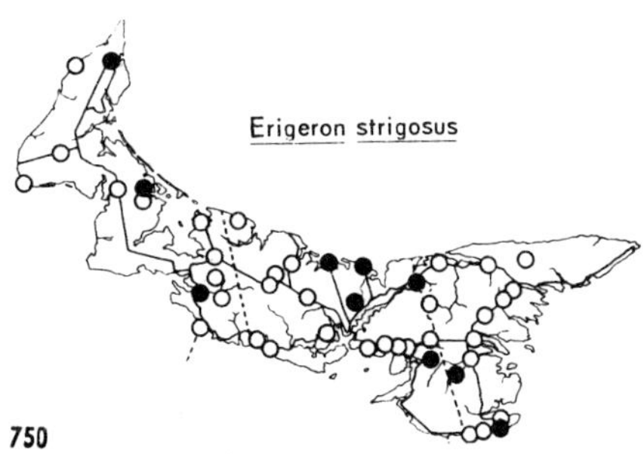

750

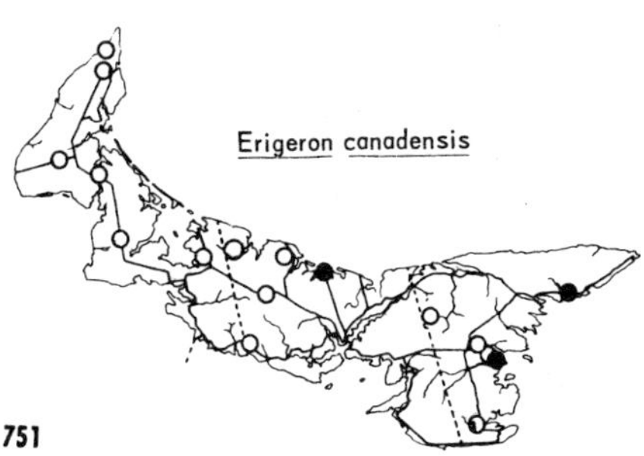

751

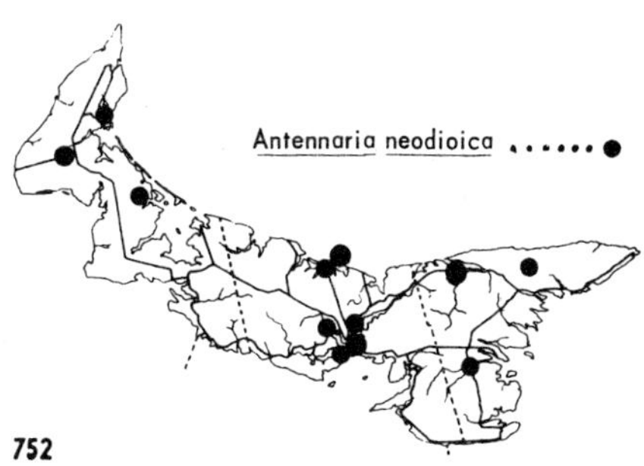

752

Antennaria petaloidea Fern. var. ***subcorymbosa*** Fern. Map 753. Scattered, along North Shore and perhaps central upland. Dry open sandy sites: old fields, thickets, banks. 23 Lots (Groh, in part). (*A. petaloidea* "and other species," Groh, Hurst, Campbell).

[*Antennaria plantaginifolia* (L.) Hook. Records are based on the two preceding species. MacSwain and Bain, MacSwain, Hurst.]

Anaphalis margaritacea (L.) C.B. Clarke, vars. Map 754. Common throughout. Dry open sites: and dunes, persisting amid shrubbery; clearings, old fields, uncultivated land. (*Antennaria margaritacea* (L.) R. Br., MacSwain and Bain, MacSwain); Hurst.

Var. ***margaritacea*** Here with broad but acute leaves dark green, with gradually deciduous cobwebby tomentum. Common.

Var. ***angustior*** (Miquel) Nakai Cape Aylesbury (*Fernald et al.*).

Var. ***intercedens*** Hara Extreme specimens are rare: Ellerslie; transitional forms common.

*****Gnaphalium macounii*** Greene Pisquid Pond near Mount Stewart, 1908 (*Warren*); abundant in southeastern uplands. Clearings, burns and poor fields. (*G. decurrens* Ives, not L., Roland).

Gnaphalium obtusifolium L. var. ***obtusifolium*** Map 755. North Shore at and around Brackley Beach; Commercial Cross (K). Dry sandy soil, laid bare by tracks or overgrazing. (*G. polycephalum,* Bain 1892, MacSwain, Hurst; *G. obtusifolium* var. *praecox,* Campbell).

Gnaphalium uliginosum L. Map 756. Throughout, common particularly in lowlands. Characteristic of recently denuded, muddy but drying soil; roadsides, cultivation, beach heads and brook banks. 40 Lots (Groh). MacSwain and Bain, MacSwain, Hurst.

Gnaphaluim sylvaticum L. Map 757. Scattered, throughout. European, said to be a naturalized species in North America. Dry open grayed sites: old fields, pastures, wood edges, banks near the shore. 5 Lots (Groh). (*G. purpureum,* MacSwain and Bain, MacSwain); Churchill 1902, Groh, Hurst.

*****Inula helenium*** L. Map 758. Becoming common in vicinity of Charlottetown: Bonshaw to Mount Stewart near Kensington. Noted only since 1943 (Messervy). European, naturalized. Old fields, roadsides.

[****Iva xanthifolia*** Nutt. Recent adventive: Bideford, 1953; Covehead, 1953. North American prairies. Hen yards, introduced in western grain.]

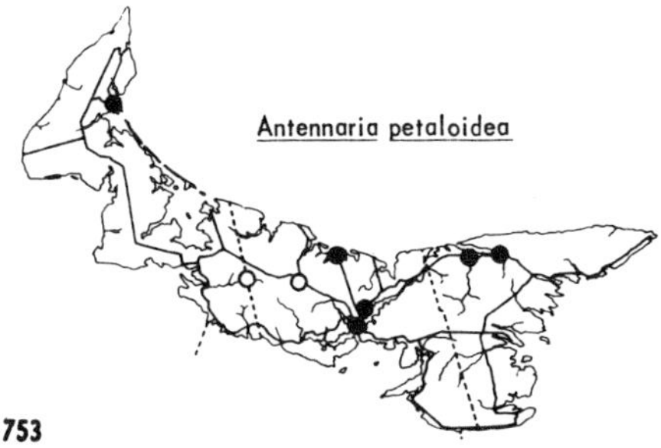

753

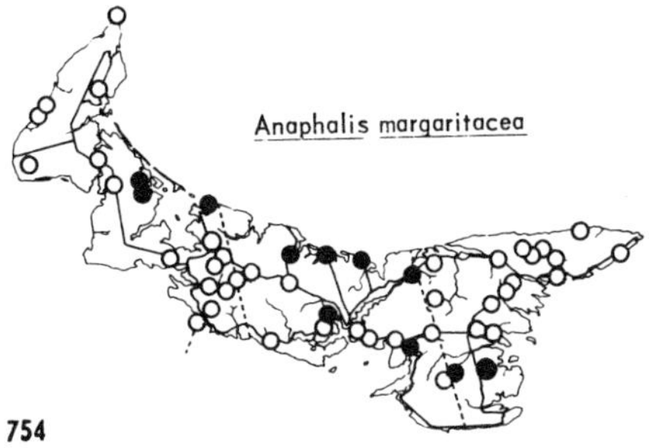

754

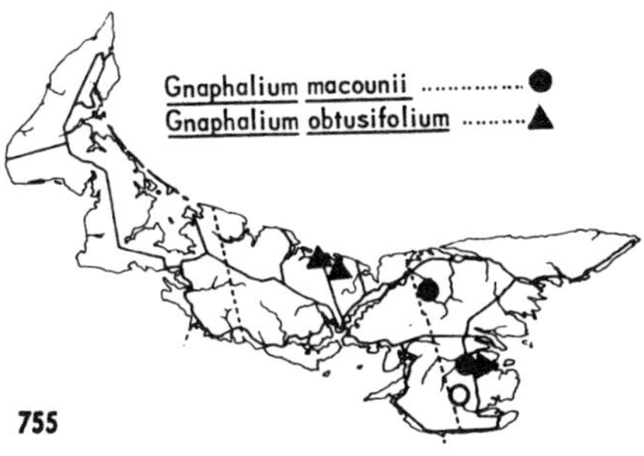

755

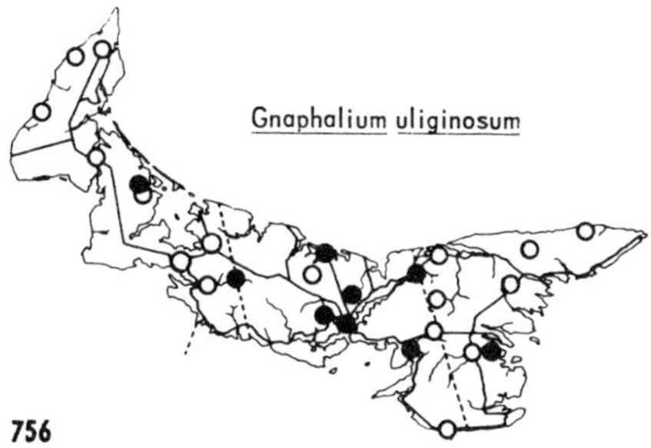

756

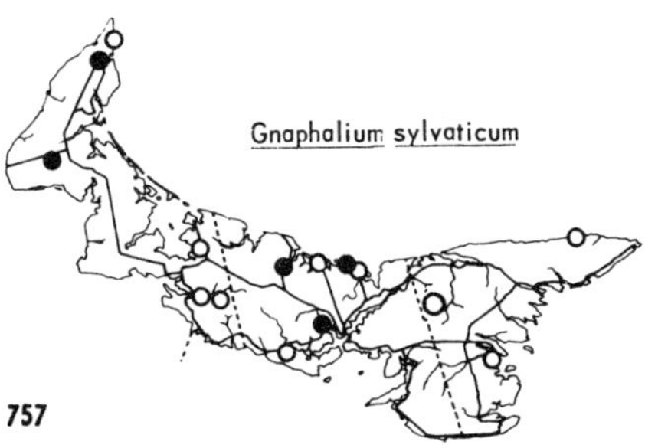

757

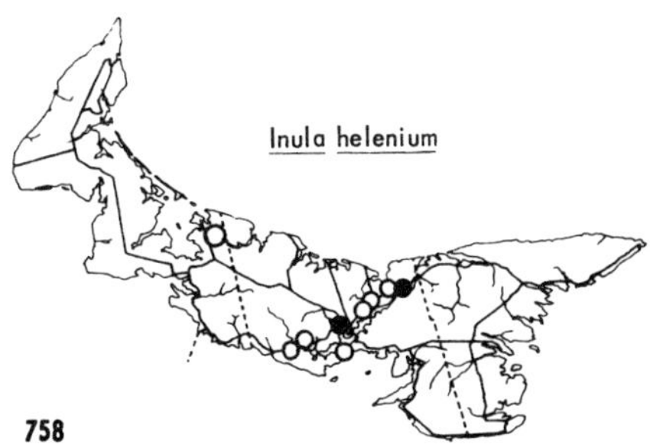

758

Ambrosia trifida L. GREAT RAGWEED Map 759. Scattered stations; serious infestation only in Charlottetown. North American, naturalized as weed of waste lots and railway sidings; often in farmyards, where adventive in western grain. 1 Lot (Groh). Most important source of allergy-inducing pollen. Prince Edward Island's resort industry depending (among other things) on its freedom from pollen that induces hay fever, determined attempts have been made in the past eleven years to locate all stations; eradication remaining, however, the responsibility of the landholder. Groh, Hurst.

Ambrosia artemisiifolia L., vars. RAGWEED Map 760. Scattered stations, in all parts. Native or naturalized from mainland. Open disturbed ground: cinders along railways; in farmyards and mown or cultivated fields, responding to mowing by flowering and fruiting at no taller than the stubble. 3 Lots (Groh).

Var. ***artemisiifolia*** Material from beach head of Waterside near Pownal is transitional to var. ***elatior***, but the site suggests native status.

Var. ***elatior*** (L.) Descourtils Common, weedy. (*A. elatior* L., MacSwain and Bain); MacSwain and Bain, MacSwain, Hurst.

[***Ambrosia psilostachya*** DC. var. ***coronopifolia*** (T. & G.) Farwell Charlottetown, 1940 (*Schurman*). Adventive from western North America.]

Xanthium italicum Moretti Charlottetown, naturalized since before 1938 along Hillsborough River front. North American. (*X. commune* Britton, Hurst 1940); Campbell.

Xanthium echinatum Murr. Map 761. Shores of Northumberland Strait east of Charlottetown; adventive at Winsloe. Upper part of sandy beaches; inland on a railway bank.

[****Heliopsis helianthoides*** (L.) Sweet Summerside, along fence and sidewalk along Water Street, 1953. Adventive from the Eastern United States.]

Rudbeckia laciniata L., vars. Map 762. Scattered stations; native or perhaps naturalized from mainland. Damp open sites; ditches and banks along roads or railways in sandy to mucky soils. 12 Lots (Groh).

Var. ***laciniata*** The usual variety here is scabrous, never soft-pubescent as in the form called var. ***gaspereauensis*** Fern., which Roland (1947) found not limited to Nova Scotia.

[Var. ***hortensis*** Bailey GOLDEN GLEAM Charlottetown, formerly at border of marsh at Experimental Farm. Cultigen, escaped. *R. laciniata* excluded by Hurst as a garden escape, on the basis of this variety.]

Rudbeckia serotina Nutt. BROWN-EYED SUSAN Map 763. Scattered throughout; very common in the southeast. Naturalized from the Great Plains before 1890. Well-drained or dry open sites; old fields, pastures. 32 Lots (Groh). (*R. fulgida*, MacSwain and Bain; *R. hirta*, MacSwain, Groh, Hurst, Campbell).

759

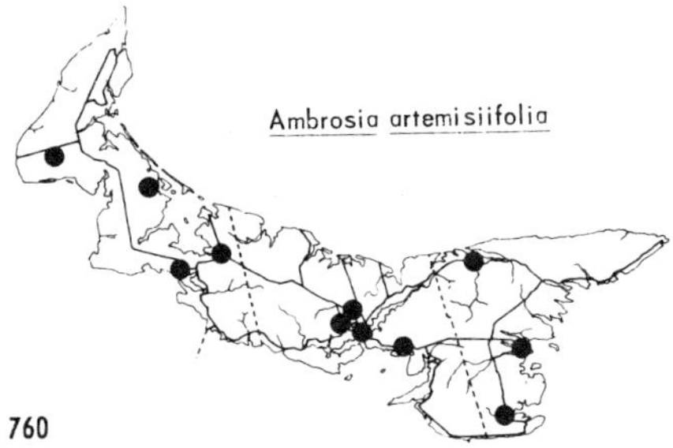

760

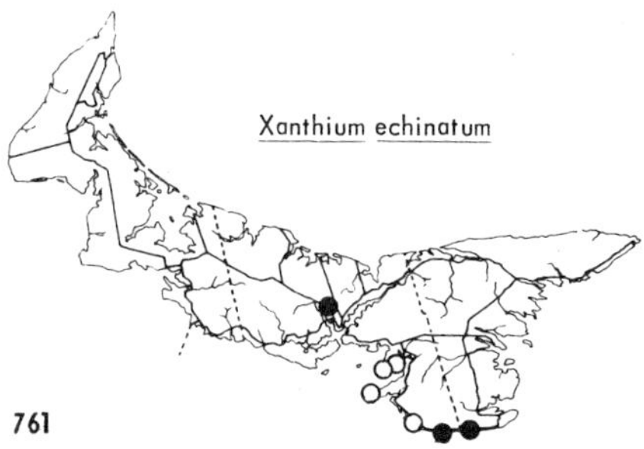

761

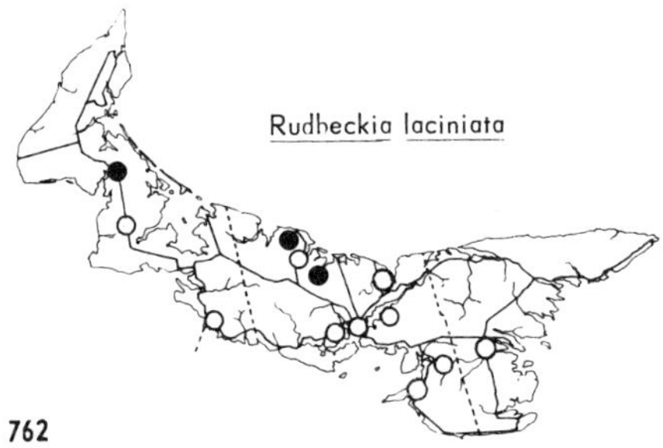

762

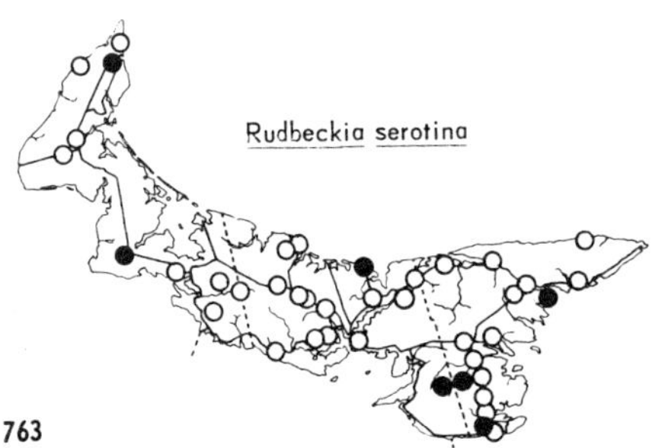

763

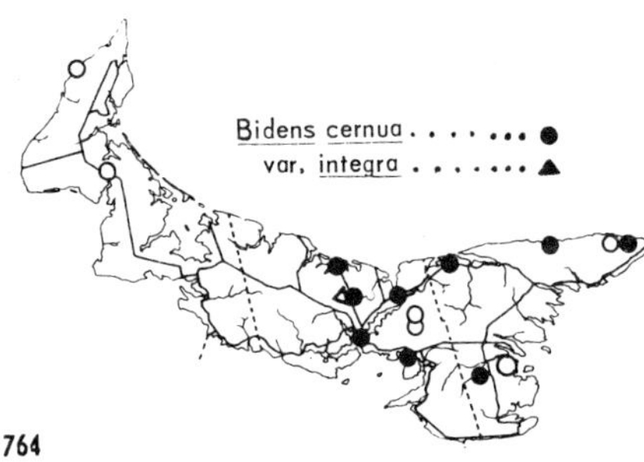

764

[*Helianthus annuus* L. SUNFLOWER Rare adventive: Kensington (1950), Covehead (1953). Plains of North America. Introduced in western grain; seldom cultivated for seeds. MacSwain, Hurst.]

[**Helianthus laetiflorus* Pers. var. ***rigidus*** (Cass.) Fern. Charlottetown, dump along railway, 1952. Introduced garden plant; prairies of North America.]

[**Helianthus maximilianii* Schrad. Southport, garden escape, 1953. Plains of North America.]

Helianthus tuberosus L. Charlottetown, northern fringes near railway; near Georgetown, Stone Arch Road. Naturalized from the Midwest. Dry ditches. 3 Lots (Groh). Hurst.

Bidens cernua L., vars. Map 764. Common in eastern lowlands; very local in West Prince. Damp muddy ground: swamps, ditches, pond margins, particularly along North Shore. Forms without ray flowers occur: North Lake (K), Dundee. At Brackley Beach, around a pond behind the dunes, dwarfed by exposure to an inch or two in height with very smooth, succulent, nearly untoothed leaves. MacSwain and Bain, MacSwain, Hurst.

Var. ***dentata*** (Nutt.) Boivin, in Nat. Can. 87: 28. 1960. Common.

Var. ***oligodonta*** Fern. & St. John North Lake, wet brackish sand, cited in the original description, Rhodora 17: 25. 1915.

Var. ***integra*** Wieg. Brackley Point Road, margin of shallow pool (*Fernald et al.*). Fernald 1950.

Var. ***elliptica*** Wieg. Southport (*Fernald et al.*).

[*Bidens connata* Muhl. If present, very rare in damp places. MacSwain and Bain, MacSwain, Groh, Hurst, Campbell.]

Bidens heterodoxa (Fern.) Fern. & St. John Map 765. East-central lowland, local, the original collection from Bunbury; marsh, at borders of salt marsh or ponds. The form with ray flowers (Rhodora 17: 24. 1915). Fernald 1950.

Bidens frondosa L., vars. BEGGAR'S TICKS Map 766. Scattered, throughout. Damp muddy swales or pond shores; weed in gardens. 12 Lots (Groh). MacSwain and Bain, MacSwain, Hurst.

Var. ***frondosa*** Common.

Var. ***stenodonta*** Fern. & St. John Bunbury, border of salt marsh (*Fernald et al.*). Cited in original description, Rhodora 17: 22. 1915.

**Galinsoga ciliata* (Raf.) Blake Charlottetown, since 1952. Cinders in alleyways; likely to become a garden weed. Naturalized by rail from central Canada; South American (Philadelphia 1836, New Hampshire 1866, Montreal 1893, Halifax 1937), as may be seen from its remaining green till first frost, then blackening and wilting.

Achillea ptarmica L. Map 767. Scattered; abundant around old settlements, more common westward. European, planted as ornamental and herb; naturalized. Roadsides, borders of field. 5 Lots (Groh). Lochhead 1904, MacSwain; Hurst.

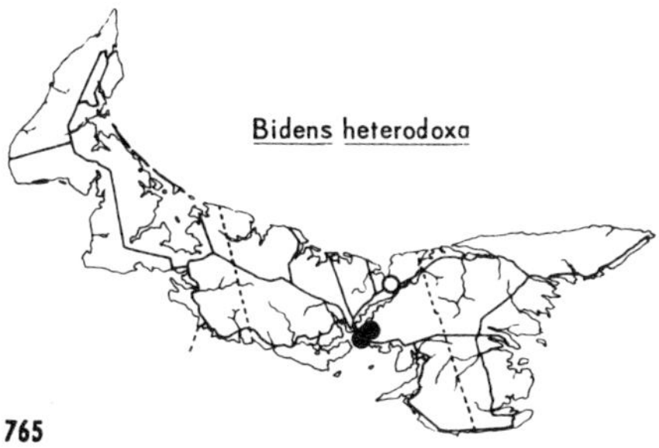

765

766

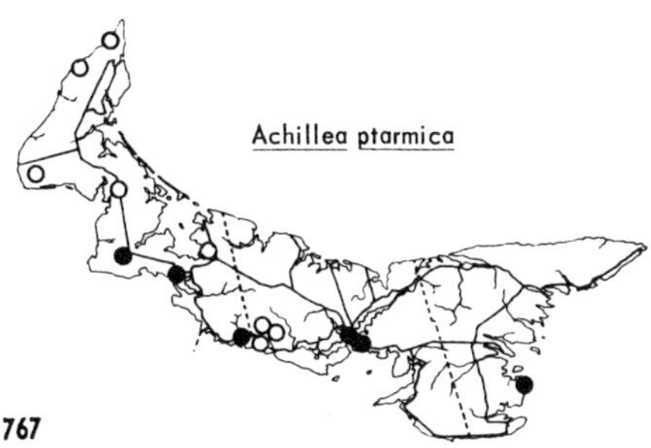

767

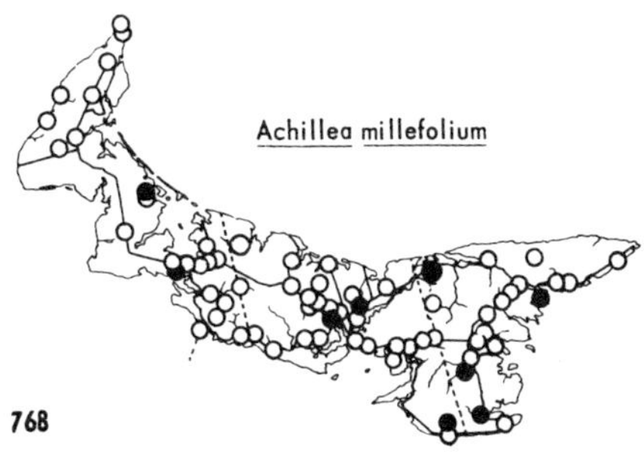

768

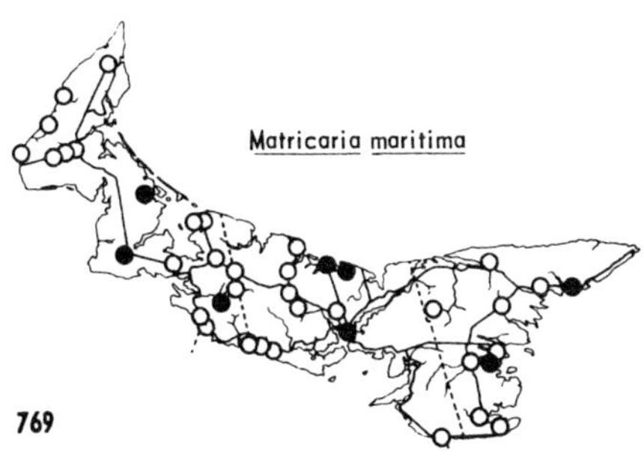

769

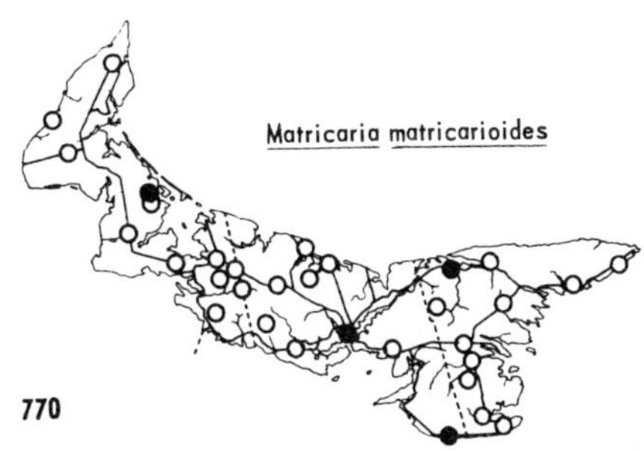

770

Achillea millefolium L. YARROW Map 768. Common throughout. Naturalized, European. Fields and roadsides. 55 Lots (Groh). Forma ***rosea*** Rand & Redfield, frequent: collected at Wood Islands, Kilmuir. MacSwain and Bain, Lochhead 1904, MacSwain, Hurst. The related native species, ***A. lanulosa*** Nutt., might be expected on outlying headlands.

[*Anthemis cotula* L. Rare: Summerside, 1914; perhaps elsewhere. Not seen in course of survey (1952-3). Weed of roadsides. 1 Lot (Groh). (***Maruta cotula*** (L.) DC., MacSwain and Bain with doubt); MacSwain, Hurst.]

[*Anthemis arvensis* L. var. *agrestis* (Wallr.) DC. Village Green, roadside, 1912 (*Fernald et al.*). MacSwain and Bain.]

Matricaria maritima L. ssp. *inodora* (L.) Clapham Map 769. BARNYARD DAISY, "TANSY" Quite common, throughout. Waste places: farmyards, cultivated or pastured fields, railway sidings, roadsides. 27 Lots (Groh). (***M. inodora*** L., MacSwain and Bain, MacSwain, Hurst; ***M. maritima*** (Knaf) Wilmott, Campbell).

Matricaria matricarioides (Less.) Porter PINEAPPLEWEED Map 770. Scattered throughout, around settlements. Naturalized, Pacific North American. Weed of farmyards, roadsides. Though not noted by Groh, was collected in Charlottetown and in the Magdalens by 1912 and in Nova Scotia by 1920. (***M. suaveolens*** (Pursh) Buchenau, not L., Hurst); Campbell.

Chrysanthemum leucanthemum L., vars. DAISY Map 771. Naturalized. European. (***Leucanthemum vulgare*** Lam., MacSwain and Bain); MacSwain.

Var. *leucanthemum* Rare: Souris, pasture.

Var. *pinnatifidum* Lecoq & Lamotte Common throughout, in fields. Establishing itself in cultivation, persistent in fallow or hayfields, forming solid masses of white in uplands in June and July; pastures and old fields. "One of the worst weeds"—Lochhead 1904. 52 Lots (Groh). Hurst, Campbell.

[*Chrysanthemum segetum* L. Wood Islands, 1946, "determined at Ottawa" (Hurst, annotation in his "New Flora"). Unconfirmed report. European, adventive.]

[*Chrysanthemum parthenium* (L.) Bernh. Introduced ornamental or herb, planted around houses. Not established. 2 Lots (Groh). Hurst.]

Tanacetum vulgare L. Map 772. Occasional: scattered in western Prince and central upland; St. Peters. European, introduced as a herbal remedy. Dry sandy sites: roadsides, about house yards. Forma ***crispum*** (L.) Hayek: Charlottetown, St. Peters. MacSwain and Bain, MacSwain, Hurst; (***Chrysanthemum vulgare*** (L.) Bernh.).

Cotula coronopifolia L. Map 773. Summerside, mud at border of salt marsh. Churchill 1902, Adams 1937, Hurst.

[*Artemisia canadensis* Michx. MacSwain and Bain, record dropped from later list.]

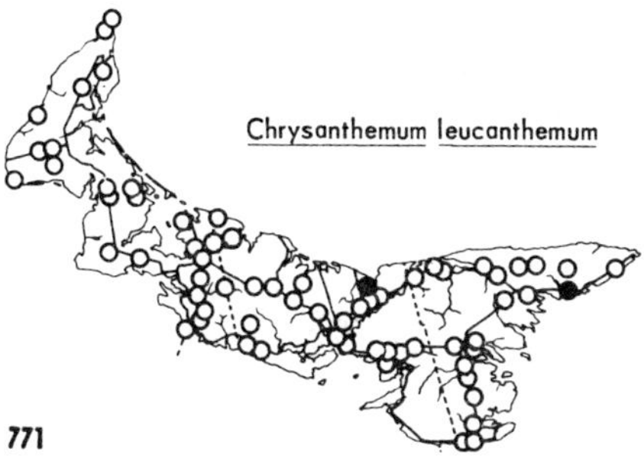

771

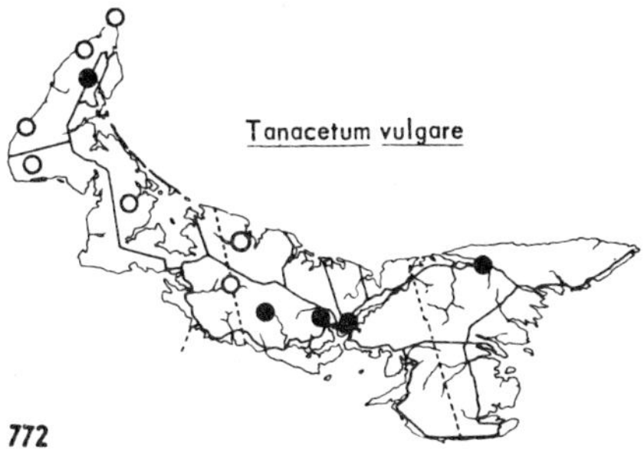

772

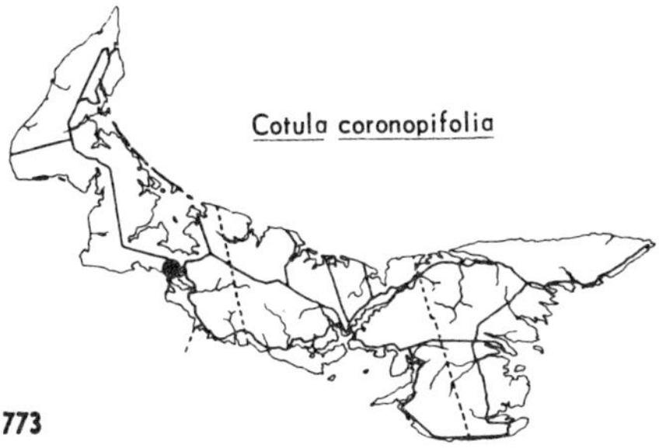

773

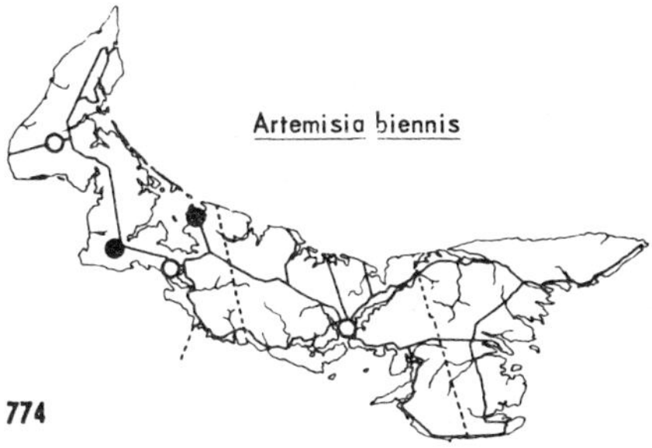

774

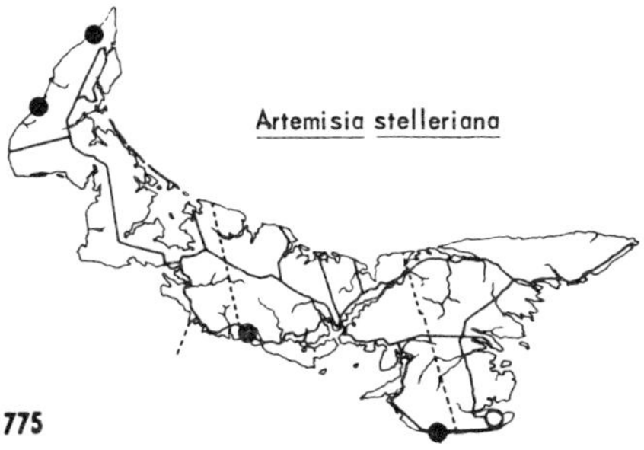

775

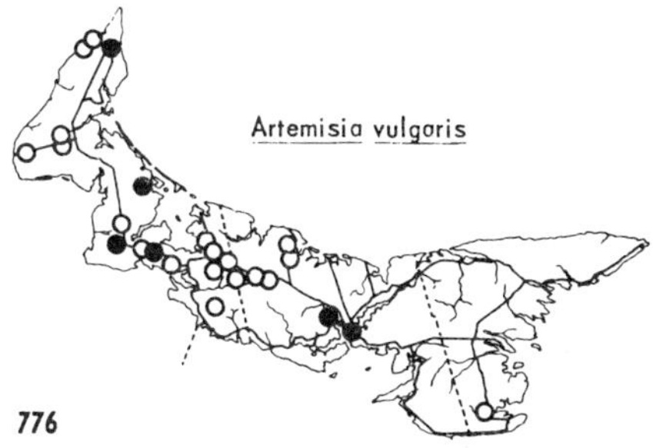

776

Artemisia biennis Willd. Map 774. Charlottetown, scattered in Prince. Naturalized from Rocky Mountains since before 1912. Waste places along railways or in sandy soil. 4 Lots (Groh). Hurst.

[*Artemisia annua* L. Charlottetown, railway yards, 1912. (*Fernald et al.*). Adventive, European. Groh, Can. Weed Surv. 2: 44. 1944.]

Artemisia stelleriana Bess. Map 775. Local: Nail Pond, Campbellton, Victoria (1924), Wood Islands, Murray harbor (1926). Naturalized, Siberian. Sandy, maritime sites: dunes, beach tension line. 1 Lot (Groh). Hurst.

Artemisia vulgaris L. var. *vulgaris*. Map 776. Common in Prince and western Queens to Charlottetown; in garden at Murray River. Naturalized, European. Open sites, light soil. Roadsides, railway banks, stony beach head. 4 Lots (Groh). Hurst.

[*Artemisia ludoviciana* Nutt. var. *gnaphalodes* (Nutt.) T. & G. Montague, 1926 (*Groh & Malte*). Adventive from Great Plains. 1 Lot (Groh). (*A. gnaphalodes* Nutt., Groh, Hurst); Campbell.]

[*Artemisia absinthium* L. var. *insipida* Stechmann Charlottetown, railway yards, 1912 (*Fernald et al.*). 1 Lot (Groh). Adventive; elsewhere introduced, European. Groh, Hurst.]

Tussilago farfara L. Charlottetown (1933), persisting. The danger of neglecting it shown by two notes: "Extirpated, 1940" (Hurst 1940); "persisting after attempt at eradication by R.R. Hurst" (Dore, on label, 1946), as it does yet (1956). Weedy, European. Hurst.

Petasites palmatus (Ait.) Gray Map 777. Local: Prince County; North Shore streams; Southport. Wet, mossy or open woods. MacSwain, Hurst.

Erechtites hieracifolia (L.) Raf. Map 778. Rare, never persistent at one station: near Tignish and in the southeast: Bunbury, Eldon, Wood Islands. Open sites: clearings and roadsides. Ours the var. *intermedia* Fern. (*Senecio hieracifolius* L., MacSwain and Bain).

Senecio vulgaris L. Map 779. Scattered. European, naturalized. Waste ground about settlements: gardens, railways, etc. Alberton, 1902. 22 Lots (Groh). MacSwain and Bain, MacSwain, Hurst.

Senecio sylvaticus L. Map 780. Local, scattered throughout. Naturalized by 1888, European. Clearings, damp open woods, shores. Macoun 1890, Churchill 1902, Hurst, Groh 1944.

Senecio viscosus L. Map 781. Local: most frequent in West Prince, where naturalized early (Alberton 1902, Tignish 1912). European. Railway cinders. 11 Lots (Groh). Hurst.

Senecio jacobaea L. Map 782. RAGWORT (incorrectly "RAGWEED"), TANSY RAGWORT, STAGGERWORT, STINKING WILLY, "BAUGHLAN" (Prince County). Common throughout. Naturalized from Europe, probably between 1850

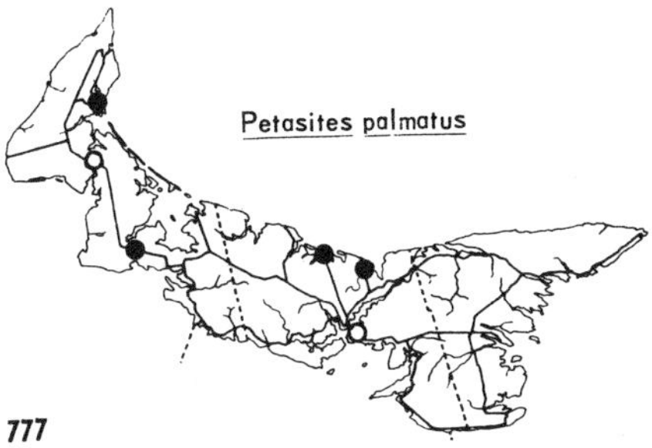

777

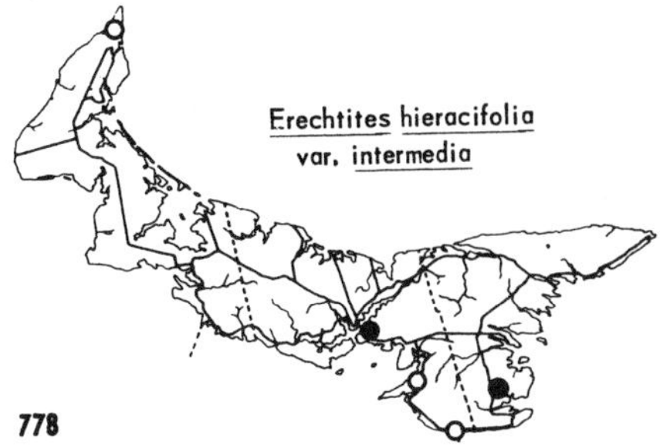

778

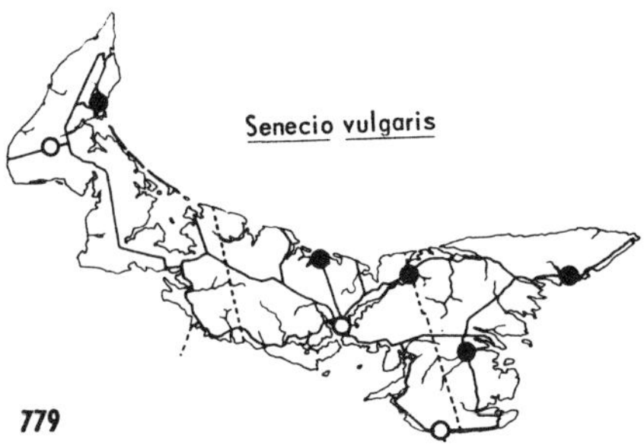

779

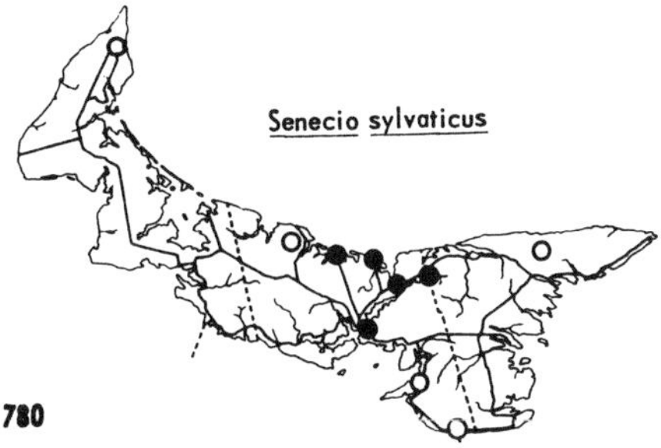

780

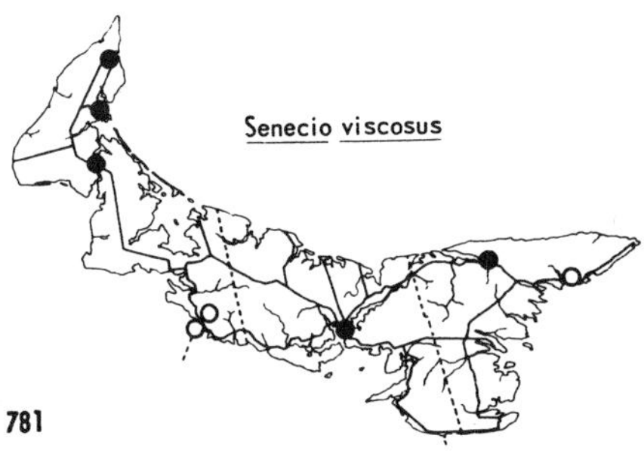

781

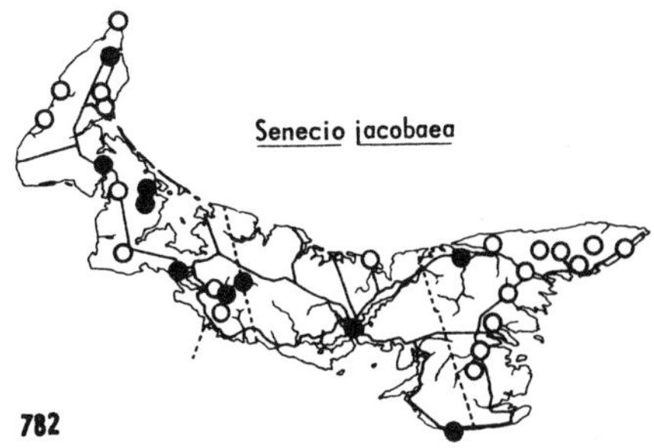

782

(collected only from Pictou County, N.S., by Mrs. Heath Haviland in 1853, presumably as interesting) and 1880. Focus of introduction perhaps Kildare (as locally reputed, though the story of its introduction in the bedtick of one Ryan may be just a story; oldest collection: Tignish 1888. By 1900 common in central Queens and northeast Kings. In 1904, 25 Lots (Lochhead). In 1926, 29 Lots (Groh). By 1953, recorded from 40 Lots. L.W. Watson 1900, Lochhead 1904, MacSwain, Hurst.

Senecio pauperculus Michx. var. ***pauperculus*** Map 783. Summerside (*Fyles*). Woods.

Senecio robbinsii Oakes Map 784. Local: Winsloe Road (*Macoun*); St. Charles (*Fernald & St. John*). Damp acid barrens. (*S. aureus* var. *Balsamitae*, MacSwain and Bain, MacSwain; *S. Balsamitae*, Hurst; *S. pauperculus* var. *Balsamitae*, Campbell).

Senecio aureus L., vars. Map 785. Vicinity of Tignish. Common, thickets along streams; cedar swamps. Bain, MacSwain and Bain, MacSwain, Hurst.

Var. ***intercursus*** Fern. Most material seems transitional between this and the var. *aquilonius*.

Var. ***aquilonius*** Fern. Specimen from larch-cedar swamp, Tignish, cited in original description by Fernald (Rhodora 45: 501. 1943).

Arctium minus (Hill) Bernh. BURDOCK Map 786. Scattered throughout. European, naturalized. Weed of houseyards and waste places. 35 Lots (Groh). (*A. Lappa*, MacSwain, Hurst with doubt); Groh, Hurst.

Forma ***pallidum*** Farwell Indian River.

Forma ***laciniatum*** Clute Kensington, Charlottetown.

Cirsium vulgare (Savi) Tenore BULL THISTLE Map 787. Common, throughout. Naturalized, European. Open places: clearings, border of woods, roadsides. A form known in Europe as *C. lanceolatum* var. *hypoleucum* DC. with the leaves white-woolly beneath occurs in streamside thicket at Peakes Station. 26 Lots (Groh). (*Cnicus lanceolatus* (L.) Willd., MacSwain and Bain with doubt, MacSwain; *Cirsium lanceolatum* (L.) Scop., not Hill, Hurst); Campbell.

[***Cirsium discolor*** (Muhl.) Spreng. Records referable to *C. muticum* (q.v.), the "biennial thistle of wild lands" (Bain) here.]

Cirsium muticum Michx. Map 788. Local: most frequent in West Prince and the northeast. Damp woods, thickets. 6 Lots (Groh). (*Cnicus discolor*, MacSwain and Bain; *C. altissimus* var. *discolor*, MacSwain; *Cirsium discolor*, Hurst, Campbell); Groh, Hurst.

Cirsium arvense (L.) Scop. Map 789. Very common throughout. Naturalized from Europe. Weed of fields and roadsides. 52 Lots (Groh). (*Cnicus arvensis* (L.)Hoffm., MacSwain and Bain, MacSwain); Groh, Hurst. Forma ***albiflora*** (Rand & Redfield) W. Hoffm. A common albino: Charlottetown, Souris, Tignish, and elsewhere.

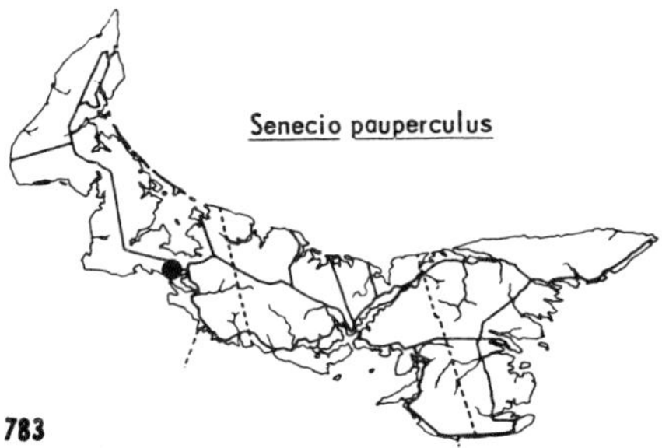

783

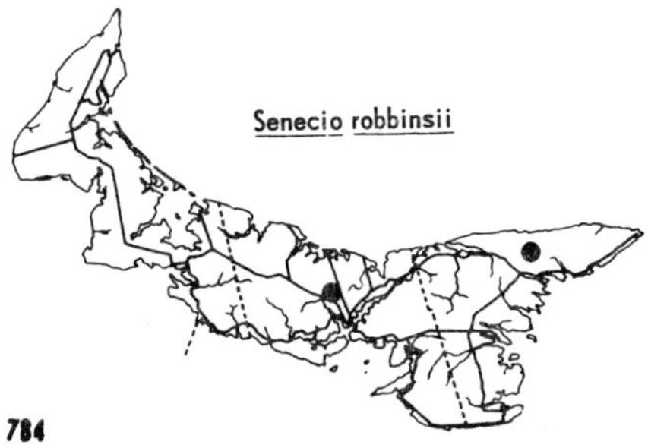

784

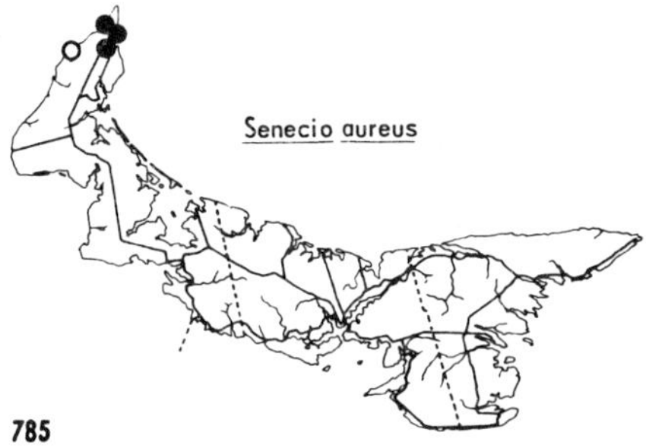

785

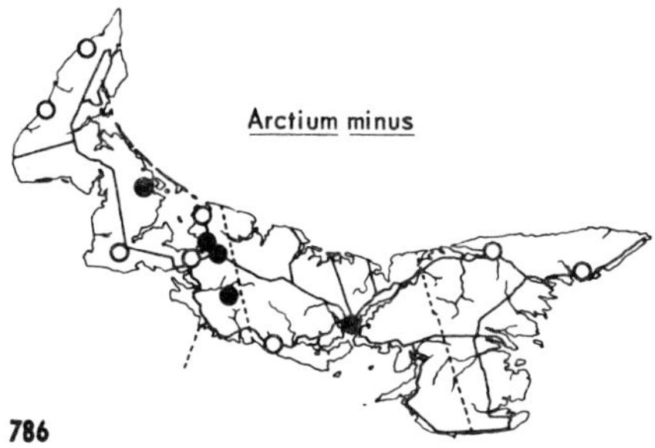

786

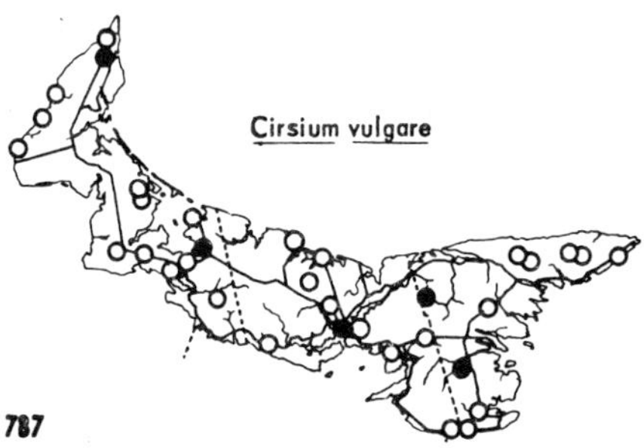

787

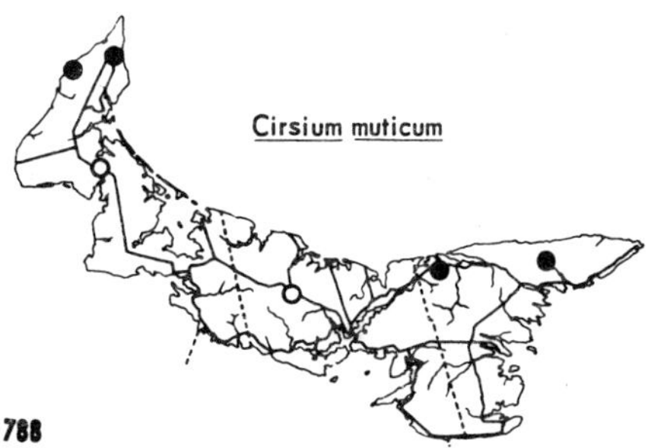

788

Centaurea nigra L. Map 790. Local: Carleton Siding; around Bonshaw; from Winsloe to Brackley Stations (1912-53); Bridgetown (1912-53) (Warren); Souris. 1 Lot — Royalty — (Groh). European, naturalized. MacSwain, Hurst.

[*Centaurea cyanus* L. CORNFLOWER (Eng.) Planted around Hurst's, Brackley Point; spreading to fallow part of garden. Introduced, European garden flower.]

Cichorium intybus L. Map 791. Restricted to towns: Summerside, North Rustico, Charlottetown, Montague, Souris. Naturalized, European (very common in Quebec Province). Waste lots. 5 Lots (Groh). MacSwain and Bain, MacSwain, Hurst.

[**Arnoseris minima** (L.) Schweigg. & Körte Charlottetown, ship's ballast dump on Hillsborough River front, 1936 (*R.R. Hurst*). European, adventive. Hurst 1940.]

Tragopogan pratensis L. Map 792. Charlottetown to O'Leary, spreading along railway since discovery at Kensington in 1943 (Hurst in *Guardian*, July 24). Naturalized, European. Railway banks, roadsides, waste lots. Campbell.

Leontodon autumnalis L. Map 793. Common, throughout. European, naturalized. Weed of fields; occasionally at roadsides. 53 Lots (Groh). MacSwain, Hurst.

Var. **pratensis** (Link) Koch Cove Head, Stanhope, Montague.

Sonchus arvensis L., vars. Map 794. Throughout; most common in central uplands, rare in Kings. Naturalized, European. Weed of cultivation, of waste places and roadsides; spreading to damp habitats: pond shores, thickets and dune slacks. Leaves variable in shape from pinnatifid to entire and in color from green to glaucous-green. MacSwain and Bain, MacSwain, Hurst.

Var. **arvensis** Common, not replaced by the variety.

Var. **glabrescens** Günth., Grab. & Wimm. Map 795. Frequent in West Prince; Summerside (1912); Charlottetown; Souris. European, recently naturalized from central Canada. Waste or cultivated fields, sands or roadsides, sometimes with var. ***arvensis*** and not intergrading. 1 Lot (Groh); not bearing out Hurst (1940's) "Found commonly by Groh," unless on 1934 visit. Hurst.

Sonchus oleraceus L. Map 796. Occasional in center, from Summerside to Dunstaffnage, mostly in towns. Waste ground, cinders and dumps. European. 4 Lots (Groh). Leaf form variable: the f. **oleraceus** usual, the f. **lacerus** (Willd.) G. Beck also present in Charlottetown. MacSwain and Bain, MacSwain, Hurst.

Sonchus asper (L.) Hill Map 797. Not common: Nail Pond, Richmond, Charlottetown, Morell, Souris. Naturalized, European. Waste ground by wharves and railways. 1 Lot (Groh). Hurst.

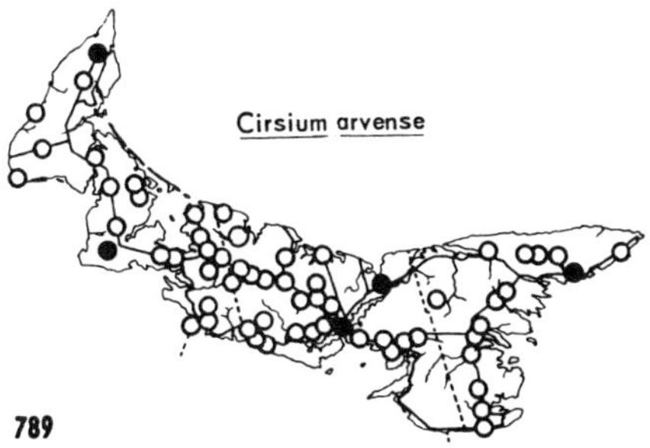

789

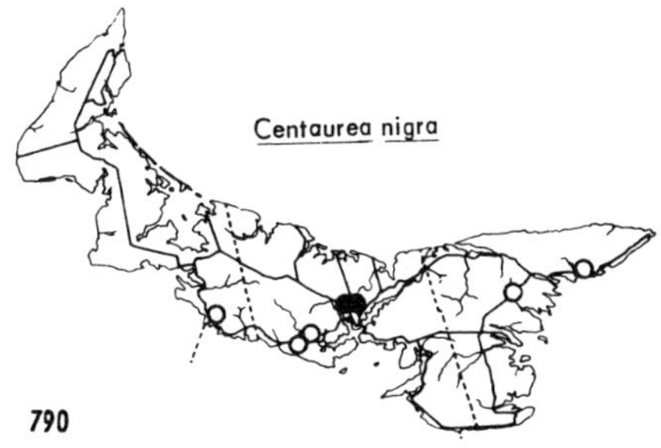

790

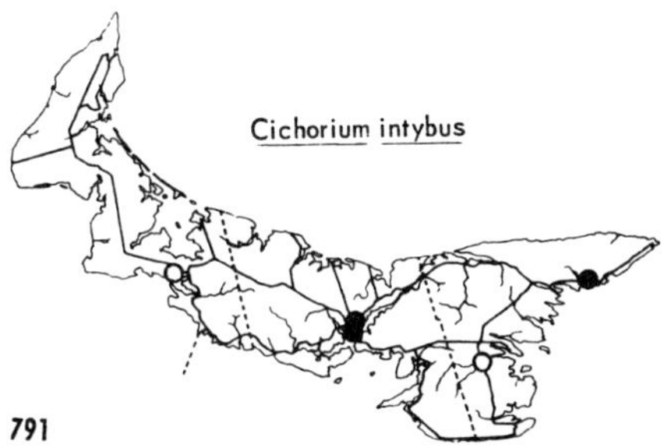

791

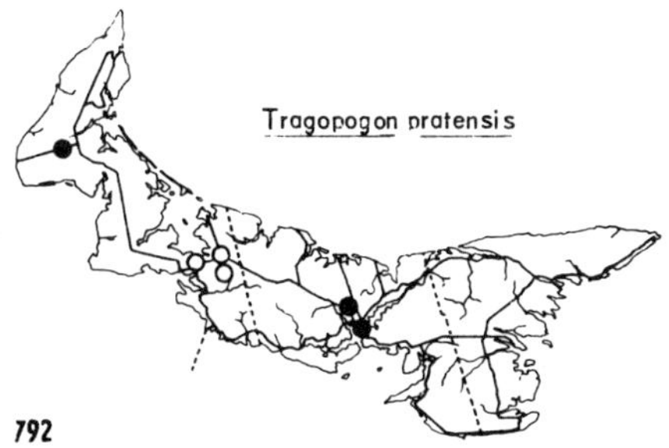

792

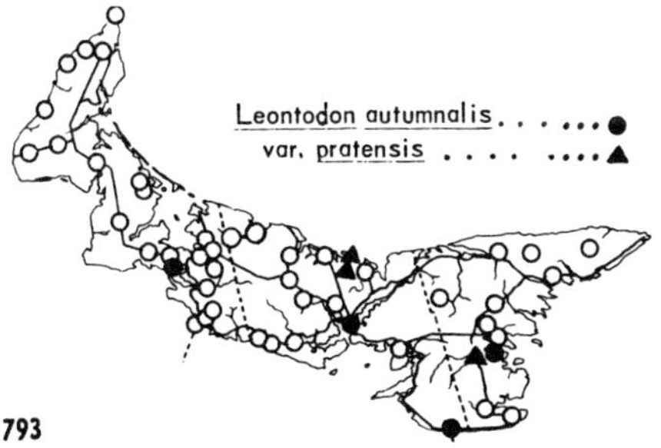

793

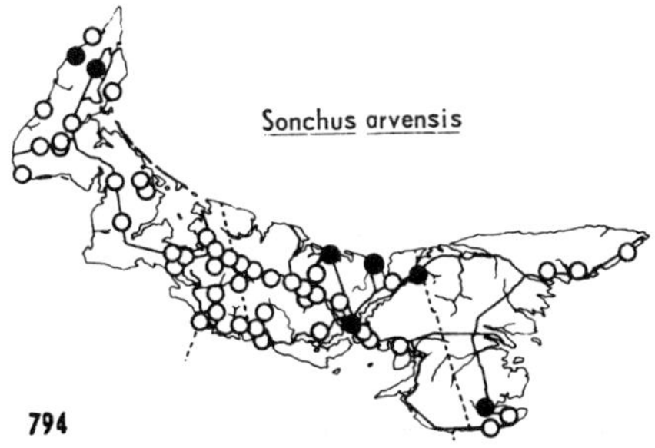

794

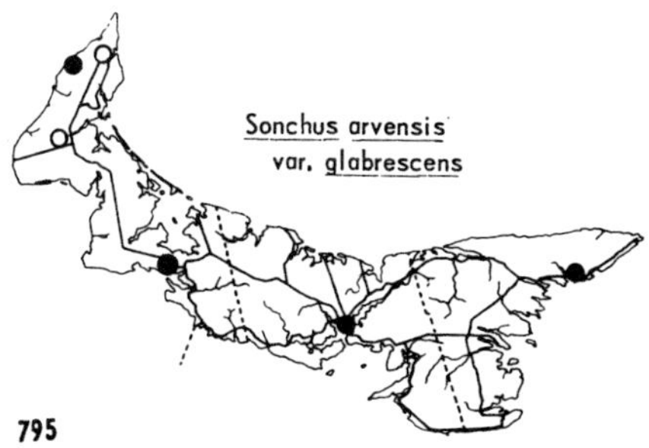

795

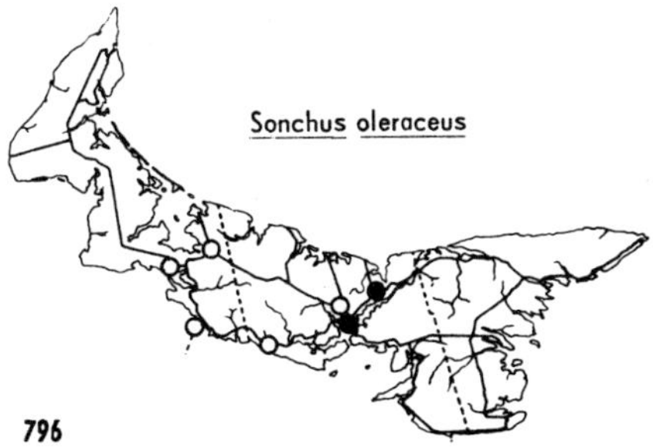

796

Hieracium pilosella L. Map 798. Common, central uplands and parts east; North Point. European, naturalized by 1888. Dry sandy open sites: old fields, clearings, dunes, roadsides. 34 Lots (Groh). (*H. Pilosella* var. *Peleterianum* Mérat, Lochhead 1904); Groh, Hurst.

Hieracium flagellare Willd. Rocky Point, 1912 (*Fernald et al.*), as weed in open woods, perhaps a shade development of *H. pilosella*. Fernald 1950.

Hieracium aurantiacum L. DEVIL'S PAINT-BRUSH Map 799. Common in central upland and uplands of Kings, local in Western Prince. Naturalized from Europe. J.A. Clarke (1904) noted that an untended cemetery in Kinkora had acted as a focus for its naturalization in East Prince; it is even now more abundant there than anywhere else in the county. Open grassy places: fields, old pastures and borders of woods. 11 Lots (Groh). Intermediates with *H. floribundum* in field with both abundant: Hunter River. MacSwain, Groh, Hurst.

Hieracium floribundum Wimm. & Grab. Map 800. Scattered throughout Prince and Queens. Naturalized, European. Open sites, sandy or disturbed soil: in old fields often persistent from last fallow stage. 25 Lots (Groh). Luxuriant forms in this and the next have upcurving secondary scapes from the base: Alberton. Intermediates with *H. aurantiacum* (q.v.) have ligule colour ranging from orange to yellow though externally orange, the foliage as setose as in *H aurantiacum* but so glaucous as to approach *floribundum*. Groh, Hurst.

Hieracium pratense Tausch Map 801. Common in Prince; scattered eastward, but general. Naturalized, European. 4 Lots (Groh). Open to partially shady sites, dry but not necessarily light or recently disturbed soils: roadsides, old fields, edges of spruce woods. Groh, Hurst.

Hieracium scabrum Michx. Map 802. Scattered; most common in central and northeastern uplands. 2 Lots (Groh). Well-drained sites: clearings or open woods. MacSwain and Bain, MacSwain, Hurst.

Hieracium canadense Michx. var. ***hirtirameum*** Fern. Map 803. Local: vicinity of Summerside; New Glasgow; Southport. 1 Lot (Groh). Dry open shade, borders of woods and clearings. Groh, Hurst.

Hieracium scabriusculum Schwein., Lepage in Can. Nat. 87: 64. 1960. Mount Vernon in southeastern upland. Steep gully in coarse soil.

Prenanthes trifoliolata (Cass.) Fern. Map 804. Common, throughout. Open, dry sites: borders of woods, clearings, thickets. (*P. Serpentaria*, Bain; *Nabalus albus*, MacSwain and Bain; *N. crepidineus*, Bain 1892; *P. alba*, MacSwain); Hurst.

Prenanthes altissima L. Map 805. Local near North Shore; on the Dunk and the Montague. Open woods along the larger streams; dry thickets at Malpeque (*Fernald et al.*). Forma ***hispidula*** Fern. Portage. Hurst, 1940.

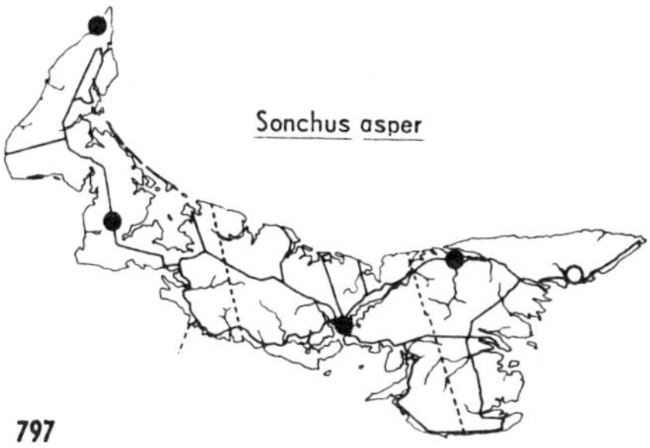

797

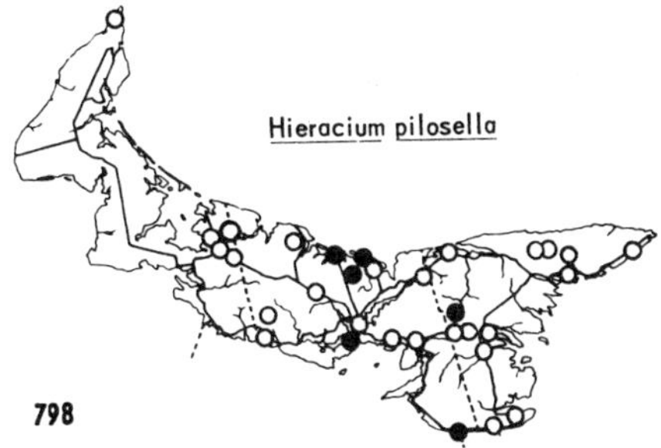

798

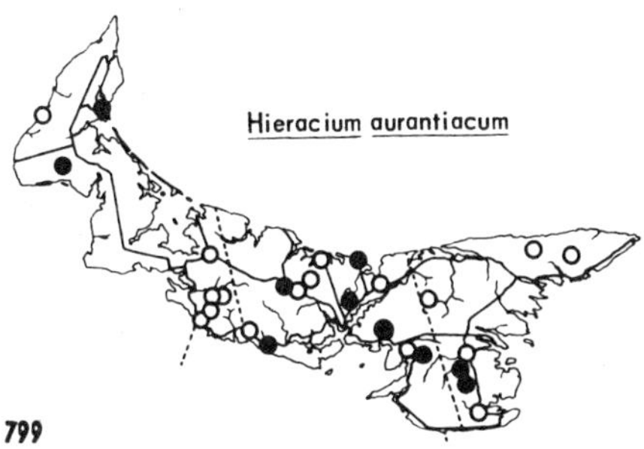

799

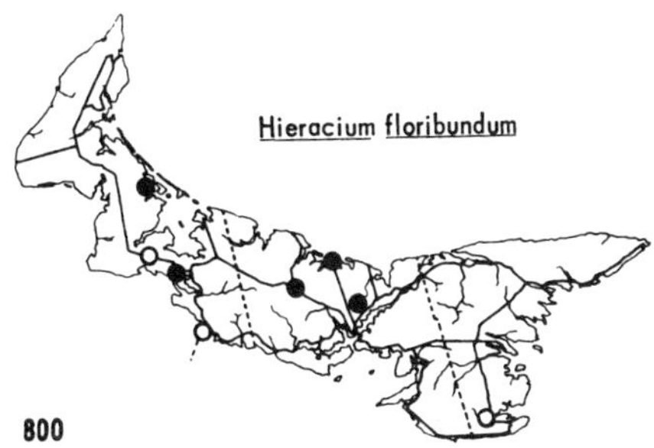

800

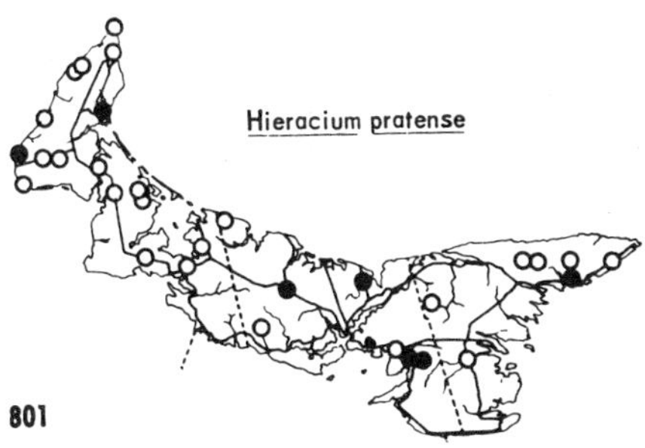

801

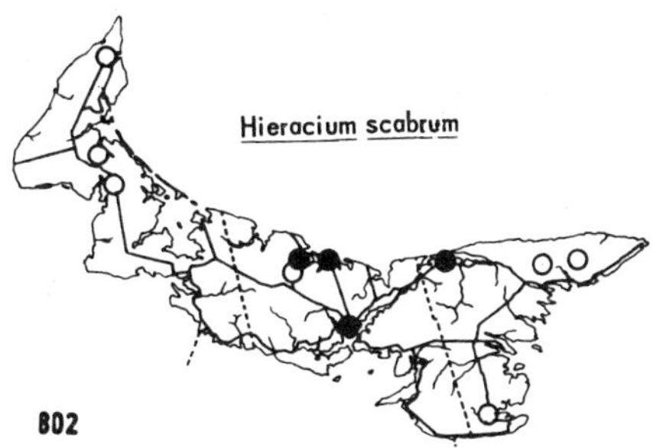

802

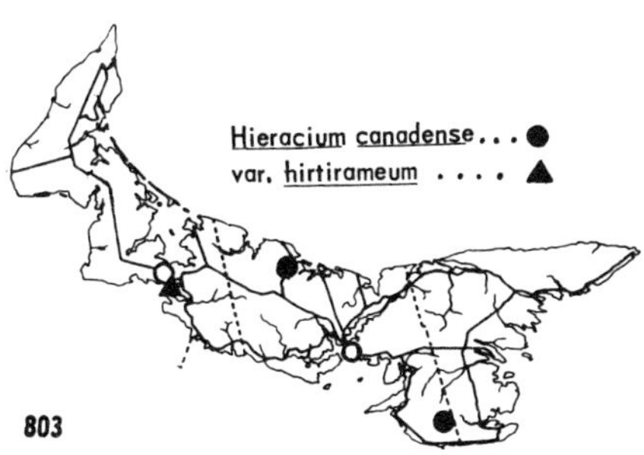

803

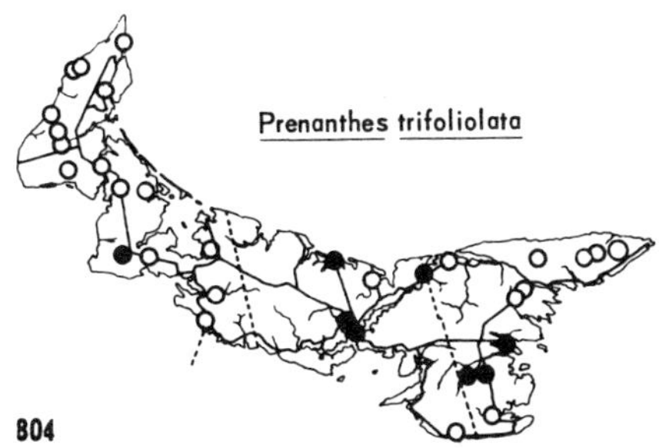

804

*_Lactuca scariola_ L. As yet local, spreading along railway; Murray River, 1951 (first collection), Charlottetown, Kensington, Summerside. European, naturalized from central Canada (in grain ?). Sidings and adjacent waste ground.

Lactuca canadensis L., vars. Map 806. Local, in the uplands. Open sites: borders of woods, clearings, tension line of salt marsh, cliff banks. MacSwain and Bain, MacSwain, Hurst.

Var. **_canadensis_** Collections from Tignish, Southport, Bothwell (tending toward var. *obovata* Wieg. in leaf form). (*L. integrifolia* Bigel., MacSwain, Hurst).

Var. **_longifolia_** (Michx.) Farwell Summerside, "rich soil in thicket."

Var. **_latifolia_** O. Ktze. Tignish, Mount Stewart.

Lactuca hirsuta Muhl. var. **_sanguinea_** (Bigel.) Fern. Map 807. Southport and Mount Stewart (*Fernald et al.*). Dry roadside thicket. MacSwain, Hurst.

Lactuca biennis (Moench) Fern. Map 808. Throughout, fairly common. Partially open sites: damp or dry thickets, borders of woods, not in organically poor soil. (*Mulgedium leucophaeum* (Willd.) DC., MacSwain and Bain; *L. leucophaea* (Willd.) Gray, MacSwain; *L. spicata*, Hurst); Campbell.

[**_Lapsana communis_** L. Determined by Macoun from a Watson collection; "roadsides, very rare" (MacSwain); Queen Street, Charlottetown, 1914. In Nova Scotia, waste ground near houses. MacSwain, Hurst.]

Taraxacum officinale Weber DANDELION Map 809. Common, particularly in towns and settlements. European, naturalized. 39 Lots (Groh). Open, disturbed or waste ground: gardens, lawns, roadsides, cinders. Its stout taproots bud below the surface of the soil, making eradication by hoe very difficult. (*T. dens-leonis* Desf., MacSwain and Bain); MacSwain, Hurst.

[**_Crepis tectorum_** L. Abrams Village, fields, 1912 (*S.E. Gallant*). Adventive, European. Groh, Can. Weed Surv. 3: 57. 1946.]

805

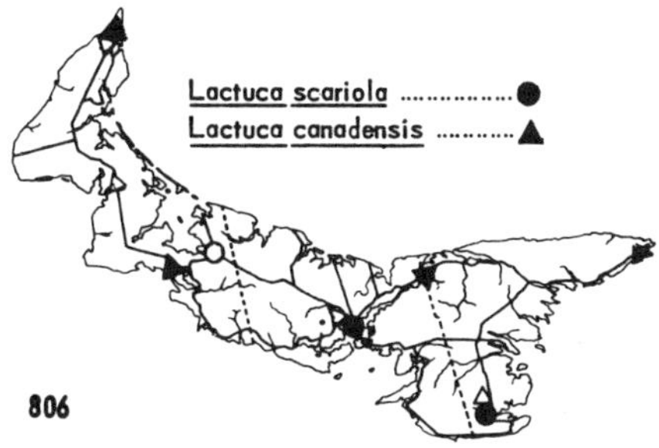

806

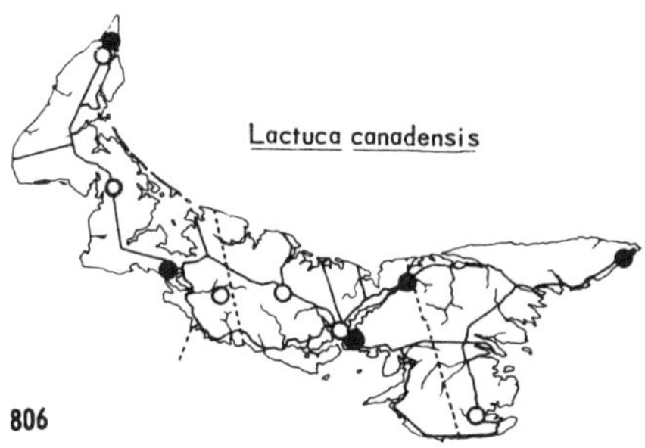

806

807

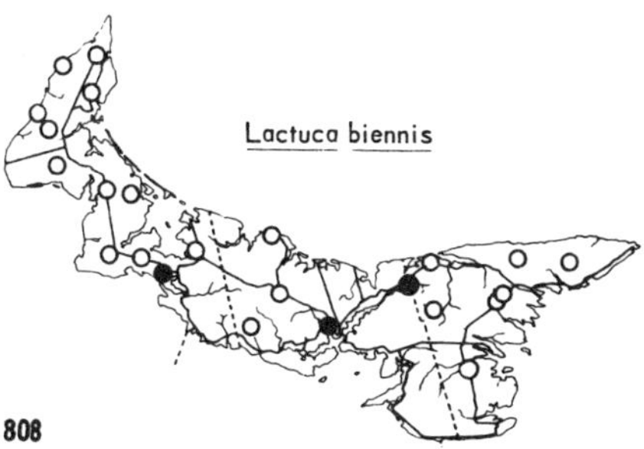

808

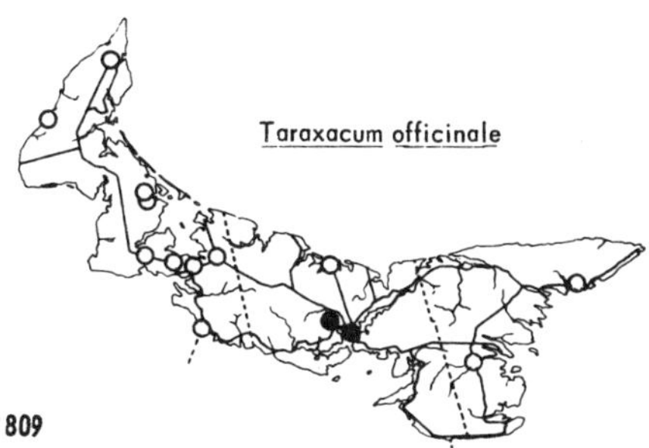

809

INDEX